Lecture Notes in Artificial Intelligence 6726

Edited by R. Goebel, J. Siekmann, and W. Wahlster

Subseries of Lecture Notes in Computer Science

Carlos Agon   Moreno Andreatta
Gérard Assayag   Emmanuel Amiot
Jean Bresson   John Mandereau (Eds.)

# Mathematics and Computation in Music

Third International Conference, MCM 2011
Paris, France, June 15-17, 2011
Proceedings

Springer

Volume Editors

Carlos Agon
Moreno Andreatta
Gérard Assayag
Jean Bresson
UMR STMS: IRCAM/CNRS/UPMC
1, place I. Stravinsky, 75004 Paris, France
E-mail: carlos.agon@ircam.fr, moreno.andreatta@ircam.fr
gerard.assayag@ircam.fr, jean.bresson@ircam.fr

Emmanuel Amiot
1, rue du Centre, 66570 St. Nazaire, France
E-mail: manu.amiot@free.fr

John Mandereau
Università di Pisa, Dipartimento di Matematica "L. Tonelli"
Largo Bruno Pontecorvo 5, 56127 Pisa, Italy
E-mail: mandereau@mail.dm.unipi.it

ISSN 0302-9743 e-ISSN 1611-3349
ISBN 978-3-642-21589-6 e-ISBN 978-3-642-21590-2
DOI 10.1007/978-3-642-21590-2
Springer Heidelberg Dordrecht London New York

Library of Congress Control Number: 2011928967

CR Subject Classification (1998): H.5.5, J.5, I.1, I.6, G.2

LNCS Sublibrary: SL 7 – Artificial Intelligence



*Typesetting:* Camera-ready by author, data conversion by Scientific Publishing Services, Chennai, India

Printed on acid-free paper

Springer is part of Springer Science+Business Media (www.springer.com)

# Preface

As in the case of the first two conferences (which took place in 2007 in Berlin and in 2009 at Yale University), the Third International Conference on Mathematics and Computation in Music (MCM 2011) aimed to provide a multi-disciplinary platform dedicated to the communication and exchange of ideas amongst researchers involved in mathematics, computer science, music theory, composition, musicology, or other related disciplines. MCM 2011 took place during June 15–17, 2011 at IRCAM, the Institute for Research and Coordination in Acoustics and Music in Paris, France. According to the mission of the Society of Mathematics and Computation in Music (SMCM), we welcomed original and high-quality contributions—including research papers, invited sessions or panels and tutorials—in all areas dealing with the relationships between music and mathematics. These areas include the formalization and geometrical representation of musical structures and processes, mathematical models for music improvisation and gestures theory, set-theoretical and transformational approaches, computational analysis, and cognitive musicology as well as more general discussions on the history, philosophy, and epistemology of music and mathematics.

These proceedings comprise 36 double-blind refereed papers that were accepted for presentation at the conference. Of 62 submissions received, 24 were accepted as long papers to be presented during the different thematic sessions of the conference and are included in the first section of the proceedings. As the reviewers considered some of the remaining submissions to be high-quality contributions, we proposed to the authors to reduce their papers and present them as posters. The second part of the proceedings comprises the 12 short papers selected and presented during the two poster sessions.

As in the previous conferences, we also solicited proposals for panel discussions and tutorials/workshops. One panel and three tutorials were selected and included in the conference program. The panel, entitled "Bridging the Gap: Computational and Mathematical Approaches in Music Research" was organized by Anja Volk (Department of Information and Computing Sciences, Utrecht University Institute for Logic) and Aline Honingh (Language and Computation, University of Amsterdam), with the participation of Alan Marsden (Lancaster Institute for the Contemporary Arts at Lancaster University and editor of *Journal of New Music Research*), Guerino Mazzola (University of Minnesota), and Geraint Wiggins (Goldsmiths College, University of London). Three tutorials were also organized, entitled "Reinforcement Learning and Computational Methods in Music Cognition" (by Nori Jacoby, Interdisciplinary Center for Neural Computation, Hebrew University of Jerusalem / Department of Music, Bar-Ilan University), "Maximal Even Sets" (by Jack Douthett, Central New Mexico Community College, Richard Plotkin, State University of New York at Buffalo, Richard Krantz, Metropolitan State College of Denver and Peter Steinbach, Central New Mexico

Community College), and "From Circle to Hyperspheres: When the Tonnetze go 4D" (by Gilles Baroin, University of Toulouse).

Thanks to the joint collaboration between IRCAM's artistic and scientific directions, MCM 2011 was integrated into the institute's most important artistic event of the season, the *Agora Music Festival*, running from June 8 to 18, 2011. The conference was thus accompanied by a series of large public scientific and artistic events organized in collaboration with some of the most important cultural and educational centers of France, such as the Centre Pompidou and Universcience (Palais de la Découverte). As a *prelude* to the conference, Universcience organized for the first time a large public conference on mathematics and music that marked the beginning of a hopefully fruitful collaboration between the researchers working at IRCAM and the scientific team of the Palais de la Découverte. The actual opening of MCM 2011 was characterized by a dialogue between two outstanding figures of music and mathematics, the renowned composer and conductor Pierre Boulez, founder and honorary director of IRCAM, and the Fields medalist Alain Connes. This dialogue, focusing on the creative process in music and mathematics, was coordinated by Gérard Assayag, director of the IRCAM/CNRS Lab and was followed by a Mathematics-Music concert organized by IRCAM's artistic direction, featuring pieces by Daniele Ghisi (*abroad*, for soprano, ensemble, and electronics), Karim Haddad (*Ce qui dort dans l'ombre sacrée...*, for bass and electronics), György Ligeti (*Monument. Selbstportrait. Bewegung*, for two pianos), and Karlheinz Stockhausen (*Kontakte*, for piano, percussion, and tape).

The second day of the conference hosted a plenary talk by the French philosopher Alain Badiou on mathematics and esthetics, whereas computer scientist Stephen Wolfram was the invited speaker of the third day of MCM 2011, with a videoconference streamed from Boston (Massachusetts, USA) entitled "Music from the Computational Universe." As a *postlude* to MCM 2011, a round table took place at the Palais de la Découverte around creativity in mathematics and arts, with the participation of Jean-Marc Lévy Leblond (physicist and essayist), Claude Bruter (mathematician and president of the ESMA, the European Society for Mathematics and Arts), Yves Hellegouarch (mathematician), Jean-Paul Allouche (mathematician), Jean-Claude Risset (physicist and composer), Tom Johnson (composer), and Jacques Mandelbrojt (painter and physicist). The multidisciplinary enlargement around the relationships between mathematics and other artistic disciplines also took profit from the exhibition of the French artist Franois Morellet at the Centre Pompidou (from March 2 to July 4) as well as the "Mathematics and Arts" exhibition organized by ESMA at the Palais de la Découverte. As a musical accompaniment of the "Mathematics and Arts" exhibition, some interactive platforms on computer-aided models in music analysis and composition were conceived and presented by Thomas Noll (ESMuC), Martin Carlé (Humboldt University), Gilles Baroin (University of Toulouse), Jérémie Garcia (IRCAM / In Situ University of Paris XI), Pierre Beauguitte (IRCAM / University Paris VI), and Benjamin Lévy (IRCAM / University of Paris VI).

MCM 2011 also inaugurated a new thematic session around a freshly published or forthcoming book whose content contributes to the understanding of the two music-theoretical and mathematical traditions that are representative of our society. This year we selected Dmitri Tymoczko's book *A Geometry of Music* (Oxford University Press), which was initially presented by the author and was then taken as the starting point for a more general philosophical and epistemological discussion on the foundational aspects of European and American maths/music-theoretical traditions.

MCM 2011 was organized under the auspices of the SMCM (Society for Mathematics and Computation in Music), the SMF (French Mathematical Society), CiE (Computability in Europe), ESMA (European Society for Mathematics and Arts), and with the financial support of the CNRS (French National Center for Scientific Research), the French Ministry of Culture, the Université Pierre et Marie Curie (UPMC), the AFIM (French Society of Music Informatics), and the SFAM (French Society of Music Analysis).

We wish to acknowledge the generous support of all IRCAM's departments in the organization of MCM 2011. We are grateful to all the invited speakers who accepted our invitation to join the conference: Pierre Boulez and Alain Connes for accepting the challenge of confronting their mutual perspectives on mathematics and music; Alain Badiou, Stephen Wolfram and all the scientists and artists who accepted to take part in the final round table at the Palais de la Découverte.

A special thanks to Sylvie Benoit of IRCAM's scientific department for helping the Program Committee with all the organizational aspects related to MCM 2011.

The Third International Conference on Mathematics and Computation in Music is dedicated to the memories of Milton Babbitt (1916–2011) and André Riotte (1928–2011), composers and music-theorists, for their outstanding contributions to the field of relationships between music and mathematics. Their ground-breaking ideas, compositional as well as theoretical, influenced generations of students and researchers up to present days.

June 2011

Carlos Agon
Emmanuel Amiot
Moreno Andreatta
Gérard Assayag
Jean Bresson
John Mandereau

# MCM 2011 Organization

MCM 2011 was organized by the Music Representations team of IRCAM (STMS Lab CNRS/UPMC) in collaboration with the Centre Pompidou and Universcience.

## Executive Committee

| | |
|---|---|
| Conference Chairs | Carlos Agon<br>Moreno Andreatta<br>Gérard Assayag<br>Jean Bresson |
| Coordination | Sylvie Benoit |
| Organizing Staff | Pierre Beauguitte<br>Louis Bigo<br>Arnaud Dessein<br>Philippe Esling<br>Jérémie Garcia<br>Fivos Maniatakos<br>Jérôme Nika |

## Program Committee

| | |
|---|---|
| Program Chairs | Carlos Agon<br>Emmanuel Amiot<br>Moreno Andreatta<br>Gérard Assayag<br>Jean Bresson<br>John Mandereau |

## Scientific Board

| | |
|---|---|
| Jean-Paul Allouche | CNRS, Université Paris 6, France |
| Christina Anagnostopoulou | University of Athens, Greece |
| John Baez | Centre for Quantum Technologies, Singapore |
| Chantal Buteau | Brock University, Canada |
| Norman Carey | CUNY - Graduate Center, New York, USA |
| Carmine Emanuele Cella | Università di Bologna, Italy |
| Marc Chemillier | EHESS, France |
| Elaine Chew | University of Southern California, USA |

| | |
|---|---|
| Adrian Childs | University of Georgia, USA |
| Ching-Hua Chuan | University of North Florida, USA |
| David Clampitt | Ohio State University, USA |
| Richard Cohn | Yale University, USA |
| Eoin Coleman | University of East Anglia, Norwich, UK |
| Darrell Conklin | Universidad del País Vasco, San Sebastián, Spain |
| Arshia Cont | IRCAM/CNRS/UPMC, France |
| Shlomo Dubnov | UCSD, USA |
| Morwaread Mary Farbood | New York University, USA |
| Davide L. Ferrario | Università di Milano-Bicocca, Italy |
| Thomas Fiore | University of Michigan-Dearborn, USA |
| Alexandre Francois | Harvey Mudd College, USA |
| Harald Fripertinger | Karl-Franzens-Universität Graz, Austria |
| Daniele Ghisi | Composer, IRCAM, France |
| Jean-Louis Giavitto | IRCAM/CNRS/UPMC, France |
| Rachel Hall | Saint Joseph's University, USA |
| Xavier Hascher | Université de Strasbourg, France |
| Francisco Herrera | Escuela Normal Superior de México, Mexico |
| Keiji Hirata | NTT Communication Science Laboratories, Kyoto, Japan |
| Aline Honingh | University of Amsterdam, The Netherlands |
| Ozgur Izmirli | Connecticut College, USA |
| Franck Jedrzejewski | CEA, France |
| Christian Kassel | CNRS/IRMA, Université de Strasbourg, France |
| Catherine Losada | University of Cincinnati, USA |
| Guerino Mazzola | University of Minnesota, USA |
| Teresa Nakra | The College of New Jersey, USA |
| Catherine Nolan | University of Western Ontario, Canada |
| Thomas Noll | ESMuC, Barcelona, Spain / TU-Berlin, Germany |
| Angelo Orcalli | Università di Udine, Italy |
| Yann Orlarey | Grame, Lyon, France |
| Athanase Papadopoulos | CNRS/IRMA, Université de Strasbourg, France |
| Richard Parncutt | Karl-Franzens-Universität Graz, Austria |
| Robert Peck | Louisiana State University, USA |
| Alberto Pinto | Università degli Studi di Milano, Italy / Stanford University, USA |
| Miller Puckette | UCSD, USA |
| Ian Quinn | Yale University, USA |
| John Rahn | University of Washington at Seattle, USA |
| André Riotte | Composer, France |
| Craig Sapp | Stanford University, USA |

| | |
|---|---|
| Sylviane Schwer | LIPN/CNRS, France |
| Godfried Toussaint | Harvard University, USA |
| Peter Van Roy | Université catholique de Louvain, Belgium |
| Anja Volk | University of Utrecht, The Netherlands |
| Geraint Wiggins | Goldsmiths College, University of London, UK |
| Jonathan Wild | McGill University, Canada |

## Sponsoring Institutions

CNRS (Centre National de la Recherche Scientifique)
SMF (Société Mathématique de France)
French Ministry of Culture and Communication
UPMC (Université Pierre et Marie Curie)
AFIM (Association Française d'Informatique Musicale)
SFAM (Société Française d'Analyse Musicale)
CiE (Computability in Europe)
ESMA (European Society for Mathematics and Arts)

## Society for Mathematics and Computation in Music

| | |
|---|---|
| President | Guerino Mazzola<br>University of Minnesota, USA |
| Vice-President | Moreno Andreatta<br>IRCAM/CNRS/UPMC, France |
| Secretary | Elaine Chew<br>University of Southern California, USA |
| Treasurer | Ian Quinn<br>Yale University, USA |

## Journal of Mathematics and Music

| | |
|---|---|
| Editors-in-Chief | Thomas Noll<br>ESMuC, Barcelona, Spain / TU-Berlin, Germany |
| | Robert Peck<br>Louisiana State University, USA |
| Reviews Editor | Julian Hook<br>Indiana University, USA |

# Table of Contents

## Long Papers

## Short Papers (Poster Sessions)

# Subsumption of Vertical Viewpoint Patterns

Mathieu Bergeron[1] and Darrell Conklin[2]

[1] CIRMMT, McGill University, Montreal, Canada
mathieu.bergeron@mail.mcgill.ca
[2] Department of Computer Science and AI
Universidad del País Vasco, San Sebastián, Spain
IKERBASQUE, Basque Foundation for Science, Bilbao
conklin@ikerbasque.org

**Abstract.** This paper formalizes the vertical viewpoint pattern language for polyphonic pattern representation. The semantics of patterns is given in terms of a translation to a relational network form. The language supports pattern subsumption, an essential inference for pattern mining, development, and refinement. Though computed in a way entirely different to relational network matching, this paper proves that subsumption inferences are sound and complete with respect to the underlying relational language.

**Keywords:** Pattern subsumption, Polyphonic patterns, Music representation, Score slicing, Computational musicology.

## 1 Motivation

Pattern recognition and discovery are important topics in computational musicology. Patterns in music can be used to describe repetition within pieces and also recurrence within a corpus of pieces. Patterns may refer to, for example, melodic lines, chord sequences, or to more complex polyphonic structures where temporal relations exist between overlapping events.

Considering polyphony, a natural and powerful representation of polyphonic patterns is graph structures, where the nodes represent notes, or more generally features of notes, and edges represent their temporal relations. Graph representations require the computation of subgraph isomorphism to determine pattern to instance relations, and this has constrained their use in computational musicology.

The aim of this paper is to elaborate a new formalism called $\mathcal{VVP}$ (vertical viewpoint patterns) for the representation of polyphonic patterns. The representation combines the viewpoint formalism [1] with the idea of segmenting the polyphonic texture into slices [2]. As a slice can contain the continuation of a musical event, a variety of temporal relations are supported. While this representation does not have the full power of graphs, it is suitably expressive for many types of musical patterns, specifically patterns where notes are temporally compact and occur in a fixed number of voices.

The method assumes that patterns have the same number of voices than the musical source from which instances are drawn. This can require the extraction

C. Agon et al. (Eds.): MCM 2011, LNAI 6726, pp. 1–12, 2011.

of fixed voice textures from actual musical data (e.g. [3] considered voice pairs extracted out of the four-voice texture of Bach chorale harmonizations). A similar technique is also required when using Humdrum, a well-known toolkit for pattern matching in symbolic music data [4,5]. Although Humdrum supports polyphony, it can be difficult to use for even simple patterns [6]. For example, Figure 1 shows some Humdrum data (after preprocessing to obtain explicit event continuations and extra spines with musical features) and a Humdrum pattern that captures the idea of a contrapuntal suspension. The regular expressions in each line of a Humdrum pattern do two things : i) match the beginning of events or the continuation of events (this is the purpose of brackets in the pattern of Fig. 1, a bracket matches the continuation of an event and the absence of a bracket matches the beginning of an event); and ii) match features of those events by matching corresponding values in additional columns (this is the purpose of the tokens ending the lines of the pattern, a dissonance feature is matched on the first line and, on the second line, features encoding the melodic contour of a step down and a consonance are matched).

| **bass** | **tenor** | **cont** | **qual** |
|---|---|---|---|
| D | B | +s | cons |
| BB | (B) | . | cons |
| FF# | (B) | . | diss |
| (FF#) | A# | -s | cons |

| | | | | | | | |
|---|---|---|---|---|---|---|---|
| [a-g A-G]+ | [- # n]* | [^)]* | [(] | [a-g A-G]+ | [- # n]* | .* | diss$ |
| [a-g A-G]+ | [- # n]* | [)] | [^(]* | [a-g A-G]+ | [- # n]* | -s | cons$ |

**Fig. 1.** A musical fragment in Humdrum (top) and a suspension pattern (bottom). The musical data is preprocessed to obtain explicit event continuations, denoted by brackets, and to add extra spines containing musical features: **cont** for melodic contour and **qual** for harmonic interval.

In addition to the required expertise in regular expressions, more fundamental difficulties with developing Humdrum patterns arise due to their opacity and lack of a denotational semantics. It is in general not possible for a computational musicologist to inspect a pattern and from that deduce what fragments will be matched. Lacking a denotational semantics, it may be intractably difficult to develop and refine patterns. For example, simply adding another melodic feature to a pattern requires both the existence of a spine containing that feature and, to reference the values of that feature, the modification of every line of the pattern in a way that respects the spine ordering.

This paper formally defines the semantics of $\mathcal{VVP}$ showing how it expresses relations between temporally related events in compact regions of time. The method offers a clear syntax for patterns that are very similar to the patterns that Humdrum can express. In addition, it is possible to formally define the notion of pattern subsumption in $\mathcal{VVP}$, indicating when the pieces matched by one pattern are always a superset of the pieces matched by another. With a formal subsumption relation the computational musicologist may develop patterns by progressive pattern specialization and generalization, bringing in fewer or more matching database pieces as desired. Furthermore, with pattern subsumption it is possible to define data mining algorithms that search refinement graphs of the pattern space [7].

This paper is organized as follows. Section 2 defines $\mathcal{R}$, a relational pattern language we replicate and extend from [8]. In that language, temporal relations between musical events are explicit and can be visualized as a network; and the notion of pattern subsumption is straightforward and intuitive. Section 3 develops a semantics for $\mathcal{VVP}$ in terms of $\mathcal{R}$, that is, in terms of the relations a $\mathcal{VVP}$ pattern expresses between musical events. Section 4 defines pattern subsumption for $\mathcal{VVP}$ and shows that it is sound and complete with respect to subsumption in $\mathcal{R}$. This is the key contribution of this paper as it establishes that the denotational semantics we give to $\mathcal{VVP}$ correctly captures the meaning of patterns as they are being specialized or instantiated. This extends the work presented in [8], where the instantiation of patterns was not considered. Finally, Section 5 further discusses the applications and limitations of the current work.

## 2 Relational Patterns

A *relational pattern* expresses temporal relations over event variables. The temporal relations we consider here are: i) $\texttt{m}(a, b)$ ($a$ **m**eets $b$: $a$ finishes when $b$ begins), ii) $\texttt{st}(a, b)$ ($a$ and $b$ **s**tart **t**ogether), and iii) $\texttt{sw}(a, b)$ ($a$ **s**tarts **w**hile $b$ is sounding). Formally, the relations are defined in Fig. 2 (see [8] for a comparison between these and the well-known temporal relations between intervals [9]).

This paper extends the relational pattern language of [8] by including two new types of atomic formulas: i) atoms of the form $\texttt{voice}(\varepsilon, \gamma)$, which indicate that some event $\varepsilon$ belongs to voice $\gamma$; and ii) atoms of the form $\tau(\varepsilon, \ldots, \varepsilon, \mu)$, which either record some basic musical information about an event, for example $\texttt{pitch}(a, \mathsf{A4\#})$, or some musical relation occuring between many events, for example $\texttt{interval}(a, b, \mathsf{M3})$. This yields the language $\mathcal{R}$, defined as follows:

**Definition 1.** $r \in \mathcal{R} ::= \omega, \ldots, \omega$ **with** $\omega ::= \texttt{m}(\varepsilon, \varepsilon)$

$$\begin{aligned} &\mid \texttt{st}(\varepsilon, \varepsilon) \\ &\mid \texttt{sw}(\varepsilon, \varepsilon) \\ &\mid \texttt{voice}(\varepsilon, \gamma) \\ &\mid \tau(\varepsilon, \ldots, \varepsilon, \mu) \end{aligned}$$

Relational networks are used to visualize the temporal relations of a pattern in $\mathcal{R}$. Figure 3 exemplifies a relational network (left) and a relational pattern (middle; the pattern is one of the many discovered by the music mining approach developed in [3]). Musical information is given, for example, by the atom $\texttt{voice}(a, \mathsf{alto})$ which specifies that event $a$ is in the alto voice; or the atom $\texttt{cont}(b, \mathsf{-s})$, which specifies

$\texttt{m}(a, b)$ **when:**
  $\texttt{offset}(a, \mu)$,
  $\texttt{onset}(b, \mu)$

$\texttt{st}(a, b)$ **when:**
  $\texttt{onset}(a, \mu)$,
  $\texttt{onset}(b, \mu)$

$\texttt{sw}(a, b)$ **when:**
  $\texttt{onset}(a, \mu_1)$,
  $\texttt{onset}(b, \mu_2)$,
  $\texttt{offset}(b, \mu_3)$,
  $\mu_1 > \mu_2$,
  $\mu_1 < \mu_3$

**Fig. 2.** Relational definition of the three temporal relations considered in this article

that event $b$ is reached by a downward melodic step. An event has the musical feature $\texttt{cont}(\varepsilon, \mu)$ which describes melodic contour and takes the possible values leap up: +l; leap down: -l; step up: +s; step down: -s; repeat: r. The other musical feature used in this paper is $\texttt{qual}(\varepsilon, \varepsilon, \mu)$, the harmonic interval (non-compound diatonic interval between lower and higher pitches in the slice), taking the values cons (intervals P1, m3, M3, P5, m6, M6) and diss (all other intervals).

Subsumption in $\mathcal{R}$ can be defined as a special case of $\theta$-subsumption [10]: a pattern $r_1$ *subsumes* a pattern $r_2$ iff there exists a mapping $\theta$ of the event variables of $r_1$ to the event variables of $r_2$ such that $r_1\theta$ is a subset of $r_2$. For example, the following patterns are in a subsumption relation, denoted by $\succeq_{\mathcal{R}}$:

*Example 1.*

$$\begin{array}{l} \texttt{st}(c, d), \\ \texttt{voice}(c, \text{alto}) \end{array} \quad \succeq_{\mathcal{R}} \quad \begin{array}{l} \texttt{st}(a, b), \\ \texttt{voice}(a, \text{alto}), \\ \texttt{voice}(b, \text{tenor}), \\ \texttt{qual}(a, b, \text{cons}) \end{array}$$

**with** $c\theta = a$ **and** $d\theta = b$

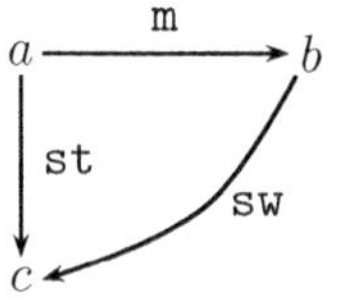


$\mathcal{R}$

$\texttt{m}(a, b)$,
$\texttt{st}(a, c)$,
$\texttt{sw}(b, c)$,
$\texttt{voice}(a, \text{alto})$,
$\texttt{voice}(b, \text{alto})$,
$\texttt{voice}(c, \text{tenor})$,
$\texttt{qual}(a, c, \text{diss})$
$\texttt{qual}(b, c, \text{cons})$

**Instances**

**Fig. 3.** Patterns in $\mathcal{R}$ and instances from Bach chorale harmonizations

# 3 Vertical Viewpoints

The vertical viewpoint method [3] represents polyphonic music as slices. The idea is to segment the musical source into a series of time-spans, or slices. A new slice is created whenever a new event appears. Events that are still sounding at the onset time of that new event have a "continuation" recorded. This results in a representation of polyphonic music that is thoroughly sequential, as the definition of $\mathcal{VVP}$ makes clear.

**Definition 2.** $v \in \mathcal{VVP} ::= [\alpha, \alpha, \ldots, \alpha]$ **with** $\alpha ::= \{\delta, \delta, \ldots, \delta\}$
$\delta ::= \tau[\gamma] \ldots [\gamma] : \mu$

According to Definition 2, a $\mathcal{VVP}$ *pattern* $v$ is a sequence of *feature sets*. Each feature set $\alpha$ encodes a slice. A *feature* $\tau[\gamma] \ldots [\gamma] : \mu$ includes a *feature name* $\tau$, *voice labels* $\gamma$ and a *feature value* $\mu$. For example, consider Fig. 4. Note that the sequence is displayed without the brackets and that commas are omitted within the sets (henceforth, we display $\mathcal{VVP}$ patterns that way to simplify presentation). The pattern contains two slices. In each slice, a feature of the form `starts`$[\gamma] : \mu$ indicates whether, for some voice $\gamma$, the slice concerns the beginning of a new event (`starts`$[\gamma]$ : t) or the continuation of some previously existing event (`starts`$[\gamma]$ : f). To relate slices and events, which is necessary to give a relational semantics to $\mathcal{VVP}$, we label these features with event variables (not part of the syntax of $\mathcal{VVP}$). In Figure 4, the first slice concerns the continuation of some event $a$ and the beginning of a new event $b$, while the second slice also concerns the continuation of $a$ and the beginning of new event (this time labelled $c$). This yields the relational network shown in the figure. In addition, the pattern specifies musical features, such as melodic contours in voice 1 (e.g. feature cont[1] : -l) and consonant intervals between voice 1 and voice 0 (e.g. feature qual[1][0] : cons).

In general, the labelling of events in a $\mathcal{VVP}$ pattern is done as follows. For the first feature set of the pattern, every voice is labelled with a fresh event variable. Then for the following feature sets, every voice exhibiting a `starts`$[\gamma_i]$ :t feature is labelled with a fresh event variable. Any other voice is labelled with its correponding event variable from the previous set. Figure 5 and the example below illustrate the labelling process.

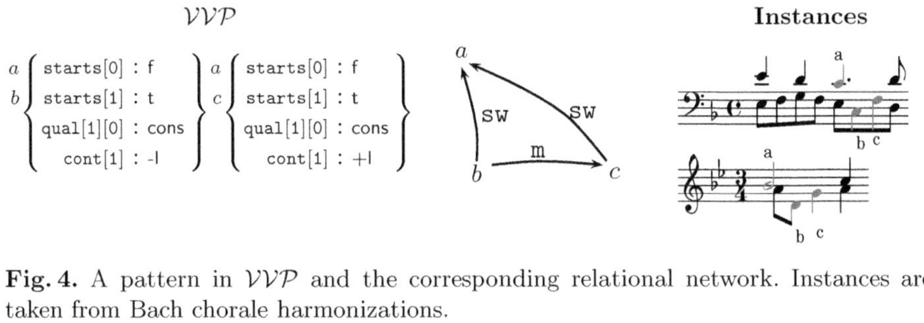


**Fig. 4.** A pattern in $\mathcal{VVP}$ and the corresponding relational network. Instances are taken from Bach chorale harmonizations.

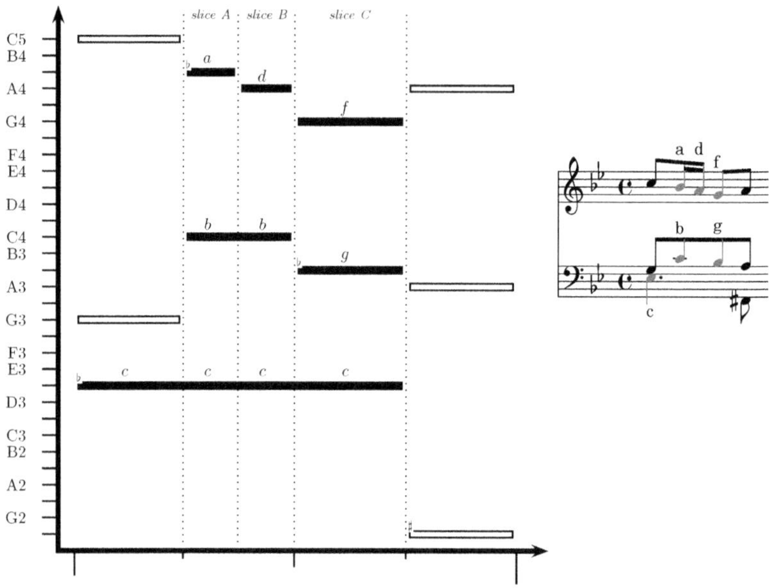


**Fig. 5.** Slicing of the polyphonic texture and labelling of events

*Example 2.*

$$\begin{array}{ccc} \textit{slice } A & \textit{slice } B & \textit{slice } C \\ \begin{matrix} a \\ b \\ c \end{matrix}\left\{\begin{matrix} \texttt{starts}[0] : \texttt{t} \\ \texttt{starts}[1] : \texttt{t} \\ \texttt{starts}[2] : \texttt{f} \end{matrix}\right\} & \begin{matrix} d \\ b \\ c \end{matrix}\left\{\begin{matrix} \texttt{starts}[0] : \texttt{t} \\ \texttt{starts}[1] : \texttt{f} \\ \texttt{starts}[2] : \texttt{f} \end{matrix}\right\} & \begin{matrix} f \\ g \\ c \end{matrix}\left\{\begin{matrix} \texttt{starts}[0] : \texttt{t} \\ \texttt{starts}[1] : \texttt{t} \\ \texttt{starts}[2] : \texttt{f} \end{matrix}\right\} \end{array}$$

The semantics of $\mathcal{VVP}$ patterns is defined in terms of a translation to $\mathcal{R}$. We do this via the five rules presented in Fig. 6. The first three rules concern temporal relations and all refer to the start of a new event (indicated by a feature of the form $\texttt{starts}[\gamma] : \texttt{t}$). The first rule concerns the case where two events start in the same slice and a $\texttt{st}(a, b)$ temporal relation is created. The second rule specifies when to create a $\texttt{sw}(a, b)$ relation. The event variable $a$ corresponds to an event that is starting in the current slice, while variable $b$ corresponds to an event that has started in the past and is being prolonged into the current slice. The third rule assigns a $\texttt{m}(a, b)$ temporal relation whenever an event starts. The event $b$ has to be in the same voice as the event $a$ (Fig. 6 and henceforth, we indicate this by horizontally aligning the labels). Finally, rules four and five simply translate the information about voices and musical features into their relational form. In rule 5, for every voice present in the musical feature $\tau[\gamma_i] \ldots : \mu$, the atom $\tau(a, \ldots, \mu)$ must refer to the event variable that labels that voice.

| $\mathcal{VVP}$ | $\mathcal{R}$ |
|---|---|
| 1) $\left\{\begin{array}{rl} & \dots \\ a & \texttt{starts}[\gamma_i] : \texttt{t} \\ & \dots \\ b & \texttt{starts}[\gamma_j] : \texttt{t} \\ & \dots \end{array}\right\}$ | $\rightsquigarrow \texttt{st}(a,b)$ |
| 2) $\left\{\begin{array}{rl} & \dots \\ a & \texttt{starts}[\gamma_i] : \texttt{t} \\ & \dots \\ b & \texttt{starts}[\gamma_j] : \texttt{f} \\ & \dots \end{array}\right\}$ | $\rightsquigarrow \texttt{sw}(a,b)$ |
| 3) $a\left\{\ \right\} b\left\{\begin{array}{l} \dots \\ \texttt{starts}[\gamma_i] : \texttt{t} \\ \dots \end{array}\right\}$ | $\rightsquigarrow \texttt{m}(a,b)$ |
| 4) $a\left\{\begin{array}{l} \dots \\ \texttt{starts}[\gamma_i] : \texttt{t} \\ \dots \end{array}\right\}$ | $\rightsquigarrow \texttt{voice}(a,\gamma_i)$ |
| 5) $\left\{\begin{array}{rl} & \dots \\ a & \texttt{starts}[\gamma] : \texttt{t} \\ & \dots \\ & \tau[\gamma]\dots : \mu \\ & \dots \end{array}\right\}$ | $\rightsquigarrow \tau(a,\dots,\mu)$ |

**Fig. 6.** Interpretation of $\mathcal{VVP}$ in terms of $\mathcal{R}$

*Example 3.* The following $\mathcal{VVP}$ pattern is translated to a relational pattern by applying the rules of Fig. 6:

$$\left\{\begin{array}{rl} a & \texttt{starts[0]} : \texttt{t} \\ b & \texttt{starts[1]} : \texttt{t} \\ & \texttt{qual[0][1]} : \texttt{cons} \end{array}\right\} \left\{\begin{array}{rl} a & \texttt{starts[0]} : \texttt{f} \\ c & \texttt{starts[1]} : \texttt{t} \end{array}\right\} \rightsquigarrow \begin{array}{ll} \text{rule 1)} & \texttt{st}(a,b) \\ \text{rule 2)} & \texttt{sw}(c,a) \\ \text{rule 3)} & \texttt{m}(b,c) \\ \text{rule 4)} & \texttt{voice}(a,0),\ \texttt{voice}(b,1), \\ & \texttt{voice}(c,1) \\ \text{rule 5)} & \texttt{qual}(a,b,\texttt{cons}) \end{array}$$

Note that the semantics presented here concerns a restricted form of $\mathcal{VVP}$, where for example the $\texttt{starts}[\gamma] : \mu$ features are always present to explicitly indicate the temporal configuration of a slice. The restriction is defined by the three well-formedness rules expressed below.

**Definition 3.** A *well-formed* $\mathcal{VVP}$ *pattern* contains only feature sets of the following form and satisfying the following rules:

$$\left\{\begin{array}{l} \texttt{starts}[\gamma_0] : \mu \\ \texttt{starts}[\gamma_1] : \mu \\ \dots \\ \texttt{starts}[\gamma_n] : \mu \\ \dots \\ \tau[\gamma_i]\dots : \mu \\ \dots \end{array}\right\}$$

- Every feature set refers to the same voices and for every voice it contains a feature of the form $\texttt{starts}[\gamma] : \mu$
- There is at least one feature $\texttt{starts}[\gamma] : \texttt{t}$ per set
- Features of the form $\tau[\gamma_i]\dots : \mu$ only appear in a set where the feature $\texttt{starts}[\gamma_i] : \texttt{t}$ appears

The purpose of the first rule is to ensure that patterns can be interpreted as compact in time. The second rule is necessary as the opposite yields a set where no new event is starting, which does not have a clear interpretation in terms of events. The third is necessary as $\tau[\gamma_i] \ldots : \mu$ is interpreted as assigning a feature to the current event in voice $\gamma_i$. To give a clear interpretation to $\mathcal{VVP}$ in terms of $\mathcal{R}$, we must avoid cases where a feature would be assigned to the continuation of an event (as this is impossible to represent in $\mathcal{R}$). Notice how a well-formed $\mathcal{VVP}$ pattern coincides with the typical use of regular expressions in a Humdrum pattern as described in Sect. 1: a slice contains information about the start or continuations of events, and about musical features.

## 4 Subsumption for Vertical Viewpoints

In this section, the relation of subsumption for $\mathcal{VVP}$ is defined. This relation is entirely different than the $\theta$-subsumption of $\mathcal{R}$ defined in Sect. 2 and it directly corresponds to how pattern matching is implemented in $\mathcal{VVP}$. We show that subsumption in $\mathcal{VVP}$ is sound and complete with respect to subsumption in $\mathcal{R}$, taking into account the $\mathcal{VVP}$ to $\mathcal{R}$ translation defined in the previous Sect. 3. Subsumption of vertical viewpoints is defined as follows.

**Definition 4.** Given two $\mathcal{VVP}$ patterns $v_1$ and $v_2$, we say that $v_1$ *subsumes* $v_2$, written $v_1 \succeq_{\mathcal{VVP}} v_2$, iff there is a mapping from the sets of $v_1$ to the sets of $v_2$ such that consecutive sets in $v_1$ are mapped to consecutive sets in $v_2$; and that that every set in $v_1$ is mapped to a superset in $v_2$.

Informally, we say that $v_1$ can be *aligned* with $v_2$ such that aligned sets are in a superset relation. Figure 7 illustrates the notion of subsumption in $\mathcal{VVP}$. The figure shows successive refinements of a suspension pattern, starting from a two-event pattern that does not contain any musical relations and culminating with a three-event pattern that expresses the appropriate temporal relations, dissonance and consonance, and resolution by a melodic step down.

**Claim 1.** Subsumption for $\mathcal{VVP}$ is sound and complete with respect to subsumption in $\mathcal{R}$. That is, given $v_1$ and $v_2$, and their corresponding relational patterns $\mathcal{R}(v_1)$ and $\mathcal{R}(v_2)$, we have $v_1 \succeq_{\mathcal{VVP}} v_2$ iff $\mathcal{R}(v_1) \succeq_{\mathcal{R}} \mathcal{R}(v_2)$.

*Case* $v_1 \succeq_{\mathcal{VVP}} v_2$. Suppose that it is not true that $\mathcal{R}(v_1) \succeq_{\mathcal{R}} \mathcal{R}(v_2)$, then for all variable substitution $\theta$, it must be the case that $\mathcal{R}(v_1)\theta$ contains at least one atom that $\mathcal{R}(v_2)$ does not contain. Then there must be one set in $v_1$ that contains a feature that its corresponding set in $v_2$ does not contain. This contradicts $v_1 \succeq_{\mathcal{VVP}} v_2$, so clearly we must have $\mathcal{R}(v_1) \succeq_{\mathcal{R}} \mathcal{R}(v_2)$. □

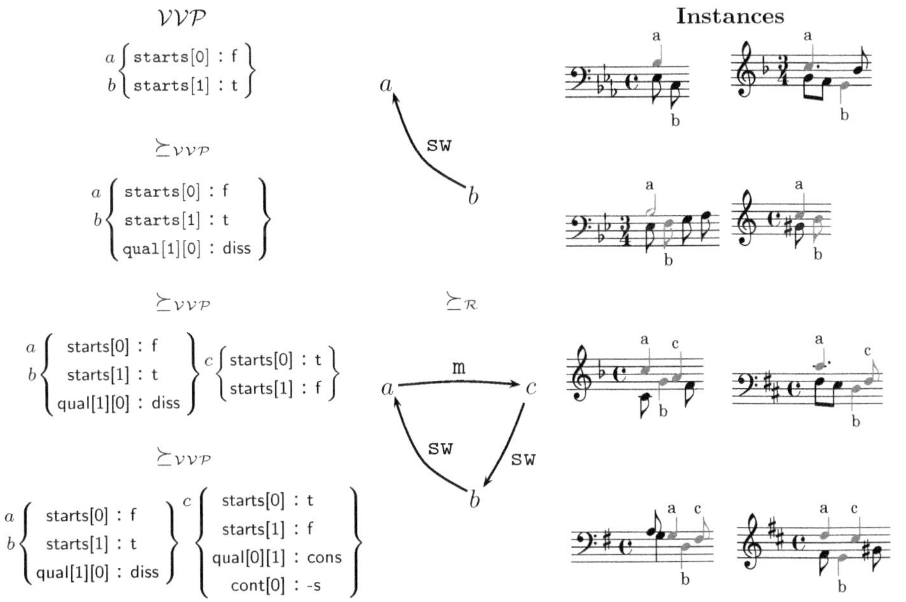


**Fig. 7.** Subsumption in $\mathcal{VVP}$ and the subsumption of corresponding relational networks. Instances are selected from Bach chorale harmonizations.

*Case* $\mathcal{R}(v_1) \succeq_{\mathcal{VVP}} \mathcal{R}(v_2)$. The proof rests on three observations: i) given the labelling of events in $v_1$ and $v_2$, the substitution $\theta$ implied by $\mathcal{R}(v_1) \succeq_{\mathcal{R}} \mathcal{R}(v_2)$ induces a mapping from the feature sets of $v_1$ to the feature sets of $v_2$; ii) any two consecutive sets in $v_1$ are mapped to consecutive sets in $v_2$; and iii) any set in $v_1$ is necessarily mapped to a superset in $v_2$. From these conditions, it follows that $v_1 \succeq_{\mathcal{VVP}} v_2$.

*First observation.* Consider any set in $v_1$. By definition, there is at least one $\texttt{starts}[\gamma_i]$ : t feature, and it is assigned a fresh label $a$:

$$v_1\!: \quad \ldots \quad \left\{ \begin{array}{rl} & \ldots \\ a & \texttt{starts}[\gamma_i] : \texttt{t} \\ & \ldots \\ b & \texttt{starts}[\gamma_j] : \texttt{t} \\ & \ldots \end{array} \right\} \quad \ldots \qquad \begin{array}{l} a\theta = a' \\ b\theta = b' \end{array}$$

$$v_2\!: \quad \ldots \quad \ldots \quad \left\{ \begin{array}{rl} & \ldots \\ a' & \texttt{starts}[\gamma_i] : \texttt{t} \\ & \ldots \\ b' & \texttt{starts}[\gamma_j] : \texttt{t} \\ & \ldots \end{array} \right\} \quad \ldots$$

The label $a$ is is mapped to some $a'$ by the substitution $\theta$ implied by $\mathcal{R}(v_1) \succeq_{\mathcal{R}} \mathcal{R}(v_2)$. As every $\mathtt{starts}[\gamma_i] : \mathtt{t}$ feature in $v_2$ is also assigned a fresh label, the $a'$ unambiguously refers to exactly one set in $v_2$. That is, there is a single set in $v_2$ where the label $a'$ is associated with a $\mathtt{starts}[\gamma_i] : \mathtt{t}$ feature. Note that because the feature $\mathtt{starts}[\gamma_i] : \mathtt{t}$ implies an atom $\mathtt{voice}(a', i)$ in $\mathcal{R}(v_1)\theta$, it follows that $\mathcal{R}(v_2)$ must also contain that atom and that the $\mathtt{starts}[\gamma_i] : \mathtt{t}$ features in $v_1$ and $v_2$ are in the same voice.

In the case where a second fresh label exists (e.g. $b$ in the example above), it must also map to the same set in $v_2$. Otherwise, we would have $\mathtt{st}(a', b')$ in $\mathcal{R}(v_1)\theta$, but $\mathcal{R}(v_2)$ would not contain $\mathtt{st}(a', b')$ as the $\mathtt{starts}[\gamma] : \mathtt{t}$ features associated with $a'$ and $b'$ would appear in different sets, which would contradict $\mathcal{R}(v_1) \succeq_{\mathcal{R}} \mathcal{R}(v_2)$. This holds for any other fresh label and ensures that we have a proper mapping: any set in $v_1$ is coupled with exactly one set in $v_2$.

*Second observation.* Consider two consecutive sets in $v_1$ and their corresponding sets in $v_2$:

$$v_1: \quad {}^{a}\Big\{\ \Big\}\ {}^{b}\Big\{\mathtt{starts}[\gamma_i] : \mathtt{t}\Big\} \qquad\qquad \begin{array}{l} a\theta = a' \\ b\theta = b' \end{array}$$

$$v_2: \quad {}^{a'}\Big\{\mathtt{starts}[\gamma_i] : \mathtt{t}\Big\} \cdots {}^{a'}\Big\{\mathtt{starts}[\gamma_i] : \mathtt{f}\Big\}\ {}^{b'}\Big\{\mathtt{starts}[\gamma_i] : \mathtt{t}\Big\}$$

Clearly, $\mathcal{R}(v_1)\theta$ and $\mathcal{R}(v_2)$ both contain a $\mathtt{m}(a', b')$ relation. This is only possible if $v_2$ contains the appropriate sucessive sets. Now suppose as above that the sucessive sets in $v_2$ are not consecutive:

$$v_1: \quad \begin{smallmatrix} a \\ c \end{smallmatrix}\Big\{\ \Big\}\ \begin{smallmatrix} b \\ d/c \end{smallmatrix}\left\{\begin{array}{l} \mathtt{starts}[\gamma_i] : \mathtt{t} \\ \mathtt{starts}[\gamma_j] : \mu \end{array}\right\} \qquad\qquad \begin{array}{l} a\theta = a' \\ b\theta = b' \\ c\theta = c' \\ d\theta = d' \end{array}$$

$$v_2: \quad \begin{smallmatrix} a' \\ c' \end{smallmatrix}\Big\{\ \Big\}\ \begin{smallmatrix} a'' \\ c'' \end{smallmatrix}\left\{\begin{array}{l} \\ \mathtt{starts}[\gamma_j] : \mathtt{t} \end{array}\right\}\ \begin{smallmatrix} b' \\ d' \end{smallmatrix}\left\{\begin{array}{l} \mathtt{starts}[\gamma_i] : \mathtt{t} \\ \\ \end{array}\right\}$$

Consider one intervening feature set in $v_2$ (shown above with labels $a''$, $c''$). That set must have at least one $\mathtt{starts}[\gamma_j] : \mathtt{t}$ feature that we suppose here in voice $j$. Both values of $\mathtt{starts}[\gamma_j] : \mu$ in the second set of $v_1$ lead to a contradiction. If $\mathtt{starts}[\gamma_j] : \mathtt{t}$, then $\mathcal{R}(v_1)\theta$ contains a $\mathtt{m}(c', d')$ relation that $\mathcal{R}(v_2)$ clearly does not contain. If $\mathtt{starts}[\gamma_j] : \mathtt{f}$, then the label $c$ is prolonged in $v_1$ and $\mathcal{R}(v_1)\theta$ contains a $\mathtt{sw}(b', c')$ relation that $\mathcal{R}(v_2)$ cannot contain as the equivalent label $c'$ is not prolonged by virtue the $\mathtt{starts}[\gamma_j] : \mathtt{t}$ feature in the intervening set, which creates the fresh label $c''$. A similar reasoning applies for more than one intervening set and hence the observation holds.

*Third observation.* The mapping induced by $\mathcal{R}(v_1) \succeq_{\mathcal{R}} \mathcal{R}(v_2)$ necessarily maps every set in $v_1$ to a superset in $v_2$. Suppose this is not true. Then some set in $v_1$ contains a feature that its corresponding set in $v_2$ does not. If that feature is of the form $\texttt{starts}[\gamma_i] : \mu$, then clearly $\mathcal{R}(v_1)\theta$ contains some temporal relation that $\mathcal{R}(v_2)$ does not contain. If that feature is of the form $\tau[\gamma_i]\ldots : \mu$, then by definition the set in $v_1$ also contains a $\mathsf{starts}[\gamma_i] : \mathsf{t}$ feature and by rule 5 of Fig. 6, $\mathcal{R}(v_1)\theta$ will contain an atom of the form $\tau(\varepsilon, \ldots, \varepsilon, \mu)$ that $\mathcal{R}(v_2)$ will not contain. In both cases, this contradicts $\mathcal{R}(v_1) \succeq_{\mathcal{R}} \mathcal{R}(v_2)$ and we must have that the sets of $v_1$ are aligned with supersets in $v_2$. By the three observations above, we must have that $v_1 \succeq_{\mathcal{VVP}} v_2$. □

The result is significant as it establishes that the semantics given in Sect. 3 is fully consistent with the implementation of pattern matching in $\mathcal{VVP}$, which is a direct application of subsumption. The steps that were taken in order to establish the result (e.g. the well-formedness constraints of Sect. 3) reveal the capacities and limitations of $\mathcal{VVP}$ patterns. As pointed out in Sect. 1, the analysis presented here provides insight, by analogy, for understanding Humdrum patterns and, in general, polyphonic patterns based on the idea of vertical slices.

## 5 Discussion

This paper has formalized the vertical viewpoint language $\mathcal{VVP}$ for expressing polyphonic patterns. This language inspired by the Humdrum pattern representation and the eventual goal of this work is to fully incorporate $\mathcal{VVP}$ within a Humdrum search tool.

A semantics for the pattern language was provided by developing a translation of $\mathcal{VVP}$ patterns to a more powerful relational network representation $\mathcal{R}$, itself a subset of predicate logic. The $\mathcal{VVP}$ language is restricted to patterns that are compact in time, in particular patterns where successive events are necessarily contiguous. Pattern languages such as Humdrum and $\mathcal{SPP}$ are more general as they allow configurations that express gaps between successive events [8].

The $\mathcal{VVP}$ language supports the important inference of subsumption, and this is done in a way entirely different to the underlying $\mathcal{R}$ representation, which requires the computation of a mapping between the variables in two relational networks. In $\mathcal{VVP}$ subsumption is computed in a manner similar to string matching, with the exception that characters are now sets and that instead of testing for equality, we test for the subset relation. In the future, we intend to detail how efficient string matching algorithms can be adapted to compute subsumption in $\mathcal{VVP}$.

Subsumption is an essential notion of any pattern representation, as it allows the development of patterns by progressive specialization and generalization thereby matching fewer or more instances. Subsumption is also essential within any data mining application that employs the pattern representation as it permits a refinement search of a subsumption taxonomy of patterns [1]. A main result of this paper is that subsumption $\mathcal{VVP}$ is sound and complete with respect to subsumption in $\mathcal{R}$. In the future we intend to explore how subsumption taxonomies in $\mathcal{VVP}$ can be used to index large score databases in Humdrum format for rapid retrieval to musicological queries, in a way similar to the use of concepts in Description Logics [11]. Finally, we also intend to analyze other representation of polyphonic patterns [12,13,14,15] by similarly expressing them in terms of a relational language. As was the case for $\mathcal{VVP}$, we anticipate that it will be only possible for subsets of the languages, and that our efforts to find such subsets will help to shed light on the strengths and weaknesses of existing polyphonic pattern languages.

## References

1. Conklin, D., Bergeron, M.: Representation and discovery of feature set patterns in music. Computer Music Journal 32(1), 60–70 (2008)
2. Maxwell, J.: An Artifical Intelligence Aproach to Computer-Implemented Analysis of Harmony in Tonal Music. PhD thesis, Indiana University (1984)
3. Conklin, D., Bergeron, M.: Discovery of contrapuntal patterns. In: ISMIR 2010: 11th International Society for Music Information Retrieval Conference, Utrecht, The Netherlands, pp. 201–206 (2010)
4. Huron, D.: Music information processing using the Humdrum Toolkit: Concepts, examples, and lessons. Computer Music Journal 26, 11–26 (2002)
5. Wild, J.: A review of the Humdrum Toolkit: Unix tools for musical research, created by David Huron. Music Theory Online 2 (1996)
6. Jan, S.: Meme hunting with the Humdrum toolkit: Principles, problems, and prospects. Computer Music Journal 28(4), 68–84 (2004)
7. De Raedt, L.: A perspective on inductive databases. ACM SIGKDD Explorations Newsletter 4(2), 69–77 (2002)
8. Bergeron, M., Conklin, D.: Temporal patterns in polyphony. In: Chew, E., Childs, A., Chuan, C.H. (eds.) MCM 2009. Communications in Computer and Information Science, vol. 38, pp. 32–42. Springer, Heidelberg (2009)
9. Allen, J.: Maintaining knowledge about temporal intervals. Communications of the ACM 26(11), 832–843 (1983)
10. Nienhuys-Cheng, S.-H., de Wolf, R.: Foundations of Inductive Logic Programming. LNCS, vol. 1228. Springer, Heidelberg (1997)
11. Baader, F.: Description logics. In: Tessaris, S., Franconi, E., Eiter, T., Gutierrez, C., Handschuh, S., Rousset, M.-C., Schmidt, R.A. (eds.) Reasoning Web. LNCS, vol. 5689, pp. 1–39. Springer, Heidelberg (2009)
12. Bergeron, M., Conklin, D.: Structured polyphonic patterns. In: Ninth International Conference on Music Information Retrieval, Philadelphia, USA, pp. 69–74 (2008)
13. Meredith, D., Lemström, K., Wiggins, G.A.: Algorithms for discovering repeated patterns in multidimensional representations of polyphonic music. Journal of New Music Research 31(4), 321–345 (2002)
14. Meudic, B., Saint-James, E.: Automatic extraction of approximate repetitions in polyphonic MIDI files based on perceptive criteria. In: Wiil, U.K. (ed.) CMMR 2003. LNCS, vol. 2771, pp. 124–142. Springer, Heidelberg (2004)
15. Ukkonen, E., Lemström, K., Mäkinen, V.: Geometric algorithms for transposition invariant content based music retrieval. In: International Conference on Music Information Retrieval (ISMIR), Baltimore, Maryland, USA, pp. 193–199 (2003)

# Building Topological Spaces for Musical Objects

Louis Bigo[1,2], Jean-Louis Giavitto[2], and Antoine Spicher[1]

[1] LACL/Université Paris-Est Creteil
louis.bigo@lacl.fr, antoine.spicher@u-pec.fr
[2] UMR CNRS STMS 9912/IRCAM
jean-louis.giavitto@ircam.fr

**Abstract.** The development of spatial representations of musical objects allows for a reformulation of algorithmic problems arising in musical theory, fosters novel classifications and provides new computational tools. In this paper, we show how a topological representation for $n$-note chords associated with the degrees of the diatonic scale and for the All-Interval Series (AIS) can be automatically built using MGS, a rule-based spatial programming language. Then, we suggest a new categorization for AIS based on their spatial construction.

**Keywords:** Spatial Computing, Algebraic Topology, All-Interval Series, $n$-chords and Diatonic Scale, Self-assembly, $n$-Hamiltonian Path.

## 1 Introduction

The algebraic nature of many musical formalizations has been very early assessed: from the equal temperament to canon, algebraic objects have been used to study combinatorial properties and classify musical structures. Recently, a fresh look on these structures has emerged focusing on topological or geometrical representations. For example, one can characterize harmonic paths in orbifolds [1,2,3] or build topological spaces embedding musical relationships in their neighborhood relationships [4].

In this paper we will follow this line of research by using tools developed in *spatial computing*. Spatial computing is an emergent domain in computer science that enlightens the notion of space in computations either as a *resource*, a *result* or a *constraint* [5,6]. Spatial computing has proven to be a fruitful paradigm for the (re-)design of algorithms tackling problems embedded in space or having a spatial extension.

In the following, we propose two studies of paradigmatic theoretical music problems from a spatial computing perspective. This paper is organized as follows. Section 2 provides a brief introduction to MGS, a domain specific programming language designed to investigate the spatial computing approach. Subsection 2.3 illustrates the MGS concepts on a self-assembly algorithm for the generic computation of a topological representation of chord series initially proposed by Guérino Mazzola. In Section 3 we propose a spatial representation of the constraints defining an All-Interval Series (AIS). This representation enables

C. Agon et al. (Eds.): MCM 2011, LNAI 6726, pp. 13–28, 2011.

us to enumerate the AIS and to classify them from a topological perspective. The paper ends with a conclusion and a discussion about future works.

## 2 Presentation of the MGS Programming Language

MGS is an experimental domain specific language dedicated to spatial computing [7,8]. MGS concepts are based on well established notions in algebraic topology [9] and relies on the use of rules to compute declaratively with spatial data structures.

In MGS, all data structures are unified under the notion of *topological collection*: an *abstract combinatorial complex* (ACC) labeled with arbitrary values. The ACC builds a space in a combinatorial way through more simple objects called *cells* [10]. It acts as a container and the values as the elements of the data structure. *Transformations* of topological collections are defined by rewriting rules [11] specifying replacement of sub-collections that can be recursively performed to build new spaces.

### 2.1 Topological Collections

An *abstract combinatorial complex* $K = (\mathsf{C}, \prec, [\cdot])$ is a set $\mathsf{C}$ of abstract elements, called *cells*, provided with a partial order $\prec$, called the *boundary relation*, and with a *dimension* function $[\cdot] : \mathsf{C} \to \mathbb{N}$ such that for each $c$ and $c'$ in $\mathsf{C}$, $c \prec c' \Rightarrow [c] < [c']$. We write $c \in K$ when a cell $c$ is a cell of $\mathsf{C}$.

A cell of dimension 0 corresponds to a point, a 1-dimensional cell corresponds to a line (an edge), a cell of dimension 2 is a surface (*e.g.* a polygon), etc. A cell of dimension $p$ is called a $p$-cell. For example, a graph is an ACC built only with 0- and 1-cells. An other example is pictured in Fig. 1.

In the context of this paper, we write $\partial_K c$ for the sub-ACC made of the cells of $K$ lower than $c$ for the relation $\prec$: $\partial_K c = (\mathsf{C}', \prec \cap\, \mathsf{C}' \times \mathsf{C}', [\cdot])$ where $\mathsf{C}' = \{c' \mid c' \prec c\}$. This ACC is called the *boundary* of $c$. The *faces* of a $p$-cell $c$

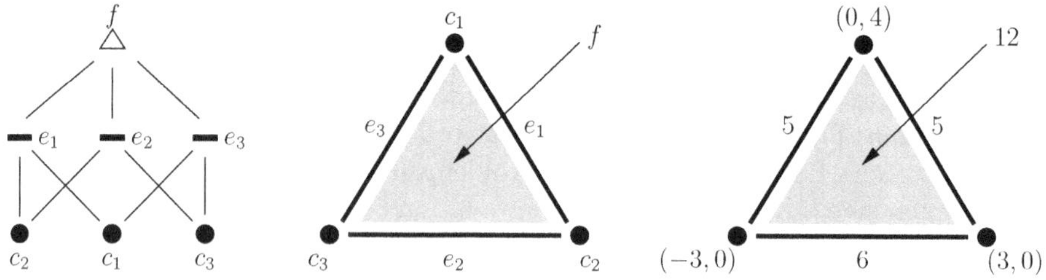


**Fig. 1.** On the left, the Hasse diagram of boundary relationship of the ACC given in the middle: it is composed of three 0-cells ($c_1$, $c_2$, $c_3$), of three 1-cells ($e_1$, $e_2$, $e_3$) and of a single 2-cells ($f$). The three edges are the faces of $f$, and therefore $f$ is a common coface of $e_1$, $e_2$ and $e_3$. On the right, a topological collection associates data with the cells: positions with vertexes, lengths with edges and area with $f$.

are the $(p-1)$-cells $c'$ of $\partial_K c$ and we write $c > c'$ or $c' < c$ and $c'$ is called a *coface* of $c$.

Two cells $c$ and $c'$ are *$q$-neighbor* either if they have a common border of dimension $q$ or if they are in the boundary of a $q$-cell (of higher dimension). If the two cells are of dimension $p$, we say that they are *$(p,q)$-neighbor*. A *$(p,q)$-path* is a sequence of $p$-cells such that two consecutive cells are *$q$-neighbor*. For example, the usual notion of path in a graph (a sequence of vertexes such that from each of its vertexes there is an edge to the next vertex in the sequence) corresponds to the notion of $(0,1)$-path.

*Topological Collections.* A topological collection $C$ is a function that associates a value with a cell in an ACC, see Fig. 1. Thus the notation $C(c)$ refers to the value of $C$ on cell $c$.

We write $|C|$ for the set of cells for which $C$ is defined. The collection $C$ can be written as a formal sum $\sum_{c \in |C|} v_c \cdot c$ where $v_c \stackrel{\mathrm{df}}{=} C(c)$. With this notation, the underlying ACC is left implicit but can usually be recovered from the context. By convention, when we write a collection $C$ as a sum $C = v_1 \cdot c_1 + \cdots + v_p \cdot c_p$, we insist that all $c_i$ are distinct. Notice that this addition is associative and commutative. This notation is directly used in MGS to build new topological collections on arbitrary ACC of any dimension.

## 2.2 Transformations

The mechanics of rewriting systems are familiar to anyone who has done arithmetic simplifications: an arithmetic expression can be simplified by repeatedly replacing parts of the term (*subterms*) with other subterms. For example, $\frac{1}{2} \cdot \frac{2}{3} \cdot \frac{3}{4} \Rightarrow \frac{1}{3} \cdot \frac{3}{4} \Rightarrow \frac{1}{4}$. The rule that is applied here is: $\frac{M}{N} \cdot \frac{N}{P} \Rightarrow \frac{M}{P}$, where $M$, $N$ and $P$ are pattern variables representing arbitrary non-null numbers. A transformation generalizes this process to topological collections.

Topological collections are transformed using sets of rules called *transformations*. A rule is a pair `pattern => expression`. When a rule is applied on a topological collection, a sub-collection matching with the `pattern` is replaced by the topological collection computed by the evaluation of `expression`. There exist several ways to control the application of a set of rules on a collection but these details are not necessary for the comprehension of the work presented here. A formal specification of topological rewriting is given in [11]. We sketch here only the specification of patterns.

A *pattern variable* specifies a cell to be matched in the topological collection together with some (optional) guard. For example the expression `x / x == 3` matches a cell labeled with the value 3. The guard is the predicate after the symbol `/`. The variable `x` can be used in the guard (and elsewhere in the rule) to denote the value of the matched cell or the cell itself, following the context (in case of ambiguity, the variable always denotes the associated value).

A pattern is a *composition* of pattern variables. There are three composition operators:

1. The composition denoted by a simple juxtaposition (e.g., "x y") does not constraint the arguments of the composition.
2. When two pattern variables are composed using a comma (e.g., "x, y"), it means that the cells matched by x and y must be $p$-neighbors. The default value for $p$ is 1 and can be explicitly specified during the application of the transformation if needed.
3. The last composition operator corresponds to the face operator: a pattern "x < y" (resp. "x > y") matches two cells $c_{\mathtt{x}}$ and $c_{\mathtt{y}}$ such that $c_{\mathtt{x}} < c_{\mathtt{y}}$ (resp. $c_{\mathtt{x}} > c_{\mathtt{y}}$).

Patterns are *linear*: two distinct pattern variables always refer to two distinct cells.

### 2.3 Self-assembly

In this subsection, we illustrate the notions of topological collection and transformation with a self-assembly mechanism used to compute a spatial representation of a chord set. Guérino Mazzola presents in [12] a topological representation of the seven trichords associated with the degrees of the diatonic scale (for example $C$ major) which appears to be a Möbius strip. We propose here a generic self-assembly process that achieves the same construction for any set of chords.

*Spatial Representation of a Chord.* We represents a $n$-note chord by a $(n-1)$-simplex. A *simplex* is a $p$-cell that has exactly $p+1$ faces. For instance, a bounded line is a simplex but a cube is not (a cube is a 3-cell but has 6 faces). Simplexes are often represented geometrically as the convex hull of their vertexes as shown in Fig. 2 for $p$-simplexes with $p \in \{0, 1, 2, 3\}$. In the simplicial representation of chord, a 0-cell represents a single note.

The first step of a spatial representation of a set of chords, consists in giving a simplicial representation (as presented on Fig. 2) of each chord. As an example, for the spatial representation of the $C$ major tonality, the seven degrees

$$I_C = \{C, E, G\} \quad II_C = \{D, F, A\} \quad III_C = \{E, G, B\}$$
$$IV_C = \{F, A, C\} \quad V_C = \{G, B, D\} \quad VI_C = \{A, C, E\} \quad VII_C = \{B, D, F\}$$

are associated with seven 2-simplexes. The complex associated with a chord is thus represented by one 2-cell (a triangular surface) whose boundary is composed of three 1-cells (edges) and three 0-cells (vertexes) respectively associated with three 2-note chords and a single note (see Fig. 2).

*The Space of Chords.* To gather in the same ACC all the considered chord simplexes and their respective boundaries, we must identify the various simplexes representing the same chord. For instance, merging chords $I_C$ and $III_C$ requires us to identify three elements: the vertexes $E$ and $G$, and the two edges $\{E, G\}$ as shown on Fig. 3.

We propose to unify these elements by using a self-assembly growing process [13] based on the identification of the cells boundaries. This operation is

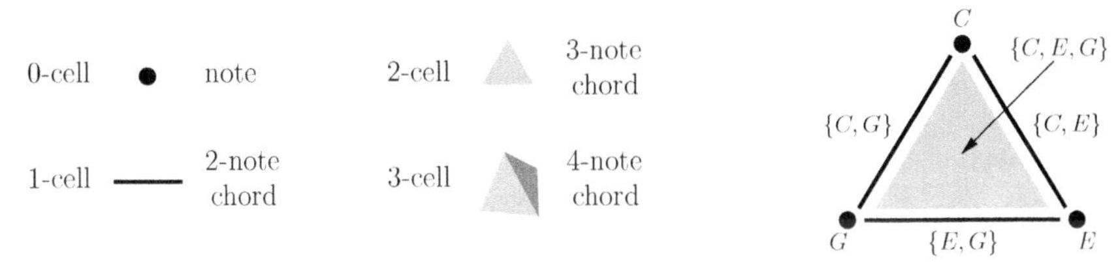


**Fig. 2.** A chord represented as a simplex. The complex on the right corresponds to first degree $I_C$ of the $C$ major tonality and all 2-note chords and notes included in it.

not elementary because the identification must occur at every dimensions. A simple way to compute the identification is to iteratively apply, until a fixed point is reached, the merge of topological cells that exactly have the same faces. The corresponding topological surgery can be expressed in the MGS syntax as follows:

```
transformation identification = {
    s1 s2 / (s1 == s2 & faces(s1) == faces(s2))
    => let c = new_cell (dim s1)
                        (faces s1)
                        (union (cofaces s1) (cofaces s2))
       in  s1·c
}
```

The primitive `new_cell` $p\ f\ cf$ returns a new $p$-cell with faces $f$ and cofaces $cf$. The rule specifies that two elements `s1` and `s2`, having the same label and the same faces in their boundaries, merge into a new element `c` (whose cofaces are the union of the cofaces of `s1` and `s2`) labeled by `s1` (which is also the label of `s2`).

In Fig. 3, the transformation `identification` is called twice. At the first application (from the left complex to the middle), vertexes are identified. The two topological operations are made in parallel. At the second application (from the complex in the middle to the right), the two edges from $E$ to $G$ that share the

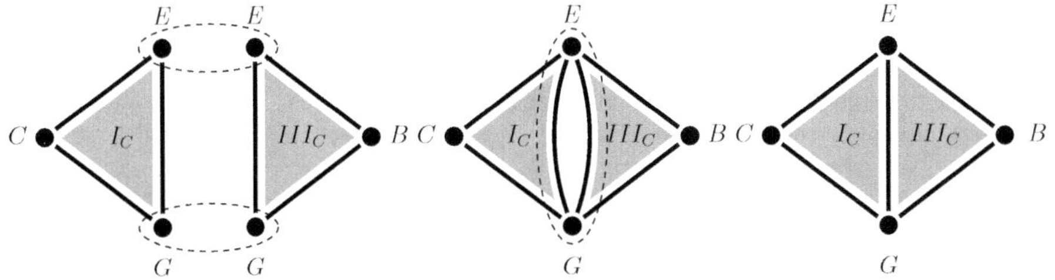


**Fig. 3.** Identification of boundaries

same boundary, are merged. The cofaces of the resulting edge are the 2-simplexes $I_C$ and $III_C$ corresponding to the union of the cofaces of the merged edges. Finally (on the right), no more merge operation can take place and the fixed point is reached.

As expected, the fixed point application of `identification` to the triadic chords $I_C$, ..., $VII_C$ builds the Möbius strip given in Figure 4. Notice that in [12], the dual complex is described: in this complex, a vertex represents a 3-note chord, an edge links two vertexes that share two notes and a face corresponds to a note common to three vertexes. Obviously, we can adapt the construction presented here to self-assemble this complex but our representation makes the dimension of a note independent of the number of notes in the considered chords.

*Higher Dimensional Spaces.* The previous transformation is generic with respect to the dimension of the simplexes. So, it can be applied without modification for the $p$-note chord for any $p$. For example, the geometrical representation of a four-note chord is a 3-simplex, that is a tetrahedron (for example $I_C = \{C, E, G, B\}$). The elaboration of the cellular complex associated with the 4-note chords of the seven degrees is difficult by hand. Thanks to the dimension-free definition of the transformation `identification`, the cellular complex is automatically computed by MGS. After computing the Euler characteristic and the orientability coefficient of the complex, its topology appears to be a *toroid* (the volume bounded by a torus). The one dimensional boundary of the Möbius strip exhibits the cycle of fifths whilst the two dimensional boundary of the toroid exhibits a surface $S$ composed of triangles representing 3-note chords. It is easier to interpret the dual $D$ of this surface: each vertex in $S$ corresponds to a polygonal surface in $D$ and conversely each surface in $S$ correspond to a vertex in $D$ (edges remain edges). The dual $D$ can be visualized as an hexagonal lattice interpreted as a kind of *tonnetz*. This lattice is generated by the intervals of third, fifth and seventh, cf. Fig. 5. These intervals can be perfect or diminished (for the fifths) and minor or major (for thirds and sevenths) in order to generate only notes in the $C$ major scale. This lattice includes the part of the circle of fifths composed by the notes of the $C$ major tonality.

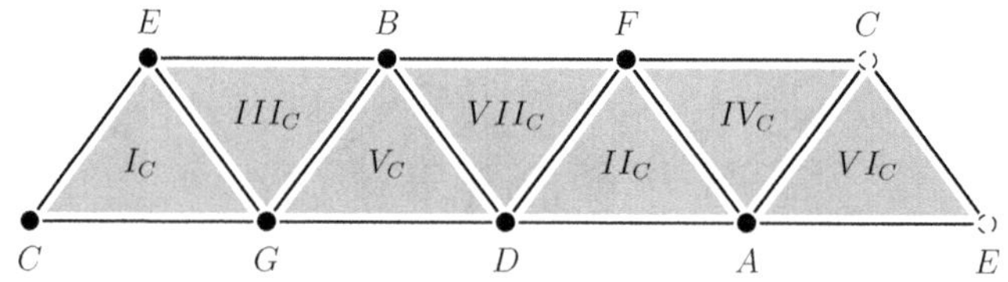


**Fig. 4.** Representation of tonality $C$ major. The edge and the vertexes at one end must be merged with the edge and the vertexes with the same label at the other end. The resulting surface is a Möbius strip.

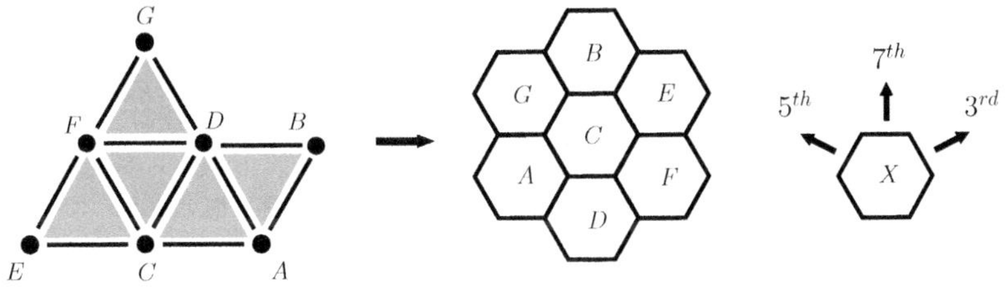


**Fig. 5.** The toroid boundary is the surface specified by the triangular lattice $S$ (unfolded on the left). Its dual is the hexagonal latice $D$ (on the center) where notes are organized following the intervals of fifth, seven and third (on the right).

# 3 All-Interval Series

In this section we illustrate a general methodology of spatial computing, using the enumeration and classification of *All-Interval Series* (AIS) as an example. We first make a short presentation of AIS. Then we propose a spatial specification of AIS. This specification is further used to characterize some noticeable AIS and to classify them.

## 3.1 Presentation of AIS

The enumeration and the classification of AIS is a widely known problem in the computing music community. An AIS is a twelve-tone series including eleven different intervals reduced in $\mathbb{Z}_{12}$. Such series contain many notable properties [14]. One of them is that the first and the last notes are always separated by a tritone interval. One of the most known use of this particular kind of series is probably on the *Lyric Suite* of Alban Berg:

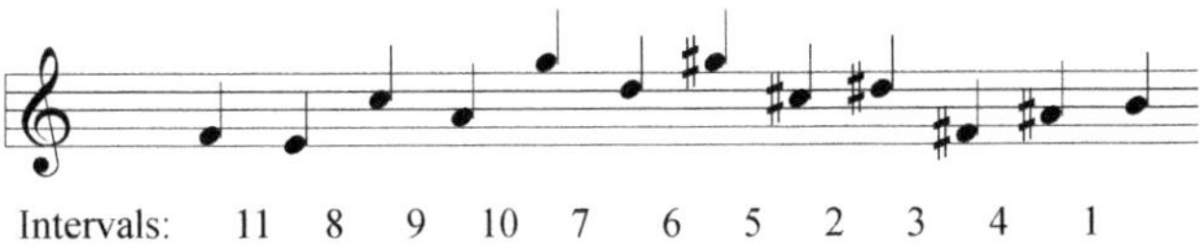


Other composers like Luigi Nono (e.g., *Il canto sospeso*) or Karlheinz Stockhausen (e.g., *Grüppen*, *Klavierstück IX*) used this material in their compositions. Different computing approaches, increasingly optimized, have been used to enumerate the totality of the AIS. One of the first enumeration was done by André Riotte [15] with the help of a FORTRAN program. This enumeration problem has quickly become a classical problem in Constraint Programming [16] and is now part of the 50 problems of the CSPLib [17]. Some previous works have been based on the enumeration of the All-Interval Chords, which is a similar problem [18].

Beyond the enumeration of the 3 856 AIS, composers and music analysts have been interested in finding pertinent criterions to classify them. André Riotte proposed a classification considering the harmonic content of the AIS. It consists for example in grouping together AIS containing a sub-sequence corresponding to the notes of particular scales or chords [15]. We will propose an example motivated by a similar goal in the next section. Elliot Carter investigated a classification enumerating all AIS containing in sequence the complete set of notes included in the All-Triad Hexachord (this 6-note chord is the only one containing the twelve possible triads) [19]. We can also mention an original classification from Franck Jedrzejewski based on knot theory [20].

### 3.2 Enumeration of All-Interval Series

We propose a new approach for the old problem of AIS enumeration. Our solution is based on the construction of a topological space adapted for their combinatorial analysis.

AIS can be seen as objects that inhabit a specific and well designed abstract space. Let's respectively call this space and the objects the *support space* and the *solutions* of the computation. The computation of the AIS consists in finding all the solutions in the support space. Space is considered as a *constraint* to guide the search: the more structured the space, the more efficient the computation. Thus, the challenge lies in the construction of a relevant support space together with an efficient search algorithms to find the solutions in the support space.

The notion of *search space* in the domain of search algorithms corresponds to the special case of support space where the possible solutions are the points of this space. In our example, instead of building a search space of $12! \simeq 479.10^6$ points and looking for the relevant solutions, we will build a much smaller support space composed of 12 points and we will look for paths in this space exhibiting some relevant properties: here, the paths associated with AIS are *Hamiltonian*.

A first naive method, corresponding to a brute force search, is given to explain our approach. Finally by adding more spatial structure to the support space, we increase the efficiency of the computation.

*A Brute Force Approach.* A brute force approach starts by considering the set of the twelve notes, and consists in computing all the possible permutations and keeping those having exactly eleven different intervals.

The spatial expression of this algorithm is straightforward. The support space of the computation is here the set of the twelve notes. From a spatial point of view, a set is a topological collection corresponding to a complete graph: each set element is associated with a vertex labelled by the element itself. The complete graph topology, meaning that there is an edge between each pair of vertexes, specifies that there is no predefined order between the elements. Fig. 6 presents the support space we consider for the computation of the AIS.

In this space, a permutation is an Hamiltonian path where each note appears only once. Using the definitions given section 2, these paths exactly correspond to the $(0, 1)$-paths of maximal length: the elements are the 0-cells that have to be neighbor by dimension 1 (that means linked by a 1-cell). An example of a

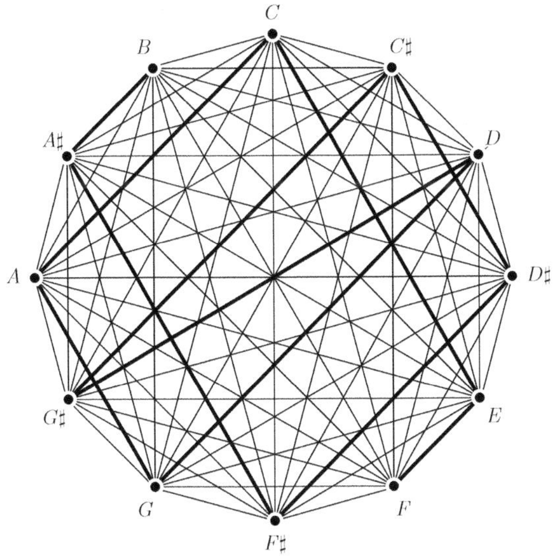


**Fig. 6.** Spatial representation of Alban Berg's AIS (in bold in the complete graph)

(0, 1)-path is given in bold on Fig. 6. Such path is easily expressed using an MGS pattern:

```
n0, n2, ..., n11 / ais(n0,...,n11)
```

In this pattern, the comma stands for the (0, 1)-neighborhood. The additional guard `ais(n0,...,n11)` checks if the Hamiltonian path matched by the n*i* is an AIS or not: the predicate `ais` holds if its arguments correspond exactly to 11 distinct intervals. Thus, this pattern specifies exactly the solutions of our computation.

Optimizing a spatial search algorithm can be done at the level of the solution specification and at the level of the support space definition. We propose in the following both optimizations.

*Optimization of the Solution Specification.* One can easily notice that the largest part of the computations involved by the brute force approach is useless. Indeed, it is possible in most of the cases to detect that a series is not all-interval before having determined the twelve notes. For example, if the interval between `n0` and `n1` is the same as the interval between `n1` and `n2`, there is no need to look further. In other words, predicate `ais` can be distributed along the path specification in the pattern.

Suppose that the edges of the graph pictured in Fig. 6 are labeled by the corresponding intervals (*e.g.* the edge between $E$ and $G$ is labeled by 3). The following pattern distributes the evaluation of the predicate `ais` along the matching of the path:

```
n0 < i1  > n1
   < i2  > n2  / (i2!=i1)
   < i3  > n3  / (i3!=i1) / (i3!=i2)
   ...
   < i11 > n11 / (i11!=i1) / ... / (i11!=i10)
```

This pattern is quite similar to the previous one. One of the major differences stands in the use of < i*p* > instead of the comma to express the neighborhood. The pattern variable i*p* corresponds to the interval between the notes n*p*′ and n*p* with $p' = p - 1$. Assuming that the edges are labelled by the corresponding interval, the path specifies an AIS if all the labels i*p* are distinct[1]. This property is ensured by the guards (i*p*!=i*q*) that are distributed along the pattern as required by the optimization. The search algorithm induced by this pattern corresponds to the algorithm of the FORTRAN program given in [14]. This new pattern speeds-up the computation of the AIS by a factor of 30.

*Optimization of the Support space.* There are two different kinds of constraint in the solution specification: *spatial constraints* and *logical constraints.* Spatial constraints specify which sub-parts of the support space are structural candidates for being solutions without taking labels into account. On the other hand, logical constraints are used to determine the solutions among the structural candidates by checking some formulas on labels. The two previous patterns are spatially equivalent. In fact, the previous optimization does not decrease the number of structural candidates but only reduces the number of visited candidates from 12! to about $9.10^6$ during the computation by avoiding the whole construction of wrongly started paths.

From a spatial point of view, it is more interesting to optimize the search by reducing the number of structural candidates. Ideally, we are looking for a totally spatial solution specification (without any logical constraint).

In our example, the proliferation of structural candidates comes from the lack of spatial distinction between intervals. Indeed, each interval is instantiated several times: for example, the semitone corresponds to the edge between $C$ and $C\sharp$, to the edge between $C\sharp$ and $D$, etc. Therefore, we propose, for each interval $i$, to add in the support space, a 2-cell whose faces are exactly the 1-cells incident to two notes separated by the interval $i$. This cell represents the class of all the intervals $i$. Figure 7 illustrates the boundary relationships defined for the classes of the fourth and the minor third. The 2-cells are filled in light gray. Notice that the topologies of the classes differs: in the given two examples, the minor third exhibits two holes while the fourth has no hole.

This new space is easily built following the method exposed in section 2.3. In this space, the specification of the AIS is completely structural. We are

[1] We do not enter in the technical details of the correct handling of the edges orientation. We assume that the pattern `c < e > c'` means that `c` (resp. `c'`) is negatively (resp. positively) oriented with respect to `e`.

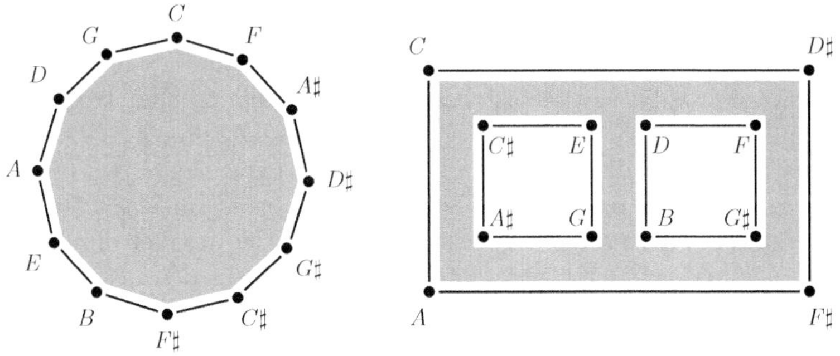


**Fig. 7.** Spatial representation of the 2-cells (and their boundaries) of the interval classes associated with the fourth (on the left) and with the minor third (on the right)

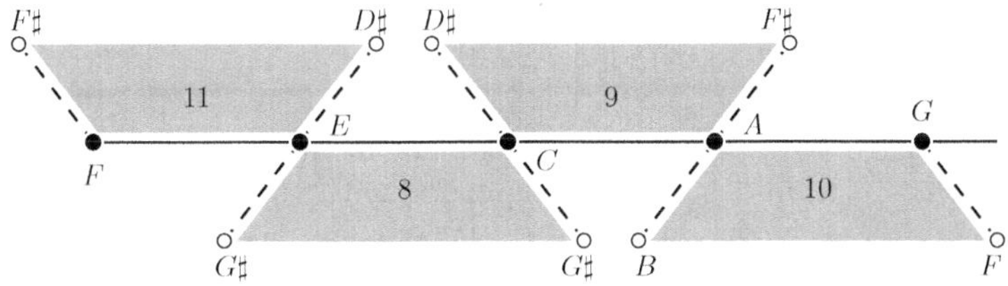


**Fig. 8.** Spatial representation of the five first notes of Alban Berg's AIS

looking for a path, made of 0, 1 and 2-cells, which is 0-Hamiltonian and also 2-*Hamiltonian*[2]:

```
n0 < i1  < I1  > i1  > n1
   < i2  < I2  > i2  > n2
   ...
   < i11 < I11 > i11 > n11
```

Figure 8 illustrates the instantiation of this path on the five first elements of Alban Berg's AIS. It is easy to see on this figure that the spatial constraint consists in matching an additional 2-cell. More precisely, two consecutive notes n$p'$ and n$p$ with $p' = (p-1)$ have to be $(0,1)$-neighbors by i$p$ and $(0,2)$-neighbors by I$p$ and such that cells I$p$ and i$p$ are incident.

Finally, the AIS computation time decreases by a factor of 4 with this solution specification applied on the original support space extended with 2-cell intervals. The speed-up obtained by tailoring the support space w.r.t. the brute force approach is greater than 120.

[2] A path in a graph which visits every edge exactly once is called *Eulerian*. However there is no general name for path in a complex which visits the $p$-cells exactly once. We name such path a $p$-Hamiltonian path.

### 3.3 A Spatial Classification of AIS

By definition, each AIS visits only one 1-cell in the boundary of each 2-cell. A natural idea to classify the AIS is then to look further in the structure of the 2-cells boundary. As we can see on Fig. 7, the boundary of a 2-cell is composed of a variable number of disconnected cycles (here a cycle is a closed $(0,1)$-path). Each cycle is one of the orbits of the action of the corresponding interval on the set of notes. For an interval $i$, it is well known that the number of orbits, hence cycles in the boundary of the associated 2-cell, is $d_i = \gcd(i, 12)$. For instance, there are $\gcd(3, 12) = 3$ cycles in the boundary of the minor third class 2-cell:

$$(C - D\sharp - F\sharp - A) \quad (C\sharp - E - G - A\sharp) \quad (D - F - G\sharp - B)$$

These cycles can be uniquely identified by an integer between 0 and $d_i - 1$ corresponding to the least note of the orbit (with the usual identification between pitch classes and numbers: $C = 0, C\sharp = 1$, etc.).

We can associate with an AIS a vector $\mathbf{V} = (v_1, \ldots, v_i, \ldots, v_{11})$ of 11 numbers giving, for each interval class, the visited cycle. This vector is called the *cyclic vector* of the AIS. Beware that the $i^{\text{th}}$ element in this vector is associated with the interval $i$ which does not usually coincide with the $i^{\text{th}}$ interval in the AIS. For example, the cyclic vector $\mathbf{V}_B$ of the AIS in the *Lyric Suite* is

| interval class $i$ | 1 | 2 | 3 | 4 | 5 | 6 | 7 | 8 | 9 | 10 | 11 |
|---|---|---|---|---|---|---|---|---|---|---|---|
| $d_i$ | 1 | 2 | 3 | 4 | 1 | 6 | 1 | 4 | 3 | 2 | 1 |
| $\mathbf{V}_B$ | 0 | 1 | 0 | 2 | 0 | 2 | 0 | 0 | 0 | 1 | 0 |

A *cluster* is the set of AIS sharing the same cyclic vector.

*Construction of a Cyclic Vector Compliant with a Given Scale.* As previously mentioned, some composers and analysts are interested in building some AIS including a sub-sequence of notes having a particular meaning [15,19].

For instance, André Riotte has analyzed AIS containing a sub-series of notes belonging to a particular scale. This problem can be formalized in various ways. Based on our representation, we propose the following formulation: given a scale $S$ containing some whole orbits for some intervals, compute the AIS whose cyclic vectors includes these orbits. In other words, the two consecutive notes in the AIS separated by such an interval belong to $S$.

For example, the harmonic $C$ minor scale $C\ D\ E\flat\ F\ G\ A\flat\ B$ contains two orbits $(D - F - A\flat - B)$ and $(E\flat - G - B)$ for the interval of minor third and of major third (and the two corresponding retrograde orbits for major sixth and for minor sixth). Following the definition of the cyclic vector, these two orbits are respectively identified by the integers $v_3 = 2$ and $v_4 = 3$. In the same way, we arbitrarily propose to use, for the tritone class interval, the cycle $(F - B)$ whose notes are included in the scale too. This choice reaches to define $v_6 = 5$. Interval classes of minor second $(i = 1)$, perfect fourth $(i = 5)$, perfect fifth $(i = 7)$ and major seventh $(i = 11)$ correspond for each to a single orbit (identified by the integer 0). We can summarize the choice of the orbits for each interval in the following table:

| interval class $i$ | 1 | 2 | 3 | 4 | 5 | 6 | 7 | 8 | 9 | 10 | 11 |
|---|---|---|---|---|---|---|---|---|---|---|---|
| $d_i$ | 1 | 2 | 3 | 4 | 1 | 6 | 1 | 4 | 3 | 2 | 1 |
| **V** | 0 | * | 2 | 3 | 0 | 5 | 0 | * | * | * | 0 |

Among the two possible orbits of the major second class interval, we choose the cycle ($C\sharp - D\sharp - F - G - A - B$) because it contains four of the seven notes of the scale, i.e. $v_2 = 1$. A check of the remaining possible combinations for the cyclic vector, gives the following vectors:

| | | | | | | | | | | | |
|---|---|---|---|---|---|---|---|---|---|---|---|
| $\mathbf{V_1}$ | 0 | 1 | 2 | 3 | 0 | 5 | 0 | 0 | 0 | 0 | 0 |
| $\mathbf{V_2}$ | 0 | 1 | 2 | 3 | 0 | 5 | 0 | 0 | 1 | 0 | 0 |
| $\mathbf{V_3}$ | 0 | 1 | 2 | 3 | 0 | 5 | 0 | 2 | 0 | 0 | 0 |
| $\mathbf{V_4}$ | 0 | 1 | 2 | 3 | 0 | 5 | 0 | 2 | 1 | 0 | 0 |

Each vector is associated with a different non-empty set of AIS. Because of the constraints on the orbits we have imposed for some intervals, we know that all these AIS include several times in their structure two consecutive notes included in the harmonic $C$ minor scale. A research of AIS containing the longer subsequence of notes belonging to the scale reaches to the following series associated with the vector $\mathbf{V_2}$:

$$E\ D\flat\ G\flat\ \boxed{F\ B\ D\ C\ A\flat\ E\flat\ G}\ A\ B\flat$$

As we can observe, this AIS contains in sequence the seven notes of the harmonic $C$ minor scale.

*Geometry of the Clusters.* There are about $\prod_i d_i = 3\,456$ possible clusters where the 46 272 AIS are scattered (some clusters are empty). We introduce here some preliminary idea to study how they are related with each other in the light of the standard algebraic operations sending an AIS to another one [14]. Given an AIS $(n_0, \ldots, n_j, \ldots, n_{11})$ with corresponding cyclic vector $\mathbf{V} = (v_1, \ldots, v_i, \ldots, v_{11})$, the application of an operation $\varphi$ computes an AIS $(n'_0, \ldots, n'_j, \ldots, n'_{11})$ with corresponding cyclic vector $\mathbf{V}' = (v'_1, \ldots, v'_i, \ldots, v'_{11})$ defined by:

– for the *transposition* $\varphi = T_b$

$$n'_j \equiv n_j + b \mod 12 \qquad \text{and} \qquad v'_i \equiv v_i + b \mod d_i;$$

– for the *homothety* $\varphi = H_p$ ($p$ relatively prime with 12)

$$n'_j \equiv p\, n_j \mod 12 \qquad \text{and} \qquad v'_i \equiv p\, v_{p^{-1} i} \mod d_i;$$

– for the *retrograde* $\varphi = R$

$$n'_j \equiv n_{-j-1} \mod 12 \qquad \text{and} \qquad v'_i \equiv v_{-i} \mod d_i;$$

– for the *circular shift* $\varphi = Q$ ($w$ is the position of the tritone)

$$n'_j \equiv n_{j+w} \mod 12 \qquad \text{and} \qquad v'_i \equiv \begin{cases} n_0 & \mod d_i \quad \text{if } i = 6 \\ v_i & \mod d_i \quad \text{otherwise} \end{cases} .$$

We aim at simplifying the study of the AIS distribution within the space of clusters by studying these operations.

Previous studies considered only the 3 856 *normalized* AIS (i.e., beginning with $C$); the others AIS can be reached by the 11 transpositions. Unfortunately this reduction is not relevant in the context of cluster classification which relies on the interval classes but not on the notes positions. In other words, the normalized AIS are not localized in a specific subset of clusters. Nevertheless, a normalized AIS always ends with $F\sharp$, so the circular shift of a normalized AIS produces an other AIS

$$Q(0, \ldots, n_w, n_{w+1}, \ldots, 6) = (n_{w+1}, \ldots, 6, 0, \ldots, n_w)$$

where the visited tritone class is $(C - F\sharp)$: thus, it belongs to a cluster with $v_6 = 0$. We can only consider the $\prod_{i \neq 6} d_i = 576$ clusters with $v_6 = 0$. Since these clusters also include the AIS of the form $(\ldots, 0, 6, \ldots)$, the symmetries induced by the retrograde $R$ and the transposition $T_6$ can be used to reduce a step further the set of clusters. The 156 remaining clusters are completely identified and *exactly* include 3 856 different AIS (equivalent to the normalized one using $R$, $T_6$ and $Q$). Finally, the last operations, namely the homotheties, decrease this number to 72. This reduction process, as well as the proofs, are detailed in a companion technical report [21].

## 4 Conclusion and Future Work

This paper aims at presenting the first results in the application of the spatial computing paradigm to musical theory problems. The two problems illustrating our approach are the structure of the $n$-note chords of the diatonic scale and the classification of the AIS. They are hardly new, but the framework presented here, based on the topological notions supported by the MGS programming language, is generic. For instance, one can study systematically the simplicial complex associated with an $n$-note chord series for a wide range of $n$, even for $n > 4$ when it cannot be graphically visualized. The topological formalization of the AIS enumeration relies on the hamiltonicity of a path in a well-designed space, a notion already used elsewhere as a compositional tool (e.g., Robert Morris in *Hamiltonian Cycle: Saxophone*, Michael Winter in *Maximum Changes* or Giovanni Albini in *Corale #4, preludio for cello and string orchestra*) to express various musical constraints [22].

We believe that this preliminary work shows the interest of a framework enabling the systematic building and processing of abstract spaces that appear in musical analysis. Our framework is based on spatial notions developed and studied in algebraic topology, and then amenable to a computer implementation due to their algebraic nature. Initially developed for the modeling and the simulation of dynamical systems, it appears well suited for the musicologist. As a matter of facts, one of the benefits of the spatial approach is the expressiveness and the concision of the constraint formulation. Only one rule is used to build the underlying space and only one rule is enough to specify the path to search.

We are currently working on spatial approach for the enumeration of Hamiltonian paths in various chord networks. We are also investigating the detailed geometry of the clusters of AIS. Future works include the validation of the approach on more examples and on larger scale problems, as well as the development of additional spatial constructions that may be necessary to handle further musical formalizations.

## Acknowledgements

The authors are very grateful to M. Andreatta, C. Agon, G. Assayag and J. Bresson from the IRCAM RepMus team, M. Malt, J.-M. Chouvel and to O. Michel from the LACL Lab at University of Paris Est for endless fruitful discussions. This research is supported in part by the IRCAM and the University Paris Est-Créteil Val de Marne.

## References

1. Tymoczko, D.: The geometry of musical chords. Science 313, 72 (2006)
2. Callender, C., Quinn, I., Tymoczko, D.: Generalized voice-leading spaces. Science 320, 346 (2008)
3. Bigo, L., Spicher, A., Michel, O.: Spatial programming for music representation and analysis. In: Spatial Computing Workshop 2010, Budapest (2010)
4. Mazzola, G., et al.: The topos of music: geometric logic of concepts, theory, and performance. Birkhäuser, Basel (2002)
5. De Hon, A., Giavitto, J.L., Gruau, F. (eds.): Computing Media and Languages for Space-Oriented Computation, Dagsthul. Dagsthul Seminar Proceedings, (06361) (2006), `http://www.dagstuhl.de/en/program/calendar/semhp/?semnr=2006361`
6. SCW: The Spatial Computing Workshops series and related events. List on the Spatial Computing Home Page, `http://www.spatial-computing.org/doku.php?id=events:start` (accessed in January 2011)
7. Giavitto, J.L., Michel, O.: MGS: a rule-based programming language for complex objects and collections. In: van den Brand, M., Verma, R. (eds.). Electronic Notes in Theoretical Computer Science, vol. 59. Elsevier Science Publishers, Amsterdam (2001)
8. Giavitto, J.L.: Topological collections, transformations and their application to the modeling and the simulation of dynamical systems. In: Nieuwenhuis, R. (ed.) RTA 2003. LNCS, vol. 2706, pp. 208–233. Springer, Heidelberg (2003)
9. Munkres, J.: Elements of Algebraic Topology. Addison-Wesley, Reading (1984)
10. Tucker, A.: An abstract approach to manifolds. Annals of Mathematics 34, 191–243 (1933)

11. Spicher, A., Michel, O., Giavitto, J.L.: Declarative mesh subdivision using topological rewriting in mgs. In: Ehrig, H., Rensink, A., Rozenberg, G., Schürr, A. (eds.) ICGT 2010. LNCS, vol. 6372, pp. 298–313. Springer, Heidelberg (2010)
12. Mazzola, G.: La vérité du beau dans la musique. Delatour (2007)
13. Giavitto, J.L., Spicher, A.: Simulation of self-assembly processes using abstract reduction systems. In: Systems Self-Assembly: Multidisciplinary Snapshots, pp. 199–223. Elsevier, Amsterdam (2008), doi:10.1016/S1571-0831(07)00009-3
14. Morris, R., Starr, D.: The structure of all-interval series. Journal of Music Theory 18, 364–389 (1974)
15. Riotte, A., Mesnage, M.: Formalismes et modèles musicaux. Delatour (2006)
16. Truchet, C., Codognet, P.: Musical constraint satisfaction problems solved with adaptive search. Soft Computing - A Fusion of Foundations, Methodologies and Applications 8, 633–640 (2004)
17. Gent, I.P., Walsh, T., Hnich, I., Miguel, I.: CSPLib: a problem library for constraints, web page at `http://www.csplib.org` (accessed in January 2011)
18. Otterström, T.: A theory of Modulation. Da Capo press, Inc. (1935)
19. Schiff, D.: The Music of Elliott Carter. Faber and Faber, London (1983)
20. Jedrzejewski, F.: Mathematical Theory of Music. Delatour (2006)
21. Spicher, A.: Spatial computation and classification of all-interval series. Technical report, LACL, Univ. of Paris-Est (2011)
22. Albini, G., Antonini, S.: Hamiltonian cycles in the topological dual of the tonnetz. In: Chew, E., Childs, A., Chuan, C.H. (eds.) MCM 2009. Communications in Computer and Information Science, vol. 38, pp. 1–10. Springer, Heidelberg (2009)

# A Model for Collective Free Improvisation

Clément Canonne[1] and Nicolas Garnier[2]

[1] Université de Lyon, Ecole Normale Supérieure de Lyon
clement.canonne@ens-lyon.fr
[2] Université de Lyon, Laboratoire de Physique de l'ENS-Lyon, CNRS UMR 5672
nicolas.garnier@ens-lyon.fr

**Abstract.** This paper presents a model for Collective Free Improvisation (CFI), a form of improvisation that can be defined as *referent-free*. While very simple, it captures some interesting mechanisms of CFI. We use two variables: the *intention* and the *objective*. Both variables are used to describe the production and organization of the improvisers' signals. Using a system of Landau equations, we propose a non-linear dynamics for the intention evolving on a short time-scale while the objective evolves on a long time-scale. In this paper, the model is used to determine if, and within which conditions, a collective structure can emerge from CFI.

**Keywords:** Free Improvisation, Cognitive Model, Non-linear Dynamic Systems, Emergent Structure.

## 1 Introduction

Collective Free Improvisation (CFI) is a musical phenomenon produced by at least two persons improvising simultaneously and freely, *i.e.* trying to leave undecided every compositional aspects until the very moment of the performance.

When talking about "free" improvisation, one should carefully distinguish between two time scales. CFI is not deprived of all the automatized behaviors that can generate the improvised musical output on a short-term time scale: Embodied patterns and learned gestures are present as much as in other kinds of improvisation. In this regard, free improvisation is not to be confused with an illusory "pure" improvisation, which would account for instantaneous *ex nihilo* creation.

In return, CFI can be defined as a referent-free improvisation. According to Pressing [1], a referent is an underlying formal scheme or guiding image specific to a given piece, used by the improviser to facilitate the generation and editing of improvised behavior on an intermediate time scale. In CFI, as opposed to referent-based improvisation (like straightforward jazz), there is no founding act (like the common choice of a standard) that confers a given set of musical or extra-musical data the status of *common knowledge* in a group.

In a broader perspective, we are not either denying the importance for CFI of cultural backgrounds or musical knowledge, especially if they are shared among

C. Agon et al. (Eds.): MCM 2011, LNAI 6726, pp. 29–41, 2011.

the group. CFI can include idiomatic borrowings: A given CFI can sound, at times, as a *be-bop* piece (with swing articulation, chords, tonal progression) or as a meditation on a *raga* (with a scale and a specific ornamentation) ; but a free improviser is someone that has no *pre-commitment* when the performance begins. His production is of course determined by several self-imposed restrictions, even stylistic restrictions, but he can modify these restrictions at any time.

In CFI, improvisers face two specific problems. First, the generation of improvised musical output on an intermediate time scale is not regulated. The formal unfolding is thus totally undetermined. Second, improvisers' musical coordination is not regulated and free improvisers' simultaneous production is much more difficult to control than in referent-based improvisation[1]. The fact that the way improvisers interact in CFI is not predetermined (roles and places in the ensemble can be redefined by anyone at anytime) makes it even harder.

We propose a model for CFI seen as this set of phenomenons[2]. An important inspiration for this model was the formalization of the improvisation's process proposed by Pressing [2] for a solo improviser. As CFI is a very interesting case of interaction, where shared information and pre-existing structures are almost nonexistent (each improviser can be described as "agnostic" before the interaction begins), it can be seen as paradigmatic. Besides the understanding of basic musical and cognitive processes in CFI, this model can be useful in both understanding social phenomena requiring effective coordination between agents (coordination problems) and in reinforcing the intuitive link between improvisational disposition and an agent's efficiency inside a complex system, following the steps of Borgo [3] that highlighted the numerous links that one can make between free improvisation's understanding and the study of complexity. This model can therefore be of interest in fields absolutely not related to music.

In this paper, the model is used to determine if, and within which conditions, a collective structure can emerge from CFI. This collective structure can be seen as a direct consequence of coordination's effectiveness or, to put it in another way, of group flow, as presented by Sawyer [4] and defined by some, if not all, of the following features: heightened consciousness, clear vision of a common goal, close listening, confidence in each other competency, experience of time dilatation, intuitive understanding of other musicians' intentions, sense of balance, high-enough risky situations to excite each musician's virtuosity and creativity... In the following, we consider that if a collective structure emerge, it is because group flow has happened: as the two notions are strongly correlated, we only focus in this paper on the question of CFI's collective structure.

---

1 For example, if the referent is a chord progression, everyone knows at each moment what pitches he can or can't play.

2 A lot of the music called free improvisation by the musicians, the audience or the record labels does not fall strictly into our category of CFI because it has some minimal form of referent, or because it takes place into an established group, with its own conventions and so on. Maybe Derek Bailey's *Company Week* is the most famous example of CFI in "real life".

# 2 Definitions and Model

## 2.1 Time Scales

The improvised production extends over several time scales, that we decompose as:

- The shortest time scale, which is the scale of the musical or acoustical signal, depending on the point of view from which the signal is described: either in a musical way (encoding the signal in "notes" of different durations on a score) or in an acoustical way (following the very evolution of the same timbre through time)... This scale is not explicitly used in our model.
- Short time scale $\tau_s$, of the order of seconds: it is the scale of the "clusters of events" [2].
- Long time scale $\tau_l$, of the order of minutes: it is the scale of the "sequences" [2].
- The scale of the complete improvisation piece. For a real-life improvisation, the duration is not *a priori* established, but in our model, it will be.

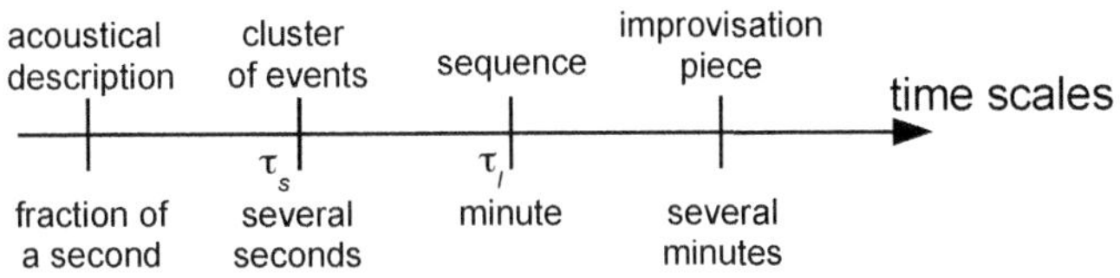


**Fig. 1.** Separation of the different time scales in the model of CFI

We detail now the two notions of *clusters of events* and *sequences* which define relevant time scales for our model.

**The cluster of events.** A cluster is a cognitive chunk that gathers a set of musical, acoustical, cinetical events that were decided at the same point in time; it can be pictured as the subsequent execution of a micro-plan, over its duration of the order $\tau_s$. Pressing [1] noted that the generation of improvised behavior on a short time scale is primarily determined by previous training and embodied patterns, and is not very piece-specific. $\tau_s$ is a short time, of the order of seconds. It is short because of the agent's cognitive limitations (one can't decide too much at the same time) and the improvisation's interactive dimension (one does not want to decide too much at the same time, to be able to react quickly to changes in the environment).

**The sequence.** The sequence is related to the long time scale $\tau_l$. A sequence is defined by a set of processes and/or a number of features (acoustical, cinetical,musical...) holding for a given length. Improvisers try to establish successive identities and stable points in musical's stream. These identities are then developed, played with or eventually negated, until new identities finally emerge. This organization in successive sequences is probably an endogenous feature of CFI [2,5,6].

## 2.2 Signal and Information

What we call "signal" in this paper is not the real musical signal produced by the musician. Its realistic description would require a huge number of variables. On the contrary, our description of the signal is extremely simplified and doesn't contain anything about the acoustic representation. In fact, we consider a real number $x$, related to the relative complexity of the signal (a cluster on the piano is more complex than a triad; a multiphonic on the clarinet is more complex than a traditionnaly-produced sound; a sub-division in septuplet is more complex than a sub-division in sextuplet; a stretch of music with very quick changes of pitches is more complex than a stretch of music with only one pitch...). As a direct consequence, our model's focus will not be on the signal *per se* but rather on more high-level phenomenas (interaction, long-term intention...). Extension of our model to more realistic signals would imply the consideration of a vector $x$ with as many components as needed (to describe, e.g., pitch, intensity, timbre, duration).

We write $N$ the number of musicians and $x^k(t)$ the temporal signal of musician $k$ ($1 \leq k \leq N$). We write $\boldsymbol{x}(t) = (x^1(t), ..., x^N(t))$ the set of signals produced at a given time $t$.

The quantity of information is directly related to the signal. We want that:

- the larger the signal, the larger its information. So we define a static information by analogy with energy in physics $I_s^k = \frac{1}{2}(x^k)^2$,
- the larger the signal varies, the larger its information. Correspondingly we define a dynamical information by analogy with kinetic energy $I_d^k = \frac{1}{2}\tau_i^2 \left(\frac{\mathrm{d}x^k}{\mathrm{d}t}\right)^2$, where $\tau_i$ is a normalization time.

Then $I^k = I_s^k + I_d^k$ is the information delivered by player $k$, whereas the total information seen by any musician of the group is:

$$I = \sum_k I^k = \frac{1}{2}\|\boldsymbol{x}\|^2 + \frac{1}{2}\tau_i^2 \left\| \frac{\mathrm{d}\boldsymbol{x}}{\mathrm{d}t} \right\|^2$$

## 2.3 Signal and Intention

Besides the signal $x^k$ produced by musician $k$, we define the intention $\omega^k$ of this musician, which represents the ideal signal that the musician would like to deliver. The intention is *a priori* more complex than the signal produced, because it contains information that the musician may not be able to actually play, due to, e.g., lack of technicality, lack of time, *etc.* The signal $x^k$ is deduced from the intention $\omega^k$ at given discrete time steps by projection, expressing the possible loss of information between the intention and its actualization in the signal:

$$x^k = g(\omega^k)$$

The function $g$ expresses the projection, and for the sake of simplicity in the present paper, we suppose that the musician is "perfect", so that $g$ is identity, *i.e.* $x^k = \omega^k$, when the projection occurs.

The intention $\omega^k$ evolves on the short time scale $\tau_s$ and we choose a continuous dynamics for it. On the contrary, we impose that $x^k$ is constant and equals $x_n^k$ during a cluster of events, *i.e.* between two projections of the intention separated by the time lag $d_n^k$ which is the duration of the cluster of events. The index $n$ labels the time. Durations $d_n^k$ of clusters are of the order of the short time scale $\tau_s$. To make our model deterministic, we impose

$$d_n^k = \tau_s - a(x_n^k)^2$$

where $a > 0$ is a constant. Clusters of events are shorter when the signal is large, *i.e.* contains more information. Conversely, if the signal is poor in information, the corresponding cluster of events is longer. We choose $a = 0.3$ in the remaining of the paper.

Because the signal $x^k$ is piecewise constant, the information $I^k$ is constant on clusters of events, while there are peaks of dynamic information at the boundaries between clusters (see Fig. 2). For coherence, we choose the normalization time $\tau_i = 2\mathrm{d}t$ where $\mathrm{d}t$ is the time step we use in numerical integration; this way, $I_d^k$ depends on the signal's amplitude variation when there is a change of cluster of events rather than on the time derivative of this amplitude which is infinite.

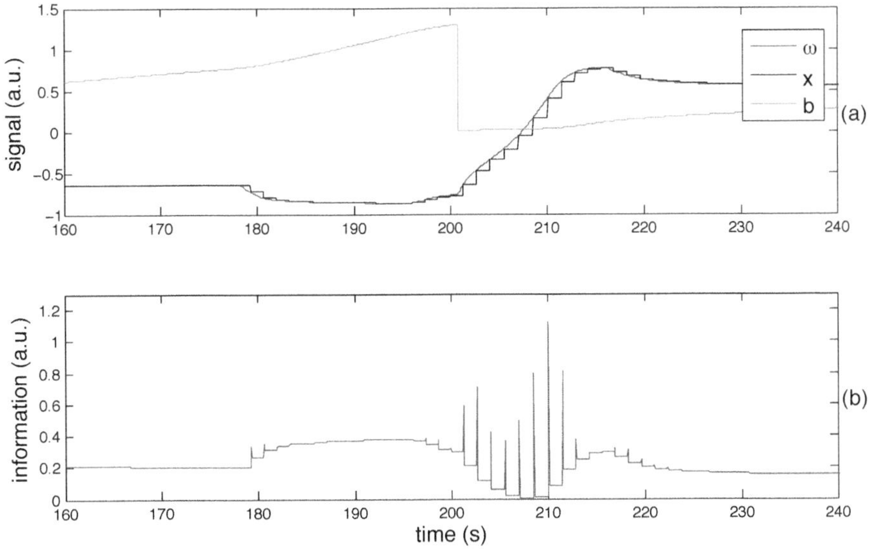


**Fig. 2.** Example of intention, signal, boreness and information for one musician. (a) the intention $\omega$ evolves continuously in time, while the signal $x$ is discrete on short time periods coresponding to cluster of events. Boreness $b$ is reset to zero when a sequence change occurs (here, around time 200 s). (b) Information $I^k$ is constant over a cluster of events, and presents a peak when a new cluster of events begins.

**Intention's dynamics.** We propose the following dynamics for $\omega^k$, inspired from dynamical systems' theory:

$$\tau_c \frac{\mathrm{d}\omega^k}{\mathrm{d}t} = \alpha^k x^k + \sum_{l \neq k} \beta^{k,l} x^l - g\|\omega_k\|^2 \omega^k \tag{1}$$

Each of these equations is a Landau equation from phase transition theory. Parameter $g > 0$ is a constant in front of a non-linear that prevents the solutions from diverging; we choose $g = 1$. Note that the evolution of intention $\omega^k$ depends on signals $\{x^l\}$, $l \neq k$, from other musicians, not from their intentions which are of course not known by musician $k$. Solutions of Eq. (1) vary on time scale $\tau_s$ and its autocorrelation function decreases to zero over this time scale. Parameter $\alpha^k > 0$ expresses the self-sensitivity of musician $k$. Parameters $\beta^{k,l}$ express the influence over musician $k$ of signals $\{x^l\}_{l \neq k}$ from other musicians and therefore quantify the interactions between musicians. $\beta^{k,l}$ are of order 1, they can have any sign or vanish.

### 2.4 Objective

We define the objective $\boldsymbol{\Omega}^k$ of the musician $k$ as the set of parameters $\alpha^k$ and $\{\beta^{k,l}\}_{l \neq k}$ that caracterizes the linear part of the equations: $\boldsymbol{\Omega}^k = (\beta^{k,l})$, where we have written $\beta^{k,k} = \alpha^k$ for $l = k$. The objective of musician $k$ is a $N$-dimensional vector that plays the role of the control parameter in the Landau theory. The set of objectives of all players defines a $N \times N$ matrix. The objective defines the value towards which the intention tends on a time scale larger than $\tau_s$. This is obvious for a solo improvisation ($N = 1$ and all $\beta^{k,l}$ are zero): We then have a Landau equation and its solution $\omega$ tends to $\sqrt{\alpha/g}$. But the objective is also a measure of the interaction with the other musicians; in a collective improvisation, any improviser interacts with the others, and the intention tends towards a value given by an appropriate combination of components of the objective. Following Pelz-Sherman [7], we distinguish some paradigmatic cases and define the corresponding coupling of musician $k$ with musician $l$:

- If $\beta^{k,l} \simeq 1$, then $\omega^k$ tends to $x^l$. Player $k$ is willing to imitate player $l$: "imitation".
- If $\beta^{k,l} \simeq -1$, then $\omega^k$ tends to -$x^l$. Player $k$ is willing to have a signal opposite from signal of player $l$: "contrast".
- If $\beta^{k,l} \simeq 0$, then player $k$ is not paying any attention to the signal from player $l$: "independency".

These are limit cases—obvious for $N = 2$ and $\alpha^k = 0$—and of course any intermediate situation can occur. Note that the couplings are non-symmetrical: $\beta^{k,l} \neq \beta^{l,k}$, *i.e.*, musician $k$ can for example "imitate" musician $l$ while $l$ is "independent" from $k$.

Contrary to classical Landau theory, we make the objective evolve in time with a specific dynamics, on the long time scale $\tau_l$. We choose a discrete dynamics, and any change in the objective of a musician defines a new sequence for this musician. This dynamics requires the introduction of the *cognitive load* and the *boreness*.

**Cognitive load.** In CFI, a musician's attention is shared between two tasks: generating his own signal and monitoring other musicians' signals. We introduce the cognitive load to account for finiteness of a musician's attention. Cognitive load bounds possible values of the objective components. We write a first part of the cognitive load as the part devoted to monitoring the signals:

$$C^k_{\text{monitor}} = \frac{1}{2}\left(\alpha^k x^k\right)^2 + \sum_{l \neq k} \frac{1}{2}\left(\beta^{k,l} x^l\right)^2 = \frac{1}{2}\left\|\boldsymbol{\Omega}^k.\boldsymbol{x}\right\|^2$$

Another component of the cognitive load is related to the production of the signal. We suppose that difficulty for musician $k$ to produce a signal is proportional, with a proportionality coefficient $(a^k)^2$, to the quantity of static information in his signal:

$$C^k_{\text{prod}} = \frac{1}{2}(a^k)^2 I^k_s = \frac{1}{2}(a^k x^k)^2$$

The total cognitive load of the musician $k$ will be noted $C^k = C^k_{\text{monitor}} + C^k_{\text{prod}}$, and we require that this variable be bounded from above by a constant $C^k_{\max}$ representing the maximal cognitive capacity of the musician $k$.

**Boreness.** When a sequence is lasting too long, the musician gets bored and ultimately breaks it. We define the boreness $b^k(t)$ of the musician $k$ to quantify this effect. Boreness grows in time, until it reaches a limit $b^k_{\max}$; then a change of objective occurs (see Fig. 2), which is also a change of sequence. We simply choose:

$$\frac{\mathrm{d}b^k}{\mathrm{d}t} = C^k$$

with initial condition $b^k = 0$ at $t = 0$. When the objective is changed, at the end of a sequence, boreness is reset to 0. Maximal boreness $b^k_{\max}$ is related to maximal cognitive charge; we choose:

$$b^k_{\max} = \frac{\tau_l}{3\tau_c} C^k_{\max}$$

**Objective's dynamic.** We choose that the objective remains constant as long as $b^k(t) < b^k_{\max}$. On the contrary, when the boreness $b^k$ becomes larger than the maximal value $b^k_{\max}$, we choose a new objective such that the cognitive charge remains bounded from above. All possible choices of $\alpha^k$ and $\beta^{k,l}$ are therefore not possible. For the sake of simplicity, the new components of the objective are choosen randomly, between $-1$ and $+1$ for $\beta^{k,l}$, and between 0 and 1 for $\alpha^k$; If the new cognitive charge $C^k$ resulting from these new values is larger than $C^k_{\max}$, we apply the factor $C^k_{\max}/C^k$ to all components of $\boldsymbol{\Omega}^k$. We also decide to project $\omega^k$ into $x^k$ at the very same time such a change of sequence occurs.

# 3 Results and Discussion

## 3.1 Collective Sequences and Their Articulation

To analyse the complete production of the group, we define collective sequences, not to be confused with individual sequences simply defined above by a given objective. This is a way to probe and quantify coordination efficiency in the group. We call collective sequence a time frame during which each improviser maintain a relative musical identity (*i.e.* his intention stays more or less constant). If we find a lot of collective sequences, and if collective sequences are long enough, we will say that coordination amongst musicians in the group is good. One of the main interest of this model is to show the existence of collective sequences.

If the position of each improviser in the group stays more or less constant, *i.e.* all objectives are contant, then we expect a collective sequence. Nevertheless, a constant objective is not a sufficient condition for the occurence of a collective sequence. Also, a collective sequence can be composed of a series of individual sequences; if after a change of individual sequence, each musican's position into the group is not strongly altered, then the same collective sequence will continue.

This collective structuring in successive sequences is one of CFI's great challenges. Fig. 3 shows that it is not always possible to detect collective sequences in CFI. One can clearly see two types of local structure in our model of CFI:

- A stable solution which can be seen as a "collective sequence" (labelled 1,2,3 in Fig. 3); this corresponds to a fixed point in the phase space of the system.

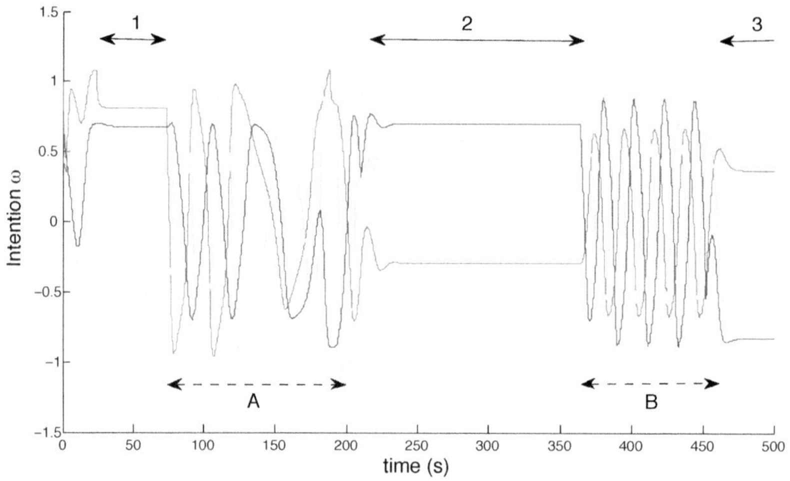


**Fig. 3.** Signals from a team of 3 musicians, all with $a^k = 0$, as produced by our model. We can discriminate 3 collective sequences (labelled 1, 2, 3). At the begining, a short transient period is observed before collective sequence 1. Between collective sequences 1 and 2, resp. 2 and 3, we observe a chaotic behavior A, resp. periodic behavior B.

- An oscillating solution which can be seen as a phase of discoordination among the musicians (labelled B in Fig. 3); this corresponds to a limit cycle, and it is obtained when some of the eigenvalues of the matrix of objectives are not real numbers but complex conjugates.

### 3.2 Contributing Factors to CFI's Structuring in Collective Sequences

The model is now used to see within which conditions the emergence of collective sequences is facilitated.

**Effects of improvisers' features.** We consider here two specific features:

- *Virtuosi* produce high-information signals at a lower cognitive cost. This is represented by a low value of $a^k$.
- *Leaders* have a superior cognitive capacity. As a direct consequence, they tend to get bored more slowly (*i.e.* from a musical perspective, they try to "work out" the different ideas and situations).

Fig. 4, where $a^k = 0.4$ for all 3 musicians, has to be contrasted with Fig. 3, where we selected $a^k = 0$ for all 3 musicians (highly virtuoso improvisers). It clearly shows the impact of this feature on improvisers' coordination.

Fig. 5 shows that the existence of leaders enhance the organization of CFI in collective sequences.

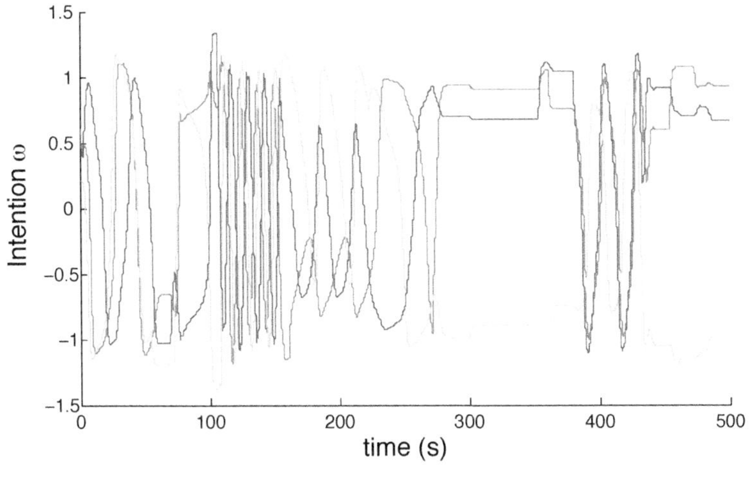


**Fig. 4.** Signals from a team of 3 musicians, all with $a^k = 0.4$. Although collective sequences still exist, they occur less often.

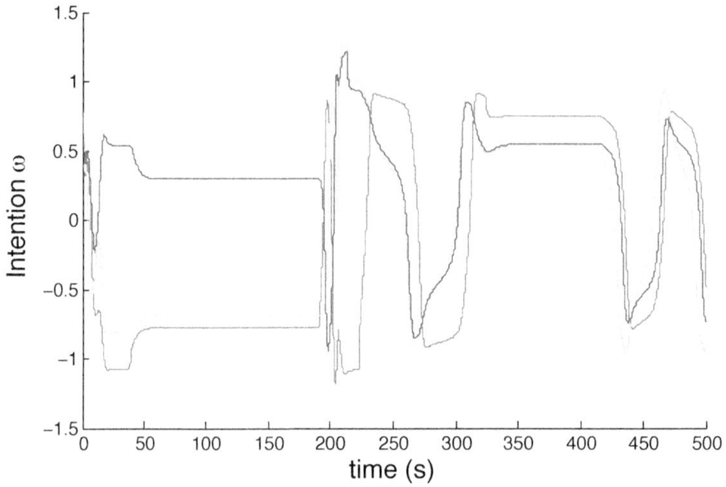


**Fig. 5.** Signals from a team of 3 virtuoso musicians, ($a^k = 0$). One of the musicians is a leader, with $C^k_{\max}$ twice larger than for regular improvisers.

**Number of improvisers.** As one could have expected, the fewer the musicians, the easier the collective organization, as shown on Fig. 6 where we increase the number of musicians to 5, and see the difficulty to define clear collective sequences.

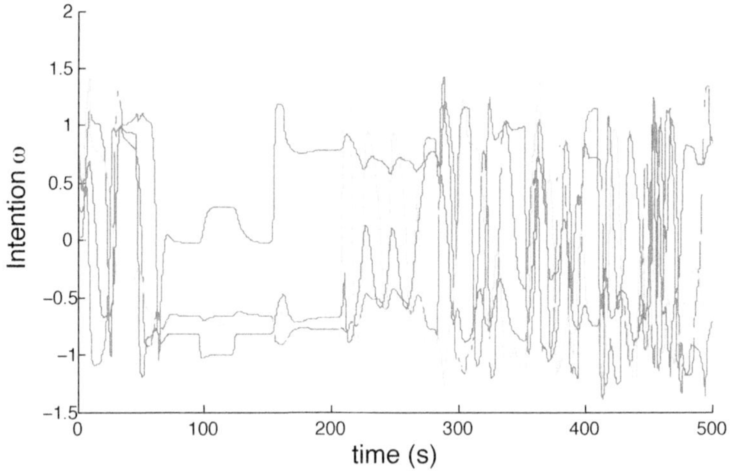


**Fig. 6.** Signals from a team of 5 virtuoso musicians ($a^k = 0$). Collective sequences are difficult to produce and/or discriminate.

**Emergence of sub-teams.** To adress the issue of obtaining collective sequences in large groups of musicians, we allow our improvisers to seek for the creation of sub-teams in CFI. This can be done in two different ways:

- Improvisers are looking for symmetrical interactions: If A is in imitation with B, B will also try to imitate A, thus unifying a sub-team by introducing *reciprocal actions*: $\beta^{k,l}$ has the same sign as $\beta^{l,k}$ and is of the same order of magnitude. Results are illustrated in Fig. 7.
- In large groups, improvisers do not interact with every other musicians. On the contrary, they focus on one or two specific musicians, so that they interact only with them. This is obtained by imposing a maximum of two non vanishing $\beta^{k,l}$ for every musician $k$. A typical case is depicted in Fig. 8.

As shown in Fig.7 and Fig. 8, this "sub-team" kind of reasoning is efficient to organize CFI in collective sequences.

### 3.3 Future Plans

The model's first results presented above are encouraging. In particular, the model is successful in showing the possibility of self-organization in CFI, despite the absence of *a priori* structures. But this self-organization depends on several features: the musicians' virtuosity, their leadership quality, their team and sub-team reasoning... and probably other features still to discover. Next step is to quantify the effects of improvisers and sub-teams' features on CFI and its organization; for this purpose, a statistical approach will be used. Some assumptions we have made can be relaxed. For example, we have tried a different projection

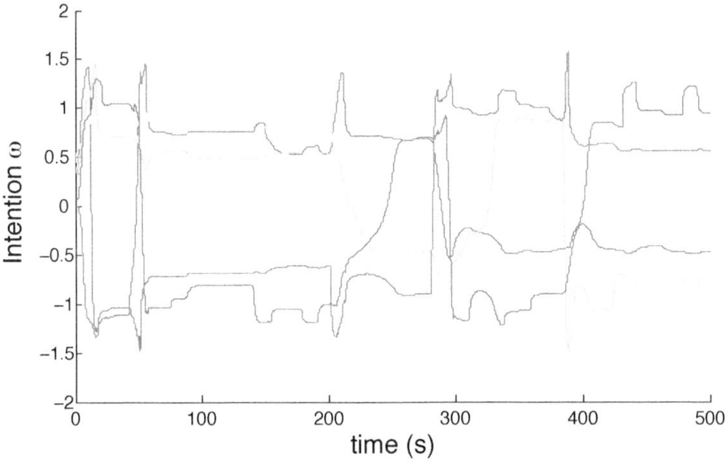


**Fig. 7.** Signals from a team of 5 musicians ($a^k = 0.4$). Here, musicians tend to have symetrical interactions.

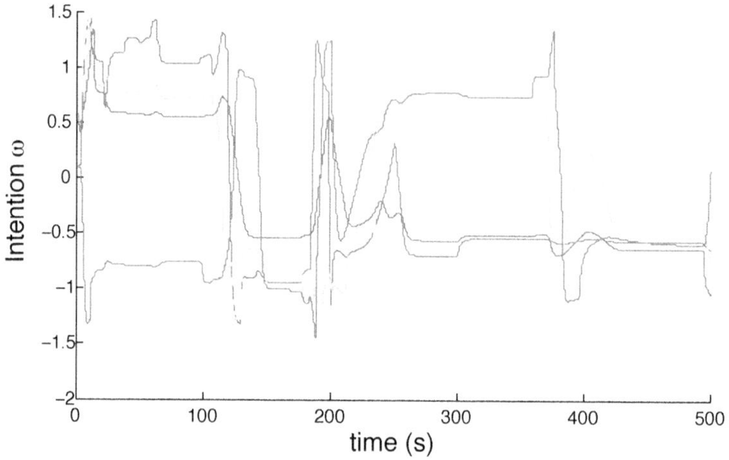


**Fig. 8.** Signals from a team of 5 virtuoso musicians ($a^k = 0$) with one leader. Here, all improvisers are interacting with at most 2 other musicians.

of intention into signal, by adding some noise, thus making the musician imperfect: this does not affect the observed behavior of the model. Future plans also include the study of more realistic dynamics for the objective, and possible inclusion of a long-term memory: in this regard, Dubnov's works [8] could be useful. This should provide a finer understanding of the way a collective structure can emerge from CFI. Confrontations of our model with laboratory human production is also under consideration, which may suggest in return some non-trivial modifications of our signal's conception.

## Acknowledgements

This work has been supported by the Fonds Recherche of ENS-Lyon.

# References

1. Pressing, J.: Cognitive Processes in Improvisation. In: Crozier, W.R., Chapman, A. (eds.) Cognitive Processes in the Perception of Art, pp. 345–363. Elsevier, Amsterdam (1984)
2. Pressing, J.: Improvisation: Methods and Models. In: Sloboda, J. (ed.) Generative Processes in Music, pp. 129–178. Clarendon, Oxford (1988)
3. Borgo, D.: Sync or Swarm: Improvising Music in a Complex Age. Continuum, New York (2005)
4. Sawyer, R.K.: Group Creativity: Music, Theater, Collaboration. Routledge, London (2003)
5. Nunn, T.: Wisdom of the Impulse: On the Nature of Musical Improvisation (1998), `http://www20.brinkster.com/improarchive/tn.htm`

6. Canonne, C.: L'improvisation Collective Libre: De l'Exigence de Coordination à la Recherche de Points Focaux. Thèse de Doctorat en Musicologie de l'Université de Saint-Etienne (2010)
7. Pelz-Sherman, M.: A Framework for the Analysis of Performer Interactions in Western Improvised Music (1998), `http://pelz-sherman.net/mpsdiss.pdf`
8. Dubnov, S.: Unified View of Prediction and Repetition Structure in Audio Signals with Application to Interest Point Detection. IEEE Transactions on Audio, Speech and Language Processing 16(2), 327–337 (2008)

# On a Class of Locally Symmetric Sequences: The Right Infinite Word $\Lambda_\theta$

Norman Carey

CUNY Graduate Center
ncarey@gc.cuny.edu

**Abstract.** The Nicomachus Triangle, a two-dimensional representation of powers of 2 and 3, provides a starting point for the development of an infinite class of right infinite *Lambda words*, $\Lambda_\theta$. The word is formed by encoding differences in the sequence $\{\mathcal{M}_\theta\}_i = \{a + b\theta\}_i$, $a, b \in \mathbb{N}$. Although the word is on an infinite alphabet, it is traversable via environments containing no more than three letters. When $\theta = \vartheta = \log_2 3$, the word encodes all well-formed scales and regions generated by the intervals octave and perfect twelfth. The study sheds additional light on the role of palindromes in musical tone structures. Regions are palindromes on two letters, and form the largest palindromes in the Lambda word, as it develops. The regions have a significant dual representation, connecting them to the palindromic prefixes of a characteristic Sturmian word. The Lambda word is rich in palindromes beyond regions. In particular, a palindrome is formed between any two successive appearances of the same letter. Although $\Lambda_\vartheta$ is of particular importance musically, Lambda words are interesting in their own right as word theoretic objects. The paper ends with a brief look at the Fibonacci Lambda word, $\Lambda_\phi$.

**Keywords:** Well-formed scale, Region palindrome, Word theory, Sturmian word, Central word, Christoffel word, Lambda word.

## 1 Introduction

Recently, a number of correspondences have been unearthed that link the theory of combinatorics on words, and the theory of well-formed scales. In particular, there are important connections to be drawn between well-formed scale theory and objects such as Sturmian words, and the conjugate class of Christoffel words. (A good introduction is in [1].) Sturmian words and Christoffel words are classes of infinite and finite words, respectively, on alphabets of two letters. Such objects have been studied extensively. (See, for example, [2,3,4,5,6].) Well-formed scales may be encoded as words on two letters, for example, the familiar pattern, WWHWWWH, of the major scale. In other words, the connection is between the conjugate class of Christoffel words of slope $\frac{g}{N-g}$ and the *class of well-formed scales* $N_g$ as defined in [7] and [8]. Similarly, there is an equivalence between what Carey and Clampitt refer to as a *region* and the notion of central word, as described in [2]. Other connections between word theory and well-formed scale theory are to be found in an upcoming article by Clampitt and Noll [9].

C. Agon et al. (Eds.): MCM 2011, LNAI 6726, pp. 42–55, 2011.

This paper will show a hitherto unacknowledged role that palindromes play in scale theory. Palindromes and measures of palindromic complexity in the context of infinite sequences on finite alphabets have been examined in detail, including [10], [11], and [12]. This paper extends the study of palindromes into a certain class of infinite sequence on infinite alphabets.

To develop the word, we begin with a new look at the Nicomachus Triangle, a structure that was significant in the original development of well-formed scales. The Nicomachus Triangle is shown in Fig. 1. This is, of course, a two dimensional representation of powers of 2 and 3 forming a multiplication table. Connecting the elements of this set in order induces a path on the grid, as shown in Fig. 1. Each vector along the path is color coded.

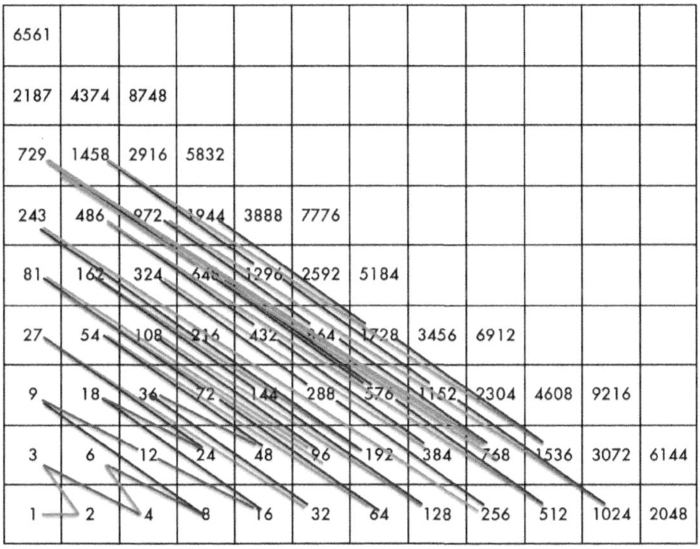


**Fig. 1.** A Path of Ordered Elements

Figure 2 unfolds the path and shows it as a sequence. Figure 3 shows the result of taking ratios between adjacent elements of Fig. 2. The ratios are then encoded as letters of an infinite alphabet, to create the beginning of the word, which will be called $\Lambda_{(\log_2 3)}$.

| |
|---|
| 1, 2, 3, 4, 6, 8, 9, 12, 16, 18, 24, 27, 32, 36, 48, 54, 64, 72, 81, 96, 108, 128 144, 162... |

**Fig. 2.** Powers of 2 and 3 and their multiples in linear order

$$\begin{array}{cccccccccccccc} \frac{2}{1} & \frac{3}{2} & \frac{4}{3} & \frac{3}{2} & \frac{4}{3} & \frac{9}{8} & \frac{4}{3} & \frac{4}{3} & \frac{9}{8} & \frac{4}{3} & \frac{9}{8} & \frac{32}{27} & \frac{9}{8} & \frac{4}{3} \ldots \\ \mathrm{P8} & \mathrm{P5} & \mathrm{P4} & \mathrm{P5} & \mathrm{P4} & \mathrm{M2} & \mathrm{P4} & \mathrm{P4} & \mathrm{M2} & \mathrm{P4} & \mathrm{M2} & \mathrm{m3} & \mathrm{M2} & \mathrm{P4} \ldots \\ a & b & c & b & c & d & c & c & d & c & d & e & c & d \ldots \end{array} \tag{1}$$

**Fig. 3.** The beginning of a right infinite word $\Lambda_{\log_2 3}$

abcbcdccdcdedcdeddededdeddfddeddfdddfddfdddfddfgfddfddfgfdd
fgfdfgfddfgfdfgfgfdfgfdfgfgfdfgfgffgfgfdfgfgffgfgfgffgfgffg
fgfgffgfgffhffgfgffgfgffhffgfgffhffgffhffgfgffhffgffhffhffg
ffhffgffhffhffgffhffhfffhffhffgffhffhfffhffhffhfffhffhfffhf
fhffhfffhffhffhihffhffhfffhffhffhihffhffhffhihffhffhihffhff
hffhihffhffhihfhihffhffhihffhffhihfhihffhffhihfhihffhihfhih
ffhffhihfhihffhihfhihfhihffhihfhihffhihfhihfhihffhihfhihfhi
hihfhihfhihffhihfhihfhihihfhihfhihfhihihfhihfhihihfhihfhihf
hihihfhihfhihihhihihfhihfhihihfhihfhihihhihihfhihfhihihhihi
hfhihihhihihfhihfhihihhihihfhihihhihihhihihfhihihhihihfhihi
hhihihhihihfhihihhihihhihihihhihihhihihfhihihhihihhihihihhi
hihhihihhihihihhihihhihihihhihihhihihhihihihhihihhihihh**jhhi**
**hihhihihhihihihhihihhihihhj**hhihihhihihhihihhjhhihihhihihhjh
hihihhihihhihihhjhhihihhihihhjhhihhjhhihihhihihhjhhihihhihi
hhjhhihhjhhihihhihihhjhhihhjhhihihhjhhihhjhhihihhihihhjhhih
hjhhihihhjhhihhjhhihhjhhihihhjhhihhjhhihihhjhhihhjhhihhjhhi
hihhjhhihhjhhihhjhhjhhihhjhhihhjhhihihhjhhihhjhhihhjhhjh

**Fig. 4.** The first 1000 letters in $\Lambda_\vartheta$

Figure 4 shows more of this word. (I will discuss the underlined segment presently.)

Although the word is on an infinite alphabet, we will see that it is comprised of local environments in which there are no more than three letters present. The word also encodes regions in which only two letters circulate. All well-formed scales and regions generated by octaves and fifths are to be found encoded in this word. For example, the so-called "structural" region is encoded in letters 5-10 as *cdccdc*.

The path described in Fig. 1, then, travels through all well-formed scale regions. There is much to study in the Lambda word concerning the palindromic nature of regions, however, another glance at Fig. 4 allows us to see that palindromes are everywhere in the word. Although the encoded regions are always palindromes on two letters, palindromes in the Lambda word may be found on one, two, or three, different letters. There are no palindromes on four or more letters, supporting the assertion that the word may be parsed into local environments containing no more than three letters. Some palindromes are to be found in the areas between regions, namely, where there are three different letters. A 31-letter palindrome in Fig. 4 that is not a region is underlined beginning on the sixth line from the bottom, *jhhihihhihihhihihihhihihhihihhj*. In fact, by inspection, between any two successive appearances of the same letter there is a palindrome, such as the one here between successive $j$s.

In this paper, I will formally define Lambda words, of which the one we have been discussing is a single instance, and show the following:

1. A Lambda word is a right infinite word on an infinite alphabet.
2. Each letter appears a calculably finite number of times.
3. Regions are encoded as central words. Any encoded region is longer than any previous palindrome in the Lambda word.
4. There exists, for any Lambda word, a unique (non-trivial) factorization into palindromes.
5. Between any two successive appearances of the same letter appears a palindrome.

All of these statements have been proven. This paper reports on results of this study: we do not give complete proofs here, but will provide proof sketches. (Complete proofs will appear in subsequent papers.)

## 2 A Right Infinite Word on an Infinite Alphabet

Each Lambda word is generated by an irrational value $\theta$, with $1 < \theta < 2$.

Let $M_\theta = \{a + b\theta\}$ where $a, b, \in \mathbb{N}$. Thus, the elements of $M_\theta$ may be represented as points $\geq 0$ on the real number line. The word $\Lambda_\theta$ is generated in three stages:

1. Order the elements of the monoid $\{\mathcal{M}_\theta\}_i = \{a + b\theta\}_i$.
2. Take differences between adjacencies in the ordered monoid.
3. Encode the differences as letters.

We illustrate with $\theta = \log_2 3$. (Here and onward we let the symbol $\vartheta = \log_2 3$.) The ordering of $\mathcal{M}_\vartheta$ yields a $\log_2$ representation of Fig. 2.

$$\{a+b\vartheta\}_i = (0+0\vartheta), (1+0\vartheta), (0+1\vartheta), (2+0\vartheta), (1+1\vartheta), (3+\vartheta), (0+2\vartheta), \ldots \quad (2)$$

(The coefficients $a$ and $b$ indicate coordinate positions in Fig. 1.)

Taking differences in (2):

$$\{a+b\vartheta\}_{i+1} - \{a+b\vartheta\}_i = (1-0\vartheta), (-1+1\vartheta), (2-1\vartheta), (-1+1\vartheta), (2-1\vartheta), (-3+2\vartheta), \ldots \quad (3)$$

Finally, using the encoding $(1 - 0\vartheta) \rightarrow a$, $(-1 + 1\vartheta) \rightarrow b$, $(2 - 1\vartheta) \rightarrow c$, $(-3 + 2\vartheta) \rightarrow d$, etc., we have

$$\Lambda_\vartheta = abcbcd \ldots \quad (4)$$

Note that the coefficients in (3) may be expressed in the form $\pm(A - B\vartheta)$.

Carey and Clampitt [13] have proven that the value $A/B$ is always a convergent or semi-convergent of $\theta$, which, in the example, is $\log_2 3$.

Therefore,

**Theorem 1.** *If* $\pm(A - B\theta)$ *is a difference between adjacent elements of* $\{\mathcal{M}_\theta\}_i$, *then* $A/B$ *is a convergent or semi-convergent of* $\theta$.

Because differences encode to letters in $\Lambda_\theta$, and because the number of convergents of an irrational value $\theta$ is infinite, we have immediately,

**Theorem 2.** $\Lambda_\theta$ *is a right infinite word on an infinite alphabet.*

## 3 Each Letter Appears a Calculably Finite Number of Times

Let $|\Lambda_\theta|_x$ represent the number of occurrences of the letter $x$ in $\Lambda_\theta$. Let $A/B$ represent the convergent or semi-convergent such that $\pm(A - B\theta) \rightarrow x$ in the encoding. Let $p/q$ represent the following convergent if $A/B$ itself is a convergent, or let $p/q$ represent the previous convergent if $A/B$ is a semi-convergent.

**Theorem 3.** $|\Lambda_\theta|_x = pq$.

| x | (A,B) | pxq | = | $\vert\Lambda_\vartheta\vert_x$ |
|---|---|---|---|---|
| a | **(1,0)** | 1x1 | | 1 |
| b | **(1,1)** | 2x1 | | 2 |
| c | **(2,1)** | 3x2 | | 6 |
| d | **(3,2)** | 8x5 | | 40 |
| e | (5,3) | 3x2 | | 6 |
| f | **(8,5)** | 19x12 | | 228 |
| g | (11,7) | 8x5 | | 40 |

*If* $A/B$ *is a full convergent of* $\vartheta$, *then* $p/q$ *is defined as the next full convergent.*
*If* $A/B$ *is a semi-convergent of* $\vartheta$, *then* $p/q$ *is defined as the previous full convergent.*

**Fig. 5.** $|\Lambda_\theta|_x$, The number of occurrences of $x$ in $\Lambda_\theta$

Figure 5 illustrates the theorem with $\Lambda_\vartheta$ and confirms that the letter $a$ only occurs once, as we have seen earlier. Theorem 3 further implies that lexicographically early letters are rarer than later ones, which is generally true, taking into account the fact that letters determined by semi-convergents are rarer than those determined by nearby convergents.

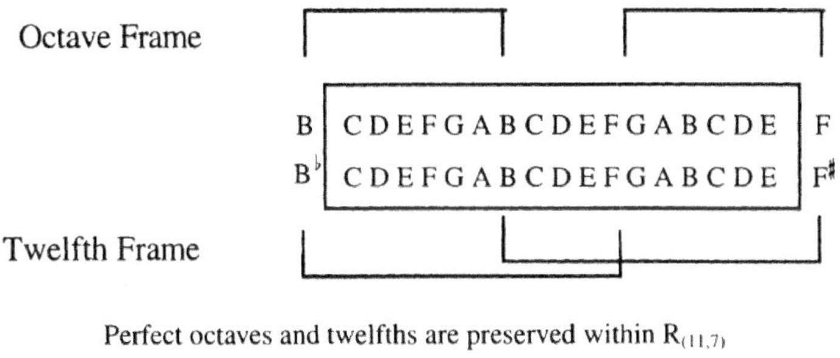


**Fig. 6.** The diatonic Octave-Twelfth region: Fig. 8 from "Regions" [14]

# 4 Palindromes in $\Lambda_\theta$

## 4.1 Regions

A region, as defined in [14], is a symmetrical tone-space that contains well-formed scales. The pattern of step intervals in a region conforms to the notion of a central word. The Lambda word encodes the step-interval pattern of each region as a central word. An important property of central words is that of double periodicity, as determined by the Fine-Wilf theorem [4]. The notion of the two periods of a central word is implicit in Carey and Clampitt's [14] description:

> The intervals [octave and twelfth] play symmetrical roles in the construction of regions, and either (or neither) may be chosen as a modular frame. Within the region, the intervals remain in a kind of balance. To begin to understand the nature of this balance, consider the diatonic region $R_{(11,7)}$. The region allows for 10 perfect octaves from C–c to e–e′ and 6 perfect twelfths from C–g to A–e′. This space is maximal in this respect, as [Fig. 6] shows. Either interval is extended outside the region only at the expense of the other[1].

The brackets above and below Fig. 6 illustrate the limits of the two periods of a central word. The symmetries of three encloded regions along the path that connects the elements in $\{\mathcal{M}_\vartheta\}_i$ are shown in Fig. 7. The first of these is the structural region, *cdccdc*, which is the region of the Scholica Enchriadis as described in [14]. The letter *c* represents P4, and *d* represents M2. The Pentatonic region contains the letters *d* and *e*, with *e* representing m3. Next comes the diatonic region, where *f* is m2. Regions are always palindromes, as the three examples suggest.

Only two letters occur in an encoded region. Any two consecutive regions share one interval, or letter. Thus, the Structural and Pentatonic regions share the letter *d*, as do Pentatonic and Diatonic, whereas the Diatonic and Chromatic regions share the letter *f*, corresponding to the diatonic half step.

[1] [14] pp. 122–123.

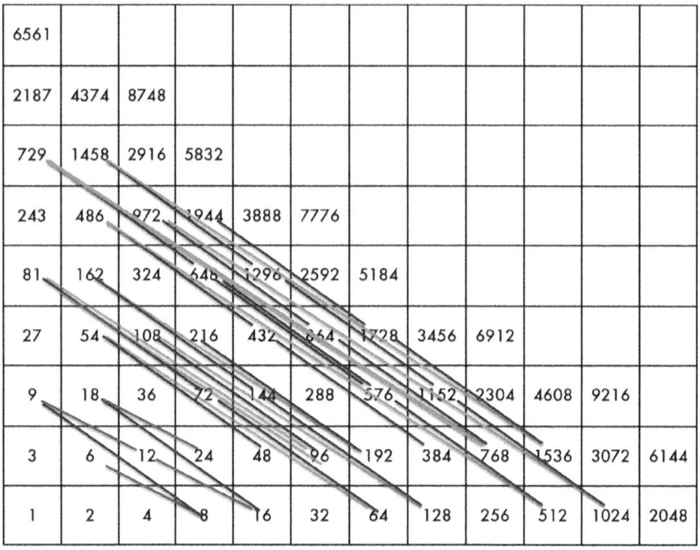


| Structural | Pentatonic | Diatonic |
|---|---|---|
| cdccdc | deddededded | ddfdddfddfdddfdd |
| Pal(*cdc*) 3/5 | Pal(*dede*) 5/8 | Pal(*ddfdf*) 5/13 |

**Fig. 7.** Three regions in $\Lambda_\vartheta$: $c = (2, -1)$, $d = (-3, 2)$, $e = (5, -3)$, $f = (8, -5)$ Letters: $\alpha, \beta$: Slope: $(|\Lambda_\vartheta|_\alpha + 1)/(|\Lambda_\vartheta|_\beta + 1)$

Another representation of the regions is shown in Fig. 8, which helps to clarify the connection between central words and the encoded regions. Note, however, that the staircase pattern of letters $x$ and $y$ here does not represent steps, as the zig-zag pattern of letters in Fig. 7 does. Instead, $x$ encodes the interval of a descending octave, and $y$, an ascending twelfth[2]. The pattern on $\{x, y\}$ is, in effect, the cutting sequence of slope $-B/A$, where $A/B$ is the determining continued fraction approximation. The stepwise orderings depicted in Fig. 7 are "dual" to the orderings on $(x, y)$ shown in Fig. 8, as described in [15]. (The same duality is applied to well-formed modes in [9].) Under this representation, the three encoded regions from Fig. 7 are shown as central words with slope approximately equal to $-(\log_2 3)^{-1} = -\log_3 2$. Here, the duality may be noted in that the arguments of the right iterated palindromic closure for each of the stepwise regions in Fig. 7 are (abstractly) reversed in the $x, y$ patterns in Fig. 8. (To illustrate with the diatonic, map $d \to y, f \to x$, then reverse.) Whereas the Structural and Pentatonic regions are both self-dual, the Diatonic produces different patterns in the two diagrams.

[2] The vectors are (-1,0) and (0,+1). Alternatively, the direction could be reversed so that the progression moves southeasterly.

The infinite word $\boldsymbol{s}$ associated with $\alpha = \log_3 2$ is found by discretizing the line of slope $\alpha/(\alpha+1)^3$. Then $\boldsymbol{s} = xxyxxyxyxxyxyxxyxxy\ldots$. Letting $\boldsymbol{c}_\alpha$ denote the infinite word such that $\boldsymbol{s} = x\boldsymbol{c}_\alpha$ gives us the *characteristic Sturmian word* $\boldsymbol{c}_\alpha = xyxxyxyxxyxyxxyxxy\ldots$. The infinite set of palindromes are the duals encoded central regions including the three depicted in Fig. 8: $\epsilon$, $x$, $xyx$, $xyxxyx$ $xyxxyxyxxy$, $xyxxyxyxxyxyxxyx$, etc. A clear exposition of these dual palindromes is found in [16].

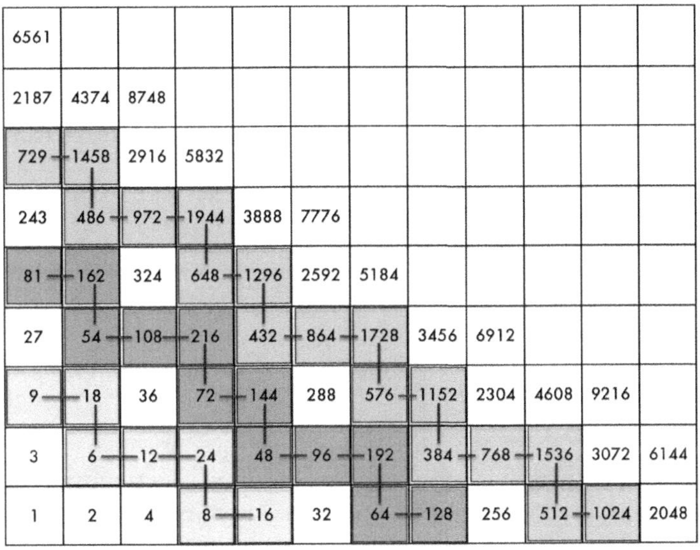


| Structural | Pentatonic | Diatonic |
|---|---|---|
| xyxxyx | xyxxyxyxxyx | xyxxyxyxxyxyxxyx |
| Pal(*xyx*) -3/5 | Pal(*xyxy*) -5/8 | Pal(*xyxyy*) -7/11 |

**Fig. 8.** Regions in $\Lambda_\vartheta$ as central words: $x = (+1, 0)$, $y = (0, -1)$. Slope $= (x+1)/(y+1)$

**Regions and Non-regions.** Figure 9 provides a partitioning of $\Lambda_\vartheta$ by central words. The central words are in bold, the segments between central words are in italics. Between regions there are, in general, three letters in circulation, namely the three letters that form the union of the letters in the two regions that lie on either side. As intervals, one of these three is the sum of the other two. The fact that no more than three letters can exist in these spaces may be recognized in the connections between well-formed scale theory and the "Three-Gap" Theorem. (See [17], [18], and [19].)

In the encoded (italicized) non-regions, the last (rightmost) letter makes its final appearance in the Lambda word. The second letter of each encoded non-region marks that letter's first appearance in the Lambda word. For example,

[3] See [2], 134.

a **bcb** **cdccdc** *dedc* **deddededded** *dfdde* **ddfdddfddfdddfdd**
*fgfddfddfgfddfgfdfgfddfgfdfgfgfdfgfdfgfgfdfgfgffgfgfd*
**fgfgffgfgfgffgfgffgfgfgffgfgf**
*fhffgfgffgfgffhffgfgffhffgffhffgfgffhffgffhffhffgffhffgffhf*
*fhffgffhffhfffhffhffg*
**ffhffhfffhffhffhfffhffhfffhffhffhfffhffhff**
*hihffhffhfffhffhffhihffhffhffhihffhffhihffhffhffhihffhffhih*
*fhihffhffhihffhffhihfhihffhffhihfhihffhihfhihffhffhihfhihff*
*hihfhihfhihffhihfhihffhihfhihfhihffhihfhihfhihihfhihfhihffh*
*ihfhihfhihihfhihfhihfhihihfhihfhihihfhihfhihfhihihfhihfhihi*
*hhihihfhihfhihihfhihfhihihhihihfhihfhihihhihihfhihihhihihfh*
*ihfhihihhihihfhihihhihihhihihfhihihhihihfhihihhihihhihihfhi*
*hihhihihhihihihhihihhihihf*
**hihihhihihhihihihhihihhihihhihihihhihihhihihihhihihhihihhih**
**ihihhihihhihih**
*hjhhihihhihihhihihihhihihhihihhjhhihihhihihhihihhjhhihihhih*
*ihhjhhihihhihihhihihhjhhihihhihihhjhhihhjhhihihhihihhjhhihi*
*hhihihhjhhihhjhhihihhihihhjhhihhjhhihihhjhhihhjhhihihhihihh*
*jhhihhjhhihihhjhhihhjhhihhjhhihihhjhhihhjhhihihhjhhihhjhhih*
*hjhhihihhjhhihhjhhihhjhhjhhihhjhhihhjhhihihhjhhihhjhhihhjhh*
*jhhihhjhhihhjhhihhjhhjhhihhjhhihhjhhjhhihhjhhihhjhhihhjhhjh*
*hihhjhhihhjhhjhhhjhhjhhihhjhhihhjhhjhhihhjhhihhjhhjhhhjhhjh*
*hihhjhhihhjhhjhhhjhhjhhihhjhhjhhhjhhjhhihhjhhihhjhhjhhhjhhj*
*hhihhjhhjhhhjhhjhhhjhhjhhihhjhhjhhhjhhjhhihhjhhjhhhjhhjhhhj*
*hhjhhihhjhhjhhhjhhjhhhjhhjhhjhhhjhhjhhhjhhjhhi*
**hhjhhjhhhjhhjhhhjhhjhhjhhhjhhjhhhjhhjhhhjhhjhhjhhhjhhjhhhjh**
**hjhhjhhhjhhjhhhjhhjhhhjhhjhhjhhhjhhjhhhjhhjhh**

**Fig. 9.** Factoring $\Lambda_\vartheta$ into central words (bold) and non-central words (italics)

the encoded non-region *dedc* contains the first appearance of $e$ and the last of $c$. Although the role of these first and last appearances were acknowledged as limits, the space between regions has been otherwise unexplored territory in well-formed scale theory.

Figure 9 gives evidence that each central word is longest palindrome up to that point in the word $\Lambda_\vartheta$. Although there are palindromes in the segments between central words, and others that contain material that may straddle a region and adjoining non-region, none of these palindromes is longer than the previous central word. Note that the italicized interregnums soon become longer than the encoded regions that punctuate them, even as the central words themselves grow longer. It is also possible to say that each central word transforms systematically into the following one, however I will not develop that perspective here.

### 4.2 Successive Letter Palindromes

Although the encoded regions play an important role in the Lambda word and are, relatively speaking, the longest of its palindromes there are another category of palindrome that has escaped notice. Beginning with any letter and moving forward in the word, a palindrome is formed when that same letter is reached a second time. In Fig. 9, in the final non-region shown on $h$, $i$, and $j$, the final two appearances of $i$ are the sixth and final letter of the line. The 41-letter word, *ihhjhhjhhhjhhjhhhjhhjhhjhhhjhhjhhhjhhjhhi*, is a palindrome.

**Theorem 4.** *Let $W$ be a word in $\Lambda_\theta$ formed between two successive appearances of some letter $x$. Then $xWx$ is a palindrome.*

After some preliminaries, a brief proof sketch of Theorem 4 follows.

### 4.3 A Factoring by Nuclear Palindromes

The factoring shown in Fig. 9 is alternately by palindromes and non-palindromes, after some successive central words at the beginning. (Although a palindrome, the letter $a$ is in a class by itself.) Any Lambda word may also be factored entirely by palindromes. An arbitrary word may be trivially be factored into palindromes by taking every letter as a one-element palindrome. Figure 10 shows a non-trivial factoring of this Lambda word.

```
a b cbc d cc dcd edcde dd ede ddedd fddeddf ddd fddfdddfddf
g fddfddf gfddfg fdfgfddfgfdf gfg fdfgfdf gfgfdfgfg
ffgfgfdfgfgff gfgfg ffgfgffgfgfgffgfgff
```

**Fig. 10.** Partitioning $\Lambda_\vartheta$ into palindromes

Although the factorization necessarily contains one-letter words, it is non-trivial because the one-letter words are greatly outnumbered and, although there will continue to be one-letter words as the factorization develops, the length of multi-letter words grows without bound. These palindromes are a subset of the set of *nuclear palindromes*, which form the basis of the proof of Theorem 4. To help provide an intuition for their structure, consider again the set of pitch classes formed by octave and fifth. Take a set of $N$ pitch classes, put into scalar order. Then, look at the segments between pitch class 0 and pitch class $N-1$. Figure 11 shows the case for $6 \leq N \leq 9$.

Any of the segments that begin and end on whole notes encloses a nuclear palindrome. Note that all of these segments are, indeed palindromes, as the

**Fig. 11.** Partitioning $\Lambda_\vartheta$ into palindromes

parenthesized words above each segment confirm. Note too, that there is a constant sum between symmetrically arranged elements in each segment. The symmetries shown here illustrate what Carey and Clampitt [20] say about the one degree of symmetry always present in any generated collection, well-formed or not.

Most nuclear palindromes may be expandable outwards to form larger, overlapping palindromes. A *maximal palindrome* is a nuclear palindrome expanded maximally. The first 16 nuclear palindromes and their maximal expansions are shown in Fig. 12. Neither of the first two nuclear palindromes in Fig. 10, $a$, $b$, may be expanded, but the next, $c$ may be expanded into $bcb$. The fourth nuclear palindrome, $cbc$ cannot be expanded, but the next two can: $d$ into $cdc$ and $cc$ into $cdccdc$, a central word.

```
λ{C1+}  a------------------------------------------------
λ{C2-}  -b-----------------------------------------------
λ{C3+}  -bcb---------------------------------------------
λ{C4+}  --cbc--------------------------------------------
λ{C5-}  ----cdc------------------------------------------
λ{C6+}  ----cdccdc---------------------------------------
λ{C7-}  --------dcd--------------------------------------
λ{C8+}  --------dcdedcd----------------------------------
λ{C9+}  ----------dedcded--------------------------------
λ{C10-} --------------dedded-----------------------------
λ{C11+} --------------deddededded------------------------
λ{C12-} ---------------------ddedd-----------------------
λ{C13+} ---------------------ddeddfddedd-----------------
λ{C14+} ------------------------ddfddeddfdd--------------
λ{C15-} ------------------------------ddfdddfdd----------
λ{C16+} ------------------------------ddfdddfddfdddfdd--
```

**Fig. 12.** Maximal Palindromes and their Nuclei in $\Lambda_\vartheta$

A proof sketch of Theorem 4 follows: Let $xWx$ be a word in $\Lambda_\theta$ such that $x \notin W$. Then $W$ is a word on at most two letters, $y$ and $z$. If $W = \emptyset$, the theorem is proved, as $xx$ is clearly a palindrome. If $W$ consists of iterations of a single letter, say, $y$, the theorem still holds, as $xy^n x$ is a palindrome. The interesting case is if $W$ is on two letters, $y$ and $z$.

The proof begins by showing that $xWx \subset C_N$, where $C_N$ is some maximal palindrome. The term *median* of a palindrome signifies its center of symmetry. If a palindrome is of odd cardinality, its median corresponds to its central letter. If even, then its median is found between its central pair of letters. Let $P$ be a palindrome. Then a *medial subword of* $P$ is a connected subword that has the same median as $P$. Then $edcde$ is a *medial subpalindrome* of $dedcded$. On the other hand, $ded$ appears twice as a subpalindrome of $dedcded$, but is not a medial subword. Clearly, any medial subword of a palindrome is also a palindrome.

Thus, if $xWx$ is a medial subword of some maximal palindrome, $C_N$, then, by the above it, too, is a palindrome. Theorem 4 is proven by means of a recursive mapping $\rho$ that maps any successive letter subword $xWx$ of $C_N$ either to its reversal $(\widetilde{xWx})$ in $C_N$, or to itself in $C_{N-1}$. The process necessarily terminates at a medial subword in $C_{N-k}$, with $0 \leq k < N$.

The successive letter palindrome property is rare. Take, for example, the usual diatonic in just intonation. This may be encoded as a word on three letters, as shown in Fig. 13.

| | | | | | | | | | | | | | | |
|---|---|---|---|---|---|---|---|---|---|---|---|---|---|---|
| C | | D | | E | | F | | G | | A | | B | | D |
| 24 | | 27 | | 30 | | 32 | | 36 | | 40 | | 45 | | 48 |
| | $\frac{9}{8}$ | | $\frac{10}{9}$ | | $\frac{16}{15}$ | | $\frac{9}{8}$ | | $\frac{10}{9}$ | | $\frac{9}{8}$ | | $\frac{16}{15}$ | |
| | $a$ | | $b$ | | $c$ | | $a$ | | $b$ | | $a$ | | $c$ | |

**Fig. 13.** The diatonic scale in just intonation as the word *abcabac*

Considering this word as circular, a reasonable one in the case of a musical scale, the words formed by successive appearances of the same letter are shown in Fig. 14:

$$abca\ bcab\ \underline{cabac}\ \underline{aba}\ \underline{bacab}\ \underline{aca}\ cabc$$

**Fig. 14.** Successive letter palindromes in *abcabac*

The four underlined words are palindromes framed by successive appearances of the same letter. Although a majority of such words are palindromes in this word, nearly half are not. The property is not universal, then, even in this highly structured seven-letter word on three letters. By contrast, such words in the infinite Lambda word—an infinite word on an infinite alphabet—are always palindromes.

## 5 The Fibonacci Lambda Word

We conclude with a reminder that a Lambda word may be built on any irrational value $\alpha$ with $1 < \alpha < 2$. Letting $\alpha = \phi = \frac{\sqrt{5}+1}{2}$,

$\Lambda_\phi = abcbcdccdcdedcdeddededdede\underline{fededef}\ldots$.

Comparing this with $\alpha = \vartheta = \log_2 3$,

$\Lambda_\vartheta = abcbcdccdcdedcdeddededdedd\overline{fddeddf}\ldots$.

The two words share a 25-letter prefix, due to the fact that the difference between $\phi$ and $\vartheta$ is relatively small. Like the Lambda word based on $\log_2 3$, $\Lambda_\phi$ will display regions, as well as successive letter palindromes. The regions in $\Lambda_\phi$ are all self-dual, suggesting that this word would be worthy of further investigation.

## 6 Conclusions

Our study of the Lambda word puts emphasis on the role of symmetry in musical structures. The Lambda word based on $\log_2 3$ encodes a series of intervals that generally decrease in value, but not monotonically. Given successive intervals $pq$ in the sequence, it is not possible to say whether $p > q$, $p = q$, or $p < q$, although, over the course of the sequence, the size of intervals will, invariably, diminish. Within this asymmetry, pockets of symmetry are found everywhere. There is an analogy to musical structures in time. Time is asymmetrical. The second law of thermodynamics normally allows us to tell the difference between a movie that runs forward or backward. Musical structures, such as arch forms, rhythmic palindromes, and regions, stamp symmetries upon the asymmetrical flow of time.

## Acknowledgments

The author wishes to thank Emanuel Amiot, and David Clampitt for invaluable help and long emails in working out this project. Great appreciation is also due to Thomas Noll for his important insights.

## References

1. Domínguez, M., Clampitt, D., Noll, T.: Well-formed scales, maximally even sets and Christoffel Words. In: Klouche, T., Noll, T. (eds.) MCM 2007. Communications in Computer and Information Science, vol. 37, pp. 477–488. Springer, Heidelberg (2009)
2. Berstel, J., Lauve, A., Reutenauer, C., Saliola, F.: Combinatorics on Words: Christoffel Words and Repetition in Words. American Mathematical Society CRM Monograph Series, vol. 27 (2008)
3. Berthé, V., de Luca, A., Reutenauer, C.: On an involution of Christoffel words and Sturmian morphisms. European Journal of Combinatorics 29(2), 535–553 (2008)
4. Lothaire, M.: Combinatorics on Words. Cambridge Math. Lib. Cambridge Univ. Press, Cambridge (1997)
5. Lothaire, M.: Algebraic Combinatorics on Words. Encylopedia Math. Appl., vol. 90. Cambridge Univ. Press, Cambridge (2002)
6. de Luca, A.: Sturmian words: Structure, combinatorics, and their arithmetics. Theoretical Computer Science 183(1), 45–82 (1997)
7. Carey, N.: On coherence and sameness, and the evaluation of scale candidacy claims. Journal of Music Theory 46, 1–56 (2002)
8. Carey, N.: Coherence and sameness in well-formed and pairwise well-formed scales. Journal of Mathematics and Music 1(2), 79–98 (2007)
9. Clampitt, D., Noll, T.: Modes, the height-width duality, and Handschin's Tone Character. Music Theory Online 17(1), (forthcoming), `http://user.cs.tu-berlin.de/%7Enoll/HeightWidthDuality.pdf`
10. Allouche, J.-P., Baake, M., Cassaigne, J., Damanik, D.: Palindrome complexity. Journal of Theoretical Computer Science 292(1), 9–31 (2003)

11. Brlek, S., Hamel, S., Nivat, M., Reutenauer, C.: On the palindromic complexity of infinite words. International Journal of Foundations of Computer Science 15(2), 293–306 (2004)
12. Droubay, X., Pirillo, G.: Palindromes and Sturmain words. Theoretical Computer Science 223, 73–85 (1999)
13. Carey, N., Clampitt, D.: Structural properties of musical scales (unpublished manuscript)
14. Carey, N., Clampitt, D.: Regions: A theory of tonal spaces in early medieval treatises. Journal of Music Theory 40, 113–147 (1996)
15. Carey, N., Clampitt, D.: Self-similar pitch structures, their duals, and rhythmic analogues. Perspectives of New Music 34(2), 62–87 (1996)
16. Singler, F.: Zur Dualität zwischen doppelter Periodizität und binärer Intervall-Struktur in der Theorie der Tonregionen. Thesis. Hochschule für Musik und Theater „ Felix Mendelssohn Bartholdy", Leipzig (2008) `http://www.qucosa.de/fileadmin/data/qucosa/documents/2536/Dualit%C3%A4t_Tonregionen_Sept09.pdf`
17. Carey, N.: Distribution Modulo 1 and Musical Scales. Ph.D. Dissertation. University of Rochester (1998)
18. Slater, N.B.: Gaps and steps for the sequence $n\theta$ mod 1. Proceedings of the Cambridge Philosophical Society 63, 1115–1122 (1967)
19. Sós, V.T.: On the distribution mod 1 of the sequence $n\alpha$. Annales Universitatis Scientarium Budapestinensis de Rolando Eötvös Nominatae. Sectio Mathematica 1, 127–134 (1958)
20. Carey, N., Clampitt, D.: Aspects of well-formed scales. Music Theory Spectrum 11(2), 187–206 (1989)

# Sensitive Interval Property for Scales as Words in the Free Group F2

David Clampitt

The Ohio State University, School of Music
Columbus, USA
clampitt.4@osu.edu

**Abstract.** The sensitive interval property is a special feature of musical scales that generalize the diatonic Ionian (major) and Aeolian (minor) modes: specifically, ascending authentic Ionian and descending plagal Aeolian. This discussion is situated in a music-theoretic interpretation of algebraic combinatorics on words over two-letter alphabets. The present paper provides an introduction to this approach, but relies on results from a number of recent papers in this area. While previous studies have restricted attention to the free monoid of words on two letters, the present one extends consideration to F2, the free group with two generators. This permits treatment of ascending and descending modal varieties of musical scales, together with rising or falling circle-of-fifths presentations (or their generalizations), within a unified mathematical framework. The special property investigated herein positions the diatonic major third (and its generalizations) as of structural significance within the theory.

**Keywords:** Scale theory, Word theory, Algebraic combinatorics on words, Modes, Sensitive interval, Divider incidence, Well-formed scales, Christoffel words, Standard and anti-standard words, Central words, Sturmian morphisms, Free group F2.

## 1 Introduction: Scale Theory and Word Theory

The term *scale theory* refers to the mathematical music theory subfield that considers the structures of pitch sets from which musical repertoires draw their tonal material. A *scale* is an ordering of pitches in terms of their pitch heights (correlated with logarithms of fundamental frequencies). That is, the scale is a (conceptually infinite) sequence of real numbers assembled in ascending order, $\ldots < h_0 < h_1 < \ldots$. Generally, this sequence is periodic modulo some value, and one introduces an equivalence relation, typically octave equivalence. If the scale has period $N$, elements whose indices are congruent mod $N$ are deemed equivalent. We therefore consider a finite sequence $h_0 < h_1 < \ldots < h_{N-1}$ and an associated set of indices $\mathbb{Z}_N = \{0, 1, \ldots, N-1\}$, which may be understood as a cyclic group with addition mod $N$. The sequence of pitch heights $h_k$ may similarly be reduced to a set $H$ of real numbers modulo 1 (i.e., a finite subset of $\mathbb{R}/\mathbb{Z}$), but we may not assume that $H$ is a group with addition mod 1.

C. Agon et al. (Eds.): MCM 2011, LNAI 6726, pp. 56–68, 2011.

Scale theorists have studied the relationships between the generic level of description—the set of indices $\mathbb{Z}_N$ and derived differences, subsets, and subsequences—and the specific level of description–the set $H$ and derived differences, subsets, and subsequences. For example, $H$ has Clough and Myerson's [1] *Myhill's Property* (MP) if for every non-zero index $k \in \mathbb{Z}_N$ the set of distinct differences $D_k = \{h_j - h_i | (j-i) \equiv k \bmod N\}$ has cardinality 2. The usual diatonic set has MP: generic diatonic steps have specific major and minor sizes, as do generic thirds, and so forth. A set H is said to be *generated* if every element $h \in H$ can be written as $g_k = (g_0 + k\theta) \bmod 1$, for some $g_0$ in $H$ and some *generating element* or *generating interval* $\theta > 0$. This sets up an alternative ordering of $H$, a permutation $\pi : \mathbb{Z}_N \to \mathbb{Z}_N$ such that $g_k = h_{\pi(k)}$, for $k = 0, 1, \ldots, N-1$. Carey and Clampitt [2] defined a *well-formed* (WF) scale to be one where $\pi$ is a linear automorphism of $\mathbb{Z}_N$, that is, $\pi(z) = mz \bmod N$ for some $m$, $\gcd(m, N) = 1$. If a well-formed scale has MP, it is said to be *non-degenerate*. In [3], it is shown that non-degenerate WF and MP are equivalent properties. The canonical musical example of a non-degenerate WF scale is the usual (white-key) diatonic scale, where the generating element has the size of a perfect fifth (in Clough and Myerson's treatment, a base-2 logarithm value of $\frac{7}{12}$; in Carey and Clampitt's treatment, $\log_2 \frac{3}{2}$ or some real number approximation). The usual (black-key) pentatonic is another such example.

Similar treatments based on the usual diatonic scale are presented by Agmon in [4] and [5]. Regener's [6] discussion of pitch notation and equal temperament foreshadows these developments, and Wilson's *Moments of Symmetry* (MOS), defined in a 1964 letter to Chalmers and described in [7], are equivalent to MP/non-degenerate WF scales. Lest music theorists be overly concerned with questions of originality, however, it is worth noting that mathematicians, e.g., the number theorist Sós in [8], had made comparable definitions in settings of greater generality.

The combinatorial properties discussed above are robust with respect to changes in $\theta$ within a well-defined range. For example, all non-degenerate WF scales with $\frac{3}{5} < \theta < \frac{4}{7}$ belong to the diatonic class wfs$(N, 2)$ (defined in generality in [9]), consisting of scales with 5 larger step intervals of size $a$ and 2 smaller step intervals of size $b$, in some rotation of the sequence $\langle aaabaab \rangle$.

These properties are, then, robust in particular with respect to (or, insensitive to) cyclic scale orderings or musical *modes*. Music theorists would say that they are dependent upon the pitch-class set, and not upon a choice of beginning or ending scale element, a privileged final (*finalis*) or tonic. And yet, in music theory and in the history of music, the particular scale orderings that follow from the choice of a privileged element are and have been of the greatest importance. In the case of common-practice period music, the major scale ordering is especially privileged, in a dichotomy with the minor scale ordering (the definition of the latter slightly less stable in the music theory literature). These two scale orderings were excluded in the classification of medieval chant. According to medieval and early renaissance theorists, modes were defined both by octave species (ordering of step intervals) and by the authentic/plagal distinction:

in the authentic form, the first note is the final and the octave that the scale spans is divided into a perfect fifth above the final, followed by a perfect fifth; in the plagal form, the final is a perfect fourth above the first note of the scale, and the octave that the scale spans is divided into a perfect fourth below the final, followed by a perfect fifth. Thus, representing major steps by $a$ and minor steps by $b$, the sequence $abaa|aba$ is distinguished from the sequence $aba|aaba$ by the placement of the divider $|$. The former represents (authentic) Dorian, the latter (plagal) hypo-Mixolydian, in modern terminology; modes 1 and 8, respectively in the medieval classification.

## 2 Word Theory and Modes

This is where word theory enters the picture. In recent studies (e.g., [10,11,12,13]), modes of MP scales have been represented as words over two-letter alphabets (Chapter 2 of [14] is a standard reference on this aspect of word theory). One considers the alphabet $A = \{a, b\}$ and the free monoid over $A$, $A^* = \{w = w_1 \dots w_n | w_i \in A, n \in \mathbb{N}\}$, with concatenation of words. One understands that the empty word $\epsilon$ is in $A^*$, and for all words $w$ in $A^*$, $\epsilon w = w = w\epsilon$. We say that words $w = uv$ and $w' = vu$ are *conjugate* or *conjugates* of each other. The following five endomorphisms of $A^*$ are important in the theory and its musical application, where each maps $A^*$ to itself by replacing each letter of $w \in A$ by: $G(a) = a, G(b) = ab$; $G\tilde{}(a) = a$, $G\tilde{}(b) = ba$; $D(a) = ba, D(b) = b$; $D\tilde{}(a) = ab$, $D\tilde{}(b) = b$; and $E(a) = b, E(b) = a$.

The mappings defined by these substitutions and compositions of them form the monoid $St$ of *Sturmian* morphisms, under composition of mappings. (A subset of three morphisms, e.g., $G$, $D\tilde{}$, and $E$, suffice to generate $St$, but for present purposes it is useful to have the five defined above available.) If $F \in St$, $F$ is a morphism by construction: for any words $u, v \in A^*$, $F(uv) = F(u)F(v)$, since in particular for any word $w = w_1 \dots w_n$ for letters $w_i \in A$, $F(w) = F(w_1)F(w_2)\dots F(w_n)$ by definition.

The submonoid of $St$ that excludes the $E$ morphism (or, what is the same thing, that includes compositions containing an even number of $E$'s) is $St_0$, the special Sturmian morphisms. The submonoid of $St_0$ generated by $\{G, D\}$ is special *standard*, the one generated by $\{G\tilde{}, D\tilde{}\}$ is special *anti-standard*, and the one generated by $\{G, D\tilde{}\}$ is special *Christoffel*. The words in $A^*$ resulting from applying a morphism from the standard, anti-standard, and Christoffel submonoids to the root-word $ab$ are defined to be *standard, anti-standard, Christoffel* words, respectively, and the set of all images of $ab$ under $St$ are called *morphic conjugates of Christoffel words.* (For each conjugacy class, there exists an *amorphic* word, or "bad conjugate," that is not in the image of $ab$.) Often one keeps track of the images of the initial $a$ and $b$ by including the divider symbol "|". Everything to the left of the divider is in the image of $a$, everything to the right is in the image of $b$. For example, $G\tilde{}GD(a|b) = G\tilde{}G(ba|b) = G\tilde{}(aba|ab) = abaa|aba$. As mentioned at the end of paragraph 5 of the Introduction, this is the pattern of the authentic Dorian mode, where final $a$'s and $b$'s represent major and minor steps, respectively. The divider symbol pays a music-theoretic dividend by detecting the authentic-plagal distinction. The word theory context, moreover, pays its dividend by distinguishing mode in the sense of octave species. It is demonstrated in [13] that the various compositions $GGD, G\tilde{}GD, G\tilde{}G\tilde{}D, GGD\tilde{}, G\tilde{}GD\tilde{}, G\tilde{}G\tilde{}D\tilde{}$, applied to $(a|b)$, yield the authentic modes as described in Glarean's 1547 *Dodecachordon*, with finals C, D, E, F, G, A, respectively. The following compositions from the complement of $St_0$, $EDDG, ED\tilde{}DG, ED\tilde{}D\tilde{}G, EDDG\tilde{}, ED\tilde{}DG\tilde{}, ED\tilde{}D\tilde{}G\tilde{}$, applied to $(a|b)$, yield the corresponding plagal modes, with corresponding finals. (NB: $G$ and $G\tilde{}$ commute, as do $D$ and $D\tilde{}$, so there are only 12 distinct compositions, corresponding to Glarean's 12 modes. The modes that Glarean excluded, the Locrian varieties, which do not permit authentic or plagal division, are also excluded on word-theoretic grounds: these are the "bad conjugates" of the word theorists.)

To reiterate a nice point respecting the music-theoretic interpretation: note that the use of the divider symbol permits a refinement of the subset of morphic conjugates of a Christoffel word. For example, we have seen that $G\tilde{}GD(a|b) = abaa|aba$, which is the pattern for authentic Dorian, whereas $ED\tilde{}DG\tilde{}(a|b) = ED\tilde{}D(a|ba) = ED\tilde{}(ba|bba) = E(bab|bbab) = aba|aaba$, which is the pattern for hypo-Mixolydian. As undivided words, they collapse to the same conjugate, abaaaba, corresponding to the same octave species, but as divided words they reflect the authentic/plagal distinction.

Space does not permit further musical interpretation of these results here. Extensive and intensive efforts in this direction are found in, e.g., [10] and [13]. From the point of view of mathematical music theory, one may understand applications of Sturmian morphisms as akin to basis transforms. Indeed, each element of $St_0$ has a *commutative image* in $SL_2(\mathbb{N})$. Those in the complement of $St_0$, on the other hand, correspond to matrices of determinant $-1$. In our musical interpretation, the meanings of the letters change with each application of a non-identity morphism; what remains fixed are the musical intervals represented by the words before and after the divider symbol, to wit, perfect fifth and perfect fourth, respectively. Thus, $a$ and $b$ in "$a|b$" represent the initial perfect fifth and perfect fourth, respectively, whose sum is the perfect octave, and for any morphism $F$ in $St_0$, the word $F(a)$ represents (or "fills the space of") the perfect fifth, the word $F(b)$ represents (or "fills the space of") the perfect fourth, and the concatenation $F(a)F(b) = F(a|b)$ is a word representing the octave: i.e., a pattern of step intervals filling in an octave. The effects of $G$ and $G\tilde{}$ are to subdivide the interval represented $b$, and similarly, mutatis mutandis, for $D$ and $D\tilde{}$. For the larger mathematical picture, see Chapter 5 of [15], which discusses the characterization of images of $a$ and $b$ under any given $F \in St$ as forming a basis of the free group F2.

What connects this modal interpretation of word theory back to earlier studies of generated scales and, in particular, of the privileged subclass of generated scales which are well-formed, is a duality on conjugates of Christoffel words called the *plain adjoint* in [13]. For the purposes of this paper, an informal presentation at the level of the musical interpretation will suffice. A purely word-theoretic definition may be found in [11], elaborated in [13]. For example, taking advantage of one's tacit understanding of the meaning of the musical note names, we may informally represent the undivided Lydian octave species and its associated word as F $a$ G $a$ A $a$ B $b$ C $a$ D $a$ E $b$ (F'). We consider that the upper octave is an excluded boundary tone, identified under octave equivalence with the initial F, but distinct from F as the terminal note in the minor step $b$ from E to F'. We know that the notes of this (and any other) "white-key" diatonic mode are generated by the perfect fifth, F–C–G–D–A–E–B, but now we ask for the pattern of upward perfect fifths and downward perfect fourths that "folds" these notes into the F–(F' ) octave. Encoding upward perfect fifths by $x$ and downward perfect fourths by $y$, we have the pattern F $x$ C $y$ G $x$ D $y$ A $x$ E $y$ B $y$ (F-sharp). Again, we have an excluded boundary tone, F-sharp, identified under "diatonic equivalence" (same uninflected note name) with the initial F, but distinct from F as the terminal note in the downward perfect fourth that yields an F-sharp within the F–(F') octave. The words $aaabaab$ and $xyxyxyy$ are said to be each other's plain adjoint, and we say that the former is the scale pattern for the Lydian mode, and the latter is the folding pattern for the Lydian mode. We can follow the same procedure to determine the folding for any mode, and, as mentioned above, the plain adjoint duality may be generalized to be applied to any conjugate of a Christoffel word (including the bad conjugate).

For some submonoids of $St$ or of $St_0$, we may lift the duality on the words to morphisms which generate them [11]. For example, if $F$ is a special Christoffel morphism, $F(a|b) = w$ is a Christoffel word, and the plain adjoint word is also a Christoffel word, $w^* = F^*(x|y)$ where $F^* = \overleftarrow{F}$, the reversal of the word in $\{G, D^{\sim}\}$ that composes $F$. $w^{**} = w$, and $F^{**} = F$. (We express the adjoint word here in the alphabet $\{x, y\}$ to make a distinction in the musical interpretation between scale pattern and folding pattern. We take it as understood that the morphisms in $St$ act analogously on any two-letter alphabet.) For example, the scale pattern for the Lydian mode is a Christoffel word: $GGD^{\sim}(a|b) = GG(ab|b) = G(aab|ab) = aaab|aab$. The plain adjoint for Lydian is $D^{\sim}GG(x|y) = D^{\sim}G(x|xy) = D^{\sim}(x|xxy) = xy|xyxyy$. The picture for the Lydian mode and its folding is shown in Fig. 1.

**Fig. 1.** Lydian mode: scale pattern and plain adjoint folding pattern

Similarly, if $F$ is a special standard morphism, $F(a|b) = w$ is a standard word, and its plain adjoint is also standard, $w^* = F^*(x|y) = \overleftarrow{F}(x|y)$. The standard word in the conjugacy class of words corresponding to diatonic modes is the musically privileged authentic Ionian mode (modern major scale): $GGD(a|b) = GG(ba|b) = G(aba|ab) = aaba|aab$. The folding pattern for the authentic Ionian mode is the plain adjoint word, $DGG(x|y) = DG(x|xy) = D(x|xxy) = yx|yxyxy$. The picture for the Ionian mode and its folding is shown in Fig. 2.

**Fig. 2.** Ionian mode: scale pattern and plain adjoint folding pattern; divider incidence

The story that is told by the pairing of the scale pattern for authentic Ionian with its plain adjoint folding pattern provides insight into the special status of this mode. As Fig. 2 shows, the *divider*—the note which marks the authentic division of the mode—is the same in the scale and in the folding. Moreover, the initial note of the scale is the *divider predecessor* in the folding, and the initial note of the folding is the divider predecessor of the scale. Finally, the penultimate note of the scale and of the folding coincide (the leading tone, or *note sensible*: sensitive tone). The totality of these conditions is called *divider incidence.* Noll's Ionian Theorem [16] proves that divider incidence holds for all standard words.

Divider incidence provides the point of departure for introduction of the sensitive interval property. But first we must extend these results from the monoid to the free group F2. The motivation to do this is because the restriction to the monoid requires that we assume a convention that interprets the letters of a two-letter alphabet as having a certain direction. It might suffice to consider ascending modes, but when we pair a mode with a folding pattern, we must choose a direction for the circle of fifths to generate the word representing the folding pattern. For musical reasons (discussed below and in [13]), we prefer to associate authentic diatonic modes with forward foldings (upwards perfect fifths, F–C–G–D–A–E–B–(F-sharp)), and plagal diatonic modes with backward foldings (downwards perfect fifths, B–E–A–D–G–C–F–(B-flat)). Thus, we would like to express both ascending and descending scale patterns and forward and backward folding patterns, within a unified mathematical framework.

Such a unification was called for in [11]. In [13], we treated each pair of letters as a separate two-letter alphabet (i.e., redefined the directional meanings of the letters), and defined the morphisms analogously on each alphabet, as we continue to do for alphabets $\{a, b\}$ and $\{x, y\}$. In this scenario, divider incidence for the plain adjoint holds only for standard words, with morphisms in $St_0$ yielding ascending authentic Ionian and descending plagal Aeolian and morphisms in the complement of $St_0$ yielding ascending plagal Ionian and descending authentic Aeolian. Because of the lack of unification, the only distinction between ascending authentic Ionian and descending hypo-Aeolian (or between ascending hypo-Ionian and descending Aeolian) is in the meaning assigned the letters of the alphabet: there is no distinction at the level of the morphisms. Similarly, there are few mathematical reasons to prefer, say, ascending authentic Ionian to its ascending plagal form. Extending from monoids over two-letter alphabets to free groups will improve this situation.

## 3 Into the Free Group F2, and Sensitive Intervals

Let $\{a, b\}$ freely generate a group, which therefore consists of all finite reduced words over an alphabet $A = \{a, b, a^{-1}, b^{-1}\}$ (all finite words over $A$ with adjacent pairs of a letter and its inverse eliminated). We keep separate another free group, with generators $\{x, y\}$, with the understanding that in all formal definitions $a$ and $x$ and $b$ and $y$ play the same roles. From group theory, the extended definition of the morphism $G$ must be $G(a) = a, G(b) = ab, G(a^{-1}) = a^{-1}, G(b^{-1}) = b^{-1}a^{-1} = (ab)^{-1}$. For example, we must have $G(aa^{-1}) = \epsilon = G(a)G(a^{-1}) = aG(a^{-1})$, so $G(a^{-1}) = a^{-1}$. $G(bb^{-1}) = \epsilon = G(b)G(b^{-1}) = abG(b^{-1})$, so $G(b^{-1}) = b^{-1}a^{-1}$.

Similar calculations yield the following extended definitions: $D(a) = ba$, $D(b) = b$, $D(a^{-1}) = a^{-1}b^{-1}, D(b^{-1}) = b^{-1}$; $G\tilde{}(a) = a, G\tilde{}(b) = ba, G\tilde{}(a^{-1}) = a^{-1}$, $G\tilde{}(b^{-1}) = a^{-1}b^{-1}$; $D\tilde{}(a) = ab, D\tilde{}(b) = b, D\tilde{}(a^{-1}) = b^{-1}a^{-1}, D\tilde{}(b^{-1}) = b^{-1}$, and obviously $E(a^{-1}) = b^{-1}, E(b^{-1}) = a^{-1}$. The morphisms $G$, $D$, etc., as extended to act on F2, are in fact members of the automorphism group, Aut(F2). They therefore not only act upon the inverse letters, they are themselves invertible. We do not need, however, to calculate their inverses here.

The result for the diatonic modes is that divider incidence is retained for standard words resulting from morphisms from $St_0$ applied to $(a|b)$ with corresponding plain adjoint standard words over $\{x, y\}$ (generalized ascending authentic Ionian modes with forward foldings, in the musical interpretation), and divider incidence is extended to anti-standard words resulting from morphisms from $St_0$ over $\{a^{-1}, b^{-1}\}$ with plain adjoint anti-standard words over $\{x^{-1}, y^{-1}\}$ (generalized descending hypo-Aeolian modes with backward foldings, in the musical interpretation). The formalism of Noll's proof [16] carries over to anti-standard words over negative letters without any problems. As we will see, anti-standard words over $\{a^{-1}, b^{-1}\}$ have the same formal structure as standard words over $\{a, b\}$.

When the plain adjoint was introduced in Sect. 2, we implicitly relied on informal knowledge of diatonic notes and intervals. In the word-theoretic environment, musical notes as such await definition, in terms of the intervals that the letters represent. Such a definition may be carried out by projecting words in F2 to trajectories in $\mathbb{Z}^2$, where the points in the lattice represent notes and vectors represent directed intervals. The definition of this projection for words in the monoid over two letters is given in [16], elaborated in [13]. The extension to F2 is straightforward, the same up to change of signs. For Christoffel words in negative letters (representing descending scales and backwards foldings), in the definition of mutually inverse affine automorphisms given by Noll in [16], substitute maxima for minima.

In the cases of divider incidence, the lifting from words in the plain adjoint relation to dual Sturmian morphisms in the appropriate submonoid holds only for the special standard and special anti-standard submonoids of $St_0$. The situation is more complicated for words representing generalizations of ascending hypo-Ionian with backward folding, and words representing generalizations of descending authentic Aeolian with forward folding. In these cases, divider incidence holds, but positive and negative alphabets are paired, and the lift pairs dual standard and anti-standard morphisms in the complement of $St_0$. A standard morphism $F$ with commutative image of determinant $-1$ (standard with $E$ contributed an odd number of times), applied to the root word $a|b$, yields a word $w$ representing a generalized ascending hypo-Ionian mode. The plain adjoint word $w^*$ representing the backward folding is the result of applying to the root word $x^{-1}|y^{-1}$ the dual anti-standard morphism $F^*$, the reversal of $F$ with each $G$ replaced by $G^{\sim}$, each $D$ replaced by $D^{\sim}$, and each $E$ left fixed. Similarly, anti-standard morphisms in the complement of $St_0$ applied to root word $a^{-1}|b^{-1}$ yield words representing generalized descending authentic Aeolian modes, and their plain adjoint forward foldings are standard words derived by applying standard morphisms in the complement of $St_0$ to root word $x|y$.

In Harrison's "renewed dualism" (cf. [17]), derived from 19th-century harmonic theorists such as Oettingen and H. Riemann and reflecting traditional associations, the major-minor modal dichotomy carries with it the following affiliations: major (Ionian), affiliated with ascending scale motion (especially ascending stepwise resolutions), authentic division, and authentic cadence (i.e., falling fifth, dominant-to-tonic cadential direction); minor (Aeolian), affiliated with descending scale motion (especially descending stepwise resolution), plagal division, and plagal cadence (i.e., rising fifth, subdominant-to-tonic cadential direction). Here and in [13], we affiliate authentic and plagal modes with anti-cadential folding directions: rising perfect fifths for authentic foldings and falling perfect fifths for plagal foldings. With these understandings, the mathematics reflects the relationships in Harrison's renewed dualism. The greater simplicity of the mathematical situation for special Sturmian morphisms privileges ascending authentic Ionian over hypo-Ionian and descending hypo-Aeolian over Aeolian (that is, in the former cases, morphisms for scale and folding are in the same submonoids, in addition to the distinction between morphisms in $St_0$ and in the complement of $St_0$). The mathematics reflects a distinction between ascending Ionian and descending hypo-Aeolian by associating the former with a special standard morphism and the latter with a special anti-standard morphism. The traditional associations of ascending/descending direction and authentic/plagal divisions with major and minor are retained. In what follows, we will confine the discussion to ascending authentic Ionian and descending plagal Aeolian (and their generalizations).

Each standard word over $\{a, b\}$ derived from a special standard morphism is of the form $uab$, where $u$ is *central*, that is, $u$ is a palindrome of length $p+q-2$ with periods $p$ and $q$, $\gcd(p, q) = 1$, $q$ the length of the divider prefix, $p$ the length of the divider suffix. We will not need the double periodicity property, only the property that $u$ is a palindrome, i.e., it coincides with its reversal [13]. Similarly, each anti-standard word over $\{a^{-1}, b^{-1}\}$ derived from a special anti-standard morphism is of the form $ua^{-1}b^{-1}$, where $u$ is central. This is proved below. Since the plain adjoints are in these cases also standard and anti-standard, respectively, they are of the forms $u^*xy$ and $u^*x^{-1}y^{-1}$, respectively, where $u^*$ is central, with periods $p'$ and $q'$, lengths of divider prefix and suffix, respectively, where $p' = p^{-1} \bmod N$, $q' = q^{-1} \bmod N$. An easy argument (see [13]) by induction on the length of standard (anti-standard) morphisms also tells us that divider prefixes in standard (anti-standard) words end in $a$ or $x$ ($a^{-1}$ or $x^{-1}$).

We wish to show that for a special anti-standard morphism $F$, $F(a^{-1}|b^{-1} = ua^{-1}b^{-1}$, where $u$ is a palindrome over $\{a^{-1}, b^{-1}\}$. Let $u$ be a palindrome, i.e., a member of $P = \{u \in \{a^{-1}, b^{-1}\} | u = u_1 \dots u_n, u_1 = u_n, u_2 = u_{n-1}, \dots\}$.

First we show that $D\tilde{}(u)b^{-1} \in P$ and $G\tilde{}(u)a^{-1} \in P$.

Proof of the lemma by induction on the length of $u$:

If $u$ is empty, the lemma holds. If $u$ is the letter $a^{-1}$, we have $D\tilde{}(a^{-1})b^{-1} = (b^{-1}a^{-1})b^{-1}$, and $b^{-1}a^{-1}b^{-1} \in P$; trivially, $G\tilde{}(a^{-1})a^{-1} = a^{-1}a^{-1} \in P$. If $u = b^{-1}$, we have $G\tilde{}(b^{-1})a^{-1} = (a^{-1}b^{-1})a^{-1}$, and $a^{-1}b^{-1}a^{-1} \in P$; trivially $D\tilde{}(b^{-1})b^{-1} = b^{-1}b^{-1} \in P$.

Assume the assertion holds for all $u$, $1 < |u| < n$. First, let $u_1 = u_n = a$. Then $D\tilde{}(u)b^{-1} = D\tilde{}(u_1 \dots u_n)b^{-1} = b^{-1}a^{-1}D\tilde{}(u_2 \dots u_{n-1})b^{-1}a^{-1}b^{-1} = b^{-1}a^{-1}D\tilde{}(u_2 \dots u_{n-1})b^{-1})a^{-1}b^{-1}$. But the length of $u' = u_2 \dots u_{n-1}$ is less than $n$ and $u' \in P$, so by the induction hypothesis, $D\tilde{}(u')b^{-1} \in P$. Therefore, $b^{-1}a^{-1}D\tilde{}(u')b^{-1})a^{-1}b^{-1} = D\tilde{}(u)b^{-1} \in P$. Second, let $u_1 = u_n = b$. Then $D\tilde{}(b^{-1}) = D\tilde{}(u_1 \dots u_n)b^{-1} = b^{-1}D\tilde{}(u_2 \dots u_{n-1})b^{-1}b^{-1} = b^{-1}D\tilde{}(u')b^{-1}b^{-1}$. Again, the length of $u'$ is less than $n$ and $u' \in P$, so by the hypothesis $D\tilde{}(u')b^{-1} \in P$. Therefore, $b^{-1}D\tilde{}(u')b^{-1}b^{-1} = D\tilde{}(u)b^{-1} \in P$. The same sort of argument holds for the case $G\tilde{}(u)a^{-1} \in P$, completing the proof of the lemma.

The proof of the principal assertion is by induction on the length of the special anti-standard morphism. For special anti-standard morphisms $f$ of length 1, $f(a^{-1}b^{-1}) = ua^{-1}b^{-1}$ where $u$ is a single letter: $G\tilde{}(a^{-1}b^{-1}) = a^{-1}a^{-1}b^{-1}$, $D\tilde{}(a^{-1}b^{-1}) = b^{-1}a^{-1}b^{-1}$. Assume for all special anti-standard morphisms $f$ of length $n > 1$ that $f(a^{-1}b^{-1}) = ua^{-1}b^{-1}$ for a palindrome $u$. Then $G\tilde{}f(a^{-1}b^{-1}) = G\tilde{}(ua^{-1}b^{-1}) = G\tilde{}(u)a^{-1}(a^{-1}b^{-1})$, and by the lemma, $G\tilde{}(u)a^{-1} \in P$.

Similarly, $D^{\sim}f(a^{-1}b^{-1}) = D^{\sim}(ua^{-1}b^{-1} = D^{\sim}(u)b^{-1}a^{-1}b^{-1}$, and by the lemma, $D^{\sim}(u)b^{-1} \in P$. Thus, for all special anti-standard morphisms of length $n+1$, the resulting anti-standard words are of the form $ua^{-1}b^{-1}$, where $u$ is a palindrome. (The argument that $u$ is in fact a central palindrome carries over from [13], but we will not require the double-periodicity property of $u$ here.)

The sensitive interval property stems from divider incidence, in that the dividers in the scale and the folding coincide, and the penultimate elements in the scale and the folding coincide. Therefore, the same interval between the divider and penultimate element is encoded in one way in the scale and another in the folding. This interval generalizes the diatonic major third, in major (ascending Ionian) the G–B rising major third from dominant to leading tone; in minor (descending hypo-Aeolian) the falling major third A–F; we call it the *sensitive interval.* Note that in Ionian, for example, the major third also extends from the initial tone of the scale, C, to E, the note that forms the upper boundary for the central word $u^*$ of the folding; and also extends from the initial note of the folding, F, to A, the note that forms the upper boundary for the central word $u$ of the scale. Fig. 3 displays the situation for authentic Ionian and its chromatic generalization, $DGGD(a|b)$, self-dual to $DGGD(x|y)$.

We will assume in the sequel a non-trivial situation, in the sense that the standard morphism $F$ contains at least one occurrence of $D$, or the anti-standard morphism $f$ contains at least one occurrence of $D^{\sim}$. Under those circumstances, as shown in [13], we also have the fact that the divider suffix $F(b)$ is a prefix

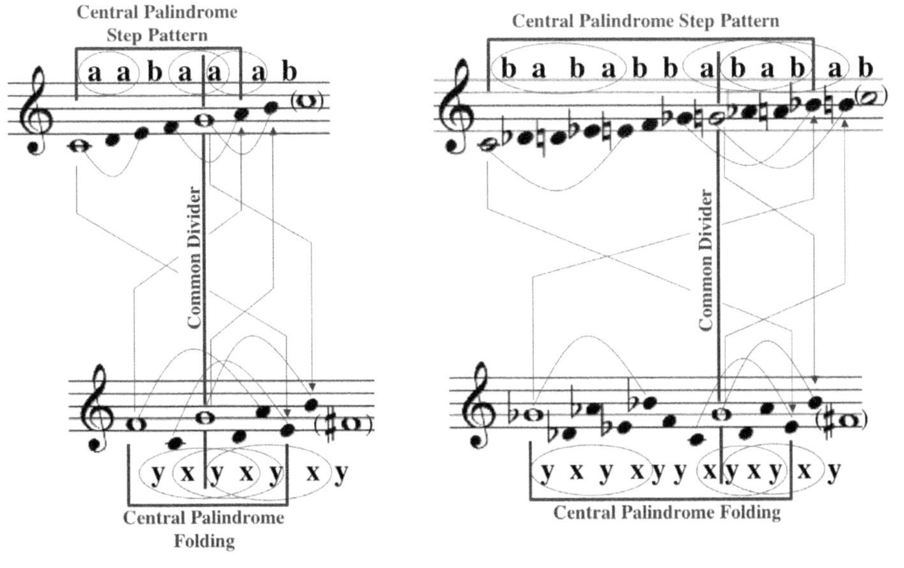


**Fig. 3.** Construction of the sensitive intervals in diatonic and chromatic modes with divider incidence. Common divider to penultimate elements, and cross connections between endpoints of adjoint central words.

of $F(a|b)$ and the divider suffix $f(b^{-1})$ is a prefix of $f(a^{-1}|b^{-1})$. An immediate corollary of this result on divider suffixes is that not only is $F(b)$ a prefix of $w = F(a|b)$, but any prefix of $F(b)$ is a prefix of $w = F(a|b)$ (and similarly for $f(b^{-1})$, $\overleftarrow{F}(y)$, and $\overleftarrow{f}(y^{-1})$).

See Figure 4. We consider prefixes $s$ and $s'$ of $F(b)$ and $\overleftarrow{F}(y)$, respectively, of lengths $|F(b)| - 1$ and $|\overleftarrow{F}(y)| - 1$, respectively. They express the interval between identical notes (from $l_q = m_{q'}$ to $l_{N-1} = m_{N-1}$), in terms of the scale step pattern, on the one hand, and of the folding pattern, on the other. (As mentioned above, given a conjugate of a Christoffel word, we may define points in the lattice to represent the notes.) By the corollary above, $s$ and $s'$ are also prefixes of $w$ and $w^*$, respectively. But by divider incidence, the initial tones of scale and folding are also divider predecessors of folding and scale, respectively. We therefore know that the prefix of length $|F(b)| - 1$ in the scale step pattern

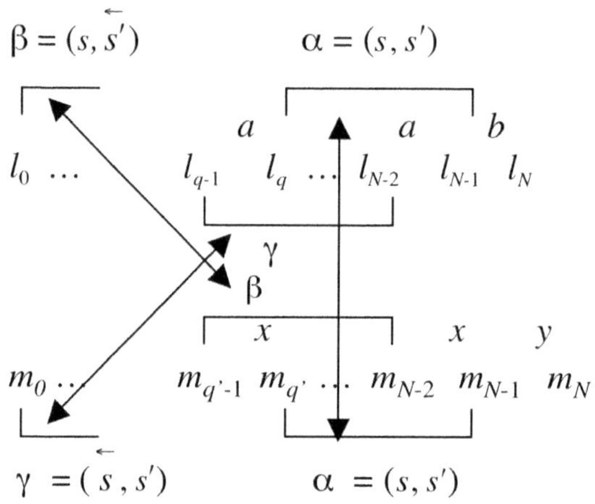


**Fig. 4.** Sensitive interval pairs, for modes/foldings associated with standard words

is $s$, and since the interval departs in the folding from initial scale tone, it is represented by a factor of length $|\overleftarrow{F}(y)| - 1$ in the folding, that is, to element $m_{N-2}$. But the antepenultimate note of the folding, $m_{N-2}$, is also evidently the endpoint of $u^*$, which is a palindrome, so the suffix of length $|\overleftarrow{F}(y)| - 1$ in the folding is the reversal of the prefix of that length of $u^*$. That prefix is, by the corollary, $s'$. Thus, the sensitive interval departing from the initial scale tone, which equals the divider predecessor in the folding, is encoded by $s$ in the word representing the scale step pattern, and by the reversal of $s'$ in the word representing the folding pattern. Similarly, the interval corresponding to prefix $s'$ in the folding departs from the initial note of the folding, which equals the divider predecessor in the scale, and extends to the antepenultimate note of the scale, the endpoint of $u$. Again, $u$ is a palindrome, so this suffix of length $|F(b)| - 1$ is the reversal of the prefix of that length of $u$, that is, the reversal of $s$. Thus, the sensitive interval departing from the initial note of the folding is encoded by the reversal of $s$ ($\overleftarrow{s}$) in the scale step pattern, and by $s'$ in the folding pattern. The picture for descending plagal modes represented by anti-standard words, over inverse letters $a^{-1}$, $b^{-1}$ and $x^{-1}$, $y^{-1}$, in the same free groups, is completely analogous to that shown in Fig. 4 for ascending authentic modes, represented by standard words over the generating letters, $a$, $b$ and $x$, $y$.

We may therefore distinguish the three appearances of the sensitive interval by defining three *sensitive pairs*, $\alpha = (s, s')$, $\beta = (s, \overleftarrow{s'})$, $\gamma = (\overleftarrow{s}, s')$. The same analysis carries over for anti-standard words on inverse letters, restricted to those derived from special anti-standard words that are non-trivial, i.e., contain at least one occurrence of $D^{\sim}$.

In conclusion, even completely neglecting the acoustical bases of tonal music (except for positing the perfect fifth and perfect octave), we see that the $\alpha$ major third sits in a structurally noteworthy position: dominant in the major scale, and subdominant in the minor scale. Moreover, the major third (or its generalization, the sensitive interval) has three instantiations, which in music theory in the diatonic environment are correlated with the functions dominant, subdominant, and tonic. The musical ramifications of this observation remain to be studied more deeply.

## References

1. Clough, J., Myerson, G.: Variety and Multiplicity in Diatonic Systems. Journal of Music Theory 29, 249–270 (1985)
2. Carey, N., Clampitt, D.: Aspects of Well-formed Scales. Music Theory Spectrum 11, 187–206 (1989)
3. Carey, N., Clampitt, D.: Self-similar Pitch Structures, Their Duals, and Rhythmic Analogues. Perspectives of New Music 34(2), 62–87 (1996)
4. Agmon, E.: A Mathematical Model of the Diatonic System. Journal of Music Theory 33, 1–25 (1989)
5. Agmon, E.: Coherent Tone Systems: A Study in the Theory of Diatonicism. Journal of Music Theory 40, 39–59 (1996)
6. Regener, E.: Pitch Notation and Equal Temperament: A Formal Study. University of California Press, Berkeley (1973)
7. Chalmers, J.: Cyclic Scales. Xenharmonicon 4, 69–78 (1975)
8. Sós, V.T.: On the Distribution mod 1 of the Sequence $n\alpha$, Annales Universitatis Scientiarum Budapestinensis de Rolando Eőtvős nominatae. Sectio Mathematica, 127–134 (1958)
9. Carey, N.: Distribution modulo 1 and Musical Scales. Ph.D. dissertation, University of Rochester (1998)
10. Noll, T., Clampitt, D., Dominguez, M.: What Sturmian Morphisms Reveal about Musical Scales and Tonality. In: Arnoux, P., Bédaride, N., Cassaigne, J. (eds.) Proceedings of the Sixth International Conference on Words, pp. 230–239. Centre International de Rencontres Mathématiques, Marseille-Luminy (2007)
11. Dominguez, M., Noll, T., Clampitt, D.: Plain and Twisted Adjoints of Well-formed Words. In: Chew, E., Childs, A., Chuan, C.-H. (eds.) MCM 2009. Communications in Computer and Information Science, vol. 38, pp. 65–80. Springer, Heidelberg (2009)

12. Clampitt, D., Noll, T.: Regions and Standard Modes. In: Chew, E., Childs, A., Chuan, C.-H. (eds.) MCM 2009. Communications in Computer and Information Science, vol. 38, pp. 81–92. Springer, Heidelberg (2009)
13. Clampitt, D., Noll, T.: Modes, the Height-Width Duality, and Handschin's Tone Character. Music Theory Online 17(1) (2011)
14. Lothaire, M.: Algebraic Combinatorics on Words. Cambridge University Press, Cambridge (2002)
15. Berstel, J., Lauve, A., Reutenauer, C., Saliola, F.: Combinatorics on Words: Christoffel Words and Repetition in Words. CRM Monograph Series, vol. 27. American Mathematical Society, Providence (2008)
16. Noll, T.: Ionian Theorem. Journal of Mathematics and Music 3, 137–151 (2009)
17. Harrison, D.: Harmonic Function in Chromatic Music: A Renewed Dualist Theory and an Account of Its Precedents. The University of Chicago Press, Chicago (1994)

# Commuting Groups and the Topos of Triads

Thomas M. Fiore[1] and Thomas Noll[2]

[1] Department of Mathematics and Statistics,
University of Michigan-Dearborn
Dearborn, Michigan, USA
tmfiore@umd.umich.edu

[2] Departament de Teoria, Composició i Direcció
Escola Superior de Música de Catalunya, Barcelona, Spain
noll@cs.tu-berlin.de

**Abstract.** The goal of this article is to clarify the relationship between the topos of triads and the neo-Riemannian $PLR$-group. To do this, we first develop some theory of generalized interval systems: 1) we prove the well known fact that every pair of dual groups is isomorphic to the left and right regular representations of some group (Cayley's Theorem), 2) given a simply transitive group action, we show how to construct the dual group, and 3) given two dual groups, we show how to easily construct sub dual groups. Examples of this construction of sub dual groups include Cohn's hexatonic systems, as well as the octatonic systems. We then enumerate all $\mathbb{Z}_{12}$-subsets which are invariant under the triadic monoid and admit a simply transitive $PLR$-subgroup action on their maximal triadic covers. As a corollary, we realize all four hexatonic systems and all three octatonic systems as Lawvere–Tierney upgrades of consonant triads.

**Keywords:** $PLR$-group, Duality, Sub dual groups, Hexatonic systems, Octatonic, Topos, Topos of triads, Triadic monoid.

## 1 Introduction

In [1], Richard Cohn coined the term "the overdetermined triad" in order to draw the reader's attention away from the celebrated acoustic attractiveness of the major and minor triads to some surprising algebraic properties of the family of the consonant triads. The starting point for the present investigation is a puzzling ramification of this over-determination of the triads even within the realm of algebraic music theory. In particular, the musical and music-theoretical prevalence of three tone-collections—major/minor mixture, hexatonic, and octatonic—emerges from the study of triads in quite different algebraic ways.

In the context of a group-theoretic approach (see [2]) one obtains these collections as carrier sets, which are covered by particular families of triads, which are orbits under certain group actions. Under this perspective it is the interplay of several triads, which makes these collections prevalent among others.

C. Agon et al. (Eds.): MCM 2011, LNAI 6726, pp. 69–83, 2011.

In the context of a monoid-theoretic approach (see [3]) one obtains these collections as extensions of a single triad. The stability properties of the set $\{0, 4, 7\}$ within $\mathbb{Z}_{12}$ can be studied in terms of a monoid of 8 affine transformations. On top of this monoid one obtains three interesting Lawvere–Tierney topologies, which also imply the prevalence of the three sets $\{0, 3, 4, 7\}$, $\{0, 3, 4, 7, 8, 11\}$ and $\{0, 1, 3, 4, 6, 7, 9, 10\}$ as extensions of a single triad.

How do these two approaches fit together? The present article prepares the group-theoretical grounding for a comprehensive answer and anticipates the remaining work on the topos-theoretical side in the concrete case of $\mathbb{Z}_{12}$.

Though this article does not require any topos theory, we have included an Appendix on Topos Theory in Sect. 5 for readers who would like to recall the main notions.

## 2 Dual Groups and Cayley's Theorem

We first briefly recall the notion of dual groups in transformational analysis, give the fundamental example of left and right regular representations, and show how to construct general dual groups in view of the left and right regular representations. As an example, we reconstruct the well-known $PLR$-group as the dual group to the $T/I$-group acting on consonant triads.

**Definition 2.1 (Dual groups in the sense of Lewin).** Let $\mathrm{Sym}(S)$ be the symmetric group on the set $S$. Two subgroups $G$ and $H$ of the symmetric group $\mathrm{Sym}(S)$ are called *dual* if their natural actions on $S$ are simply transitive and each is the centralizer of the other, that is,

$$C_{\mathrm{Sym}(S)}(G) = H \quad \text{and} \quad C_{\mathrm{Sym}(S)}(H) = G.$$

**Example 2.2 (Left and right regular representations).** If $G$ is any group, the left and right regular representations $\lambda, \rho\colon G \to \mathrm{Sym}(G)$ are defined by $\lambda_g(h) = gh$ and $\rho_g(h) = hg^{-1}$. As is well known, $\lambda$ and $\rho$ are injective and the natural actions of $\lambda(G)$ and $\rho(G)$ are simply transitive. The images $\lambda(G)$ and $\rho(G)$ clearly commute, since $\lambda_{g_1}\rho_{g_2}(h) = g_1 h g_2^{-1} = \rho_{g_2}\lambda_{g_1}(h)$. In fact, $\rho(G)$ is the centralizer of $\lambda(G)$ in $\mathrm{Sym}(G)$, for if $f\colon G \to G$ is a bijection satisfying $f(gh) = gf(h)$ for all $g$ and $h$ in $G$, then $f(h) = f(he) = hf(e)$ and $f$ is the same as $\rho(f(e)^{-1})$. Similarly, $\lambda(G)$ is the centralizer of $\rho(G)$ in $\mathrm{Sym}(G)$. The groups $\lambda(G)$ and $\rho(G)$ provide the most basic example of dual groups. All other examples of dual groups are isomorphic to this one, as we see in the following construction.

**Construction 2.3 (Construction of the dual group).** Given a group $G$ acting simply transitively on a set $S$, we may construct the dual group as follows. We consider $G$ as a subgroup of $\mathrm{Sym}(S)$ via the homomorphism $G \to \mathrm{Sym}(S), g \mapsto (s \mapsto gs)$, which is an injection by simple transitivity. We suggestively call this subgroup $\lambda(G)$, although it is not the same as $\lambda(G)$ in Example 2.2. Fix an element $s_0 \in S$. Again by simple transitivity, the function $G \to S, h \mapsto hs_0$ is a bijection. The natural action of $\lambda(G)$ on $S$ is essentially the same as left multiplication: $g(hs_0) = (gh)s_0$. We now define a second injection $G \to \mathrm{Sym}(S)$ by $g \mapsto (hs_0 \mapsto hg^{-1}s_0)$ and call the image $\rho(G)$. The group $\rho(G)$ is the dual group to $\lambda(G)$, which we sought to construct. The bijection $G \to S, h \mapsto hs_0$ and the injection $G \to \mathrm{Sym}(S), g \mapsto (hs_0 \mapsto hg^{-1}s_0)$ depend on the choice of $s_0$, but the group $\rho(G)$ does not.

**Example 2.4 (Construction of $PLR$-group).** If in Construction 2.3 we take $G$ to be the $T/I$-group, $S$ to be the set of major and minor triads, and $s_0 = \{0, 4, 7\}$, then $\rho(G)$ is the neo-Riemannian $PLR$-group. For example, the operations parallel, leading tone exchange, and relative, denoted $P$, $L$, and $R$ respectively, correspond to right multiplication by $I_7$, $I_{11}$, and $I_4$.

$$\begin{array}{lll} P: T_n\{0,4,7\} \mapsto T_nI_7\{0,4,7\} & \text{and} & I_n\{0,4,7\} \mapsto I_nI_7\{0,4,7\} \\ L: T_n\{0,4,7\} \mapsto T_nI_{11}\{0,4,7\} & \text{and} & I_n\{0,4,7\} \mapsto I_nI_{11}\{0,4,7\} \\ R: T_n\{0,4,7\} \mapsto T_nI_4\{0,4,7\} & \text{and} & I_n\{0,4,7\} \mapsto I_nI_4\{0,4,7\}. \end{array}$$

The embedding $\rho\colon T/I \to \mathrm{Sym}(S)$ takes the generators $I_1$ and $I_8$ to $L$ and $R$ respectively (recall $I_1 \circ I_8 = T_{1-8} = T_5$). If we choose a different $s_0$, then we obtain another isomorphism of the $T/I$-group with the $PLR$-group.

## 3 Sub Dual Groups

We next give a very practical method for constructing sub dual groups from dual groups. This method minimizes computation: instead of checking commutativity of all functions on every element, it suffices to evaluate a putative commuting function on the single element $s_0$ and check if the output lies in $S_0$.

**Theorem 3.1 (Construction of sub dual groups).** *Let $G, H \leq \mathrm{Sym}(S)$ be dual groups, $G_0$ a subgroup of $G$, and $s_0$ an element of $S$. Let $S_0$ be the orbit of $s_0$ under the action of $G_0$. Then the following hold.*

*(i) The group $G_0$ acts simply transitively on $S_0$.*
*(ii) If $g \in G$ and $gs_0$ is in $S_0$, then $g \in G_0$. In particular, if $g \in G$ and $gs_0$ is in $S_0$, then $g$ preserves $S_0$ as a set, that is, $g(S_0) \subseteq S_0$.*
*(iii) Let $H_0$ denote the subgroup of $H$ consisting of those elements $h \in H$ with $hs_0 \in S_0$. Then $H_0$ acts simply transitively on $S_0$.*
*(iv) Restriction from $S$ to $S_0$ embeds $G_0$ and $H_0$ in $\mathrm{Sym}(S_0)$. We denote their images in $\mathrm{Sym}(S_0)$ by $G_0|_{S_0}$ and $H_0|_{S_0}$.*
*(v) If $g \in \mathrm{Sym}(S_0)$ and $g$ commutes with $H_0|_{S_0}$, then $g$ admits a unique extension to $S$ which belongs to $G$. This extension necessarily belongs to $G_0$ by (ii). Similarly, if $h \in \mathrm{Sym}(S_0)$ commutes with $G_0|_{S_0}$, then $h$ admits a unique extension to $S$ which belongs to $H$. This extension necessarily belongs to $H_0$.*
*(vi) The groups $G_0|_{S_0}$ and $H_0|_{S_0}$ are dual in $\mathrm{Sym}(S_0)$.*

*Proof.* (i) By the definition $S_0 = G_0 s_0$, we have $gs \in S_0$ for all $g \in G_0$ and $s \in S_0$. The orbit of $s_0$ is all of $S_0$, so the action of $G_0$ on $S_0$ is transitive. Simplicity of the $G_0$-action on $S_0$ follows from the simplicity of the $G$-action on $S$.

(ii) Suppose $g \in G$ and $gs_0 = s \in S_0$. By the simple transitivity of $G_0$ acting on $S_0$ from (i), there is a unique $\overline{g} \in G_0$ such that $\overline{g}s_0 = s$. But on the other hand, $g$ and $\overline{g}$ are in $G$, which acts simply transitively on $S$, so $g = \overline{g} \in G_0$.

(iii) Suppose $h \in H$ and $hs_0 \in S_0 = G_0 s_0$. An arbitrary element of $S_0$ has the form $gs_0$ for some $g \in G_0$, and $h(gs_0) = g(hs_0) \in G_0 s_0 = S_0$. Thus $h(S_0) \subseteq S_0$, and the action of $H_0$ on $S$ restricts to an action on $S_0$. The orbit of $s_0$ under $H_0$ is $S_0$, for if $gs_0 \in S_0$ is an arbitrary element, then there exists $h \in H$ such that $hs_0 = gs_0 \in S_0$, which implies $h \in H_0$. Simplicity follows from the simplicity of the $H$-action on $S$.

(iv) Suppose $g, g' \in G_0$ and $g|_{S_0} = g'|_{S_0}$. Then $gs_0 = g's_0$ and $g = g'$ by the simple transitivity of the $G$-action on $S$. The injectivity of the $H_0$-embedding is similar.

(v) Suppose $g \in \mathrm{Sym}(S_0)$ and $g$ commutes with $H_0|_{S_0}$. By the simple transitivity of the $G$-action on $S$, there is a unique $\overline{g} \in G$ such that $\overline{g}s_0 = gs_0$. By (iii) a general element of $S_0$ has the form $hs_0$ with $h \in H_0$. We have $\overline{g}hs_0 = h\overline{g}s_0 = h|_{S_0}gs_0 = gh|_{S_0}s_0 = ghs_0$. Thus $\overline{g}$ and $g$ agree on $S_0$ and $\overline{g}$ is a true extension of $g$. The proof for the analogous statement concerning elements in $\mathrm{Sym}(S_0)$ which commute with $G_0|_{S_0}$ is similar.

(vi) We already have simply transitive group actions of $G_0|_{S_0}$ and $H_0|_{S_0}$ on $S_0$ by (i) and (iii). The groups $G_0|_{S_0}$ and $H_0|_{S_0}$ commute because $G$ and $H$ do. Suppose $h \in \mathrm{Sym}(S_0)$ commutes with $G_0|_{S_0}$. Then by (v), the bijection $h$ admits an extension to $\overline{h} \in H_0 \subseteq \mathrm{Sym}(S)$, which precisely means $h \in H_0|_{S_0}$. This proves $C_{\mathrm{Sym}(S_0)}(G_0|_{S_0}) = H_0|_{S_0}$. The proof of $C_{\mathrm{Sym}(S_0)}(H_0|_{S_0}) = G_0|_{S_0}$ is similar. □

**Remark 3.2.** In Theorem 3.1, the subgroups $G_0$ and $H_0$ of $\mathrm{Sym}(S)$ commute, since they are subgroups of the dual groups $G$ and $H$, respectively. However, $G_0$ and $H_0$ are *not* dual groups in $\mathrm{Sym}(S)$ when $G_0 \lneqq G$, though the restrictions $G_0|_{S_0}$ and $H_0|_{S_0}$ are dual in $\mathrm{Sym}(S_0)$. First of all, $G_0$ and $H_0$ do not act simply transitively on $S$. Secondly, the centralizer of $G_0$ is larger than $H_0$, as the centralizer contains at least all of $H$. The isomorphism type of $C_{\mathrm{Sym}(S)}(G_0)$ may be found using the methods of [4].

**Corollary 3.3 (Transforming the orbits of $G_0$).** *Let $G, H \leq \mathrm{Sym}(S)$ be dual groups, $G_0$ a subgroup of $G$, and $s_0$ an element in $S$. Let $S_0$ be the orbit of $s_0$ under the action of $G_0$, and $H_0$ the subgroup of $H$ consisting of those elements $h \in H$ with $hs_0 \in S_0$. Then the following hold.*

*(i) If $k \in H$, then the $G_0$-orbit of $ks_0$ is $kS_0$. Thus, if we transform $s_0$ to $ks_0$ in Theorem 3.1, then $S_0$ transforms to $kS_0$.*

*(ii) If $k \in H$, then $kH_0k^{-1}$ is precisely the subgroup of $H$ consisting of those $h \in H$ such that $h(ks_0) \in kS_0$. Thus, if we transform $s_0$ to $ks_0$ in Theorem 3.1, then $H_0$ transforms to $kH_0k^{-1}$.*

*(iii) If $k \in H$, then the restrictions of $G_0$ and $kH_0k^{-1}$ to $kS_0$ are dual in* $\mathrm{Sym}(kS_0)$.
*(iv) Each orbit of $G_0$ can be written as $kS_0$ for some $k \in H$. Moreover, the group $H$ acts transitively on the orbits of $G_0$ by $h(kS_0) = (hk)S_0$.*

*Proof.* (i) The element $k \in H$ commutes with each element in $G_0$, so $G_0ks_0 = kG_0s_0 = kS_0$.
(ii) If $kh_0k^{-1} \in kH_0k^{-1}$, then $kh_0k^{-1}(ks_0) = kh_0s_0 \in kS_0$ since $h_0s_0 \in S_0$. On the other hand, if $h \in H$ and $hks_0 = gks_0$ for some $g \in G_0$, then $k^{-1}hks_0 = gs_0$ and $k^{-1}hk \in H_0$, so $h \in kH_0k^{-1}$.
(iii) This follows directly from Theorem 3.1 (vi).
(iv) This follows directly from the transitivity of the $H$-action on $S$ and (i). □

## 3.1 Hexatonic Systems

To see the utility of Theorem 3.1, we construct the hexatonic systems of Cohn [5] and the dual pair of Clampitt [6]. In Theorem 3.1, we take $S$ to be the set of major and minor triads, $G$ to be the neo-Riemannian $PLR$-group, $H$ to be the $T/I$-group, and $G_0$ to be the $PL$-group, which is generated by $P$ and $L$, is isomorphic to $S_3$, and therefore has only 6 elements.

Let $s_0$ be $E\flat$ (we write major chords with upper-case letters and minor chords with lower-case letters). Beginning with $E\flat$, and evaluating $P$ and $L$ alternately, we obtain the 6 chords $\{E\flat, e\flat, B, b, G, g\}$. We have now found the complete orbit $S_0$ of $E\flat$ and may stop: by Theorem 3.1 (i), $G_0$ acts simply transitively on $S_0$, which means $|S_0| = |G_0| = 6$ by the Orbit-Stabilizer Theorem.

To find $H_0$, it suffices to find those transpositions and inversions which map $E\flat$ into $S_0$, or in other words, it suffices to solve the six equations

$$\begin{aligned} T_i E\flat &= E\flat & \qquad I_\ell E\flat &= e\flat \\ T_j E\flat &= G & I_m E\flat &= g \\ T_k E\flat &= B & I_n E\flat &= b. \end{aligned}$$

Clearly, $i$, $j$, and $k$ are 0, 4, and 8 respectively, and less clearly, $\ell$, $m$, and $n$ are 1, 5, and 9. In cycle notation, the group $G_0|_{S_0}$ is

$$G_0|_{S_0}: \quad \begin{aligned} \mathrm{Id} &= () & \qquad LP|_{S_0} &= (E\flat\ G\ B)(e\flat\ b\ g) \\ P|_{S_0} &= (E\flat\ e\flat)(G\ g)(B\ b) & PL|_{S_0} &= (E\flat\ B\ G)(e\flat\ g\ b) \\ L|_{S_0} &= (E\flat\ g)(G\ b)(B\ e\flat) & PLP|_{S_0} &= (E\flat\ b)(G\ e\flat)(B\ g) \end{aligned}$$

and the group $H_0|_{S_0}$ is

$$H_0|_{S_0}: \quad \begin{aligned} T_0|_{S_0} &= () & \qquad I_1|_{S_0} &= (E\flat\ e\flat)(G\ b)(B\ g) \\ T_4|_{S_0} &= (E\flat\ G\ B)(e\flat\ g\ b) & I_5|_{S_0} &= (B\ b)(G\ e\flat)(E\flat\ g) \\ T_8|_{S_0} &= (E\flat\ B\ G)(e\flat\ b\ g) & I_9|_{S_0} &= (G\ g)(E\flat\ b)(B\ e\flat), \end{aligned}$$

exactly as in Clampitt's article [6]. The groups $G_0|_{S_0}$ and $H_0|_{S_0}$ are dual in $\mathrm{Sym}(S_0)$ by Theorem 3.1 (vi).

We may now use Corollary 3.3 and the above computation to find the other three hexatonic systems and the corresponding duals to the restricted $PL$-groups. In Corollary 3.3, we take $k = T_1$, $T_2$, and $T_3$. Starting with $T_1$, we find the $PL$-orbit

$$T_1 S_0 = \{E, e, C, c, A\flat, a\flat\}$$

and (extended) dual group

$$T_1 H_0 T_1^{-1} = \{T_0, T_4, T_8, I_3, I_7, I_{11}\}.$$

With $T_2$ we obtain the $PL$-orbit

$$T_2 S_0 = \{F, f, C\sharp, c\sharp, A, a\}$$

and (extended) dual group

$$T_2 H_0 T_2^{-1} = \{T_0, T_4, T_8, I_5, I_9, I_1\}.$$

With $T_3$ we obtain the $PL$-orbit

$$T_3 S_0 = \{G\flat, f\flat, D, d, B\flat, b\flat\}$$

and (extended) dual group

$$T_3 H_0 T_3^{-1} = \{T_0, T_4, T_8, I_7, I_{11}, I_3\}.$$

We know we have all orbits and may now stop because the union of the orbits thus far gives us all of $S$. For a direct computation of all four hexatonic systems, see the exposition [7].

### 3.2 Octatonic Systems

As another application of Theorem 3.1 and Corollary 3.3, we construct the three octatonic systems. In Theorem 3.1, we take $S$ to be the set of major and minor triads, $G$ to be the neo-Riemannian $PLR$-group, $H$ to be the $T/I$-group, and $G_0$ to be the $PR$-group, which is generated by $P$ and $R$, is isomorphic to $D_4$, and therefore has only 8 elements[1].

Let $s_0$ be the $C$ chord. Beginning with $C$, and evaluating $P$ and $R$ alternately, we obtain the 8 chords $\{C, c, E\flat, e\flat, G\flat, g\flat, A, a\}$. We have now found the complete orbit $S_0$ of $C$ and may stop: by Theorem 3.1 (i), $G_0$ acts simply transitively on $S_0$, which means $|S_0| = |G_0| = 8$ by the Orbit-Stabilizer Theorem.

[1] To see that the $PR$-group is dihedral of order 8, we note that alternately applying $P$ and $R$ to $C$ gives $C, c, E\flat, e\flat, G\flat, g\flat, A, a, C$, so that $s := RP$ has order 4 (recall $P$ and $R$ commute with $T_1$). The element $t := P$ has order 2, and the equation $tst = s^{-1}$ holds true. Thus the $PR$-group is generated by $\{s, t\}$ and $s$ and $t$ satisfy the relations of $D_4$.

To find $H_0$, it suffices to find those transpositions and inversions which map $C$ into $S_0$, or in other words, it suffices to solve the eight equations

$$\begin{aligned} T_iC &= C & I_mC &= c \\ T_jC &= E\flat & I_nC &= e\flat \\ T_kC &= G\flat & I_oC &= g\flat \\ T_\ell C &= A & I_pC &= a. \end{aligned}$$

Clearly, $i$, $j$, $k$, and $\ell$ are 0, 3, 6, and 9 respectively, and with a little more work, $m$, $n$, $o$, and $p$ are 7, 10, 1, and 4. The groups $G_0|_{S_0}$ and $H_0|_{S_0}$ are dual in $\mathrm{Sym}(S_0)$ by Theorem 3.1 (vi).

We may now use Corollary 3.3 and the above computation to find the other two octatonic systems and the corresponding duals to the restricted $PR$-groups. In Corollary 3.3, we take $k = T_1$ and $T_2$. Starting with $T_1$, we find the $PR$-orbit

$$T_1S_0 = \{D\flat, d\flat, E, e, G, g, B\flat, b\flat\}$$

and (extended) dual group

$$T_1H_0T_1^{-1} = \{T_0, T_3, T_6, T_9, I_9, I_0, I_3, I_6\}.$$

With $T_2$ we obtain the $PR$-orbit

$$T_2S_0 = \{D, d, F, f, A\flat, a\flat, B, b\}$$

and (extended) dual group

$$T_2H_0T_2^{-1} = \{T_0, T_3, T_6, T_9, I_{11}, I_2, I_5, I_8\}.$$

We know we have all orbits and may now stop because the union of the orbits thus far gives us all of $S$. For a direct computation of all three octatonic systems, see [7].

## 4 The Topos of Triads and the neo-Riemannian *PLR*-Group

We first recall the topos of triads $\mathbf{Set}^{\mathcal{T}}$ from [3] and then present the enumeration. In the present article we use the semi-tone encoding of pitch classes, and not the circle-of-fifths encoding used in [3]. Consequently, some formulas will differ, though the objects under consideration are exactly the same.

The *triadic monoid* $\mathcal{T}$ of [3] is the collection of affine maps $\mathbb{Z}_{12} \to \mathbb{Z}_{12}$ which preserve the $C$-triad $\{0, 4, 7\}$ as a set. A map $\psi\colon \mathbb{Z}_{12} \to \mathbb{Z}_{12}$ is *affine* if it has the form $z \mapsto mz + b$ for some $m, b \in \mathbb{Z}_{12}$. There are 144 affine maps $\mathbb{Z}_{12} \to \mathbb{Z}_{12}$, the invertible ones are characterized by $m \in \{1, 5, 7, 11\}$. A function $\psi$ *preserves* $\{0, 4, 7\}$ *as a set* if $\psi(\{0, 4, 7\}) \subseteq \{0, 4, 7\}$. This is weaker than preserving $\{0, 4, 7\}$ *pointwise*. The set $\mathcal{T}$ forms a *monoid* under function composition, that is, function composition is an associative, unital operation on $\mathcal{T}$ (functions in $\mathcal{T}$ are not necessarily invertible). The triadic monoid $\mathcal{T}$ has 8 elements and is generated by the affine maps $f(z) := 3z + 7$ and $g(z) := 8z + 4$. The other maps in $\mathcal{T}$ are $f \circ f(z) = 9z + 4$, $g \circ g(z) = 4z$, the three constant maps 0, 4, 7, and the identity $\mathrm{Id}_{\mathbb{Z}_{12}}$.

If $\mathcal{S} \subseteq \mathbb{Z}_{12}$ is closed under the action of $\mathcal{T}$, that is $\mathcal{T}\mathcal{S} \subset \mathcal{S}$, then $\mu|_{\mathcal{S}} \colon \mathcal{T} \times \mathcal{S} \to \mathcal{S}$ is called a *subaction of* $\mu$ and $\mathcal{S}$ is called the *carrier set of the subaction* $\mu|_{\mathcal{S}}$.

The *topos of triads* is the category $\mathbf{Set}^{\mathcal{T}}$, its objects are sets equipped with a $\mathcal{T}$-action and its morphisms are $\mathcal{T}$-equivariant maps. Like any presheaf category, this category is a *topos*. See [3] for a complete investigation of the triadic monoid and the topos of triads. An exposition of [3] may be found in [8]. See the Appendix of the present paper for a recollection of topos-theoretic concepts used in this article.

Our aim in the present article is to clarify the relationship between the neo-Riemannian $PLR$-group, the triadic monoid, and the topos of triads, so we focus on subactions of the natural action $\mu \colon \mathcal{T} \times \mathbb{Z}_{12} \to \mathbb{Z}_{12}$ which are covered by consonant triads. A tremendous difference between the triadic monoid action and the $PLR$-group action is that the triadic monoid acts on pitch classes, whereas the $PLR$-group acts on aggregate chords; the $PLR$-group simply *does not act on individual pitch classes*. However, we may use Theorem 3.1 to induce an action of certain transpositions and inversions on the pitch classes of a monoid subaction in the case that the carrier set is covered by an orbit of a $PLR$-subgroup.

For example, consider the major-minor mixture $\{0, 3, 4, 7\}$. A quick computation verifies that $\mathcal{T}\{0, 3, 4, 7\} \subseteq \{0, 3, 4, 7\}$, as it suffices to check if the two generators $f$ and $g$ of $\mathcal{T}$ preserve $\{0, 3, 4, 7\}$. The carrier set $\{0, 3, 4, 7\}$ of this monoid subaction is covered by the consonant triads $C$ and $c$, while $\{C, c\}$ is an orbit of the $PLR$-subgroup $\{\mathrm{Id}, P\}$. Theorem 3.1 now implies that the dual subgroup in $\mathrm{Sym}(\{C, c\})$ is $\{\mathrm{Id}, I_7\}$. The identity and the inversion operator $I_7$ *do act on individual pitch classes*, and since Id and $I_7$ preserve the set $\{C, c\}$, they will also preserve the set of underlying pitch classes. This is how we move from $PLR$-subgroup actions on chords to $T/I$-subgroup actions on pitch classes. It is a coincidence in this very small example that $I_7$ and $P$ agree on the set $\{C, c\}$.

We may now enumerate all $\mathcal{T}$-sets $\mathcal{S}$ which are covered by triads and admit an action of a $PLR$-subgroup on the maximal cover $\mathbb{S}$, see Table 1 and Table 2. In Table 1, the operation $Q_6$ transposes major chords up by 6 semitones and minor chords down by 6 semitones (this is the same as $T_6$, but we call it $Q_6$ to emphasize that we think of it as element of the $PLR$-group). The notation $S\ell$ indicates the *slide transformation* on major and minor chords, which holds the third constant and moves the root and the fifth in parallel by a semitone.

**Theorem 4.1 (Enumeration).** *Let $\mathcal{T}$ be the triadic monoid, that is, $\mathcal{T}$ is the monoid of affine functions $\mathbb{Z}_{12} \to \mathbb{Z}_{12}$ which preserve the $C$-chord $\{0, 4, 7\}$ as a set. Suppose $\mathcal{S} \subseteq \mathbb{Z}_{12}$ is closed under the natural $\mathcal{T}$-action in the sense that $\mathcal{T}\mathcal{S} \subseteq \mathcal{S}$. Assume further that $\mathcal{S}$ is covered by consonant triads, and let $\mathbb{S}$ be the collection of consonant triads $Y$ with $Y \subseteq \mathcal{S}$. If $\mathbb{S}$ admits a simply transitive action of a $PLR$-subgroup $G$, then $\mathcal{S}$ must be one of the sets enumerated in Table 1.*

**Table 1.** Enumeration of all $\mathcal{T}$-Subactions with Maximal Cover Admitting a Simply Transitive Action of a $PLR$-Subgroup

| Carrier Set $\mathcal{S}$ | Type | Maximal Cover $\mathbb{S}$ | $PLR$-Subgroup |
|---|---|---|---|
| $\{0,4,7\}$ | Major Chord | $C$ | $\{\text{Id}\}$ |
| $\{0,3,4,7\}$ | Major-Minor Mixture | $C,c$ | $\{\text{Id},P\}$ |
| $\{0,3,4,7,8,11\}$ | Hexatonic | $C,c,A\flat,a\flat,E,e$ | $\langle P,L\rangle$ |
| $\{0,1,3,4,6,7,9,10\}$ | Octatonic | $C,c,E\flat,e\flat,G\flat,g\flat,A,a$ | $\langle P,R\rangle$ |
| $\{0,1,4,6,7,10\}$ | Maj. Triad Tritone Mix. | $C,G\flat$ | $\{\text{Id},Q_6\}$ |
| $\{0,1,2,4,6,7,8,10\}$ | Prometheus Tritone Mix. | $C,d\flat,G\flat,g$ | $\{\text{Id},Q_6,\text{S}\ell,Q_6\text{S}\ell\}$ |
| $\mathbb{Z}_{12}$ | Chromatic Scale | All Consonant Triads | $PLR$-group |

**Table 2.** Carrier Sets of Table 1

| Carrier Set $\mathcal{S}$ in $\mathbb{Z}_{12}$ | Carrier Set $\mathcal{S}$ in Musical Notation |
|---|---|
| $\{0,4,7\}$ | $\{C,E,G\}$ |
| $\{0,3,4,7\}$ | $\{C,E\flat,E,G\}$ |
| $\{0,3,4,7,8,11\}$ | $\{C,E\flat,E,G,A\flat,B\}$ |
| $\{0,1,3,4,6,7,9,10\}$ | $\{C,D\flat,E\flat,E,G\flat,G,A,B\flat\}$ |
| $\{0,1,4,6,7,10\}$ | $\{C,D\flat,E,G\flat,G,B\flat\}$ |
| $\{0,1,2,4,6,7,8,10\}$ | $\{C,D\flat,D,E,G\flat,G,A\flat,B\flat\}$ |
| $\mathbb{Z}_{12}$ | $\{C,D\flat,D,E\flat,E,F,G\flat,G,A\flat,A,B\flat,B\}$ |

*Proof.* Suppose $G$ is a $PLR$-subgroup which acts simply transitively on $\mathbb{S}$.

**Case 1: $G$ contains $P$.** Assume $P \in G$. Our case-by-case method is to consider the orbit of $C = \{0,4,7\}$ under each $PLR$-subgroup which contains $P$, and to test if the underlying pitch-class set is closed under $\mathcal{T}$. By Theorem 3.1 (i), each $PLR$-subgroup acts simply transitively on its $C$-orbit.

We may quickly determine all $PLR$-subgroups containing $P$ via the explicit structure of the $PLR$-group, described for instance in [9] and [2]. The $PLR$-group has two generators, $Q_1$ of order 12 and $P$ of order 2, where $Q_1$ transposes major chords up by 1 and minor chords down by 1. The bijection $Q_k$ is defined similarly, and satisfies $Q_k = (Q_1)^k$. The 24 elements of the $PLR$-group are $\{Q_k | k \in \mathbb{Z}_{12}\} \cup \{PQ_k | k \in \mathbb{Z}_{12}\}$, for example $L = PQ_4$ and $R = PQ_9$. For these reasons, any $PLR$-subgroup containing $P$ is generated by $P$ and some $Q_i$. The group $\{Q_k | k \in \mathbb{Z}_{12}\}$ is cyclic, hence all of its subgroups are cyclic, and for our computation of the $PLR$-subgroups containing $P$ it suffices to check $\langle P, Q_i\rangle$ where $Q_i$ is a generator of a cyclic subgroup, that is, $Q_i \in \{\text{Id}, Q_1, Q_2, Q_3, Q_4, Q_6\}$.

The smallest $PLR$-subgroup containing $P$ is $\{\text{Id}, P\}$, which produces the $C$-orbit $\{C, c\}$ with underlying pitch-class set $\{0,3,4,7\}$. This is closed under the $\mathcal{T}$-action, as $f(3) = 4 = g(3)$.

The next larger subgroup is $\langle P, Q_6\rangle = \{\text{Id}, Q_6, P, PQ_6\}$, which produces the $C$-orbit $\{C, c, G\flat, g\flat\}$ with underlying pitch-class set $\{0,1,3,4,6,7,9,10\}$, the octatonic. The group $\langle P, Q_6\rangle$ does not act simply transitively on the set of all consonant triads in the octatonic, so a hypothesis of the theorem is not satisfied, and there is nothing to check.

We arrive next at $\langle P, Q_4 \rangle = \{\mathrm{Id}, Q_4, Q_8, P, PQ_4, PQ_8\}$, which is the same as $\langle P, PQ_4 \rangle = \langle P, L \rangle$.

The $C$-orbit is a hexatonic system of Cohn $\{C, c, A\flat, a\flat, E, e\}$, see [5] and Sect. 3.1 of the present paper. The underlying pitch-class set $\{0, 3, 4, 7, 8, 11\}$ is closed under $\mathcal{T}$, since $f(8) = 7$, $f(11) = 4$, $g(8) = 8$, and $g(11) = 8$.

The group $\langle P, Q_3 \rangle = \{\mathrm{Id}, Q_3, Q_6, Q_9, P, PQ_3, PQ_6, PQ_9\}$ is the same as $\langle P, PQ_9 \rangle = \langle P, R \rangle$. The $C$-orbit is an octatonic system

$$\{C, c, E\flat, e\flat, G\flat, g\flat, A, a\},$$

see Sect. 3.2. The underlying pitch-class set $\{0, 1, 3, 4, 6, 7, 9, 10\}$ is closed under $\mathcal{T}$, since $f(1) = 10$, $f(6) = 1$, $f(9) = 10$, $f(10) = 1$, and $g(1) = 0$, $g(6) = 4$, $g(9) = 4$, $g(10) = 0$.

The group $\langle P, Q_2 \rangle$ has $\mathbb{Z}_{12}$ as the underlying pitch-class set for its $C$-orbit because $\{0, 3, 4, 7\}$ has a representative from each coset $\mathbb{Z}_{12}/6\mathbb{Z}_{12}$. That is, iterated applications of $Q_2$ to $\{0, 4, 7\}$ and $\{0, 3, 7\}$ will reach all of $\mathbb{Z}_{12}$. However, the group $\langle P, Q_2 \rangle$ does not act simply transitively on the set of all consonant triads in $\mathbb{Z}_{12}$, so a hypothesis of the theorem is not satisfied, and there is nothing to check.

The group $\langle P, Q_1 \rangle$ is the entire $PLR$-group, and its $C$-orbit is the entire collection of major and minor triads. Its underlying pitch-class set is of course $\mathbb{Z}_{12}$, which is clearly closed under the $\mathcal{T}$-action.

We have considered all $PLR$-subgroups containing $P$ and determined which ones produce a $C$-orbit with underlying pitch-class set closed under the $\mathcal{T}$-action.

**Case 2: $G$ does not contain $P$.** Assume $P \notin G$ and $G$ acts simply transitively on $\mathbb{S}$.

If $G$ is trivial, then the $C$-orbit is simply $\{C\}$ with underlying pitch-class set $\{0, 4, 7\}$, which is closed under the $\mathcal{T}$-action by definition of $\mathcal{T}$.

For the rest of Case 2, we assume $G$ is nontrivial and argue using the subgroup $H \leq T/I$ which maps $C = \{0, 4, 7\}$ to triads in $\mathbb{S}$. The subgroup $H$ was called $H_0$ in Theorem 3.1, and is defined via the action on consonant triads. We will use the $H$-action on individual pitch classes to conclude that the only two possibilities for $\mathcal{S}$ are the Major Triad Tritone Mixture $\{0, 1, 4, 6, 7, 10\}$ and the Prometheus Tritone Mixture $\{0, 1, 2, 4, 6, 7, 8, 10\}$ whenever $P \notin G$.

We first remark $3 \notin \mathcal{S}$, for if 3 were in $\mathcal{S}$, then $\{0, 3, 7\}$ and $\{0, 4, 7\}$ would be in $\mathbb{S}$, which would necessitate $P \in G$. Based on this observation, we may conclude that the only transposition in $H$ is $T_6$, as follows. Since $3 \notin \mathcal{S}$, we have $T_3 \notin H$, and consequently $T_9 \notin H$. Similarly, $T_8 \notin H$, for if $T_8$ were in $H$, then $T_8\{0, 4, 7\} = \{8, 0, 3\}$ would be a subset of $\mathcal{S}$, a contradiction. Consequently, $T_2, T_4, T_{10} \notin H$. Clearly, $T_1, T_5, T_7, T_{11} \notin H$, for if any of these were in $H$, all transpositions would be in $H$. Thus, the only transposition that can be in $H$ is $T_6$. The transposition $T_6$ must be in $H$, since $H \leq T/I$ is nontrivial by Theorem 3.1 (iii), the composite of two inversions is a transposition, and the only allowable transposition is $T_6$.

At this point, we note that $\{\mathrm{Id}, T_6\}$ acts simply transitively on $\{C, G\flat\}$, and its dual group $\{\mathrm{Id}, Q_6\}$, determined by Theorem 3.1, is a possibility for $G$. The underlying pitch-class set is the Major Triad Tritone Mixture $\{0,1,4,6,7,10\}$, which is closed under the $\mathcal{T}$-action because $f(1)=10$, $f(6)=1$, $f(10)=1$, and $g(1)=0$, $g(6)=4$, $g(10)=0$, so the Major Triad Tritone Mixture is a possibility for $\mathcal{S}$.

Let us return to general $G$ with $P \notin G$. We claim $9 \notin \mathcal{S}$. Since $T_6 \in H$ and $H$ preserves $\mathbb{S}$, so also $\mathcal{S}$, we have $T_6\{0,4,7\} = \{6,10,1\} = G\flat \subseteq \mathcal{S}$. If 9 were in $\mathcal{S}$, then $g\flat = \{6,9,1\}$ would be a subset of $\mathcal{S}$, which would necessitate $P \in G$, a contradiction. Thus $9 \notin \mathcal{S}$.

We also claim $5 \notin \mathcal{S}$. If 5 were in $\mathcal{S}$, then $f \circ f(5) = 1$ and $g(5) = 8$ must also belong to $\mathcal{S}$. Then we have the parallel triads $\{1,4,8\}$ and $\{1,5,8\}$ both in $\mathbb{S}$, which is not allowed by $P \notin G$. Hence, $5 \notin \mathcal{S}$.

Suppose for the remainder of the proof that $H$ properly contains $\{\mathrm{Id}, T_6\}$. Our next goal is to show that the other elements of $H$ are exactly $I_2$ and $I_8$. The other elements of $H$ must be inversions, as the only transposition in $H$ is $T_6$ as we proved above. Recall that the inversion $I_{k+\ell}$ is the unique inversion which interchanges $k$ and $\ell$. Since $3,5,9 \notin \mathcal{S}$ as we saw above, we can exclude from $H$ all inversions that interchange 0, 4, or 7 with 3, 5, or 9. Thus, $I_3, I_7, I_{10}, I_5, I_9, I_0, I_9, I_1, I_4 \notin H$. The composite of $T_6$ with any of these cannot be in $H$, so we also have $I_{11}, I_6 \notin H$. The only two inversions that can be in $H$ are $I_2$ and $I_8$. But if either is in $H$, then so is the other, so both $I_2$ and $I_8$ must be in $H$. Thus $H = \{\mathrm{Id}, T_6, I_2, I_8\}$.

At this point, we note that $\{\mathrm{Id}, T_6, I_2, I_8\}$ acts simply transitively on the chord set $\{C, d\flat, G\flat, g\}$ and the dual subgroup is $\{\mathrm{Id}, Q_6, \mathrm{S}\ell, Q_6\mathrm{S}\ell\}$ by Theorem 3.1. The underlying pitch-class set $\{0,1,2,4,6,7,8,10\}$ is closed under the $\mathcal{T}$ action, since $f(1)=10$, $f(2)=1$, $f(6)=1$, $f(8)=7$, $f(10)=1$, and $g(1)=0$, $g(2)=8$, $g(6)=4$, $g(8)=8$, $g(10)=0$. We conclude that the Prometheus Tritone Mixture $\{0,1,2,4,6,7,8,10\}$ is the final possibility for $\mathcal{S}$. □

We may now consider upgrades by the Lawvere–Tierney topologies

$$j_{\mathcal{P}}, j_{\mathcal{L}}, j_{\mathcal{R}} \colon \Omega \to \Omega$$

in the topos $\mathbf{Set}^{\mathcal{T}}$, found on page 117 of [3]. Recall that if $\nu$ is a sub $\mathcal{T}$-action of a $\mathcal{T}$-action $\sigma$ and its characteristic morphism is $\chi\colon \sigma \to \Omega$, then the *j-upgrade* of $\nu$ has carrier set equal to the pre-image of the point $\mathcal{T} \in \Omega$ under $j \circ \chi$, that is, the $j$-upgrade has carrier set $(j \circ \chi)^{-1}(\mathcal{T})$.

**Corollary 4.2 (Maximal covers of Lawvere–Tierney upgrades admit $PLR$-subgroup actions).** *Let $\varphi\colon \mathbb{Z}_{12} \to \mathbb{Z}_{12}$ be an element of the $T/I$-group and let $(\mathbb{Z}_{12}, \mu^{\varphi})$ denote the $\mathcal{T}$-action given by the $\varphi$-conjugation of the natural $\mathcal{T}$-action, that is, the generators $f$ and $g$ of $\mathcal{T}$ act by $\varphi f \varphi^{-1}$ and $\varphi g \varphi^{-1}$. Then the maximal covers of the $j_{\mathcal{P}}$-, $j_{\mathcal{L}}$-, and $j_{\mathcal{R}}$-upgrades of the subaction $(\varphi(C), \mu^{\varphi}|_{\varphi(C)})$ admit simply transitive actions by $PLR$-subgroups. In fact, the upgrades and their maximal covers are the $\varphi$-images of the upgrades of $C$ and their respective covers. The $PLR$-subgroups remain the same. See Table 3.*

**Table 3.** Upgrades of the Subaction $(\varphi(C), \mu^{\varphi}|_{\varphi(C)})$ by the Lawvere–Tierney Topologies $j_{\mathcal{P}}$, $j_{\mathcal{L}}$, and $j_{\mathcal{R}}$

| Upgrade | Carrier Set | Type | $PLR$-Subgroup |
|---|---|---|---|
| $j_{\mathcal{P}}$ | $\varphi\{0, 3, 4, 7\}$ | Major-Minor Mixture | $\langle P \rangle$ |
| $j_{\mathcal{L}}$ | $\varphi\{0, 3, 4, 7, 8, 11\}$ | Hexatonic | $\langle P, L \rangle$ |
| $j_{\mathcal{R}}$ | $\varphi\{0, 1, 3, 4, 6, 7, 9, 10\}$ | Octatonic | $\langle P, R \rangle$ |

*Proof.* Let $\chi^C \colon \mathbb{Z}_{12} \to \Omega$ denote the characteristic morphism of the subaction $\{0, 4, 7\}$ of the natural $\mathcal{T}$-action on $\mathbb{Z}_{12}$. Then we have the following two pullbacks (see the Appendix in Sect. 5 for the notion of pullback).

$$\begin{array}{ccccc} \varphi(C) & \xrightarrow{\varphi^{-1}} & C & \longrightarrow & \{\mathcal{T}\} \\ \downarrow & \text{pullback} & \downarrow & \text{pullback} & \downarrow \\ \mathbb{Z}_{12} & \xrightarrow[\varphi^{-1}]{} & \mathbb{Z}_{12} & \xrightarrow[\chi^C]{} & \Omega \end{array}$$

The composite of these two pullback squares is also a pullback, so the characteristic morphism of $(\varphi(C), \mu^{\varphi}|_{\varphi(C)})$ is $\chi^C \circ \varphi^{-1}$. Thus, the $j$-upgrade of $(\varphi(C), \mu^{\varphi}|_{\varphi(C)})$ is $\varphi$ of the $j$-upgrade of $(C, \mu^{\mathrm{Id}})$, that is,

$$(j \circ \chi^{\varphi(C)})^{-1}(\mathcal{T}) = (j \circ \chi^C \circ \varphi^{-1})^{-1}(\mathcal{T}) = \varphi\left((j \circ \chi^C)^{-1}(\mathcal{T})\right).$$

We recall from page 123 of [3] that the $j_{\mathcal{P}}$-, $j_{\mathcal{L}}$-, and $j_{\mathcal{R}}$-upgrades of $C$ are the major-minor mixture, hexatonic, and octatonic appearing in Table 1. The maximal cover of the $\varphi$-image is clearly the $\varphi$-image of the maximal cover, so we are precisely in the situation of Corollary 3.3. Sections 3.1 and 3.2 further illustrate the point. □

In future work, we will study triadic covers of sets which are simultaneously invariant under multiple conjugated $\mathcal{T}$-actions. Another goal is to better understand the role of the three groups $\langle P \rangle$, $\langle P, L \rangle$, $\langle P, R \rangle$ within the framework of other groups which act simply transitively on triadic covers.

## 5 Appendix on Topos Theory

We recall some basic notions from topos theory. For explanations, examples, and proofs we refer to the excellent textbook of Mac Lane–Moerdijk [10]. Only a very small fraction of this Appendix is actually used in the present paper, so readers do not need to understand this Appendix in order to follow the present paper.

A *topos* is a category $\mathbf{C}$ such that the following three axioms hold.

(i) The category $\mathbf{C}$ admits all *finite limits.*
(ii) The category $\mathbf{C}$ admits a *subobject classifier* $* \to \Omega$. This is an object $\Omega$ equipped with a universal monomorphism from the terminal object $*$ to $\Omega$, which is universal in the sense that any other monomorphism is a pullback of it in a unique way. Often, one simply calls $\Omega$ the *subobject classifier.*
(iii) The category $\mathbf{C}$ is *Cartesian closed*, that is, for any two objects $B$ and $C$ there is an object $C^B$ in $\mathbf{C}$ and for any object $A$ of $\mathbf{C}$ a bijection

$$\mathrm{Mor}_{\mathbf{C}}(A \times B, C) \cong \mathrm{Mor}_{\mathbf{C}}(A, C^B)$$

natural in $A$, $B$, and $C$.

To understand these axioms, we consider how the category **Set** is a topos. This category admits all finite limits. For instance, the diagram defined by sets $A, B, C$ and two maps $f : A \to C, g : B \to C$ has finitely many objects and morphisms, and $L := \{(a, b) \in A \times B \mid f(a) = g(b)\}$ makes the diagram

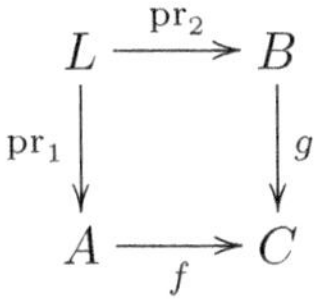

(1)

commute, and moreover there is a unique morphism from $L$ to any analogous $L'$ in such a way the two triangles involving $L$, $L'$, $A$, and $B$ also commute. This $L$ is called a *limit* of the diagram, and the limit for this special diagram is called a *pullback.* The projection $\mathrm{pr}_1$ is called a *pullback* of $g$ and the projection $\mathrm{pr}_2$ is called a *pullback* of $f$.

For the discussion of the subobject classifier of **Set**, we first point out that the monomorphisms in **Set** are precisely the injective maps. The subobject classifier in **Set** is the inclusion $\{1\} \hookrightarrow \{0, 1\}$. If $D$ is a subset of $E$, then the inclusion $D \hookrightarrow E$ is a pullback of $\{1\} \hookrightarrow \{0, 1\}$ in a unique way: there is only one function $\chi$ that makes the following diagram a pullback.

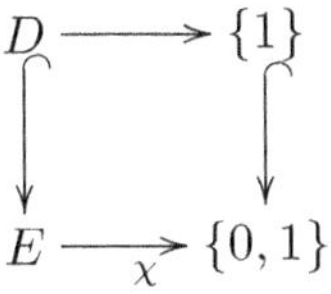


This function $\chi$ is necessarily the usual *characteristic function*, which takes the value 1 for inputs in $D$ and the value 0 for inputs in $E \backslash D$. Note that $\chi^{-1}(1) = D$.

For the Cartesian closedness of **Set**, the object $C^B$ is defined to be the set of functions $B \to C$, that is, $C^B = \mathrm{Mor}_{\mathbf{Set}}(B, C)$. The bijection in (iii) sends a function $f\colon A \times B \to C$ to the function $A \to C^B$ defined by $a \mapsto (b \mapsto f(a, b))$. Thus, **Set** is a topos.

In the topos $\mathbf{Set}^{\mathcal{T}}$, the underlying set of the subobject classifier $\Omega$ is the set of left ideals in $\mathcal{T}$. For instance, the full monoid $\mathcal{T}$ and the empty set $\emptyset$ are both left ideals in $\mathcal{T}$, so $\mathcal{T}, \emptyset \in \Omega$. To obtain the other left ideals it is convenient to inspect a Cayley graph of $\mathcal{T}$ with respect to its generators $f$ and $g$.

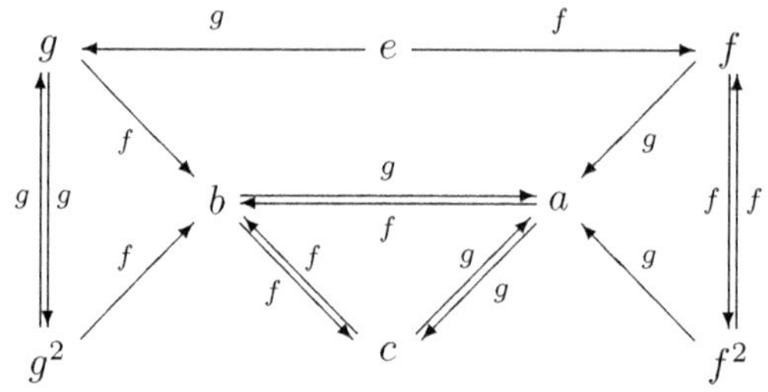


The nodes labelled $e, f, f^2, g, g^2, a, b, c$ represent the 8 elements of $\mathcal{T}$. The left ideals are node sets where no arrow starts that leads outside of this set. One may directly check by eye that there are four non-empty proper subsets $\mathcal{B} \subset \mathcal{T}$ satisfying this condition, namely

$$\mathcal{C} = \{a, b, c\},\ \mathcal{L} = \{a, b, c, f, f^2\},\ \mathcal{R} = \{a, b, c, g, g^2\},\ \mathcal{P} = \{a, b, c, f, f,^2, g, g^2\}.$$

The $\mathcal{T}$-action on the subobject classifier $\Omega = \{\emptyset, \mathcal{C}, \mathcal{L}, \mathcal{R}, \mathcal{P}, \mathcal{T}\}$ is

$$m \cdot \mathcal{B} := \{n \in \mathcal{T} \mid n \cdot m \in \mathcal{B}\}$$

for $m \in \mathcal{T}$ and $\mathcal{B} \in \Omega$, and the universal monomorphism is $\{\mathcal{T}\} \hookrightarrow \Omega$.

A *Lawvere–Tierney topology* in the topos $\mathbf{Set}^{\mathcal{T}}$ is a $\mathcal{T}$-equivariant map $\Omega \to \Omega$ such tha

(i) $j(\mathcal{T}) = \mathcal{T}$,
(ii) $j \circ j = j$,
(iii) $j(\mathfrak{r}) \cap j(\mathfrak{s}) = j(\mathfrak{r} \cap \mathfrak{s})$ for all $\mathfrak{r}, \mathfrak{s} \in \Omega$.

The six Lawvere–Tierney topologies of $\mathbf{Set}^{\mathcal{T}}$ are determined in Sect. 2.3 and 2.4 of [3]. Given a sub $\mathcal{T}$-set $D$ of a $\mathcal{T}$-set $E$, one may *upgrade* it by a Lawvere–Tierney topology $j$ in the following way. By definition of subobject classifier, the inclusion $D \hookrightarrow E$ corresponds uniquely to a morphism $\chi\colon E \to \Omega$, its characteristic morphism. The *$j$-upgrade of $D$* is then the sub $\mathcal{T}$-set of $E$ which has characteristic morphism $j \circ \chi$. The underlying set of the upgrade is $(j \circ \chi)^{-1}(\mathcal{T})$.

For example, the characteristic morphism $\chi^{\{0,4,7\}}$ of the $C$-major chord $\{0, 4, 7\}$ in $\mathbb{Z}_{12}$ with the natural $\mathcal{T}$-action can be determined by precomposing the characteristic morphism $\chi^{\{0,1,4\}}$ on page 121 of [3] with multiplication by 7.

| $t$ | 0 | 1 | 2 | 3 | 4 | 5 | 6 | 7 | 8 | 9 | 10 | 11 |
|---|---|---|---|---|---|---|---|---|---|---|---|---|
| $\chi^{\{0,4,7\}}(t)$ | $\mathcal{T}$ | $\mathcal{R}$ | $\mathcal{C}$ | $\mathcal{P}$ | $\mathcal{T}$ | $\mathcal{C}$ | $\mathcal{R}$ | $\mathcal{T}$ | $\mathcal{L}$ | $\mathcal{R}$ | $\mathcal{R}$ | $\mathcal{L}$ |

Using this, and the tables on pages 117 and 118 of [3], the upgrades of $\{0, 4, 7\}$ are found to be as follows.

| Upgrade | Carrier Set | Type |
|---|---|---|
| $j_{\mathcal{T}}$ | $\{0, 4, 7\}$ | Major Chord |
| $j_{\mathcal{P}}$ | $\{0, 3, 4, 7\}$ | Major-Minor Mixture |
| $j_{\mathcal{L}}$ | $\{0, 3, 4, 7, 8, 11\}$ | Hexatonic |
| $j_{\mathcal{R}}$ | $\{0, 1, 3, 4, 6, 7, 9, 10\}$ | Octatonic |
| $j_{\mathcal{C}}$ | $\mathbb{Z}_{12}$ | Chromatic Scale |
| $j_{\mathcal{F}}$ | $\mathbb{Z}_{12}$ | Chromatic Scale |

The numbers in this table differ from those on page 123 in [3] by multiplication with 7 because we are using the semitone encoding in the present article.

A surprising discovery of [3] is that these music-theoretic collections are the Lawvere–Tierney upgrades of the $C$-major chord. One of the goals of the present paper is to thicken this music-theoretic meaning by way of the $PLR$-group in Theorem 4.1.

**Acknowledgements.** Thomas M. Fiore thanks Ramon Satyendra for explaining David Clampitt's article [6] to him in 2004.

## References


1. Cohn, R.: Neo-Riemannian operations, parsimonious trichords, and their "Tonnetz" representations. Journal of Music Theory 41(1), 1–66 (1997)
2. Fiore, T.M., Satyendra, R.: Generalized contextual groups. Music Theory Online 11(3) (2005)
3. Noll, T.: The topos of triads. In: Colloquium on Mathematical Music Theory. Grazer Math. Ber., vol. 347, pp. 103–135. Karl-Franzens-Univ. Graz, Graz (2005)
4. Peck, R.: Generalized commuting groups. Journal of Music Theory 54(2) (2010)
5. Cohn, R.: Maximally smooth cycles, hexatonic systems, and the analysis of late-romantic triadic progressions. Music Analysis 15(1), 9–40 (1996)
6. Clampitt, D.: Alternative interpretations of some measures from Parsifal. Journal of Music Theory 42(2), 321–334 (1998)
7. Oshita, K.: The hexatonic systems under neo-Riemannian theory: an exploration of the mathematical analysis of music. REU Project, University of Chicago (2009), `http://www.math.uchicago.edu/~may/VIGRE/VIGRE2009/REUPapers/Oshita.pdf`
8. Bartlett, P.: Triads and topos theory. REU Project, University of Chicago (2007), `http://www.math.uchicago.edu/~may/VIGRE/VIGRE2007/REUPapers/FINALFULL/Bartlett.pdf`
9. Crans, A.S., Fiore, T.M., Satyendra, R.: Musical actions of dihedral groups. Amer. Math. Monthly 116(6), 479–495 (2009)
10. Mac Lane, S., Moerdijk, I.: Sheaves in geometry and logic. Universitext. Springer, New York (1994); A first introduction to topos theory, Corrected reprint of the 1992 edition

# Spelled Heptachords

Julian Hook

Indiana University, Bloomington, USA
juhook@indiana.edu

**Abstract.** This paper develops a theory of *spelled pitch classes* (spcs) and *spelled pitch-class sets* (spc sets), incorporating pitch spelling into the techniques of pitch-class set theory. The symmetries of spc space are transposition and inversion along the line of fifths. Because of the inextricable link between pitch spelling and diatonic scales, *spelled heptachords*—seven-note spc sets that include each letter name exactly once—occupy a privileged position in this theory. Spelled heptachords may be regarded as inflected diatonic scales, and possess a number of structural characteristics not shared by other spc sets. The 66 equivalence classes of spelled heptachords without enharmonic doublings or voice crossings are enumerated. A diatonic musical structure together with a spelled heptachord determine an *spc structure* in which the notes of the diatonic structure are inflected by the corresponding accidentals from the heptachord; spc structures arising in this way show promise as powerful tools in analysis of chromatic harmony.

**Keywords:** Pitch-class set theory, Diatonic set theory, Pitch spelling, Heptachords, Chromaticism.

The development of American pitch-class set theory was largely motivated by a desire to understand "the structure of atonal music" (to quote Forte's well-known title [1]). Since set theory's inception, however, its techniques have often been applied to study music exhibiting at least some characteristics of tonality, such as tertian chords and diatonic scales. The theory's traditional disregard for enharmonic distinctions proves a hindrance in these applications. Set theory regards F♯ and G♭ as the same note, an augmented second and a minor third as the same interval, and a dominant seventh and German augmented sixth as the same chord, while in tonal analysis such distinctions are indispensable.

The first two sections of this paper integrate pitch spelling into set theory, developing notions of *spelled pitch classes* (spcs) and *spelled pitch-class sets* (spc sets), and enabling distinctions such as those listed above to be respected. Because of the inextricable link between pitch spelling and diatonic scales, *spelled heptachords*—seven-note spc sets that include each letter name exactly once—occupy a privileged position in this theory. Spelled heptachords are the subject of section 3, and section 4 demonstrates the application of spelled heptachords in analysis of chromatic harmony.

C. Agon et al. (Eds.): MCM 2011, LNAI 6726, pp. 84–97, 2011.

# 1 Spelled Pitch Classes

Many approaches have been taken to the theoretical integration of chromatic and diatonic structure, but few studies have even indirectly confronted matters of spelling, for instance by representing a note as an ordered pair consisting of both mod-12 and mod-7 pitch-class numbers (see for example [2–4]). The strategy adopted here is rather different; it relates to constructions found in [5], [6], [7], and [8], though none of these sources explore the set-theoretic connections pursued here.

The name *spelled pitch class* reflects our understanding of an spc as a pitch class in a particular spelling, but in fact an spc is nothing more than a *note name*: a letter A–G possibly with any number of sharps or flats attached. Table 1 shows a portion of *spc space*. As the table shows, the spcs may be arranged along the line of fifths and identified with the integers. (Regener [5] calls this space the *quint group*; Temperley [9] refers to spcs as *tonal pitch classes*.) Assigning spc 0 to the note D proves advantageous because of the central position of D among the seven white keys. Also shown for each spc $n$ are a corresponding mod-12 pc number $\pi(n)$ in *pc space*; a mod-7 *diatonic pitch class* (*dpc*) $\delta(n)$, identified with a white-note letter name in *diatonic pitch-class space* (*dpc space*); and an integer *accidental index* $\alpha(n)$, indicating the number of sharps (+) or flats (–) attached to that letter. For consistency with common conventions, pc and dpc numbers are reckoned from C = 0. The table shows all spcs with $-2 \le \alpha(n) \le 2$, from F♭♭ to B×, corresponding to $-17 \le n \le 17$.

Calculation of $\pi$, $\delta$, and $\alpha$ is straightforward:

$$\pi(n) \equiv 7n + 2 \pmod{12}\ ; \tag{1}$$

$$\delta(n) \equiv 4n + 1 \pmod{7}\ ; \tag{2}$$

$$\alpha(n) = n/7 \text{ rounded to the nearest integer.} \tag{3}$$

From (1)–(3) we may deduce :

**Table 1.** A central segment of spelled pitch-class space ($-17 \le n \le 17$)

| Note name | | F♭♭ | C♭♭ | G♭♭ | D♭♭ | A♭♭ | E♭♭ | B♭♭ | F♭ | C♭ | G♭ | D♭ | A♭ | E♭ | B♭ |
|---|---|---|---|---|---|---|---|---|---|---|---|---|---|---|---|
| Spelled pitch class (spc) | $n \in \mathbb{Z}$ | –17 | –16 | –15 | –14 | –13 | –12 | –11 | –10 | –9 | –8 | –7 | –6 | –5 | –4 |
| Pitch class (pc) | $\pi(n) \in \mathbb{Z}_{12}$ | 3 | 10 | 5 | 0 | 7 | 2 | 9 | 4 | 11 | 6 | 1 | 8 | 3 | 10 |
| Diatonic pitch class (dpc) | $\delta(n) \in \mathbb{Z}_7$ | 3 | 0 | 4 | 1 | 5 | 2 | 6 | 3 | 0 | 4 | 1 | 5 | 2 | 6 |
| Accidental index | $\alpha(n) \in \mathbb{Z}$ | –2 | –2 | –2 | –2 | –2 | –2 | –2 | –1 | –1 | –1 | –1 | –1 | –1 | –1 |

| | F | C | G | D | A | E | B | F♯ | C♯ | G♯ | D♯ | A♯ | E♯ | B♯ | F× | C× | G× | D× | A× | E× | B× |
|---|---|---|---|---|---|---|---|---|---|---|---|---|---|---|---|---|---|---|---|---|---|
| $n$ | –3 | –2 | –1 | 0 | 1 | 2 | 3 | 4 | 5 | 6 | 7 | 8 | 9 | 10 | 11 | 12 | 13 | 14 | 15 | 16 | 17 |
| $\pi(n)$ | 5 | 0 | 7 | 2 | 9 | 4 | 11 | 6 | 1 | 8 | 3 | 10 | 5 | 0 | 7 | 2 | 9 | 4 | 11 | 6 | 1 |
| $\delta(n)$ | 3 | 0 | 4 | 1 | 5 | 2 | 6 | 3 | 0 | 4 | 1 | 5 | 2 | 6 | 3 | 0 | 4 | 1 | 5 | 2 | 6 |
| $\alpha(n)$ | 0 | 0 | 0 | 0 | 0 | 0 | 0 | 1 | 1 | 1 | 1 | 1 | 1 | 1 | 2 | 2 | 2 | 2 | 2 | 2 | 2 |

$$n \equiv 7\pi(n) - 2 \pmod{12}\ ; \tag{4}$$

$$n \equiv 2\delta(n) - 2 \pmod{7}\ ; \tag{5}$$

$$n \equiv 7\pi(n) - 12\delta(n) - 2 \pmod{84}\ ; \tag{6}$$

$$7\alpha(n) - 3 \le n \le 7\alpha(n) + 3\ . \tag{7}$$

The value of $n$ may therefore be recovered if any two of $\pi(n)$, $\delta(n)$, and $\alpha(n)$ are known, except that if only $\pi(n)$ and $\delta(n)$ are known then $n$ is determined only up to congruence mod 84. Spcs that differ by a multiple of 84, such as C(6♯) ($n = 40$) and C(6♭) ($n = -44$), share a letter name and are enharmonically equivalent, but occupy distinct positions on the line of fifths; this situation never arises in familiar musical contexts involving only spcs with $-2 \le \alpha(n) \le 2$.

*Transposition* $\tau_k$ (transposition by $k$ fifths) and *inversion* $\iota$ (about D) are defined on spc space as translation and reflection along the line of fifths:

$$\tau_k(n) = n + k \quad (k \in \mathbb{Z})\ ; \tag{8}$$

$$\iota(n) = -n\ . \tag{9}$$

These operations interact systematically with the familiar mod-12 $T_k$ and $I$ on pc space, and their mod-7 dpc-space counterparts $t_k$ and $i$ [10]:

$$\pi(\tau_k(n)) \equiv \pi(n) + 7k \equiv T_{7k}(\pi(n)) \pmod{12}\ ; \tag{10}$$

$$\pi(\iota(n)) \equiv -\pi(n) + 4 \equiv T_4I(\pi(n)) \pmod{12}\ ; \tag{11}$$

$$\delta(\tau_k(n)) \equiv \delta(n) + 4k \equiv t_{4k}(\delta(n)) \pmod{7}\ ; \tag{12}$$

$$\delta(\iota(n)) \equiv -\delta(n) + 2 \equiv t_2i(\delta(n)) \pmod{7}\ . \tag{13}$$

The additive constants 4 and 2 in (11) and (13) arise because $\iota$ inverts about D in spc space, while $I$ and $i$ invert about C in pc space and dpc space.

These operations satisfy the relations $\tau_k\iota = \iota\tau_{-k}$ and $\iota^2$ = identity. Together $\tau_1$ and $\iota$ generate an *infinite dihedral group*, the group of isometries of the integers.

Observe that $\tau_7$(C) = C♯, $\tau_{-7}$(C) = C♭, $\tau_{12}$(C) = B♯, and $\tau_{-12}$(C) = D♭♭. Generally $\tau_7$ (respectively $\tau_{-7}$) raises (respectively lowers) any spc chromatically, leaving its letter name unchanged; $\tau_{12}$ (respectively $\tau_{-12}$) respells any note enharmonically in the "sharpwise" (respectively "flatwise") direction.

## 2 Spelled Pitch-Class Sets

Several *spc sets*—subsets of spc space—are displayed in Table 2. The *fifth-string* of an spc set enumerates the pattern of gaps in its line-of-fifths representation; the length of the fifth-string is 1 less than the cardinality of the set. The *span* of an spc set is the difference between its largest and smallest spc numbers, or equivalently the sum of the terms in its fifth-string.

**Table 2.** Some spelled pitch-class sets

| Description | A♭ | E♭ | B♭ | F | C | G | D | A | E | B | F♯ | C♯ | G♯ | Fifth-string | Span |
|---|---|---|---|---|---|---|---|---|---|---|---|---|---|---|---|
| B♭$^7$ chord | • | | • | • | | | • | | | | | | | 213 | 6 |
| F$^7$ chord | | • | | • | • | | | • | | | | | | 213 | 6 |
| C$^7$ chord | | | • | | • | • | | | • | | | | | 213 | 6 |
| F♯$^{\varnothing 7}$ chord | | | | | • | | | • | • | | • | | | 312 | 6 |
| Ger$^{+6}$ chord | | | • | • | | | • | | | | | | • | 136 | 10 |
| F♯$^{o7}$ chord | | • | | | • | | | • | | | • | | | 333 | 9 |
| Whole-tone scale | | • | | • | | • | | • | | • | | • | | 22222 | 10 |
| Octatonic scale | | • | | • | • | | • | • | | • | • | | • | 2121212 | 11 |

Spc sets related by $\tau$-transposition share the same fifth-string and span. All dominant seventh chords, including the first three sets in the table, share the fifth-string 213 and span 6. Inversion by $\iota$ preserves the span of an spc set but retrogrades its fifth-string. The F♯$^{\varnothing 7}$, the $\iota$-inversion of the C$^7$, has fifth-string 312. The fifth-string may be used as a sort of "prime form" for a $\tau$-transposition class (hereafter called a *$\tau$-class*) of spc sets, while a pair of retrograde-related fifth-strings together define a *$\tau\iota$-class*, all of whose members are related by $\tau$ and/or $\iota$. The German augmented sixth in the table, though enharmonically equivalent to the B♭$^7$, is not $\tau$- or $\tau\iota$-related to it, and has a different fifth-string and span.

Diminished seventh chords, whole-tone scales, and octatonic scales, in their most common spellings, are symmetrical in spc space as they are in pc space. Nonstandard spellings of the same pc sets, such as {C, D♯, F♯, B♭♭} for the diminished seventh, are not $\tau$-related to the standard spellings, and may be asymmetrical.

An spc set may contain a *letter doubling* such as {F, F♯}, an *enharmonic doubling* such as {E♯, F}, or even a *(voice) crossing* such as {E♯, F♭}. Letter doublings are necessarily present in all spc sets of cardinality ≥ 8 (the octatonic scale in Table 2 contains {F, F♯}), and enharmonic doublings are necessarily present in all spc sets of cardinality ≥ 13. Spc sets without letter doublings will be called *letter-distinct*; spc sets without enharmonic doublings or crossings will be called *proper*.

If $S$ is an spc set, we write

$$\pi(S) = \{\pi(n) \mid n \in S\} . \tag{14}$$

This $\pi(S)$ is an ordinary pc set, called the *projection* of $S$ onto pc space. Clearly $|\pi(S)| \leq |S|$; if $S$ is proper, equality holds. Spc sets $S_1$ and $S_2$ are *enharmonically equivalent* if $\pi(S_1) = \pi(S_2)$. From (10) and (11),

$$\pi(\tau_k(S)) = T_{7k}(\pi(S)) ; \tag{15}$$

$$\pi(\iota(S)) = T_4 I(\pi(S)) . \tag{16}$$

If spc sets $S_1$ and $S_2$ belong to the same $\tau$-class (respectively $\tau\iota$-class) in spc space, then $\pi(S_1)$ and $\pi(S_2)$ must therefore belong to the same $T$-class (respectively $TI$-class) in pc space. If $S$ is symmetric in spc space (that is, if $S$ is $\tau$-related to its $\iota$-inversion), then $\pi(S)$ is symmetric in pc space ($T$-related to its $I$-inversion). The converses of these statements are not true in general, as the oddly spelled diminished seventh chord above illustrates.

The *diatonic projection* $\delta(S)$ of an spc set onto dpc space may be defined similarly, and satisfies similar properties. Details are left to the reader.

The *accidental index* of an spc set is the sum of the accidental indices of its elements:

$$\alpha(S) = \Sigma_{n \in S}\, \alpha(n) \ . \tag{17}$$

Because $\alpha(n)$ enumerates sharps positively and flats negatively, calculation of $\alpha(S)$ amounts to counting the set's sharps minus its flats.

## 3 Spelled Heptachords

By a *spelled heptachord* we mean a *letter-distinct* spc set of cardinality 7. Letter-distinctness ensures that each letter name appears exactly once in every spelled heptachord. The spc numbers of the notes therefore include one representative of each mod-7 congruence class; the diatonic projection $\delta(H)$ of a spelled heptachord $H$ is all of dpc space. A spelled heptachord may be regarded as an altered diatonic scale $\{C^{\lambda_1}, D^{\lambda_2}, E^{\lambda_3}, F^{\lambda_4}, G^{\lambda_5}, A^{\lambda_6}, B^{\lambda_7}\}$ whose letter names C–B are inflected by accidentals $\lambda_1$–$\lambda_7$, which may be identified with integers (e.g., ♯ = 1, ♭♭ = –2). By attaching the same accidentals to the lines and spaces of a musical staff we may regard a spelled heptachord as a sort of generalized key signature, an idea related to concepts presented in [10] and [11].

Many of the distinctive properties of spelled heptachords follow from the fact that fifth-transposition of a diatonic scale alters exactly one accidental:

$$\tau_1(\{C, D, E, F, G, A, B\}) = \{G, A, B, C, D, E, F\sharp\} \ . \tag{18}$$

Applied to a diatonic scale, $\tau_1$ has the same effect as the *signature transformation* $s_1$, which transposes a standard key signature one position "sharpwise" [10]. The above property holds, however, for the $\tau_1$-transposition of *every* spelled heptachord: whatever accidentals inflect the notes C–D–E–F–G–A of a spelled heptachord $H$ will appear unchanged on G–A–B–C–D–E in $\tau_1(H)$, but the accidental originally on B transposes to F altered by the addition of one sharp (or removal of one flat):

$$\tau_1(\{C^{\lambda_1}, D^{\lambda_2}, E^{\lambda_3}, F^{\lambda_4}, G^{\lambda_5}, A^{\lambda_6}, B^{\lambda_7}\}) = \{G^{\lambda_1}, A^{\lambda_2}, B^{\lambda_3}, C^{\lambda_4}, D^{\lambda_5}, E^{\lambda_6}, F^{\lambda_7+1}\} \ . \tag{19}$$

It follows that $\alpha(\tau_1(H)) = \alpha(H) + 1$ for every spelled heptachord $H$, and thus $\alpha(\tau_k(H)) = \alpha(H) + k$. Hence every $\tau$-class of spelled heptachords contains exactly one heptachord of every accidental index. The accidental index serves as a useful means of indexing the elements of the $\tau$-class. In particular, every $\tau$-class of spelled heptachords contains exactly one *balanced* form $H_0$ for which $\alpha(H_0) = 0$, its sharps equal in number to its flats.

These properties are illustrated in Table 3 for the $\tau$-class that we shall call HMIN, represented by fifth-string 121113, consisting of the various transpositions of the har-

**Table 3.** Five representatives of the $\tau$-class 121113 = HMIN (harmonic minor scales). The accidental index (count of sharps minus flats) of HMIN($k$) is always equal to $k$.

| | A♭ | E♭ | B♭ | F | C | G | D | A | E | B | F♯ | C♯ | G♯ | D♯ | Common name |
|---|---|---|---|---|---|---|---|---|---|---|---|---|---|---|---|
| HMIN(–2) | • | • | | • | • | • | • | | | • | | | | | C harmonic minor |
| HMIN(–1) | | • | • | | • | • | • | • | | | • | | | | G harmonic minor |
| HMIN(0) | | | • | • | | • | • | • | • | | | • | | | D harmonic minor |
| HMIN(+1) | | | | • | • | | • | • | • | • | | | • | | A harmonic minor |
| HMIN(+2) | | | | | • | • | | • | • | • | • | | | • | E harmonic minor |

monic minor scale. The balanced form of this $\tau$-class, denoted HMIN(0), is the D harmonic minor scale, which contains one flat and one sharp. A few of its $\tau$-transpositions, HMIN($k$) = $\tau_k$(HMIN(0)), are also shown, and it is readily verified that $\alpha$(HMIN($k$)) = $k$ for each $k$.

The sum of the spc numbers of HMIN(0) is $(-4) + (-3) + (-1) + 0 + 1 + 2 + 5 = 0$. In fact, the spc numbers in the balanced form of every spelled heptachord sum to 0. This is because the spc numbers of the white-key diatonic collection DIA(0) sum to 0, and any balanced spelled heptachord may be obtained from DIA(0) by adding equal numbers of sharps and flats.

These properties of spelled heptachords do not hold for spelled spc sets of other cardinalities. For example, there are not one but two balanced diminished seventh chords (F♯$^{o7}$ and C♯$^{o7}$), and in neither of them do the spc numbers sum to 0. The octatonic scale $S$ from Table 2 has no balanced $\tau$-transposition at all, for $\alpha(S) = 1$ while $\alpha(\tau_{-1}(S)) = -1$.

Twelve spelled heptachords are listed in Table 4. The balanced form of each is shown, along with the fifth-string, span, and a shorthand name. Several of these correspond to familiar scale types, but it should be remembered that a spelled heptachord contains no information to indicate which of its seven notes is the tonic. While DIA(0) is conveniently regarded as the C major scale, it could also be a modal scale on a different tonic; MMIN(0) is the ascending form of the G melodic minor scale, but is also the "acoustic" scale on C.

The remaining scales appearing in the table are as follows: HMIN(0) is a harmonic minor scale from Table 3, and HMAJ(0) is the "harmonic major" scale on G, a major scale inflected by ♭$\hat{6}$ [12]. NMIN(0) is the "Neapolitan minor" scale on A, a harmonic minor scale with ♭$\hat{2}$, while NMAJ(0) is the "Neapolitan major" scale on G, a major scale with ♭$\hat{2}$. GYP is the "gypsy" (or "Hungarian minor") scale, characterized by two augmented seconds. SWT ("super–whole-tone") is a seven-note superset of a whole-tone collection; SOCTA and SOCTB ("sub-octatonic") are subsets of an octatonic collection; and SHEXA and SHEXB ("super-hexatonic") are supersets of a hexatonic collection.

Other scales of these types are readily obtained from the balanced forms. Knowing that the G harmonic major scale is HMAJ(0) and that B♭ = $\tau_{-3}$(G), for example, we may conclude that the B♭ harmonic major scale is HMAJ(–3).

From the fifth-strings in Table 4 we see immediately that DIA, MMIN, GYP, and SWT are symmetrical in spc space, and therefore have symmetrical projections in pc space. The other eight examples form four $\iota$-related pairs: HMIN/HMAJ, NMIN/

**Table 4.** Balanced forms of twelve spelled heptachords

| | D♭ | A♭ | E♭ | B♭ | F | C | G | D | A | E | B | F♯ | C♯ | G♯ | D♯ | Fifth-string | Span |
|---|---|---|---|---|---|---|---|---|---|---|---|---|---|---|---|---|---|
| DIA(0) | | | | | • | • | • | • | • | • | • | | | | | 111111 | 6 |
| MMIN(0) | | | | • | | • | • | • | • | • | | • | | | | 211112 | 8 |
| HMIN(0) | | | | • | • | | • | • | • | • | | | • | | | 121113 | 9 |
| HMAJ(0) | | | • | | | • | • | • | • | | • | • | | | | 311121 | 9 |
| NMIN(0) | | | | • | • | • | | • | • | • | | | | • | | 112114 | 10 |
| NMAJ(0) | | • | | | | • | • | • | | • | • | • | | | | 411211 | 10 |
| GYP(0) | | | • | • | | | • | • | • | | | • | • | | | 131131 | 10 |
| SWT(0) | | | • | | • | | • | • | • | | • | | • | | | 221122 | 10 |
| SOCTA(0) | | | • | | • | • | | • | • | | • | | | • | | 212123 | 11 |
| SOCTB(0) | | • | | | • | | • | • | | • | • | | • | | | 321212 | 11 |
| SHEXA(0) | | • | | | • | • | • | | | • | • | | | | • | 311314 | 13 |
| SHEXB(0) | • | | | | • | • | | | • | • | • | | | • | | 413113 | 13 |

NMAJ, SOCTA/SOCTB, and SHEXA/SHEXB, whose projections in pc space are *I*-related.

Spelled heptachords, like other spc sets, generally may have enharmonic doublings or crossings; for instance, the "whole-tone heptachord" {C, D, E, F♯, G♯, A♯, B♯} is a spelled heptachord. All the spelled heptachords in Table 4, however, are *proper*—free of doublings or crossings—which ensures that as the seven letters proceed around the circle of dpc space, the corresponding projected pitch classes of the heptachord proceed similarly around the circle of mod-12 pc space, without pausing (for an enharmonic doubling such as B♯–C in the above example) or reversing direction (for a crossing).

A general spc set admits numerous enharmonic respellings, but for *proper* spelled heptachords the possible respellings are greatly limited. Consider, for example, the D♯ harmonic minor scale HMIN(+7) = {D♯, E♯, F♯, G♯, A♯, B, C𝄪}, and suppose we wish to respell this set as a proper spelled heptachord containing F rather than E♯. Letter-distinctness then requires that the original F♯ be respelled, and to avoid crossings it must be respelled as G♭ (rather than E𝄪 or A♭♭♭ or some more remote possibility). Continuing in this way, we find that the only possible respelling of the set is {E♭, F, G♭, A♭, B♭, C♭, D}, which is the E♭ harmonic minor scale HMIN(–5)—the original spc set transposed by $\tau_{-12}$.

If $H_1$ and $H_2$ are enharmonically equivalent proper spelled heptachords, it is always the case that $H_2 = \tau_{12k}(H_1)$ for some integer $k$. It follows that if $H_1$ and $H_2$ are proper spelled heptachords for which $\pi(H_1)$ and $\pi(H_2)$ belong to the same $T$-class in pc space, then $H_1$ and $H_2$ belong to the same $\tau$-class in spc space ("$T$-equivalence implies $\tau$-equivalence"). Consequently a proper spelled heptachord $H$ is symmetrical in spc space if and only if $\pi(H)$ is symmetrical in pc space. Moreover, $\tau$-classes of proper spelled heptachords are in one-to-one correspondence with $T$-classes of seven-note pc sets. These properties do not generally hold for spc sets (even proper letter-distinct spc sets) of any nontrivial cardinality other than 7.

The 66 $\tau$-classes of proper spelled heptachords are tabulated in an Appendix to this paper.

## 4 Applications to Chromatic Harmony

By a *structure* we mean any musical excerpt, such as a chord, chord progression, melody, contrapuntal passage, or any combination of these. Structures whose notes are spcs are *spc structures* or *spelled chromatic structures*; there are also *pc structures* (*unspelled chromatic structures*) and *dpc structures* (*diatonic structures*). Spc structures may be transposed by $\tau_k$, pc structures by $T_k$, and diatonic structures by $t_k$.

A diatonic structure is *complete* if all seven dpcs occur within it. It is *trivially complete* if the seven notes occur only simultaneously, so that the structure is invariant under every $t_k$. For a nontrivially complete diatonic structure, the seven dpc-space transpositions are necessarily all distinct.

If $X$ is a diatonic structure and $H$ is a spelled heptachord, we may obtain a spelled chromatic structure by applying the accidentals in $H$ to the corresponding dpcs in $X$ wherever they appear. This spc structure is called *X inflected by H*, denoted $X/H$. For example, if $X$ is the seventh chord {G, B, D, F} and $H$ is the spelled heptachord GYP(+1) = {G♯, A, B♭, C♯, D, E, F}, then $X/H$ is the spc set {G♯, B♭, D, F}, the German augmented sixth chord from Table 2. There are several ways in which we may regard $H$: as a "filter" through which the notes of $X$ are passed, determining their accidentals; in the terminology of [13], as a "field" of spcs from which the notes of an spc structure are selected; or in the terminology of [14], as a "macroharmony" that controls the pitch content of the music over a certain span of time.

The "C major" scale in Fig. 1(a) may be regarded as a complete diatonic structure $X$, or as an spc structure $X/H$, where $H$ is the white-key heptachord DIA(0). The remainder of Fig. 1 shows three transformations of this spc structure. In (b), $X$ is transposed by $t_1$ in dpc space, yielding $t_1(X)/H$, a scale starting on D, still within DIA(0) ("D Dorian"). In (c), $X$ is unchanged but the heptachordal field $H$ is transposed by $\tau_1$ in spc space, yielding $X/\tau_1(H)$, a scale on C in DIA(+1) ("C Lydian"). In (d), $\tau_1$ is applied not just to $H$ but to the entire spc structure, yielding $\tau_1(X/H)$, a scale on G in DIA(+1) ("G major"). (The musical notation in Fig. 1 is potentially misleading in its implication of register; in working with pcs, spcs, or

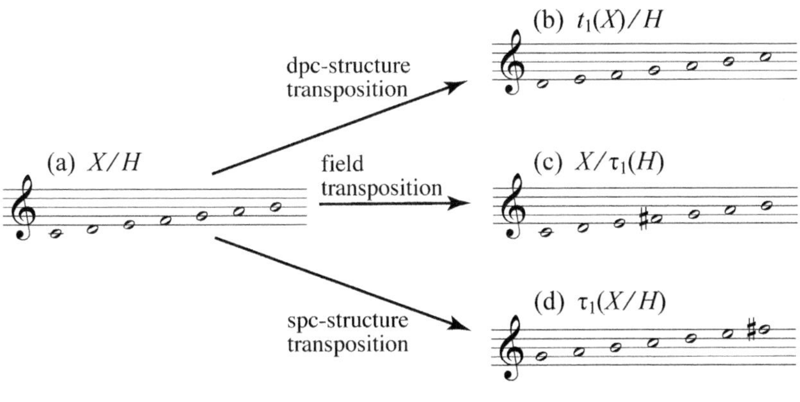


**Fig. 1.** Three transformations of the spc structure $X/H$, where $X$ is a complete dpc structure consisting of a seven-note scale, and $H$ is the proper spelled heptachord DIA(0)

dpcs, we consider the low C of (a) and the high C of (b) to be the same object.)

These three transformations—dpc-structure transposition, field transposition, and spc-structure transposition—interact in systematic ways. For example, the G major scale $\tau_1(X/H)$ of Fig. 1(d) could also have been obtained from $X/H$ by first transposing $X$ by $t_4$ (yielding $t_4(X)/H$, "G Mixolydian"), then transposing $H$ by $\tau_1$. Thus

$$\tau_1(X/H) = t_4(X)/\tau_1(H) \; ; \tag{20}$$

more generally, for any diatonic structure $X$ and spelled heptachord $H$, and for any $k$,

$$\tau_k(X/H) = t_{4k}(X)/\tau_k(H) \; . \tag{21}$$

Similarly one can show that

$$t_k(X)/H = \tau_{2k}(X/\tau_{-2k}(H)) \; ; \tag{22}$$

$$X/\tau_k(H) = \tau_k(t_{3k}(X)/H) \; . \tag{23}$$

If $X$ is nontrivially complete and $H$ is proper, then the seven dpc-space transpositions $t_k(X)$ are all distinct, as are the seven spc structures $t_k(X)/H$. The projections of these structures in pc space all exhaust the notes of $H$, but as pc structures, no two of them can be $T$-related, because a seven-note pc set cannot be invariant under any transposition other than $T_0$. Within a diatonic scale, it is a consequence of the "cardinality equals variety" principle [15] that distinct dpc-space transpositions of a nontrivially complete diatonic structure can never be related by chromatic transposition. In fact, the same property holds for intrascalar transpositions in the field determined by any proper spelled heptachord; every such heptachord, like the diatonic scale, has seven distinguishable "modes" or "species."

By (23), the field-transposed structure $X/\tau_k(H)$ is $\tau$-related to (and the same "species" as) the diatonic transposition $t_{3k}(X)/H$. If $X$ is nontrivially complete and $H$ is proper, then by the preceding paragraph, this structure $t_{3k}(X)/H$ cannot be $T$-related to $X/H$ unless $3k \equiv 0$ and hence $k \equiv 0 \pmod 7$. Therefore the pc-space projections of $X/H$ and the field-transposed $X/\tau_k(H)$ cannot be $T$-related unless $k \equiv 0 \pmod 7$. Because $\tau_7$ raises all spcs chromatically, the pc-space projections of $X/H$ and $X/\tau_{7k}(H)$ are always related by $T_k$.

We may therefore take any nontrivially complete diatonic structure and filter it through any of the 66 different proper spelled heptachords, each in seven different transpositions such as $\tau_{-3}(H)$, ..., $\tau_3(H)$. This will yield 462 different spc structures, no two of which are $\tau$-related in spc space or $T$-related in pc space.

This observation engenders a fruitful perspective on chromatic harmony. Many chromatic progressions may be analyzed as simple dpc structures inflected by a variety of spelled heptachords, particularly those appearing in Table 4 and their $\tau$-transpositions. Space does not permit extensive analytical examples here, but Figs. 2–3 should give some idea of the possibilities.

Fig. 2 shows the diatonic progression {B, D, F, A} → {C, E, G} inflected by sixteen different spelled heptachords. The diatonic structure is complete (the two chords are complements in dpc space), so each inflected form exhausts the notes of a spelled

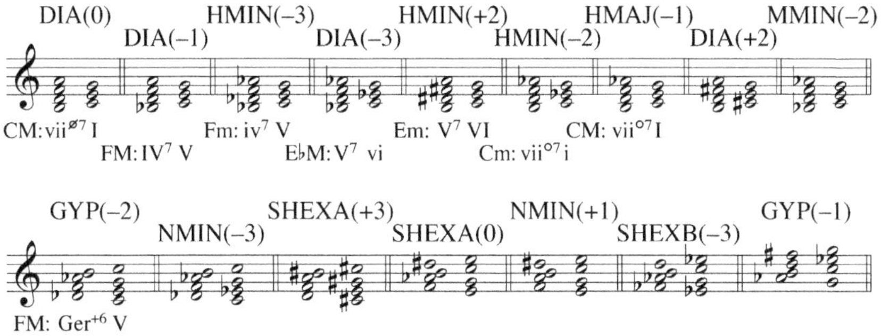


**Fig. 2.** Sixteen spc structures arising through inflection of the dpc structure {B, D, F, A} → {C, E, G} by a proper spelled heptachord

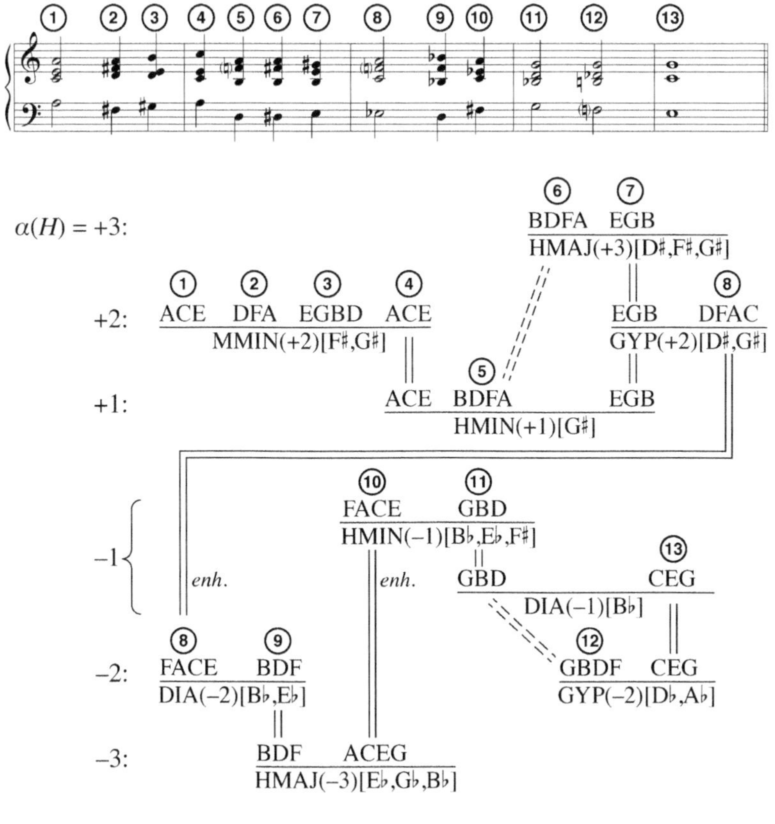


**Fig. 3.** A chromatic chord progression and an analysis using spc structures in spelled-heptachord fields. Chords are identified by circled numbers, keyed to the analysis below. Dpcs for each chord are listed above the horizontal line, and fields below (accidentals for each field given in brackets). Double lines indicate reinterpretations of one chord in two different fields, with *enh.* indicating that enharmonic respelling is involved. Dashed double lines denote inflections in which a chord is altered by a change of field.

heptachord. The first seven progressions are common, as the Roman-numeral analyses confirm, though several different keys are suggested. The remaining progressions also occur in the musical literature, some more frequently than others. Those on the second line of the figure illustrate a variety of resolutions of augmented sixths.

Fig. 3 presents a representative analysis of a chromatic chord progression, similar to some more elaborate analyses in [13] but using the notation of the present paper. All the dpc structures in this example are tertian chords (triads or seventh chords), and the spelled-heptachord fields are limited to a few of the most common types from Table 4, with accidental indices ranging from –3 to +3. Although a transformational system is not developed explicitly here, the analysis effectively describes the progression unfolding through two superimposed layers of transformations: diatonic transformations of the dpc chords (whose dpcs are shown above the horizontal lines in the analysis), overlaid on a second layer of transformations acting independently on the spelled-heptachord fields (shown below the lines).

The excerpt includes several simple successions of two or more chords within a single field, wherein only the diatonic transformations are operative. For example, the opening measures in A minor are partly in MMIN(+2) ("A melodic minor") and partly in HMIN(+1) ("A harmonic minor"). In constructing a musically plausible analysis it is desirable to choose fields that accommodate several consecutive chords whenever possible, thereby accounting for large spans of music with the fewest possible changes of field.

In highly chromatic passages, however, frequent field changes are unavoidable. In Fig. 3, movement from one field to another is often accomplished through pivot-chord–type reinterpretations of a single harmony, denoted by double lines. For example, chord 4 pivots from melodic to harmonic minor to accommodate the F♮ of chord 5, and chord 11 pivots from G minor to F major. Three different fields are shown for chord 7: the harmonic minor field HMIN(+1) relates this chord to the earlier passage in A minor, HMAJ(+3) links it to the tonicizing chord 6 (DIA(+4) is also possible, but there is no need to postulate a C♯ since none occurs in the music), and GYP(+2) is shared with the following augmented sixth (chord 8).

Two of the pivots involve enharmonic respellings (denoted *enh.*). Chord 8 is approached as an augmented sixth in A minor {D♯, F, A, C} and resolved as a dominant seventh in B♭ major {F, A, C, E♭}; chord 10 is a diminished seventh, approached in B♭ major {A, C, E♭, G♭} and resolved in G minor {F♯, A, C, E♭}. Of course, only one of the two spellings appears in the score.

The excerpt also includes two chord inflections (dashed double lines), in which one dpc structure appears successively in two different fields, at least one note of the chord being altered as a result (chord 12 also includes an added seventh). Inflections typically result in a change of chord quality. The two inflections in Fig. 3 are the only places in the excerpt in which two consecutive chords cannot be accommodated by one field. Together with cross relations, inflections account for most of the situations in which a field change is required but no pivot chord is possible.

## References


1. Forte, A.: The Structure of Atonal Music. Yale University Press, New Haven (1973)
2. Brinkman, A.R.: A Binomial Representation of Pitch for Computer Processing of Musical Data. Music Theory Spectrum 8, 44–57 (1986)
3. Agmon, E.: A Mathematical Model of the Diatonic System. Journal of Music Theory 33, 1–25 (1989)
4. Douthett, J., Hook, J.: Formal Diatonic Intervallic Notation. In: Chew, E., Childs, A., Chuan, C.-H. (eds.) MCM 2009. Communications in Computer and Information Science, vol. 38, pp. 104–114. Springer, Heidelberg (2009)
5. Regener, E.: Pitch Notation and Equal Temperament: A Formal Study. University of California Press, Berkeley (1973)
6. Pople, A.: Using Complex Set Theory for Tonal Analysis: An Introduction to the Tonalities Project. Music Analysis 23, 153–194 (2004)
7. Hook, J.: Enharmonic Systems: A Theory of Key Signatures, Enharmonic Equivalence and Diatonicism. Journal of Mathematics and Music 1, 99–120 (2007)
8. Junod, J., Audétat, P., Agon, C., Andreatta, M.: A Generalisation of Diatonicism and the Discrete Fourier Transform as a Mean for Classifying and Characterising Musical Scales. In: Chew, E., Childs, A., Chuan, C.-H. (eds.) MCM 2009. Communications in Computer and Information Science, vol. 38, pp. 166–179. Springer, Heidelberg (2009)
9. Temperley, D.: The Line of Fifths. Music Analysis 19, 289–319 (2000)
10. Hook, J.: Signature Transformations. In: Douthett, J., Hyde, M.M., Smith, C.J. (eds.) Music Theory and Mathematics: Chords, Collections, and Transformations, pp. 137–160. University of Rochester Press, Rochester (2008)
11. Tymoczko, D.: Voice Leadings as Generalized Key Signatures. Music Theory Online 11(4) (2005)
12. Riley, M.: The 'Harmonic Major' Mode in Nineteenth-Century Theory and Practice. Music Analysis 23, 1–26 (2004)
13. Hook, J.: An Integrated Transformational Theory of Diatonic and Chromatic Harmony. Presented at: Society for Music Theory, Los Angeles (2006)
14. Tymoczko, D.: A Geometry of Music: Harmony and Counterpoint in the Extended Common Practice. Oxford University Press, New York (2011)
15. Clough, J., Myerson, G.: Variety and Multiplicity in Diatonic Systems. Journal of Music Theory 29, 249–270 (1985)
16. Audétat, P., Junod, J.: The Diatonic Bell, `http://www.cloche-diatonique.ch/`

## Appendix: The 66 $\tau$-Classes of Proper Spelled Heptachords

$\beta$ is the smallest number of accidentals required to realize a heptachord class (for example, for the class MMIN, $\beta = 1$ because MMIN(+1) has only a single sharp). The table is sorted first by $\beta$, second by span, and third by fifth-string, keeping inversionally related heptachords together. The $\tau$-classes of proper spelled heptachords are in one-to-one correspondence with $T$-classes of seven-note pc sets. Prime forms and Forte numbers [1] are given for the corresponding pc sets; for asymmetrical sets, Forte's $TI$-classes split into two $T$-classes denoted A and B. "Bell no." gives the number assigned to each set in the *diatonic bell* of Audétat [8], [16].

| $\beta$ | Span | Fifth-string | Balanced form | | Prime form | Forte no. | Bell no. | Name |
|---|---|---|---|---|---|---|---|---|
| | | | Spcs | Scale | | | | |
| 0 | 6 | 111111 | −3 −2 −1 0 1 2 3 | B C D E F G A | 013568t | 7–35 | 1 | DIA |
| 1 | 8 | 211112 | −4 −2 −1 0 1 2 4 | F♯ G A B♭ C D E | 013468t | 7–34 | 2 | MMIN |
| 1 | 9 | 121113 | −4 −3 −1 0 1 2 5 | C♯ D E F G A B♭ | 0134689 | 7–32 A | 3+ | HMIN |
| 1 | 9 | 311121 | −5 −2 −1 0 1 3 4 | F♯ G A B C D E♭ | 0135689 | 7–32 B | 3− | HMAJ |
| 1 | 10 | 112114 | −4 −3 −2 0 1 2 6 | G♯ A B♭ C D E F | 0124689 | 7–30 A | 6+ | NMIN |
| 1 | 10 | 411211 | −6 −2 −1 0 2 3 4 | B C D E F♯ G A♭ | 0135789 | 7–30 B | 6− | NMAJ |
| 1 | 11 | 111215 | −4 −3 −2 −1 1 2 7 | D♯ E F G A B♭ C | 0124679 | 7–29 A | 10+ | |
| 1 | 11 | 512111 | −7 −2 −1 1 2 3 4 | E F♯ G A B C D♭ | 0235789 | 7–29 B | 10− | |
| 2 | 10 | 131131 | −5 −4 −1 0 1 4 5 | C♯ D E♭ F♯ G A B♭ | 0125689 | 7–22 | 5 | GYP |
| 2 | 10 | 221122 | −5 −3 −1 0 1 3 5 | C♯ D E♭ F G A B | 012468t | 7–33 | 4 | SWT |
| 2 | 11 | 113141 | −5 −4 −3 0 1 5 6 | G♯ A B♭ C♯ D E♭ F | 0125679 | 7–20 A | 9+ | |
| 2 | 11 | 141311 | −6 −5 −1 0 3 4 5 | B C♯ D E♭ F♯ G A♭ | 0234789 | 7–20 B | 9− | |
| 2 | 11 | 122132 | −5 −4 −2 0 1 4 6 | F♯ G♯ A B♭ C D E♭ | 0234689 | 7–28 B | 8+ | |
| 2 | 11 | 231221 | −6 −4 −1 0 2 4 5 | C♯ D E F♯ G A♭ B♭ | 0135679 | 7–28 A | 8− | |
| 2 | 11 | 212123 | −5 −3 −2 0 1 3 6 | G♯ A B C D E♭ F | 0134679 | 7–31 A | 7+ | SOCTA |
| 2 | 11 | 321212 | −6 −3 −1 0 2 3 5 | B C♯ D E F G A♭ | 0235689 | 7–31 B | 7− | SOCTB |
| 2 | 13 | 211135 | −5 −3 −2 −1 0 3 8 | A♯ B C D E♭ F G | 0124579 | 7–27 A | 14+ | |
| 2 | 13 | 531112 | −8 −3 0 1 2 3 5 | A B C♯ D E F G♭ | 0245789 | 7–27 B | 14− | |
| 2 | 13 | 311314 | −6 −3 −2 −1 2 3 7 | D♯ E F G A♭ B C | 0124589 | 7–21 A | 11+ | SHEXA |
| 2 | 13 | 413113 | −7 −3 −2 1 2 3 6 | E F G♯ A B C D♭ | 0134589 | 7–21 B | 11− | SHEXB |
| 2 | 15 | 411315 | −7 −3 −2 −1 2 3 8 | A♯ B C D♭ E F G | 0123679 | 7–19 A | 15+ | |
| 2 | 15 | 513114 | −8 −3 −2 1 2 3 7 | A B C D♯ E F G♭ | 0123689 | 7–19 B | 15− | |
| 2 | 16 | 512215 | −8 −3 −2 0 2 3 8 | A♯ B C D E F G♭ | 0124678 | 7–15 | 18 | |
| 3 | 13 | 212332 | −6 −4 −3 −1 2 5 7 | C♯ D♯ E F G A♭ B♭ | 0234679 | 7–25 A | 13+ | |
| 3 | 13 | 233212 | −7 −5 −2 1 3 4 6 | F♯ G♯ A B C D♭ E♭ | 0235679 | 7–25 B | 13− | |
| 3 | 13 | 221323 | −6 −4 −2 −1 2 4 7 | D♯ E F♯ G A♭ B♭ C | 0134579 | 7–26 A | 12+ | |
| 3 | 13 | 323122 | −7 −4 −2 1 2 4 6 | E F♯ G♯ A B♭ C D♭ | 0245689 | 7–26 B | 12− | |
| 3 | 15 | 221145 | −6 −4 −2 −1 0 4 9 | E♯ F♯ G A♭ B♭ C D | 0123579 | 7–24 A | 20+ | |
| 3 | 15 | 541122 | −9 −4 0 1 2 4 6 | D E F♯ G♯ A B♭ C♭ | 0246789 | 7–24 B | 20− | |
| 3 | 15 | 231414 | −7 −5 −2 −1 3 4 8 | F♯ G A♯ B C D♭ E♭ | 0145679 | 7–Z18A | 16+ | |
| 3 | 15 | 414132 | −8 −4 −3 1 2 5 7 | C♯ D♯ E F G♭ A B♭ | 0234589 | 7–Z18B | 16− | |
| 3 | 16 | 122155 | −6 −5 −3 −1 0 5 10 | B♯ C♯ D E♭ F G A♭ | 0123578 | 7–14 A | 25+ | |
| 3 | 16 | 551221 | −10 −5 0 1 3 5 6 | G♯ A B C♯ D E♭ F♭ | 0135678 | 7–14 B | 25− | |

(table continues on next page)

**Appendix** (continued)

| β | Span | Fifth-string | Balanced form: Spcs | Balanced form: Scale | Prime form | Forte no. | Bell no. | Name |
|---|---|---|---|---|---|---|---|---|
| 3 | 16 | 141415 | –7 –6 –2 –1 3 4 9 | E♯ F♯ G A♭ B C D♭ | 0123678 | 7–7 A | 22+ | |
| 3 | 16 | 514141 | –9 –4 –3 1 2 6 7 | D♯ E F G♯ A B♭ C♭ | 0125678 | 7–7 B | 22– | |
| 3 | 16 | 321235 | –7 –4 –2 –1 1 4 9 | E♯ F♯ G A B♭ C D♭ | 0124578 | 7–Z38 A | 21+ | |
| 3 | 16 | 532123 | –9 –4 –1 1 2 4 7 | D♯ E F♯ G A B♭ C♭ | 0134678 | 7–Z38 B | 21– | |
| 3 | 18 | 531135 | –9 –4 –1 0 1 4 9 | E♯ F♯ G A B♭ C♭ D | 0124569 | 7–Z17 | 23 | |
| 4 | 15 | 221532 | –7 –5 –3 –2 3 6 8 | G♯ A♯ B C D♭ E♭ F | 0234579 | 7–23 A | 17+ | |
| 4 | 15 | 235122 | –8 –6 –3 2 3 5 7 | B C♯ D♯ E F G♭ A♭ | 0245679 | 7–23 B | 17– | |
| 4 | 16 | 323323 | –8 –5 –3 0 3 5 8 | A♯ B C♯ D E♭ F G♭ | 0134578 | 7–Z37 | 19 | |
| 4 | 18 | 212355 | –7 –5 –4 –2 1 6 11 | F𝄪 G♯ A B♭ C D♭ E♭ | 0123568 | 7–Z36 A | 30+ | |
| 4 | 18 | 553212 | –11 –6 –1 2 4 5 7 | C♯ D♯ E F♯ G A♭ B♭♭ | 0235678 | 7–Z36 B | 30– | |
| 4 | 18 | 233235 | –8 –6 –3 0 2 5 10 | B♯ C♯ D E F G♭ A♭ | 0124568 | 7–13 A | 27+ | |
| 4 | 18 | 532332 | –10 –5 –2 0 3 6 8 | G♯ A♯ B C D E♭ F♭ | 0234678 | 7–13 B | 27– | |
| 4 | 18 | 323145 | –8 –5 –3 0 1 5 10 | B♯ C♯ D E♭ F G♭ A | 0123569 | 7–16 A | 26+ | |
| 4 | 18 | 541323 | –10 –5 –1 0 3 5 8 | G A♯ B C♯ D E♭ F♭ | 0134569 | 7–16 B | 26– | |
| 4 | 20 | 414155 | –9 –5 –4 0 1 6 11 | F𝄪 G♯ A B♭ C♭ D E♭ | 0123478 | 7–6 A | 31+ | |
| 4 | 20 | 551414 | –11 –6 –1 0 4 5 9 | C♯ D E♯ F♯ G A♭ B♭♭ | 0145678 | 7–6 B | 31– | |
| 4 | 20 | 541145 | –10 –5 –1 0 1 5 10 | B♯ C♯ D E♭ F♭ G A | 0123479 | 7–Z12 | 28 | |
| 5 | 18 | 235323 | –9 –7 –4 1 4 6 9 | E♯ F♯ G♯ A B♭ C♭ D♭ | 0134568 | 7–11 A | 24– | |
| 5 | 18 | 323532 | –9 –6 –4 –1 4 7 9 | D♯ E♯ F♯ G A♭ B♭ C♭ | 0234578 | 7–11 B | 24+ | |
| 5 | 20 | 221555 | –8 –6 –4 –3 2 7 12 | C𝄪 D♯ E F G♭ A♭ B♭ | 0123468 | 7–9 A | 33+ | |
| 5 | 20 | 555122 | –12 –7 –2 3 4 6 8 | F♯ G♯ A♯ B C D♭ E♭♭ | 0245678 | 7–9 B | 33– | |
| 5 | 20 | 235145 | –9 –7 –4 1 2 6 11 | F𝄪 G♯ A B♭ C♭ D♭ E | 0123469 | 7–10 A | 32+ | |
| 5 | 20 | 541532 | –11 –6 –2 –1 4 7 9 | C D♯ E♯ F♯ G A♭ B♭♭ | 0234569 | 7–10 B | 32– | |
| 5 | 23 | 532355 | –11 –6 –3 –1 2 7 12 | C𝄪 D♯ E F G A♭ B♭♭ | 0123567 | 7–5 A | 34+ | |
| 5 | 23 | 553235 | –12 –7 –2 1 3 6 11 | F𝄪 G♯ A B C D♭ E♭♭ | 0124567 | 7–5 B | 34– | |
| 6 | 20 | 235532 | –10 –8 –5 0 5 8 10 | A♯ B♯ C♯ D E♭ F♭ G♭ | 0234568 | 7–8 | 29 | |
| 6 | 23 | 323555 | –10 –7 –5 –2 3 8 13 | G𝄪 A♯ B C D♭ E♭ F♭ | 0123467 | 7–4 A | 35+ | |
| 6 | 23 | 555323 | –13 –8 –3 2 5 7 10 | B♯ C♯ D♯ E F G♭ A♭♭ | 0134567 | 7–4 B | 35– | |
| 6 | 25 | 541555 | –12 –7 –3 –2 3 8 13 | G𝄪 A♯ B C D♭ E♭♭ F | 0123458 | 7–3 A | 36+ | |
| 6 | 25 | 555145 | –13 –8 –3 2 3 7 12 | B C𝄪 D♯ E F G♭ A♭♭ | 0345678 | 7–3 B | 36– | |
| 7 | 25 | 235555 | –11 –9 –6 –1 4 9 14 | D𝄪 E♯ F♯ G A♭ B♭♭ C♭ | 0123457 | 7–2 A | 37+ | |
| 7 | 25 | 555532 | –14 –9 –4 1 6 9 11 | E♯ F𝄪 G♯ A B♭ C♭ D♭♭ | 0234567 | 7–2 B | 37– | |
| 8 | 30 | 555555 | –15 –10 –5 0 5 10 15 | A𝄪 B♯ C♯ D E♭ F♭ G♭♭ | 0123456 | 7–1 | 38 | CHROM |

# Fundamental Passacaglia: Harmonic Functions and the Modes of the Musical Tetractys

Karst de Jong[1] and Thomas Noll[2]

[1] Royal Conservatoire Den Haag,
Escola Superior de Música de Catalunya, Barcelona
karstdj@xs4all.nl
[2] Escola Superior de Música de Catalunya, Barcelona
noll@cs.tu-berlin.de

**Abstract.** In this paper we take the three tonal functions *tonic*, *subdominant* and *dominant* out of their usual theoretical domicile—the combinatorics of fifth-related triads enriched by a dialectical interpretation—and redeploy them within an alternative theoretical framework: the combinatorics of the modes of the musical *tetractys*, enriched by musical-theoretical interpretations of selected mathematical facts. Section 1 introduces tonal perspectives of the analysis of the fundamental bass. Section 2 provides a short overview of the combinatorics of the three modes of the musical tetractys. The concluding Section 3 binds the two strands of investigation together.

**Keywords:** Fundamental Bass, Functional Modes, Harmonic Functions, Well-Formed Scales, Well-Formed Modes, Christoffel Duality, Musical Tetractys.

With the present investigation we act on a suggestion, which has been made more than two decades ago. A side remark in Norman Carey and David Clampitt's article on well-formed scales [1] draws a link between what they name the *structural scale* and Riemann's tonic-subdominant-dominant relationship. Basically, this link interprets the roots of the tonic, subdominant and dominant triads in terms of a well-formed fifth-generated 3-tone-scale. One of the three modes of this scale is also known under the name *musical tetractys*. The tones $C - F - G - (C')$ exempify this mode in ascending order. The tetractys plays an established theoretical role in the derivation of the diatonic modes and is associated with early music rather than common practice harmonic tonality. The arguments of the present article will contribute to an augmentation of its theoretical role for the study of musical harmony.

## 1 Tonal Perspectives of the Harmonic Underworld

This section is dedicated to the introduction of *functional modes* to the analysis of fundamental bass progressions. The approach is not a theory of analysis, though. The competence to identify chord progressions and to extract a fundamental line is

C. Agon et al. (Eds.): MCM 2011, LNAI 6726, pp. 98–114, 2011.

a presupposition to the approach, not an outcome. The intended outcome is an enriched interpretation of this fundamental line through functional modes. We regard these modes as a manifestation of tonality within the fundamental bass.

## 1.1 Fifths and Minor Thirds Revisited

In our first joint paper [2] we explore the descriptive and explanatory power of a two-dimensional grid—spanned by perfect fifths and minor thirds—for the analysis of fundamental bass progressions. The idea of regarding these two intervals as basic generators of all possible progressions involves the assumption that other intervals, such as the major second or the major third, are compounds.

To begin with, the free P5/m3 grid provides a useful framework for the exploration of unlimited pathways of the fundamental bass. The fact that common fundamental bass progressions seem to have relatively simple pathways in the grid is particularly intriguing. Many passages from classical pieces and jazz standards can be simply described by progressions of descending minor thirds and falling fifths. A suggestive example is the opening bars of Victor Young's *My Foolish Heart* (see Fig. 1).

In [2] we make the terminological distinction between *fifth progressions* on the vertical axis and *minor third substitutions* on the horizontal axis, and we motivate this choice by their search for a precise equivalent to the constitution of (ordinary) melodies. In the melodic "upper world" diatonic steps and alterations (♯, ♭) serve as the basic intervals, while leaps are regarded to be compounds. In the "underworld" of the fundamental bass these roles are taken by the perfect fifth, and the minor third, respectively. However,we left it an open task to further elaborate upon this analogy.

Another issue in the article ([2]) which needs further elaboration, is a convincing treatment of tonality on the background of the P5/m3 grid. The fundament progressions are studied in this unlimited free grid while no principles are provided to regulate the functions of fundamental basses within the framework of tonality. We show in the present paper that both theoretical gaps can be filled at once, namely a clarification of the different roles of fifth and minor third as basic constituents of fundamental bass progressions and the establishment of a layer of tonal function modes on top of the P5/m3 grid.

**Fig. 1.** Opening bars of Victor Young's *My Foolish Heart*. The arrow ↓ stands for descending fifths or ascending fourths in the fundament. The arrow ← stands for descending minor thirds in the fundament. Thus the entire progression is represented by the arrow chain ←↓←↓↓↓↓↓.

## 1.2 Identifying the Fundamental Bass

In a typical harmonic progression the real bass line can be distinguished from the fundamental bass. Especially in the idoms of the western Classical and Romantic styles, the "melodization" of the bass line is a common feature. Inversions replace the root-positions, resulting in a more melodic bass line. Several theorists, such as Jean Phillipe Rameau, Simon Sechter, Arnold Schönberg and Paul Hindemith have paid considerable attention to the role of the fundamental bass in relation to the real bass in analysis and composition. Their efforts reinforce our confidence that this level of musical description deserves further investigation.

Our aim is to look at the fundamental bass as if it were a dual melody, in order to reveal its guiding role within tonality and tonal structures. The attribute "dual" refers to a mathematical concept of duality in the application to modes, where step interval patterns have fifth/fourth-foldings as their dual counterparts. In this respect we regard the prominence of linear steps in ordinary melodies, and the prominence of fifths and fourths in fundament progressions as manifestations of this duality. More details will be investigated in Sect. 2.

Prior to the theoretical challenge to decipher these melodies and place them in a larger framework of *tetractys modes*, there is the obvious practical question which needs to be addressed when analyzing the music: how can we exactly identify the fundamental bass in actual musical textures? There are situations where this task is relatively straightforward, such as in the homophonic C-major Prelude of J. S. Bach's WTK I. However, in most musical textures there is considerable room for interpretation. Generally, to identify the fundament, a basic principle is to follow the rule of the stacked thirds, according to which any chord can be rearranged into a chain of thirds. However, there are exceptions. It is beyond the scope of this article to investigate these exceptions in depth, but nevertheless we list here some guidelines to which we adhere in our own analyses:

1. Root of a Stack of Thirds: The fundamental bass is identified as the root. The prototypical examples are the root positions and inversions of major and minor triads, and the four types of diatonic seventh chords. In some cases, such as the diminished triad and the diminished seventh chord, we identify the fundament as a non-sounding tone, a major third below the root.
2. Real Bass: In the cases of the Cadential 6/4-chord, the Neapolitan sixth chord, and chords with a added sixth, the fundamental bass is identified as the real bass.
3. Augmented Sixth Chords: For all types of augmented sixth chords, such as Italian, French, German, the fundamental bass is identified as a tone, a diminished fifth below the real bass. To put it differently, in reference to the stack of thirds, for the French chord the fundament coincides with the root. For the Italian and German chords we follow the same reasoning as for the diminished chords, whereby the non-sounding fundament is a major third below the root.
4. Ambiguity: Sometimes the vertical dimension alone does not provide enough information to determine the fundamental bass properly. In such a case the

context puts the default choice into question. Such is the case for the half-diminished chord, which resolves with its root as a leading tone. Another example is the so called common-tone diminished seventh chord on $\sharp II$ in Major.

5. Upper Structures: Many late romantic and jazz chords need to be interpreted with care in their context (Tristan chord, Eulenspiegel chord, half-diminished chord as an upper structure of an altered dominant, tritone related dominants etc.)

### 1.3 Functional Modes: The $T$ – $S$ – $D$ Constellation Revisited

The compound concept *harmonic function mode* (or short: *functional mode*) is the central new concept proposed in this paper. It stands on two pillars: (1) the analysis of harmonic functions and (2) the melody formed by the progression of the fundamental bass. The proposed concept reinterprets the three tonal functions tonic ($T$), subdominant ($S$) and dominant ($D$) as scale degrees of three possible modes of an underlying three-note scale.

The generic step order of the abstract three-note scale is the pattern $T$ – $S$ – $D$ – $(T')$. The three modes with the common tonic C are thus: $C - F - G - (C')$, $C - D - G - (C')$ and $C - F - B\flat - (C')$ (see Fig. 2). The step intervals between $T$ and $S$, $S$ and $D$, $D$ and $T$ vary from one mode to the other.

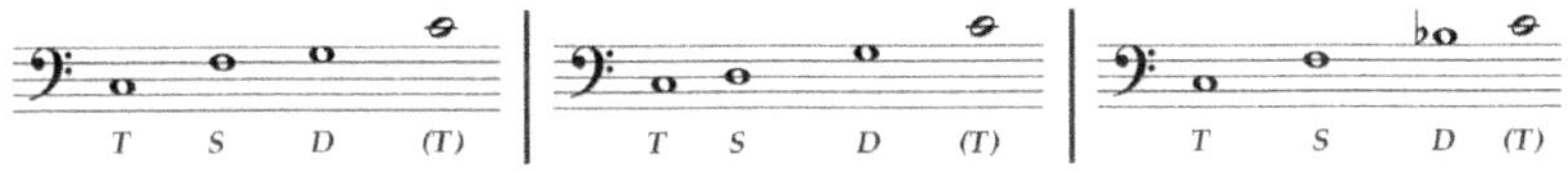


**Fig. 2.** The three modes of the Tetractys

This is a considerable expansion of the traditional concept of harmonic functions with fixed fundaments within any given key. Furthermore this reinterpretation is quite different from Moritz Haupmann's ([3]) and Hugo Riemann's ([4]) original ideas about the internal relations among the three functions, where the two dominants $S$ and $D$ are always in a perfect fifth/fourth relationship to the tonic.

We believe that we find manifestations of these modes in all tonal music. On a local level, many musical passages can be identified to be in either one of these functional modes. Figure 3 provides short examples for each of the three functional modes:

The first mode is illustrated by the first theme of Haydn's Sonata Hob. XVI No. 33. $D_1$ denotes the first mode of $D$. The complete fundament progression $D$ – $(G$ – $D$ – $A$ – $D)$ – $G$ – $A$ – $D$ can be associated with the standard cadential progression (of triads) $D$ – $G$ – $A$ – $D$ of Riemannian function theory.[1]

[1] The secondary dominant seventh chord of the subdominant $G$ strengthens the presence of the subdominant within this progression.

The second mode is illustrated by Handel's Aria *Lascia ch'io pianga.* The second mode of $F$ is denoted by the symbol $F_2$. In this example the fundament $G$ represents the subdominant function $S$. Note that it is not in a perfect fifth relationship to the tonal center $F$. Nevertheless it is regarded to be the "true" subdominant of that mode, and there is no need to interpret the $G$ as a substitute of $B\flat$. The opening of the song *An Englishman in New York* by Sting illustrates the third mode: The symbol $B_3$ denotes the third mode of $B$. In this example the fundament $A$ represents the dominant function $D$. It has a major second relationship to the tonal center $B$ and it is regarded to be the "true" dominant of that mode. There is no need to interpret the $A$ as a substitute of $F\sharp$.

**Fig. 3.** Examples illustrating the Modes 1, 2 and 3. Mode 1: Haydn, Sonata Hob. XVI No. 33, Mode 2: Handel's Aria *Lascia ch'io pianga*, Mode 3: *An Englishman in New York* by Sting.

The three modes are not of the same importance in common practice tonal music. First and second modes can be found throughout the repertoire of baroque, classical and romantic pieces. The same holds for jazz harmony, even though second modes appear here more often than first ones. The third modes however, are the odd ones out. We find them in pop music, as the Sting-example suggests.

We propose to draw a connection between differences among the three modes with regard to their presence in the music—on the one hand—and differences among them with regard to their mathematical structure, on the other. The latter shall be dealt with in Sect. 2.2.

From an analytical point of view, it is not always possible to unambiguously decide wether or not a given chord progression has to be identified as a first or as a second mode. The following example from the 2nd movement of Haydn's symphony No. 94 (see Fig. 4) contains the celebrated ambiguity between a root position IV-chord with added sixth and a $II^7$-chord in first inversion ($II^{6}_{5}$).

**Fig. 4.** Mode 1, Haydn, Symphony no.94/ii

Traditionally, the ambiguity is either attributed to the music itself (Rameau) or it remains in a frozen aggregate state between the conceptions of coexisting theoretical traditions (Riemann functions, Roman numerals). Rameau's argument about a double employment of the tones of both chords uses this ambiguity for the purpose of fundamental bass analyses. Ramists would therefore acknowledge a substitution of the fundament F by the fundament D. In the tradition of Riemann one would subsume the $IV$-chord with (or without) the added sixth as well as the $II$- and $II^7$-chord (with their inversions) into the category of the subdominant function and one would attribute a fundament $F$ to the first half of the second bar. In the traditions of scale degree theory and Roman numeral analysis one would adhere to the third-chain-structure of the $II^7$-chord and would disregard the option of a $IV$ with added sixth. Hence one would attribute a fundament $D$ to the first half of the second bar.

The functional mode approach as such does not dissolve this ambiguity. It rather amplifies it by offering three alternative analyses (see Fig. 4): The first and the second analyses both attribute a first mode $C_1$ to the music. The unusual notation $\sharp S$ in the first ("Ramist" double employment) analysis $C_1 : T\,S\,\sharp S\,D\,T$ will be explained in Sect. 3.2. Our preferred analysis in this case is the second ("Riemannian") option, namely $C_1 : T\,S\,D\,T$.

## 2 Combinatorics of the Tetractys Modes

This section familiarizes the reader with the tetractys[2] modes as instances of *well-formed modes*. It illustrates the approach taken in [5] in a situation, which is simpler—but perhaps less familiar—than the diatonic modes. Thus it is useful to refer to the diatonic case for comparison. Figure 5 opposes the structure of the first tetractys mode (left) from Fig. 2 and the structure of the authentic Ionian mode (right). Both subfigures show in the lower staff an ascending pattern of large and small steps and in the upper staff a folding pattern in ascending fifths and descending fourths. In both modes the ascending step intervals are

[2] Traditionally, the chain of integer ratios $6 : 8 : 9 : 12$, has been called the *musical tetractys*. In its ancient interpretation in terms of string length ratios the number 6 represents the highest tone. The outer ratio $6 : 12 = 1 : 2$ corresponds to a descending octave. In its modern interpretation in terms of frequency ratios the direction (of encoding the tetractys) is inverted accordingly.

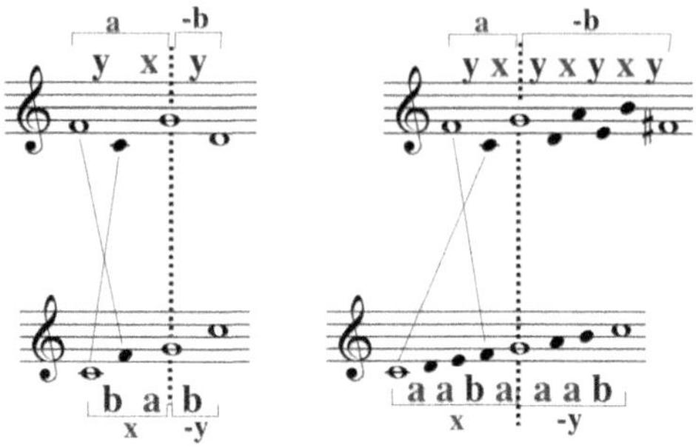


**Fig. 5.** Analogy between the tetractys and the authentic Ionian mode. The thick dotted lines as well as the crossing thin lines refer to the authentic division of the step interval patterns (see Sect. 2.2).

designated by the letters *a* and *b*, while ascending fifth and descending fourth in the folding patterns are designated by the letters *x* and *y* respectively.

In the case of the Ionian mode, the two-letter word *aabaaab* of length seven is the plain adjoint of the two-letter word *yxyxyxy* of that same length (see [6] for details). The mutual adjointness of these words is a modal refinement of the well-formedness property of the diatonic scale. The scale folding with the pattern *yxyxyxy* fills the interval of the augmented prime $F - F\sharp$ while the ascending scale with the step pattern *aaabaab* fills the interval of the octave $C - C'$. Thus, these two intervals themselves are dually related to each other in the context of the adjointness. The ascending augmented prime is the difference between the ascending major and minor step intervals. Dually, the octave is the sum of an ascending fifth and an ascending fourth, or—in order to highlight the duality we can say—it is the difference between an ascending fifth and a descending fourth.

In the case of the first tetractys mode, the two-letter word *bab* of length three is the plain adjoint of the two-letter word *yxy*. The mutual adjointness of these words is a modal refinement of the well-formedness property of the *tetractys scale* (see also Fig. 6). The scale folding with the pattern *yxy* fills the interval of the minor third $F - D$ while the ascending scale with the step pattern *bab* fills the interval of the octave $C - C'$. Thus, these two intervals themselves are dually related to each other. The descending minor third is the difference between the ascending major second (being the small tetractys step interval) and the ascending perfect fourth (being the large tetractys step interval). Dually—as above—the octave is the difference between an ascending fifth and a descending fourth. The minor third is the *alteration interval* in the world of the tetractys modes, and might therefore be called *augmented tetractys prime*. This analogy is the driving idea for the present paper.

Aside from the analogy between the tetractys and the diatonic there is a "genealogic" connection between them in a mathematical sense. This relates to the historic derivation of the diatonic modes as tetrachord fillings of the P4-steps in the tetractys within a hierarchy of modes. The following subsection recapitulates this connection on the level of scales.

## 2.1 The Structural Scale

The octave is regarded as the ambitus interval of the tetractys modes. At the same time it is the interval of periodicity for the underlying three tone scale. The theory of well-formed scales ([1]) is only concerned with the aspect of periodicity and studies generated scales as arithmetic sequences $(0, g, 2g, 3g, ...)\, mod\, 1$ on the chroma circle, where the pitch height 1 of the octave is identified with the pitch height 0 of the prime. We review some facts about the structural scale $(0, g, 2g)\, mod\, 1$, with $g = log_2(3/2)$ representing the pitch height of the fifth. The conversion from generation order $(t_0, t_1, t_2) = (0, g, 2g - 1)$ to scalar order $(t_0, t_2, t_1) = (0, 2g - 1, g)$ is a multiplication by 2 *mod* 3 on the indices 0, 1, 2 of $t_0, t_1, t_2$. As a fifth-generated well-formed scale the structural scale (= tetractys scale) is situated within a (virtually infinite) hierarchy of such scales. Figure 6 shows the first five instances of this family. Its root is the two note *division scale*, whose step intervals are the perfect fifth P5 and the perfect fourth P4.

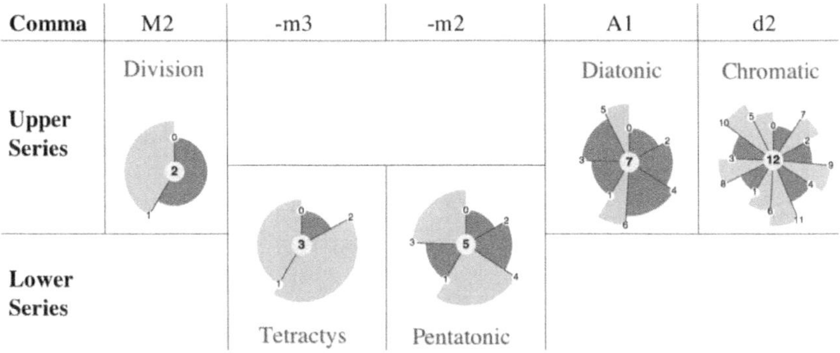


**Fig. 6.** The table lists five instances of fifth-generated well-formed scales. The pie charts represent the associated cyclic scale-step patterns. The bold numbers n = 2, 3, 5, 7, 12 in the center of these diagrams denote the cardinalities of the scales. The tiny numbers along the circumference denote the tones of the scale in generation order. Primary steps are displayed in dark and secondary steps in light gray. The difference between the two is the comma. Depending of the sign of the comma we distinguish between the upper series (positive comma) and the lower series (negative comma).

The fifth P5 is regarded the primary and the fourth P4 the secondary step interval of the division scale. This confirms the markedness of the plagal division with respect to the (unmarked) authentic division. These two exemplify the modes of the division scale. The principle of a markedness of the secondary scale step intervals, with respect to the primary one is inherited throughout the hierarchy, as explained in the next paragraph. The pie charts for each of the five scales in Fig. 6 indicate this in terms of the two different gray levels for the coloring of scale steps (dark gray = primary, light gray = secondary).

The difference between the primary and the secondary step interval is the associated augmented prime (also called the *comma*). Whenever the comma is ascending it becomes the primary step interval at the next level of the hierarchy, while the previous secondary step interval remains in this role. Whenever the comma is descending its ascending counterpart becomes the secondary step interval at the next level of the hierarchy, while the previous primary step interval remains in its role. In the case of the structural scale, the major second M2 is primary (being the ascending comma of the division scale) and the fourth P4 is secondary. The comma is a descending minor third, as already mentioned. It eventually becomes the secondary step interval of the pentatonic scale.

Modes of a scale can either be obtained by simply cutting the pie charts at some tone which then becomes the finalis. In this manner one gets the *common origin family* of modes (see Fig. 7), which differs from the *common finalis family* (as shown in Fig. 2). The first, second and third modes result from cutting at points 1, 0 and 2, respectively.

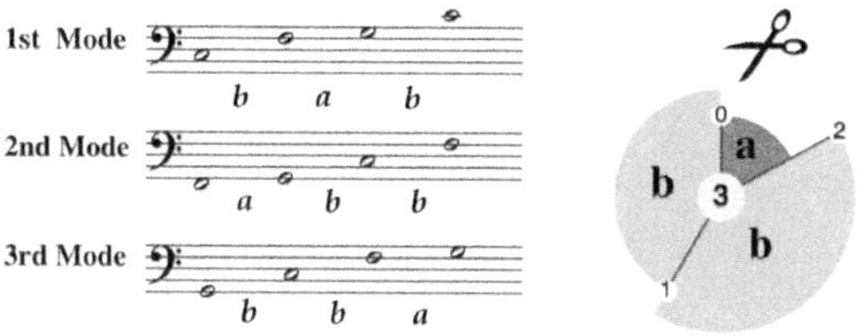


**Fig. 7.** Each tone of the structural scale serves as the finalis of an ascending mode with the step-interval patterns *bab*, *abb* and *bba*, respectively

This method, however, of cutting the pie charts and thereby rotating the scale step pattern from one mode to the next does not qualitatively distinguish between the modal varieties. It can not explain, why the third tetractys mode or the locrian diatonic mode are the odd ones out among their conjugates. Furthermore, the method does not grasp historical aspects of the emergence of modes.

## 2.2 The Tesserachordon – Authentic and Plagal Tetractys Modes

The listing of the tetractys modes in the order 1st, 2nd, and 3rd is in analogy to the familiar listing of seven diatonic modes: Ionian, Dorian, Phygian, Lydian, Mixolydian, Aeolian, and Locrian. The medieval octenary system contains only Dorian, Phrygian, Lydian and Mixolydian, each one in authentic and plagal division. Heinrich Glarean ([7]) extends this list to his famous Dodekachordon by including the authentic and plagal Ionian and Aeolian modes. The Locrian modes are explicitly rejected by Glarean. The historically motivated classification of modes into three classes can be convincingly generalized to the modal varieties of any $n$-tone well-formed scale (see [5] for details). There are $n-3$ inner modes (i.e. generalized *Guidonian modes*), all of which come in authentic and plagal division and which—in both cases—have the Guidonian affinities. In addition there are two *edge modes* (i.e. generalized *new Glarean modes*), each of which comes in authentic and plagal divisions and which have double neighbor polarity (different double-neighbor figures at finalis and confinalis). Finally, in each such modal family there is the *bad conjugate* whose divider is altered by the associated augmented prime of that scale. It is also *amorphous* in the sense, that it cannot be generated from the authentic division $(a|b)$ through the application of Sturmian morphisms (see [5]).

For the present investigation it is interesting to inspect these three classes in the case of the family of the tetractys modes. The third tetractys mode is amorphous in analogy to the Locrian mode. Closer inspection also shows that there are no (generalized) Guidonian modes. The tetractys analogue to the octenary system is utterly empty, i.e. there is no instance of identical double-neighbor figures at finalis and confinalis. Instead we find double-neighbor polarity, which is characteristic for Glarean's new modes. In other words the analogue of the Dodekachordon is a Tesserachordon. Figure 8 displays the proper authentic and plagal tetractys modes.

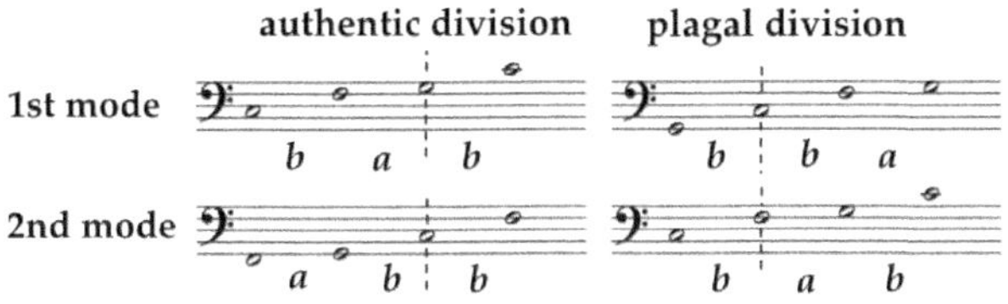


**Fig. 8.** The "Tesserachordon" as a tetractys analogue to Glarean's Dodekachordon

The authentic and plagal divisions of the 1st tetractys mode are analogous to the Ionian and Hypo-Ionian modes and the authentic and plagal divisions of the 2nd tetractys mode are analogous to the Aeolian and Hypo-Aeolian modes (see [5] for details).

## 2.3 Modulation between Tetractys Modes

In order to study of tonal modulations and chromatic deflections we may benefit from an analogy to the diatonic modes. Jay Hook's approach to *signature transforms* (see [8]) and other work can be suitably adopted to the situation of the tetractys modes. In the present paper we mention the two elementary types of parsimonious modal transformations. Figure 9 displays three tetractys modes with a common finalis $C$. The ascending step interval patterns of these modes are shown in the top systems, while the associated fifth/fourth foldings are shown at the bottom of each example. The foldings contain the active chromatic tone of each mode—the minor third alteration of its origin.

One type of parsimonious modulation is based on a transformation where the chromatic tone of the source mode is turned into an essential tone of the target mode, while the origin of the source mode loses its essential status. In a modulation from the first mode $C_1$ to the second mode $C_2$ the chromatic tone

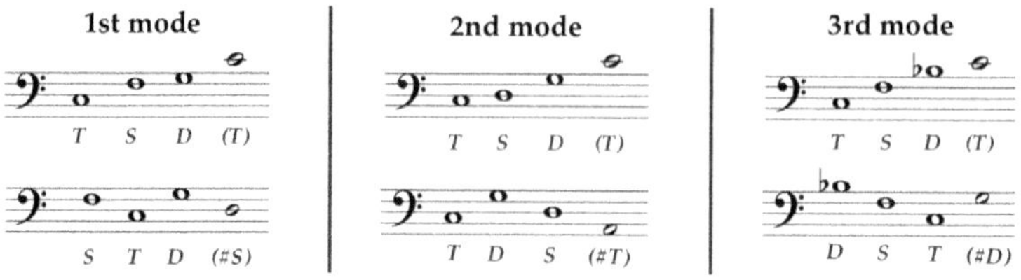


**Fig. 9.** Three tetractys modes with a common finalis

$D$ with function $\sharp S$ becomes an essential tone with the function $S$. The tone $A$ with function $\sharp T$ now becomes the new active chromatic tone of the target mode $C_2$. The tone $C$ remains in the role of the finalis. The same way, the first mode $C_1$ can be reached from the third mode $C_3$, namely by making the chromatic tone $G$ with function $\sharp D$ an essential tone with the function $D$. A second type of parsimonious modulation is dual to the first type. Hereby the origin of the mode is fixed and the finalis is shifted (see Fig. 7). It is based on transformations, where the finalis of the source mode with function $T$ is replaced by its octave, which becomes an essential tone of the target mode with function $D$. For example the first mode $C_1$ with finalis $C$ becomes a second mode $F_2$ with finalis $F$. [3]

In the analytical practice one is not forced, however, to assume a modulation for every non-tetractys fundament. In Sect. 3.2 we argue in favor of a distinction between essential and accidental alteration.

# 3 Exploring the World of Functional Modes

In this section we establish some ties between the combinatorial facts from the previous section and the study of the fundamental modes. The basic consideration involves two thoughts. Firstly we think that the fundamental bass benefits from a modal treatment. Therefore it is reasonable to connect the three tonal functions with the tetractys modes. Secondly, we think that the duality between scale step patterns and foldings is the crux in the relation between real melody and fundamental bass progression. While melody manifests itself in step proximity, fundamental bass progression manifests itself in the proximity along folding patterns.

## 3.1 Lines, Foldings and Duality

A few words should be said about the mode as a syntactic unit and about the typical order of fundaments in harmonic progressions. Thereby it is instructive

[3] As tentative terms we propose *function shift* for the second modulation type (with fixed origin), and *character shift* for the first type (with fixed finalis). Hereby the term *character* would be in accordance with Jacques Handschin's concept of *tone character* (see [5] for a discussion of Handschin's approach in the context of well-formed mode theory).

to take a music-theoretical interpretation of the duality mentioned above into consideration. In mathematics, two objects which stand in a dual relation to each other are related by the interchange of particular pairs of terms. If the objects are the scale step-interval pattern and the scale 5th/4th folding, the immediacy of the neighborhood relation of the steps in the scale is dual to the immediate kinship of perfect fifths in the folding. Since we strive to interpret the fundament progressions as a melody, the way in which Schenkerians pay attention to the immediacy of the step patterns within a line (Urlinie) is inspiring. By transfering the Schenkerian concept of immediacy, or contiguity, from the Urlinie to a space where tones are folded in fifths and fourths, we hope to find an effective way of describing and analyzing fundamental bass patterns. Looking for more sources, there are traces of a contiguity concept in the theories of Moritz Hauptmann and Hugo Riemann, who give a dialectical explanation of the order in which the tonal functions should appear: $T - S - D - T$. Compared to their musically questionable explanation, the duality approach seems now to offer a promising alternative explanation for the logical order in which the tonal functions should follow each other: Schenkerians regard tones on the Urlinie as indispensable stations on a trajectory. Similarly, the three tones of the functional modes can be regarded as indispensible stations on the folding patterns ("Urfaltung").

The Schenkerian emphasis of the bass arpeggiation $I - V - I$ as the basic manifestation of harmony is nevertheless not disregarded. In the modal perspective it relates to the authentic division of the octave, which is a mode of fifth-generated two note scale $C - G - (C')$. The mode is the common root of both, the first and second functional mode in $C$, within a hierarchy of well-formed fifth-generated modes. In each of the two cases the lower fifth interval within $C - G - (C')$ is being expanded into a fourth and a major step, respectively: $C - F - G - (C')$ and $C - D - G - (C')$ (cf. Fig. 3). This means that both modes still inherit the authentic division of the octave. This implies a concept of hierarchy different from the Schenkerian point of view: Rather than interpreting the single scale degrees $II$ and $IV$ as being subordinate to the single scale degree $V$ we interpret the entire modes $C - F - G - (C')$ and $C - D - G - (C')$ as being located at another level in the hierarchy of modes and as being active at middleground levels of musical structure.

The same line of argument illuminates the special and exotic role of the third mode. It is not a descendant of the authentic division: the dominant corresponds to a minor third alteration of the divider. This makes it analogous to the Locrian mode, which was rejected by Glareanus as *hyper-aeolius reiectus I* (see [7]). We tend to find the third mode more in the pop repertoire and modal French music of the turn of the 20th century than in the common practice period.

## 3.2 Essential vs. Accidental Alteration of Fundaments

In continuation of the considerations of Sect. 2.3, let us elaborate a little further upon the concept of alteration. We distinguish between essential and accidental alterations. Returning to the previous example from the diatonic world we would speak of an essential alteration when we modulate from the C-ionian mode to the C-mixolydian mode (character shift). The tone $B$ is essentially altered to become $B\flat$. An accidental alteration takes place, when we use the $B\flat$ as passing tone between the Ionian scale degrees $B$ and $A$. Now a similar distinction can be made in the world of the functional modes. Please refer again to Fig. 2. When modulating from a first mode $C_1$ to a second mode $C_2$ with common finalis $C$ the subdominant scale degree $F$ has to be essentially altered by a minor third to $D$ (see Sect. 2.3). However, the fundament $D$ can also occur within the first mode as an accidental alteration of the proper subdominant function. The accidental alteration serves as a theoretical alternative to the Riemannian concept of functional substitution within our approach. In the case of an accidental alteration no character shift takes place, i.e. it has no effect on the dual side of the mode. We view it as a mere surface disturbance.

A typical instance of a fundament progression with an accidental alteration in the second mode is the *turnaround* in jazz. Here the tone representing the tonic function is immediately followed by its minor third alteration downwards and then followed by three falling fifths. This fundamental bass progression subsumes a multitude of chord progressions with quite different Roman numeral analyses: $I-vi-ii-V-I$, $I-V/ii-ii-V-I$, $I-V/ii-V/V-V-I$, $i-\sharp vi-ii-V-i$, etc. Figure 10 displays a turnaround from the minor key theme *Frantonality* by Errol Garner. Note that the fundamental bass analysis is indifferent to the distinction between major and minor keys. It is therefore interesting to observe that the turnaround in a minor key involves the sharpened diatonic scale degree $\sharp vi$ rather than the pure diatonic scale degree $VI$.

**Fig. 10.** Errol Garner: *Frantonality* (1946, Pastiche of *Love Me or Leave Me*)

Chopin's $A$-major prelude Op. 28/7 is also a good illustration of an accidental alteration within a second mode. In this case the whole piece culminates in the completion of a single turnaround. At the climactic point of the piece, the tonic A is being altered (accidental alteration) to $F\sharp$ and then followed by the subdominant $B$, dominant $E$, and tonic $A$, respectively (see Fig. 11).

A striking example of a 1st mode with altered subdominant (♯S) is Ravel's *Forlane* from *Le Tombeau de Couperin* (Fig. 12). In this theme a clear first functional mode of E is accompanied by a colorful and strongly chromatic upper structure. In this example the altered subdominant $\sharp S$ occurs explicitly in the music. A different application of the concept of accidental alteration ties in with our discussion of the ambiguity between the root position $IV$-chord with added sixth and a $II^7$-chord in first inversion ($II^6_5$). In the case of the Haydn symphony (Fig. 4) the very first of the three alternative analyses is an instance

**Fig. 11.** Chopin Op. 28 No. 7, Bars 9-16

**Fig. 12.** Ravel: Forlane from *Le Tombeau de Couperin*, bars 1-4

of an accidentally altered subdominant $\sharp S$ within a first mode of $C$. This is how Rameau's idea of double employment can be re-interpreted in the functional mode approach.

### 3.3 "Double Star", Hierarchy, Modulation and Sequence

A thorough investigation of the scope and the boundaries of the functional mode approach within the investigation of harmonic tonality requires a careful analysis of a sufficiently large corpus of works as well as an extension of the paradigmatic study of the single modes towards an investigation of their combinatoriality on the syntagmatic side. For the present paper we need to limit ourselves to a couple of pre-considerations on the basis of examples.

First of all, it is reasonable to assume that the expression of tonality through the fundamental bass has different manifestations along the hierarchy of well-formed scales (cf. Fig. 6). In the present approach we concentrate on the authentic tetractys modes. They form the next higher level immediately following the authentic division mode at the root of the hierarchy, which is manifest in the Schenkerian bass arpeggiation $I - V - I$. Immediately after the tetractys modes we obtain the pentatonic modes, which we may interpret as chromatically saturated tetractys modes. This is where the scope of the tetractys approach is reaching a limit.

Complete diatonic falling fifth progressions offer themselves as manifestations of a next hierarchical level. Some theorists, however, such as François-Joseph Fètis or Hugo Riemann, argue that complete diatonic circle-of-fifth sequences have a tendency to suspend the presence of tonality. A clarification of this point on the basis of theoretical arguments and examples deserves a separate investigation. It is interesting though, to discuss at least one such example, where the

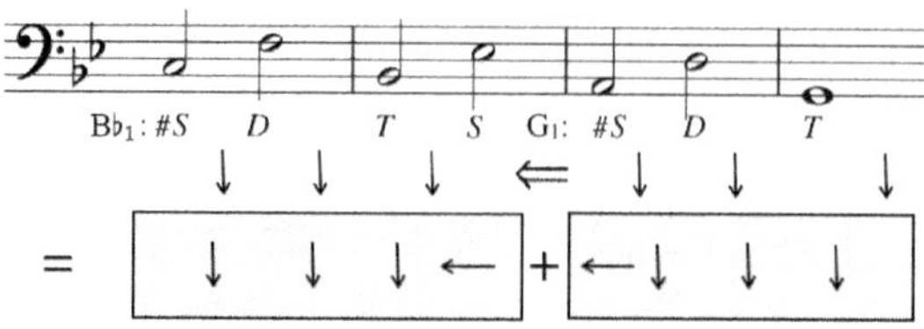


**Fig. 13.** Analysis of the complete diatonic cycle of fifths as an interaction of two functional modes. The double arrow symbolizes the interval of a diminished fifth.

assumption of a suspended tonality is no less than problematic. It concerns the fundamental bass progression of the song *Autumn Leaves* by J. Kosma (1945).

In this minor mode song the fundamental bass traverses the whole diatonic cycle of fifths in G-minor (see Fig. 13). But this song also shows a typical feature of the harmony in minor keys: a decomposition into two regions, both of which are organized as first functional modes: $G_1$ and $B\flat_1$. What prevents the suspension of tonality in this case could be based upon the constitution of the minor mode as a kind of "double star". The diminished fifth between $A$ and $E\flat$ decomposes into two minor thirds ($A - C$, $C - E\flat$), both of which are the active alteration intervals in the modes $G_1$ and $B\flat_1$, respectively.

Another type of interaction between functional modes is *hierarchical embedding.* The fundamental bass line of the first 8 bars of Benny Golson's *Whisper Not* (see Fig. 14) offers an example of a 2nd mode whose $T - S$ progression is prolonged on the foreground by a series of 2nd functional modes.

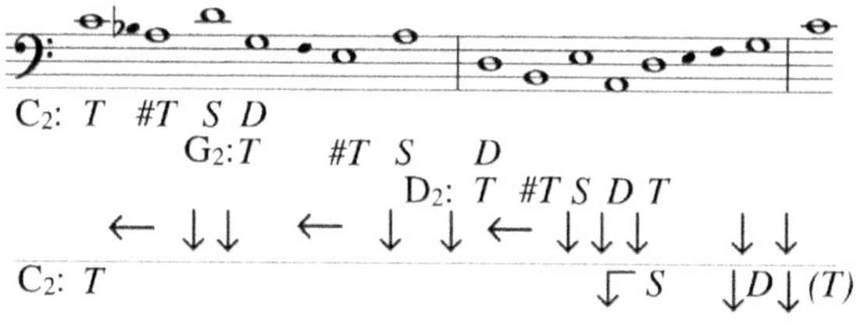


**Fig. 14.** Real Bass and Fundamental Bass (open note heads) of Golson's *Whisper Not*

An interesting modulating example is the jazz piece *Excerpt from Canonic Passacaglia* by Clare Fischer (issued on *Alone Together* in 1997). The chord progression, the bass line, and also melodic details are reminiscent of Benny Golson's *Whisper Not* (1956). Clare Fisher turns this model into a continuously modulating pattern $d$: i – ♯vi/$a$: $ii - V - i$ which traverses the entire circle of fifths. The five bass tones $D - C - B - D - A$ combine the descending line $C - C - B - A$ with the zig-zag $D - B - E - A$ in the m3/P5 lattice. The title *Passacaglia* is most likely a reference to the descending fourth-line (such as $D - C - B - A$). The deviation from the more typical descend $D - C - B\flat - A$) with $B\flat$ instead of $B$ is

$D_2$: $T \leftarrow \#T$

$A_2$: $S \downarrow D \downarrow T \leftarrow \#T$

$E_2$: $S \downarrow D \downarrow T \leftarrow \#T$

$B_2$: $S \downarrow D \downarrow T \leftarrow \#T$

**Fig. 15.** Clare Fischer's *Canonic Passacaglia* as a modulating chain of second functional modes

in solidarity with the constitution of the fundamental bass pattern with because $B$ being a minor third below $D$. Despite of the obvious similarities with *Whisper Not*, the *Canonic Passacaglia* by Clare Fischer (see Fig. 15) does not show the same kind of hierarchical organization. It is a chain of modulating 2nd modes through all twelve tonal centers, each of which provides a clear tonal anchor.

## 4 Conclusion

The present paper is a first attempt to envision the meaning of the side remark in Norman Carey and David Clampitt's article [1] in some detail. The functional mode approach to tonality is based upon the assumption that the presence of tonality is linked to the experience of a tetractys mode, and that the mode is typically expressed through its folding pattern. These tetractys modes can be chromatically extended. The active chromatic tone is already included in the associated folding pattern and constitutes a minor third (as the augmented tetractys prime) with the origin of the folding. As soon as the finalis and the framing minor third of the folding are in place, the mode in question is determined. Modulations on the tetractys level rely on the breaking-up of a mode and thereby on a moment of tonal instability. The *Canonic Passacaglia* example shows a first level of such instability, where in the model and each sequence $\leftarrow\downarrow\downarrow$ the final falling fifth is missing (in comparison to the complete folding pattern $\leftarrow\downarrow\downarrow\downarrow$). The modes follow each other in a sequence of rising perfect fifths. A further degree of instability occurs, whenever two fifths are taken away from the complete folding pattern $\leftarrow\downarrow\downarrow\downarrow$. Such is the case in the *Monte Sequence* where a sequence of ascending major seconds is established through the repetition of the pattern $\leftarrow\downarrow$ (e.g. in Bach's *Air* from the 3rd Orchstral Suite in $D$-Major, bars 13 - 14). When the third, and final fifth is taken away from the pattern, all that remains is the minor third $\leftarrow$, and tonality is on the verge of dissolving. A good example where this happens, is the *Omnibus progression*, whose underlying fundamental bass progression can be reduced to a sequence of the pattern $\leftarrow\downarrow\uparrow$. The ultimate possibility to dissolve tonality is the avoidance of the comma altogether (either explicitly or as the boundary interval). Nothing, but progressions of freely falling and rising perfect fifths remain.

## Acknowledgements

We would like to thank William Caplin, Norman Carey, David Clampitt, Clemens Kemme and three anonymous reviewers for valuable comments and suggestions.

## References

1. Carey, N., Clampitt, D.: Aspects of Well Formed Scales. Music Theory Spectrum 11(2), 187–206 (1989)
2. De Jong, K., Noll, T.: Contiguous Fundamental Bass Progressions. Dutch Journal of Music Theory (2008)
3. Hauptmann, M.: Die Natur der Harmonik und der Metrik, Leipzig, Breitkopf und Härtel (1853)
4. Riemann, H.: Musikalische Logik: Hauptzüge der physiologischen und psychologischen Begründung unseres Musiksystems. Verlag von C. F. Kahnt, Leipzig (1874)
5. Clampitt, D., Noll, T.: Modes, the Height-Width Duality, and Handschin's Tone Character. Theory Online 17(1) (2011)
6. Noll, T.: Ionian Theorem. JMM 3(3), 137–151 (2009)
7. Glareanus, H.: Dodecachordon, Basel (1547); Reprint: Hildesheim: Olms (1970)
8. Hook, J.: Signature Transformations. In: Douthett, J., Hyde, M.M., Smith, C.J. (eds.) Music Theory and Mathematics: Chords, Collections, and Transformations, pp. 137–160. University of Rochester Press, Rochester (2008)

# Musical Tonality, Neural Resonance and Hebbian Learning

Edward W. Large

Center for Complex Systems and Brain Sciences, Florida Atlantic University
large@ccs.fau.edu

**Abstract.** A new theory of musical tonality is explored, which treats the central auditory pathway as a complex nonlinear dynamical system. The theory predicts that as networks of neural oscillators phase-lock to musical stimuli, stability and attraction relationships will develop among frequencies, and these dynamic forces correspond to perceptions of stability and attraction among musical tones. This paper reports on an experiment with learning in a model auditory network. Results suggest that Hebbian synaptic modification can change the dynamic responses of the network in some ways but not in others.

**Keywords:** Tonality, Stability, Attraction, Neurodynamics, Oscillation, Network, Learning.

## 1 Introduction

Tonality is a set of stability and attraction relationships perceived among musical tones. Stability and attraction relationships are thought to function analogously to the meaning of words in language. As Zuckerkandl [1] asserts, "... musical tones point to one another, attract and are attracted" and these dynamic qualities "make melodies out of successions of tones and music of acoustical phenomena." Dynamical theories [2,3,4] aim to explain which aspects of tonality may be universal, and which are likely to be learned, based on neurodynamic principles. The current theory claims that 1) the auditory system is a complex, nonlinear dynamical system, 2) the way that the auditory system resonates to sound determines the highly structured perceptual responses that we collectively call "tonal cognition," and 3) Hebbian synaptic modification can change these responses in some ways, but not others.

## 2 Tonal Theory

Music combines individual sounds into well-formed structures, such as melodies. One feature that the melodies of most musical systems share is that they elicit tonal percepts: Listeners experience feelings of *stability* and *attraction* among tones in a melody. Stability, in this sense, means that one or more tones are perceived as points of repose. One specific tone, called the tonic, provides a

C. Agon et al. (Eds.): MCM 2011, LNAI 6726, pp. 115–125, 2011.

focus around which the other tones are organized in a hierarchy of relative stability (Fig. 1), such that some tones are perceived as more stable than others. Less stable tones provide points of dissonance or tension, more stable tones provide points of consonance or relaxation. Less stable tones are heard relative to more stable ones, such that more stable tones are said to attract the less stable tones [5]. Some theorists describe tonal attraction by analogy to physical forces (e.g., [6]), others as resolution from dissonance to consonance [7].

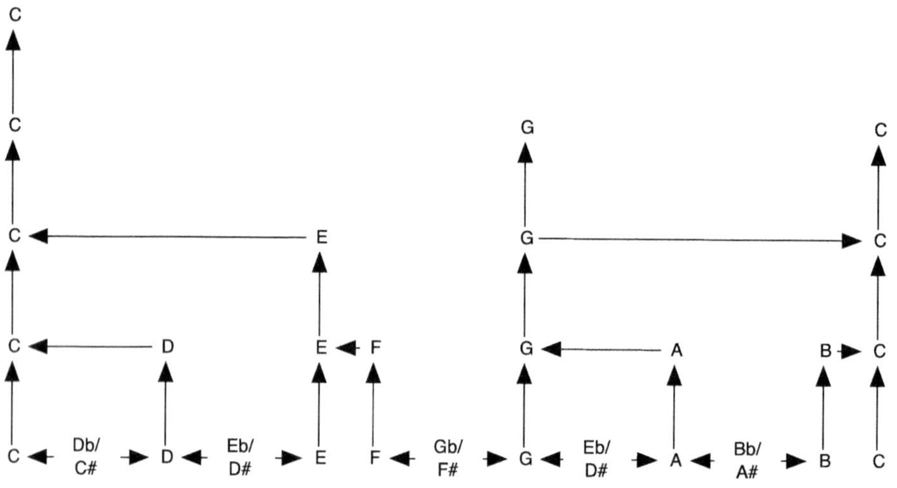


**Fig. 1.** Music theoretic depiction of tonal stability and attraction for the key of C Major. More stable tones occupy higher positions (upward pointing arrows). Attractions occur at each hierarchical level (horizontal arrows; adapted from [5]).

## 2.1 Theories of Musical Tonality

**Small Integer Ratio (SIR) Theory.** The oldest theory of musical tonality states that perceptions of consonance and dissonance are governed by ratios of whole numbers. Pythagoras used the principle of small integer ratios (SIRs) [8] to explain the musical scale that was in use in the West at the time, and Pythagoras and his successors proposed SIR systems for tuning musical instruments, such as just intonation (JI). The study of frequency ratios and tuning systems formed the basis for theoretical approaches to music cognition until the middle of the nineteenth century. Around that time, modern Western equal temperament (ET) came into use. ET intervals closely approximate JI ratios. However, apart from octaves, the intervals are not small integer ratios, they are irrational numbers.

**Helmholtz Theory.** Ohm [9] proposed that pitch was a consequence of the auditory system's ability to perform Fourier analysis—a linear transformation—on acoustical signals. Helmholtz [10] noted that from a linear processing point of view, there is nothing special about small integer ratios. He hypothesized that the dissonance of a pair of simultaneously sounding complex tones was due to the interference of its pure tone components [10]. So-called *sensory dissonance* occurs when simultaneous tones interact within an auditory critical band [11], and the interaction of pure tone components correctly predicts ratings of consonance for pairs of complex tones [12]. Helmholtz went on to observe that ET intervals (irrational ratios) are approximately as consonant as nearby SIR intervals. He used such arguments to build a strong case against SIR theory.

**Modern Tonal Theory.** Relating Helmholtz's theory of consonance to musical perceptions of stability and attraction, however, has proved problematic. *Musical* consonance and dissonance differ across cultures and styles, and have changed over the history of Western music [13]. *Musical* (as opposed to sensory) consonance and dissonance now tend to be defined in sequential terms: "...a dissonance is that which requires resolution to a consonance" [13]. Such definitions rule out Helmholtz theory as a potential explanation because the sensory dissonance phenomenon is heard only for simultaneously sounded tones, not for sequences (e.g., melodies). Mainly for these reasons, it is argued that—beyond transient sensations of roughness—particular frequency ratios do not matter *at all* in higher-level music cognition. Rather, the auditory system transforms musical notes into *abstract symbols* unrelated to frequency, in much the same way that spoken words are transformed into abstract symbols whose meaning is unrelated to their sound. Importantly, correlational evidence is interpreted to mean that stability and attraction relationships are internalized solely through long term exposure to the music of one's culture through a process of statistical learning, e.g. [14,15,16,17,18,19].

### 2.2 Some Issues with Modern Tonal Theory

According to modern tonal theory, the sounds themselves don't matter, much as the sounds associated with spoken words do not determine what they mean. However, this view does not make intuitive sense; in music, it feels like the sounds *do* matter. Modern theories displaced SIR theories based largely on early evidence that the auditory system analyses sound by linear frequency decomposition. However, this argument fails to account for at least two lines of important recent evidence.

**The auditory system is highly nonlinear.** The cochlea produces sharp mechanical frequency tuning, exquisite sensitivity and nonlinear distortion products that cannot be explained by linear systems, but can be explained as self-tuned critical oscillations of hair cells, e.g. [20]. In the central auditory system, action potentials phase-lock to acoustic stimuli at many different levels, including cochlear nucleus, superior olive, inferior colliculus (IC), thalamus and A1 [21,22]. Nonlinear response curves have been identified in the central auditory systems of animals [23,24,25]. In humans, nonlinear responses to musical intervals have been measured in the auditory brainstem response [26]. Such results provide evidence of nonlinear resonance in the auditory system all the way from the cochlea to the primary auditory cortex. In nonlinear systems there is something special about integer ratios.

**There is a cross-cultural tendency toward SIR intervals.** A recent review observed "a strong propensity for perfect consonances (2:1, 3:2, 4:3) in the scales of most cultures, the main exceptions being the ofttime highly variable scales of cultures that are either preinstrumental, or whose main instruments are of the xylophone type" [8] (p. 249). Moreover, the three largest non-Western musical traditions—Chinese, Indian, and Arab-Persian—all employ tuning systems based on small integer ratios [8]. Such tuning systems may be more than 9,000 years old [27]. While there is no one "natural scale", the strong cross-cultural tendency toward SIR frequency relationships may point toward an underlying universal.

## 3 Dynamical Model of Auditory Processing

In the current approach, nonlinear cochlear responses are modeled as nonlinear oscillations of hair cells (cf. [20]), and phase-locked responses of auditory neural populations are modeled as phase-locked neural oscillations [28]. Here, we briefly present the basic conceptual components of the model.

Recently, nonlinear models of the cochlea have been proposed to simulate the nonlinear responses of outer hair cells. It is important to recognize that outer hair cells are thought to be responsible for the cochlea's extreme sensitivity to soft sounds, excellent frequency selectivity and amplitude compression [20]. Models of nonlinear resonance that explain these properties have been based on the Hopf normal form for nonlinear oscillation, and are generic [29]. Normal form (truncated) models have the form

$$\frac{dz}{dt} = z(\alpha + i\omega + \beta|z|^2) + x(t) + h.o.t. \tag{1}$$

where $z$ is a complex-valued state variable, $\omega$ is the intrinsic oscillator frequency ($\omega = 2\pi f$, $f$ in Hz), $\alpha$ is a bifurcation parameter, and the value $\alpha = 0$ is the critical value, or bifurcation point. When $\alpha > 0$, the system can spontaneously oscillate. $\beta < 0$ is a nonlinear damping parameter, which prevents oscillation amplitude from blowing up when $\alpha > 0$. $x(t)$ denotes linear forcing by an external signal. The term *h.o.t.* denotes *higher-order terms* of the nonlinear expansion that are truncated (i.e., ignored) in normal form models. Nonlinear oscillator address behaviors that linear filters do not, such as extreme sensitivity to weak signals, amplitude compression and high frequency selectivity.

A canonical model was recently derived from a model of neural oscillation in excitatory and inhibitory neural populations [28,30]. The canonical model (Eq. 2) is related to the normal form [29] (Eq. 1), but it has properties beyond those of Hopf normal form models because the underlying, more realistic oscillator model is fully expanded, rather than truncated. The complete expansion of higher-order terms produces a model of the form

$$\frac{dz}{dt} = z(\alpha + i\omega + (\beta_1 + i\delta_1)|z|^2 + \epsilon\frac{(\beta_2 + i\delta_2)|z|^4}{1 - \epsilon|z|^2}) + cP(\epsilon, x(t))A(\epsilon, \bar{z}) \quad (2)$$

There are, of course, similarities with the normal form model. The parameters, $\omega$, $\alpha$ and $\beta_1$ correspond to the parameters of the truncated model. $\beta_2$ is an additional amplitude compression parameter, and $c$ represents strength of coupling to the external stimulus. Two frequency detuning parameters $\delta_1$ and $\delta_2$ are new in this formulation, and make oscillator frequency dependent upon amplitude. The parameter $\epsilon$ controls the amount of nonlinearity in the system. Here, $x(t)$ represents a generic input, that could represent either external input (i.e., a sound), or network input from afferent, efferent, or internal network connectivity. In the latter case, $x = \sum a_j z_j$ where $a_j$ ranges over a row of a connectivity matrix $A$ (i.e., $a_j$ is a row vector) and $z_j$ is the $j^{th}$ oscillator in a column vector representing a network state.

The canonical model (Eq. 2) is more general than the Hopf normal form (Eq. 1) and encompasses a wide variety of behaviors that are not observed in linear resonators, including compressive nonlinearities and frequency detuning. Most importantly, coupling to the stimulus is nonlinear and has a passive part, $P(\epsilon, x(t))$, and an active part, $A(\epsilon, z)$, producing higher order resonances. In such nonlinear systems, higher order resonances—also known as "distortion products"— produce neural responses at frequencies that are not physically present in the stimulus, which may correspond to harmonics ($k \times f_1$), subharmonics ($f_1/m$), summation frequencies (e.g., $f_1+f_2$), difference frequencies (e.g., $f_2 - f_1$), and integer ratios (e.g., $k \times f_1/m$), where $f_1$ and $f_2$ are stimulus frequencies, and $k$ and $m$ are integers.

Higher order resonance is a key feature of this model, leading to predictions of stability and attraction. Fig. 2 illustrates entrainment regions and stability/attraction relationships in a tonotopically arranged network of oscillators, each stimulated with a signal external frequency. Gray regions are regions in which oscillators resonate, or entrain, to the stimulus. Higher-order resonances are found at frequencies that form integer ratios with the external stimulus, and the entrainment regions increase in extent with the strength of coupling. More *stable* resonances have larger resonance regions, and these are colored in darker shades. Where regions overlap, the stronger resonance overpowers the weaker resonance, and attraction effects are observed [4]. Higher-order resonance produces the types of frequency responses that have been reported in the central auditory nervous system. Recently, a variant of this model predicted precisely the nonlinearities observed in brainstem responses to musical intervals [26], and did so with high accuracy for both consonant and dissonant intervals [31]. Such models predict ratings of consonance/dissonance [3] as well as empirical ratings of stability (e.g., [32]) for major and minor modes, with high precision [4].

We have recently developed a canonical version of Hebbian learning that can dynamically evolve connections between oscillators of different frequencies through detection of multifrequency phase coherence [4].

$$\dot{c}_{ij} = -\delta_{ij}c_{ij} + k_{ij}\frac{z_i}{1 - \sqrt{\epsilon z_i}}\frac{\bar{z}_i}{1 - \sqrt{\epsilon \bar{z}_i}} \quad (3)$$

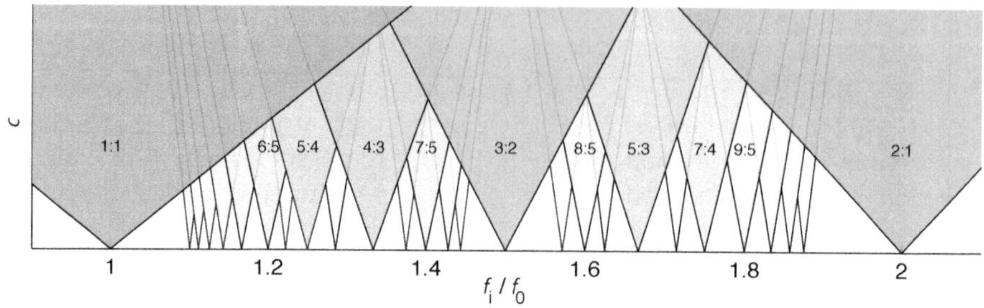


**Fig. 2.** Bifurcation diagram showing natural resonances in a tonotopic nonlinear oscillator array as a function of connection strength and frequency ratio (adapted from [4])

where $c_{ij}$ is complex, representing the magnitude and phase of the connection between any two nonlinear oscillators at a point in time [33], and $\delta_{ij}$ and $k_{ij}$ are real parameters representing the speed of change of the connection. The variables, $z_i$ and $z_j$ are the complex valued state variables of the two oscillators connected by $c_{ij}$. Previous analysis of the learning rule suggested that as a listener experiences Western musical sequences, connections within and between networks would essentially be pruned, learning the frequencies of Western equal temperament (ET). However, no experiments with learning were reported [4].

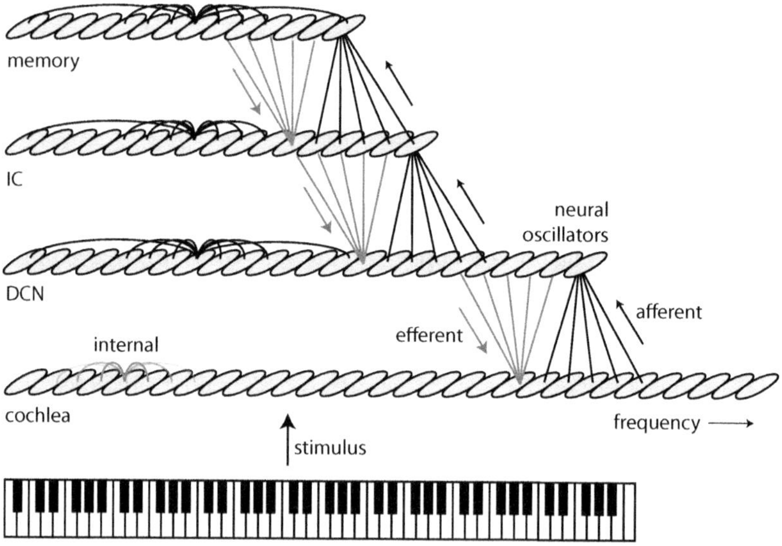


**Fig. 3.** Model architecture of phase-locking neural dynamics in the auditory system. Fixed connections are shown in gray, learned connections in black. Higher levels operate only at lower frequencies, modeling deterioration of phase-locking at higher auditory levels.

With these building blocks in place, is possible to construct complex models of phase-locking neural networks in the central auditory pathway, and make predictions about listeners with differing levels of auditory experience. Our simple model of the early auditory system includes a model cochlea, dorsal cochlear nucleus (DCN) and inferior colliculus (IC). Cochlear analysis is modeled with a network of locally coupled outer hair cell oscillators, each tuned to a distinct intrinsic frequency [34,35]. Phase-locked responses in the DCN and IC are modeled as the phase-locking of neural oscillators. To this network we add a memory network, which could be interpreted as MGB / A1, but is justified here mainly based on psychological grounds, as memory is a key component in tonal cognition and perception. High-frequency phase-locking deteriorates as the auditory pathway is ascended. To capture this, the upper frequency limit of GFNNs decreases at each processing level, as illustrated in Fig. 3. Note that the multi-layered architecture is not redundant: the transformations are nonlinear, so information is added at each level.

## 4 A Learning Experiment

To study the behavior of the learning algorithm, a stimulus was generated consisting of two complex, steady state tones, shown in Fig. 4(A). Tone 1 was a harmonic complex consisting of frequencies 500, 1000, 1500, 2000, and 2500 Hz. Tone 2 was a harmonic complex consisting of frequencies 600, 1200, 1800, 2400, and 3000 Hz. The interval was a minor third, tuned according to JI (6:5).

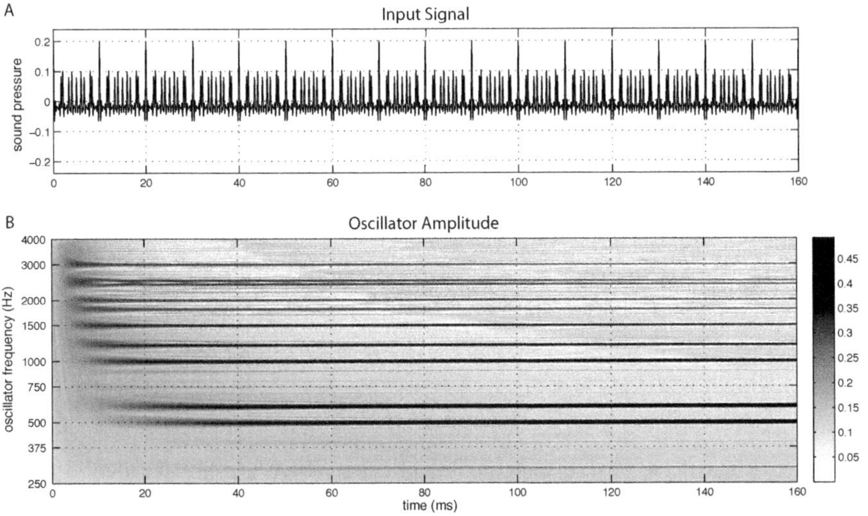


**Fig. 4.** (A) A harmonic musical interval (minor third) made up of two complex tones. (B) A spectrogram-like representation—oscillator amplitude as a function of time (horizontal) and natural frequency (vertical axis)—showing the response of the Layer 3 (IC) network to this stimulus.

The network of Fig. 3 processed the sound mixture. Layers 1 and 2 of the network of oscillators operated in the critical parameter regime (i.e., $\alpha = 0$), with the oscillators poised between damped and spontaneous oscillation, Layer 3 operated in the active parameter regime (i.e., $\alpha > 0$) resulting in low amplitude spontaneous oscillations. Layer 4 operated in a generalized Hopf (a.k.a. Bautin) parameter regime, (i.e., $\alpha < 0$, $\beta_1 > 0$, $\beta_2 < 0$), enabling a persistent memory of the stimulus frequencies (cf. [4]). Internal connectivity in the cochlea (gray) was local and fixed, simulating local basilar membrane coupling among outer hair cells. All efferent connections (gray) were set to zero. The connections in black were initialized to all-to-all connectivity at low amplitude and zero phase; these connections were learned. The response of the Layer 3 (IC) network to this stimulus (oscillator amplitude, $|z|$, as a function of time) is shown in Fig. 4(B).

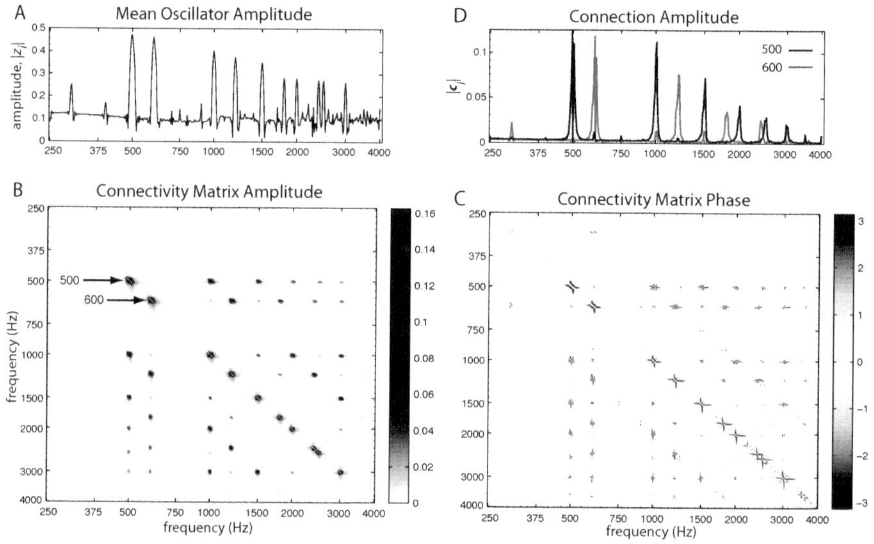


**Fig. 5.** Internal connection matrices learned in the third layer of the network after the processing the minor third. Amplitude response of the oscillator network, averaged over the last 20 ms. Panels B and C show the magnitude and phase of the connection matrix respectively. Panel D focuses on two rows of the amplitude matrix (Panel B), showing the amplitudes as a function of frequency.

The learning algorithm discussed above was run synchronously, evolving connections as the network processed the stimulus. The result of learning is shown in Fig. 5. Panel A shows the amplitude response of the oscillator network, averaged over the last 20 ms. Reading counterclockwise, Panels B and C show the amplitude and phase of the connection matrix. Note that in the amplitude matrix (Panel B) the peaks in the rows corresponding to the 500 Hz and 600 Hz oscillators are different. Connections are learned from those oscillators whose activity is phase coherent with the oscillators of interest (500 and 600 Hz) over the relevant time scale. Panel D focuses on two rows of the amplitude matrix (Panel B), showing the amplitudes as a function of frequency. Note that the strongest connections are to harmonics, revealing the components of the two different sources, Tone 1 and Tone 2. However, lower amplitude connections also exist connecting the fundamental frequencies (and some harmonics) of the minor third interval. Thus the algorithm learns appropriate connections associating the harmonics of individual tones, as well as connections associating the fundamentals of simultaneously presented intervals. The latter connections are weaker, because the 6:5 ratio is a weaker high-order resonance. This initial study used simple stimuli and short training protocols to analyze the behavior of the algorithm. Ongoing studies are using longer pieces of music and learning sequential contingencies.

## 5 Conclusions

Tonality is a universal feature of music, found in virtually every culture, but tonal "languages" vary across cultures with learning. Here, a model auditory system, based on knowledge of auditory organization and general neurodynamic principles, was described and studied. The model provides a direct link to neurophysiology and, while simplified compared to the organization and dynamics of the real auditory system, it makes realistic predictions [31]. Analysis of the model suggests that certain musical universals may arise from intrinsic neurodynamic properties (i.e., nonlinear resonance). Moreover, preliminary results of learning studies suggest that different tonal languages may be learned in such a network through passive exposure to music. In other words, this neurodynamic theory predicts the existence of a dynamical, universal grammar (cf. [36]) of music.

## Acknowledgements

This work was supported by NSF grant BCS-1027761. The author thanks Nicole Flaig for her comments on this manuscript.

## References

1. Zuckerkandl, V.: Sound and Symbol: Music and the External World. Princeton University Press, Princeton (1956)
2. Large, E.W., Tretakis, A.E.: Tonality and Nonlinear Resonance. In: Avanzini, G., Lopez, L., Koelsch, S., Majno, M. (eds.) The Neurosciences and Music II: From Perception to Performance. Annals of the New York Academy of Sciences, vol. 1060, pp. 53–56 (2005)
3. Shapira Lots, I., Stone, L.: Perception of musical consonance and dissonance: an outcome of neural synchronization. Journal of The Royal Society Interface 5(29), 1429–1434 (2008)
4. Large, E.W.: A dynamical systems approach to musical tonality. In: Huys, R., Jirsa, V.K. (eds.) Nonlinear Dynamics in Human Behavior. Studies in Computational Intelligence, vol. 328, pp. 193–211. Springer, Heidelberg (2010)

5. Lerdahl, F.: Tonal Pitch Space. Oxford University Press, New York (2001)
6. Larson, S.: Musical forces and melodic expectations: Comparing computer models and experimental results. Music Percept 21(4), 457–498 (2004)
7. Bharucha, J.J.: Anchoring effects in music: The resolution of dissonance. Cogn. Psychol. 16, 485–518 (1984)
8. Burns, E.M.: Intervals, scales, and tuning. In: Deustch, D. (ed.) The Psychology of Music, pp. 215–264. Academic Press, San Diego (1999)
9. Ohm, G.S.: Uber die Denition des Tones, nebst daran geknufter Theorie der Sirene und ahnlicher tonbildender Vorichtungen. Ann. Phys. Chem. 135(8), 513–565 (1843)
10. Helmholtz, H.L.F.: On the sensations of tone as a physiological basis for the theory of music. Dover Publications, New York (1863)
11. Plomp, R., Levelt, W.J.M.: Tonal consonance and critical bandwidth. J. Acoust. Soc. Am. 38, 548–560 (1965)
12. Kameoka, A., Kuriyagawa, M.: Consonance theory part II: Consonance of complex tones and its calculation method. J. Acoust. Soc. Am. 45, 1460–1471 (1969)
13. Dowling, W.J., Harwood, D.L.: Music Cognition. Academic Press, San Diego (1986)
14. Krumhansl, C.L.: Cognitive foundations of musical pitch. Oxford University Press, NY (1990)
15. Cuddy, L.L., Lunney, C.A.: Expectancies generated by melodic intervals: Perceptual judgements of melodic continuity. P&P 57, 451–462 (1995)
16. Schellenberg, E.G.: Expectancy in melody: Tests of the implication realization model. Cognition 58, 75–125 (1996)
17. Knopoff, L., Hutchinson, W.: An index of melodic activity. Interface 7, 205–229 (1978)
18. Temperley, D.: The cognition of basic musical structures. MIT Press, Cambridge (2001)
19. Pearce, M.T., Wiggins, G.A.: Expectation in Melody: The Influence of Context and Learning. Music Perception 23(5), 377–405 (2006)
20. Eguìluz, V.M., Ospeck, M., Choe, Y., Hudspeth, A.J., Magnasco, M.O.: Essential nonlinearities in hearing. PhRvL 84(22), 5232 (2000)
21. Langner, G.: Periodicity coding in the auditory system. Hear. Res. 60, 115–142 (1992)
22. Joris, P.X., Schreiner, C.E., Rees, A.: Neural Processing of Amplitude-Modulated Sounds. Physiol. Rev. 84(2), 541–577 (2004)
23. Escabi, M.A., Schreiner, C.E.: Nonlinear Spectrotemporal Sound Analysis by Neurons in the Auditory Midbrain. J. Neurosci. 22(10), 4114–4131 (2002)
24. Langner, G.: Temporal processing of periodic signals in the auditory system: Neuronal representation of pitch, timbre, and harmonicity. Z. Audiol. 46(1), 80–21 (2007)
25. Sutter, M.L., Schreiner, C.: Physiology and topography of neurons with multipeaked tuning curves in cat primary auditory cortex. J. Neurophysiol. 65(5), 1207–1226 (1991)
26. Lee, K.M., Skoe, E., Kraus, N., Ashley, R.: Selective Subcortical Enhancement of Musical Intervals in Musicians. J. Neurosci. 29(18), 5832–5840 (2009)
27. Zhang, J., Harbottle, G., Wang, C., Kong, Z.: Oldest playable musical instruments found at Jiahu early Neolithic site in China. Nature 401, 366–368 (1999)
28. Large, E.W., Almonte, F., Velasco, M.: A canonical model for gradient frequency neural networks. Physica D: Nonlinear Phenomena 239(12), 905–911 (2010)
29. Hoppensteadt, F.C., Izhikevich, E.M.: Weakly Connected Neural Networks. Springer, New York (1997)

30. Wilson, H.R., Cowan, J.D.: A mathematical theory of the functional dynamics of cortical and thalamic nervous tissue. Kybernetik 13, 55–80 (1973)
31. Large, E.W., Almonte, F.: Phase-locked neural oscillation predicts human auditory brainstem responses to musical intervals. PNAS Under Review (2011)
32. Krumhansl, C.L., Kessler, E.J.: Tracing the dynamic changes in perceived tonal organization in a spatial representation of musical keys. Psychol. Rev. 89(4), 334–368 (1982)
33. Hoppensteadt, F.C., Izhikevich, E.M.: Synaptic organizations and dynamical properties of weakly connected neural oscillators II: Learning phase information. Biol. Cybern. 75, 126–135 (1996)
34. Duke, T., Julicher, F.: Active traveling wave in the cochlea. PhRvL 90(15), 158101 (2003)
35. Kern, A., Stoop, R.: Essential role of couplings between hearing nonlinearities. Phys. Rev. Lett. 91(12), 128101–128104 (2003)
36. Prince, A., Smolensky, P.: Optimality: From Neural Networks to Universal Grammar. Science 275, 1604–1610 (1997)

# Interval Cycles, Affinity Spaces, and Transpositional Networks

José Oliveira Martins

Eastman School of Music
jmartins@esm.rochester.edu

**Abstract.** The paper proposes a framework that coordinates several models of pitch space whose constructive features rely on the concept of interval cycles and transpositional relations. This general model brings under a focused perspective diverse pitch structures such as *Tonnetze*, *affinity spaces*, Alban Berg's "master array" of interval-cycles, and several types of transpositional networks (T-nets). This paper argues that applying incremental changes on some of the constructive features of the generic *Tonnetz* (Cohn 1997) results in a set of coherent and analytically versatile transpositional networks (T-nets), here classified as homogeneous, progressive, and dynamic. In this context, several properties of the networks are investigated, including voice-leading and common-tone relations. The paper also explores the music-modeling potential of progressive and dynamic T-nets by attending to characteristic compositional deployments in the music of Witold Lutosławski and György Kurtág.

**Keywords:** Interval cycles, Affinity spaces, Dasian, Transposition, *Tonnetz*, neo-Riemannian theory, Network, T-nets.

## 1 Introduction

Since the mid-1990s, the two-dimentional lattice known as the *Tonnetz* has received considerable attention as a privileged pitch framework in neo-Riemannian literature[1]. The neo-Riemannian *Tonnetz* has largely adopted a mod 12-perspective, in contrast with the (theoretically infinite) nineteenth-century counterpart based on just-intonation. The modern reinterpretation of the *Tonnetz* resulted in a more versatile geometrical framework of a torus that supports a closed group-theoretic perspective. From a focus on the modeling of triadic relations, the framework of the *Tonnetz* has been expanded to accommodate other trichordal relations and other modular cardinalities. Richard Cohn [3] framed such generalization in a lattice he termed the *generic Tonnetz*. Figure 1 presents the pc-relations of the generic Tonnetz by assigning an arbitrary pc-reference 0

[1] The literature on neo-Riemannian theory is significantly extensive. For a historical overview and a variety of approaches to neo-Riemannian theory and analysis see the volume 42.2 of the Journal of Music Theory [1,2]. Particularly germane to this paper are [3] and [4].

C. Agon et al. (Eds.): MCM 2011, LNAI 6726, pp. 126–139, 2011.

| | | | | |
|---|---|---|---|---|
| -x + 3y | 3y | x + 3y | 2x + 3y | 3x + 3y |
| -x + 2y | 2y | x + 2y | 2x + 2y | 3x + 2y |
| -x + y | y | x + y | 2x + y | 3x + y |
| -x | 0 | x | 2x | 3x |
| -x - y | -y | x - y | 2x - y | 3x - y |

**Fig. 1.** Cohn's *generic Tonnetz*

and mapping the surrounding pcs according to the pc-distance from the origin 0 under the lattice grid structured by the two axes $x$ and $y$[2].

This paper argues that applying incremental changes on some of the constructive features of the generic *Tonnetz* results in a set of coherent and analytically versatile transpositional networks (T-nets), here classified as homogeneous, progressive, and dynamic. This set of T-nets captures important features of diverse pitch models such as *Tonnetze*, *affinity spaces*, Alban Berg's "master array" of interval-cycles, and other pitch constructs. The differentiation between the three types of T-nets reflects the implementation of different "rules" or constraints upon a single general framework. This framework, shown in Fig. 2(a), depicts a lattice of ordered integer pairs in $Z^2$, where the position (0,0) serves as the referential marker upon which other positions along the $x$ and $y$ coordinates are calculated. Figure 2(b) introduces a set of variables, which structure the transpositional relations involved in the various T-nets. Given a referential pc in position (0,0) and directed intervals $dx(0,0) = pc(1,0) - pc(0,0)$, and $dy(0,0) = pc(0,1) - pc(0,0)$ mod 12, the T-net can be defined by calculating the pc in position $(i,j)$, or the directed intervals $dx(i,j)$ and $dy(i,j)$ associated with $(i,j)$[3]. This calculation, however, involves the assessment of the degree of variation of directed intervals along and across the $x$- and $y$-axes. The notation $\Delta dx|x$ and $\Delta dy|y$ refers to the change of transpositional values $dx$ along the $x$-axis, and $dy$ along the $y$-axis respectively (1); whereas $\Delta dx|y$ and $\Delta dy|x$ refer to the change of transpositional values of $dx$ across the $y$-axis, and $dy$ across $x$-axis respectively (2).

[2] The figure recaptures Cohn's Figure 6, p. 10 [3]. See, for instance, Lewin [10] for the modeling of (013) trichordal relations in a tonal passage in what amounts to a lattice where $x = 1$ and $y = 2$.

[3] All the formulas throughout the paper refer to a mod-12 universe.

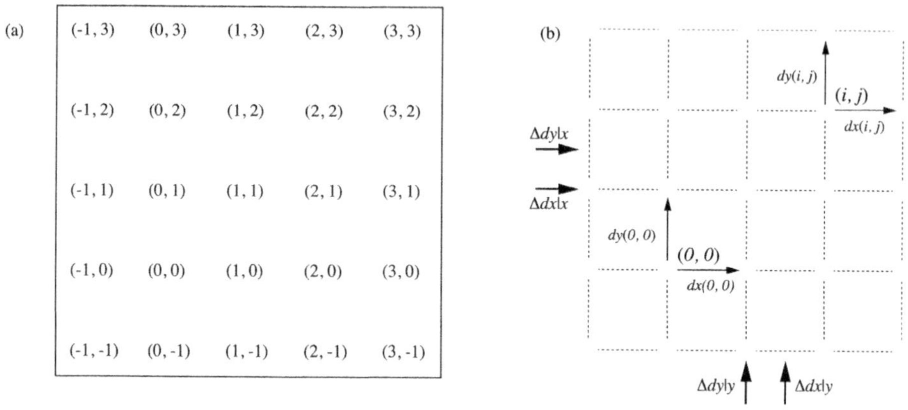


**Fig. 2.** (a) Lattice of ordered integer pairs in $Z^2$. (b) Generic layout of a T-net.

$$\Delta dx|x = dx(i+1, j) - dx(i, j) \text{ and } \Delta dy|y = dy(i, j+1) - dy(i, j). \quad (1)$$

$$\Delta dx|y = dx(i, j+1) - dx(i, j) \text{ and } \Delta dy|x = dy(i+1, j) - dy(i, j). \quad (2)$$

## 2 Homogeneous T-Nets and Affinity Spaces

### 2.1 Homogeneous T-Nets

The generic framework for T-nets introduced in Fig. 2 can now be used to examine some properties of Cohn's generic *Tonnetz*. In a *Tonnetz*, any straight (horizontal or vertical) path formed by a series of directed intervals is an interval cycle (i-cycle), and any two parallel paths project the same i-cycle. Another way of capturing the i-cycle relations of a *Tonnetz* is to say that the four rates of change ($\Delta d$) along and across the $x$ and $y$-axes are zero (3). I refer to this construct as the *homogeneous T-net* in order to reflect the lack of variation in the transpositional values in the lattice.

$$\text{In a homogeneous T-net: } \Delta dx|x = \Delta dy|y = \Delta dx|y = \Delta dy|x = 0. \quad (3)$$

Given the pc and directed intervals at the referential position (0,0), the space of a homogeneous T-net can be defined by calculating the pc, or directed intervals at an arbitrary position $(i, j)$ as stated at (4) and (5).

$$pc : Z^2 \rightarrow Z_{12}, pc(i, j) = pc(0, 0) + i(dx(0, 0)) + j(dy(0, 0)). \quad (4)$$

$$dx(i, j) = dx(0, 0) \text{ and } dy(i, j) = dy(0, 0). \quad (5)$$

The homogeneous T-net of Fig. 3 represents the familiar neo-Riemanian *Tonnetz*, where $dx = 4$ and $dy = 3$. For instance, given pc G at position (0,0), the resulting pc at position (2,3) is C, i.e., $pc(2, 3) = 7 + 2 \times 4 + 3 \times 3 = 0$.

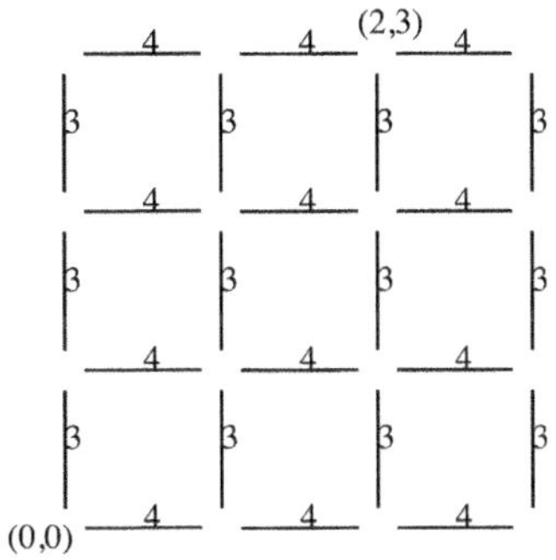


**Fig. 3.** The familiar neo-Riemanian *Tonnetz* as a homogeneous T-net

## 2.2 Affinity Spaces

Homogenous T-nets provide a useful framework to model certain structural relations of what I call *affinity spaces*[4]. Affinity spaces are (often non-octave) periodic constructs in which an interval pattern or modular unit is replicated twelve times in different parts of the space. For instance, Figure 4(a) represents the so-called *Dasian* space, a construct whose periodicity is structured by the recurrent unit of $< 2, 2, 2, 1 >$ semitones (or any of its rotations)[5]. The periodicity of the space induces local interval similarities or affinities between pcs distancing four steps or a perfect fifth, a relation akin to the medieval notion of *transpositio*[6].

As a counterpart to pattern tiling, affinity spaces can also be conceived as a set of interlocked i-cycles encompassing the entire space. Figure 4(b) partitions the Dasian space into four 7-cycles (disposed within the dotted circles), such that each of the four notes in a modular unit is assigned to a different i-cycle. Notes in the same i-cycle are related by affinities given their shared local interval context. The modular unit of affinity spaces is modeled by the formula $t = nr + s$ (mod 12), where $t$, the interval of *transpositio*, measures the affinity (or periodic transposition) of the space, $r$ measures the interval that recurs $n$ times within

[4] Martins [6] introduces the notion of affinity spaces by generalizing the concept of affinitas, which refers to the assignment of the same local scalar pattern in different parts of the medieval scale so as to preserve modal identity; the generalization afforded by the affinity spaces is used to model the music of diverse composers as Bartók, Stravinsky, Milhaud, and Lutosławski. Pesce [7] traces and studies the concept of "related tones" in general, and of "affinities" in particular, from the end of the ninth century through the mid-sixteenth century. See also [8] and [9] for approaches to the modeling of interlocked cycles that are resonant with my own.

[5] The generalization of the Dasian space [6] is based on the periodic structure of the medieval dasian scale discussed in the ninth-century *Enchiriadis* treatises. See Carey and Clampitt [10] for a discussion of the properties of the medieval scale in diatonic set-theoretic terms.

[6] Transpositio refers to the transfer of modal quality (local interval pattern) to various positions in the tonal system, see Pesce [7].

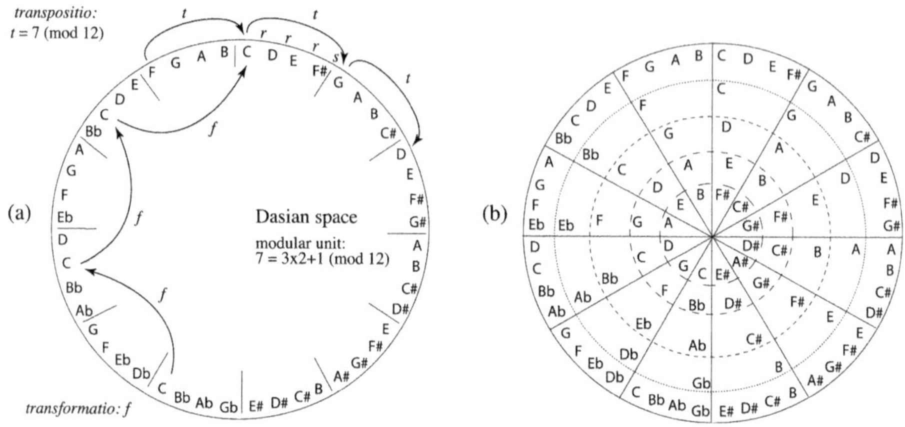


**Fig. 4.** (a) The Dasian space (affinity space): modular unit, *transpositio*, and *transformatio*. (b) Partition of the Dasian space into four interlocked 7-cycles.

a modular unit ($n$ is a positive integer), and $s$ measures the interval $t - nr$ and always represents a unique interval in the space[7]; the Dasian's modular unit is $7 = 3 \times 2 + 1$ (mod 12). Given the partition of an affinity space into $n+1$ i-cycles, there are always $n + 1$ pc replications in the space. In the Dasian space, there are four instances of each pc distancing seven clockwise steps (see the arrows connecting instances of pc C in Figure 4(a); this relation between instances of the same note in different local interval contexts is akin to the medieval notion of *transformatio* ($f$)[8].

Figure 5 presents a homogeneous T-net that coordinates $dx = 2$ and $dy = 7$ (mod 12) and serves as a framework to model some relations in the Dasian space. This framework is replicated four times in Fig. 6 along the $z$-axis. In this three-dimensional framework, $dy$ corresponds to the interval of *transpositio* $t$ (the fifth periodicity of the space) and $dz$ corresponds to the step motion $r$ within the modular unit; each vertical path thus corresponds to an i-cycle of the Dasian space, and the modular pattern is obtained by moving along the $z$- and $y$-axes. The interval $dx$ is here used to coordinate relations between transposed

[7] The number of allowable intervals in a modular unit is here restricted to two, one ($s$) iterated singly, and the other ($r$) recurrently; other, less restrictive, combinations of intervals for a modular unit are possible.

[8] *Transformatio* refers to the change of modal quality (local interval pattern) on a given pc effected by a note alteration, see Pesce [7]. Martins [6,11] conceives of *transformatio* as a function that (depending on the pc's order position within the modular unit) maps a pc $x$ into itself in a different part of the space (where the pc image of $x$ occupies an immediately lower order position than the argument within the modular unit), or maps a pc $x$ into pc $y$ ($y - x = r - s$), such that both argument and image are neighbored by the same pair of pcs in the affinity space. In the *Dasian* space, the *transformatio* operation spans intervals 0 (between replicated pcs) and 1 (between different pcs).

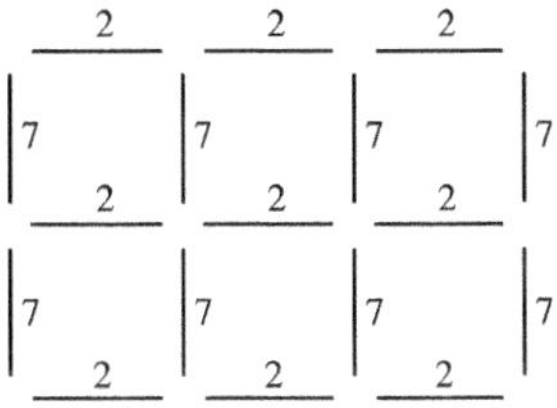


**Fig. 5.** Homogeneous T-net coordinating $dx = 2$ and $dy = 7$

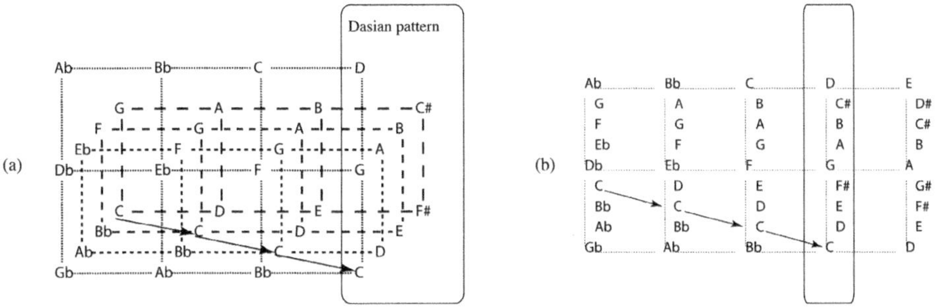


**Fig. 6.** (a) The *Dasian* space framed by the T-net in Fig. 5; vertical lines represent the *transpositio* relation and diagonal arrows represent the *transformatio* relation. (b) Framework of (a) simplified.

(or displaced) versions of the Dasian space, in particular common-tone relations between adjacent T-nets as captured by the arrows connecting instances of pc C in different i-cycles (*transformatio* relation)[9]. Figure 6(b) simplifies the layout Fig=6(a) by collapsing the T-nets replications along the vertical $dy$; this simplified version will be used in the remainder of the paper. The formulas at (6) and (7) define pitch relations in any affinity space formed by the coordination of homogeneous T-nets[10].

$$pc : Z^3 \rightarrow Z_{12}, pc(i,j,k) = pc(0,0,0) + i(dx(0,0,0)) + j(dy(0,0,0)) + k(dz(0,0,0)) \tag{6}$$

$$dx(i,j,k) = dx(0,0,0), dy(i,j,k) = dy(0,0,0), \text{ and } dz(i,j,k) = dz(0,0,0) \tag{7}$$

## 3 Progressive T-Nets

The transpositional relations that frame the homogeneous T-net are now expanded by allowing for a variation of values across both the $x$- and $y$-axis.

[9] The interval $dx$ corresponds to the step motion $r$, such that axes $z$ and $x$ are interchangeable in this representation.

[10] The value of $dz$ does not apply when $k$ is hosted at the i-cycle in order position $n$.

(a)

| | | | | | | | | | | | | |
|---|---|---|---|---|---|---|---|---|---|---|---|---|
| C | C | C | C | C | C | C | C | C | C | C | C | C |
| C | B | Bb | A | Ab | G | F# | F | E | Eb | D | C# | C |
| C | Bb | Ab | F# | E | D | C | Bb | Ab | F# | E | D | C |
| C | A | F# | Eb | C | A | F# | Eb | C | A | F# | Eb | C |
| C | Ab | E | C | Ab | E | C | Ab | E | C | Ab | E | C |
| C | G | D | A | E | B | F# | C# | Ab | Eb | Bb | F | C |
| C | F# | C | F# | C | F# | C | F# | C | F# | C | F# | C |
| C | F | Bb | Eb | Ab | C# | F# | B | E | A | D | G | C |
| C | E | Ab | C | E | Ab | C | E | Ab | C | E | Ab | C |
| C | Eb | F# | A | C | Eb | F# | A | C | Eb | F# | A | C |
| C | D | E | F# | Ab | Bb | C | D | E | F# | Ab | Bb | C |
| C | C# | D | Eb | E | F | F# | G | Ab | A | Bb | B | C |
| C | C | C | C | C | C | C | C | C | C | C | C | C |

(b)

**Fig. 7.** (a) Alban Berg's "master array" of interval cycles. (b) Berg's array represented as a progressive T-net.

A known pitch model for this construct, referred to here as the *progressive* T-net, is Alban Berg's "master array" of i-cycles[11]. Berg's array in Fig. 7(a) coordinates all i-cycles in both vertical and horizontal dimensions. Figure 7(b) recasts Berg's array as a T-net, emphasizing transpositional relations. Unlike in the homogenous T-net, however, some of the transpositional values in progressive T-nets vary throughout the network; in the case of the Berg's array, the transpositional values increase by 1 from left to right and from bottom to top in the network; in other words: $\Delta dx|y = \Delta dy|x = 1$. The constraints defining the change of transpositional values in a *generic* progressive T-net are stated at (8), and the formulas (9) and (10) calculate any pc and directed intervals at position $(i, j)$[12].

$$\text{In a progressive T-net: } \Delta dx|y \neq 0, \Delta dy|x \neq 0 \text{ and } \Delta dx|y = \Delta dy|x; \quad \text{and also, } \Delta dx|x = \Delta dy|y = 0 \tag{8}$$

$$pc: Z^2 \rightarrow Z_{12}, pc(i,j) = pc(0,0) + i(dx(0,0)) + (dy(0,0) + (\Delta dy|x)i)j, \text{ or}$$
$$pc: Z^2 \rightarrow Z_{12}, pc(i,j) = pc(0,0) + j(dy(0,0)) + (dx(0,0) + (\Delta dx|y)j)i. \tag{9}$$

$$dx(i,j) = dx(0,0) + j(\Delta dx|y) \text{ and}$$
$$dy(i,j) = dy(0,0) + i(\Delta dy|x). \tag{10}$$

Figure 8 shows a progressive T-net, whose transpositional levels are incremented by 9, i.e., $\Delta dx|y = \Delta dy|x = 9$, mod 12 (while $\Delta dx|x = \Delta dy|y = 0$). This progressive T-net provides a framework for the modeling of the closing

[11] See Perle [12]; for a contextualization of the role of Berg's array in the development of Perle's theories, see Headlam [13].

[12] Another structure for a progressive T-net not explored here is: $\Delta dx|x = \Delta dy|y \neq 0$ and $\Delta dx|y = \Delta dy|x = 0$.

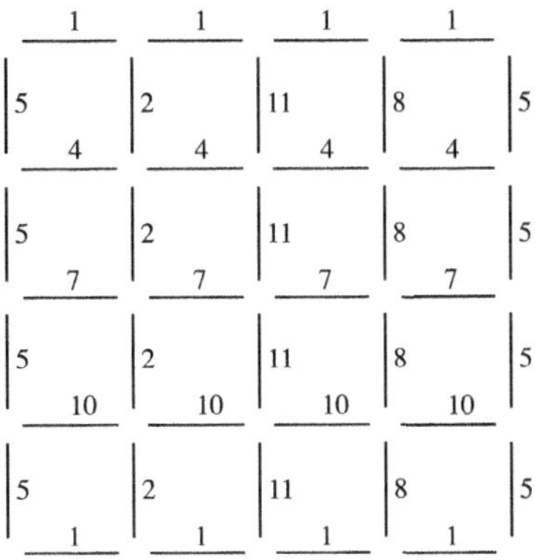


**Fig. 8.** A progressive T-net: $\Delta dx|y = \Delta dy|x = 9$; $\Delta dx|x = \Delta dy|y = 0$

section of Witold Lutosławski's song "Rycerze" (*Iłłakowicz Songs* 1956-57). The passage (mm. 181–199, reduced in Fig 9(a)) presents a series of "vertical" 12-tone chords, each being formed by the stacking of three distinct "fully diminished-seventh chords". Each 12-note chord can thus be thought of as a continuous segment of an affinity space. The five 12-note chords suggest four different affinity spaces structured by $5 = 3 \times 3 + 8$, $2 = 3 \times 3 + 5$, $11 = 3 \times 3 + 2$, and $8 = 3 \times 3 + 11 \pmod{12}$, for the first and fifth 12-tone chords, the second, the third, and the fourth respectively. The transpositional framework of Fig. 8 is "filled in" by the "diminished-seventh chords" in Fig. 9(b), thus mapping the different affinity space segments onto the progressive T-net. Lutosławski's closing passage is highlighted in the figure: while the 12-tone chord progression circles around all available $dy$-levels $< 5, 2, 11, 8, (5) >$ returning to its starting point at $dy = 5$, the stacking of the diminished-seventh chords does not cover all the theoretically available $dx$-levels $< 1, 10, 7, 4 >$, as the stacking does not use $dx = 4$; having done so, however, would create a repetition of the "lower" diminished-seventh chord and thus a tetrachordal redundancy in the 12-tone

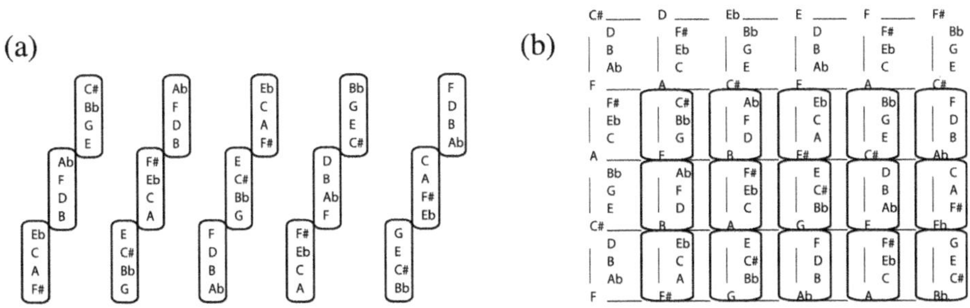


**Fig. 9.** (a) Lutosławski's 12-note chord succession in the closing section of *Rycerze* (mm. 181–199). (b) *Rycerze*'s closing section modeled by the mapping of affinity spaces within the progressive T-net.

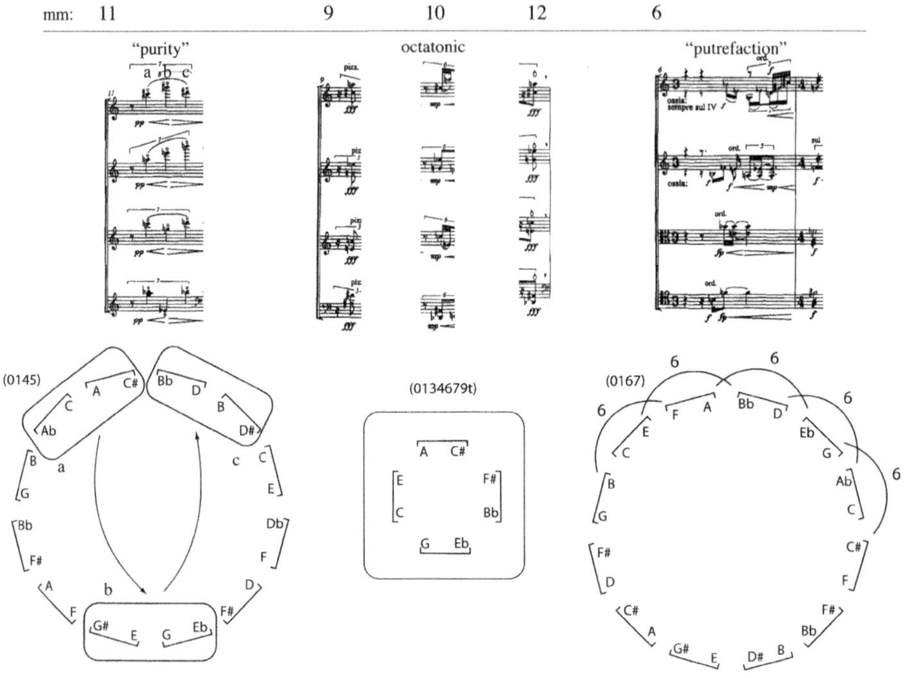


**Fig. 10.** Three affinity spaces in the first movement of György Kurtág's "Quartetto per Archi" op. 1 (1959)

chord[13]. The transpositional levels for the stacking and succession of fully diminished-sevenths in Lutosławski's passage also ensure that no common tones are retained in adjacent tetrachords both vertically (along the $y$-axis) and horizontally (along the $x$-axis).

While common-tone relations between adjacent affinity spaces are absent in Lutosławski's 12-tone chord succession in *Rycerze*, the progressive T-net suggested by the György Kurtág's "Quartetto per Archi" op. 1 (1959) maximizes common-tone relations between three distinct adjacent affinity spaces. Figure 10 presents three affinity spaces suggested by the quartet's first movement materials, modeled by the modular units: $1 = 1 \times 4 + 9$, $9 = 1 \times 4 + 5$, and $5 = 1 \times 4 + 1 \pmod{12}$[14]. It is interesting that the *transpositio* intervals of the

[13] The stacking of chordal structures that repeat some of the notes of the 12-tone chord does occur in *Zima* (*Iłłakowicz* Songs), where the lowest Aug-triad is rotated and repeated on top.

[14] Two of affinity spaces proposed here include "sonorities" and articulations which are associated with "purity" or "light" and "putrefaction," according to an autobiographical account of the composer's compositional struggles in Paris in the mid-50s (Hoffman [14]). In this sense the "octatonic" chords can be thought of as mediating between these extreme associations; Martins [15] elaborates on this reading for the first movement of Kurtág's quartet.

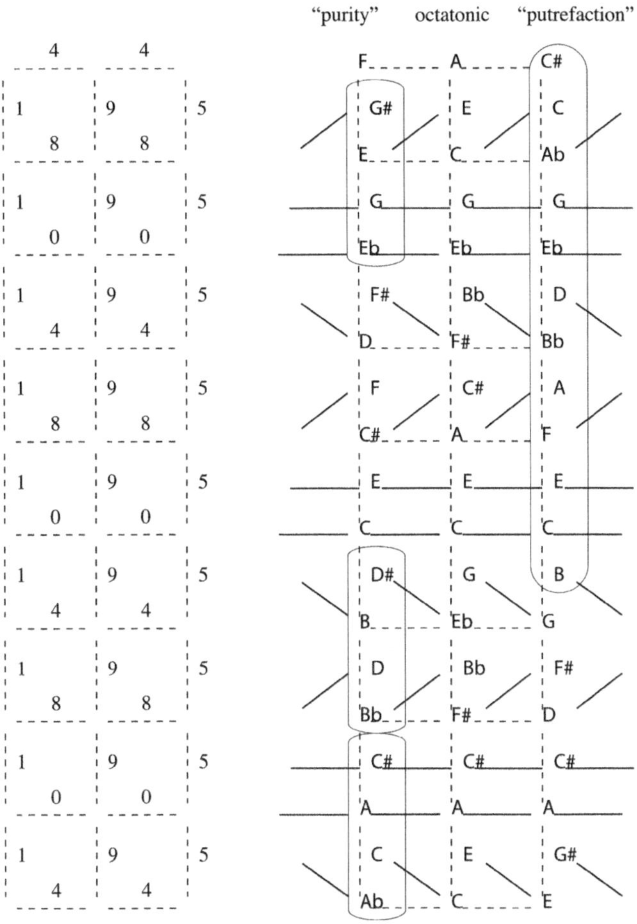


**Fig. 11.** A progressive T-net ($\Delta dx|y = \Delta dy|x = 8$ and $\Delta dx|x = \Delta dy|y = 0$) "filled in" by the three affinity spaces of Fig. 10. Horizontal and oblique lines connect common tones between adjacent affinity spaces.

three spaces $< 1, 9, 5 >$ are also present in each affinity space as the intervals $t$, $s$, and $t + r$. This coincidence allows for maximal common-tone relations between affinity spaces when framed by the progressive T-net of Fig. 10, in which $\Delta dx|y = \Delta dy|x = 8 \pmod{12}$. Unlike in homogeneous T-nets, where common-tone relations between affinity spaces (structured by *transformatio*) are represented by parallel oblique lines (cf. Fig. 6), in progressive T-nets, common-tone relations between adjacent affinity spaces are represented by either horizontal lines and/or mirroring oblique lines. In the progressive T-net of Fig. 11, common-tone relations "wrap around" the three spaces emphasizing the torus geometry of the T-net.

## 4 Dynamic T-Nets

This section expands further the versatility of T-nets by modeling a constant contraction or expansion of the network. Such networks are here referred to as *dynamic* T-nets. Whereas in progressive T-nets the transpositional values of $dx$ do not vary along the $x$-axis, and the values of $dy$ do not vary along the $y$-axis (i.e., $\Delta dx|x = \Delta dy|y = 0$), such is not the case in dynamic T-nets where the transpositional values of $dx$ and $dy$ do vary along their respective axes. This distinction is captured by the constraints (11) on the rate of change for directed intervals in a dynamic T-net; the formulas (12) and (13) calculate the values for any pc and directed intervals at an arbitrary position $(i, j)$.

$$\text{In a dynamic T-net: } \Delta dx|x = \Delta dy|y \neq 0 \text{ and } \Delta dx|y = \Delta dy|x \neq 0 \tag{11}$$

$$\begin{aligned} pc : Z^2 \to Z_{12}, pc(i,j) = pc(0,0) + \sum_{l=1}^{i}(dx(0,0) + (\Delta dx|x)(l-1)) \\ + \sum_{l=1}^{j}(dy(0,0) + (\Delta dy|x)i) + (\Delta dy|y)(l-1), \\ \text{where } l \text{ is an integer, or} \\ pc(i,j) : pc(0,0) + i(dx(0,0)) + (\Delta dx|x)(\frac{i(i-1)}{2}) \\ +((\Delta dy|x)i)j + j(dy(0,0) + (\Delta dy|y)(\frac{j(j-1)}{2}). \end{aligned} \tag{12}$$

$$\begin{aligned} dx(i,j) = dx(0,0) + j(\Delta dx|y) + i(\Delta dx|x), \text{ and} \\ dy(i,j) = dy(0,0) + i(\Delta dy|x) + j(\Delta dy|y). \end{aligned} \tag{13}$$

Figure 12 presents a dynamic T-net, which models the longer central section of Lutosławski's *Postludium I* (*Three Postludes*, 1959). The transpositional changes for this dynamic T-net are $\Delta dx|x = \Delta dy|y = \Delta dx|y = \Delta dy|x = 11$.

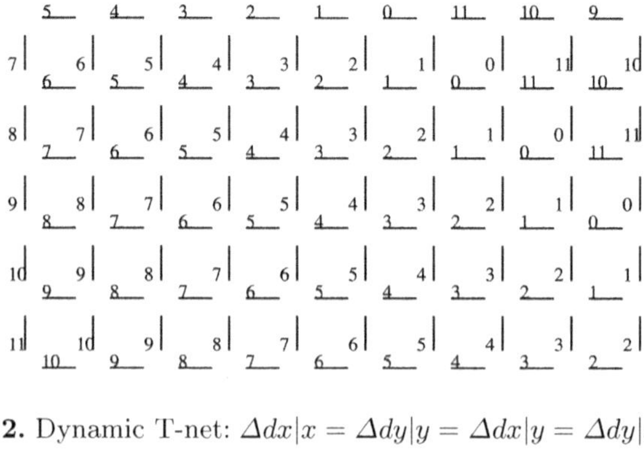


**Fig. 12.** Dynamic T-net: $\Delta dx|x = \Delta dy|y = \Delta dx|y = \Delta dy|x = 11$

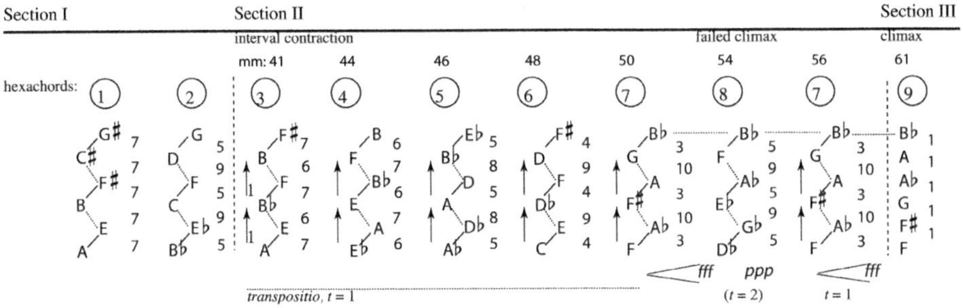


**Fig. 13.** Lutosławski's *Postludium I*, mm. 41–60 (*Three Postludes*, 1959); pitch reduction of gradually contracting affinity-space segments.

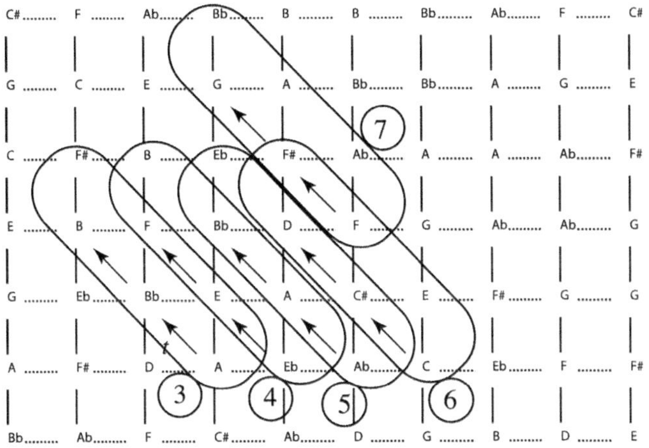


**Fig. 14.** Mapping of affinity-space segments into the dynamic T-net

Lutosławski's passage (mm. 41–60) reduced in Fig. 13 achieves a gradual contraction of pc-space by juxtaposing a series of affinity-space segments that tend towards a semitonal cluster, while retaining a constant interval of *transpositio* $t = 1 \pmod{12}$ (except hexachord 8). Figure 14 maps the succession of affinity-space segments of the section into the dynamic T- net of Fig. 12. Given the contracting aspect of the dynamic T-net from left to right and bottom to top, the affinity spaces are no longer mapped along the y-axes as in previously discussed T-nets, but are rather mapped in zig-zag lines along the southeast-northwest direction.

## 5 Conclusion

The paper proposes a framework that coordinates several models of pitch space whose constructive features rely on the concept of interval cycles and transpositional relations. This general model brings under a focused perspective diverse

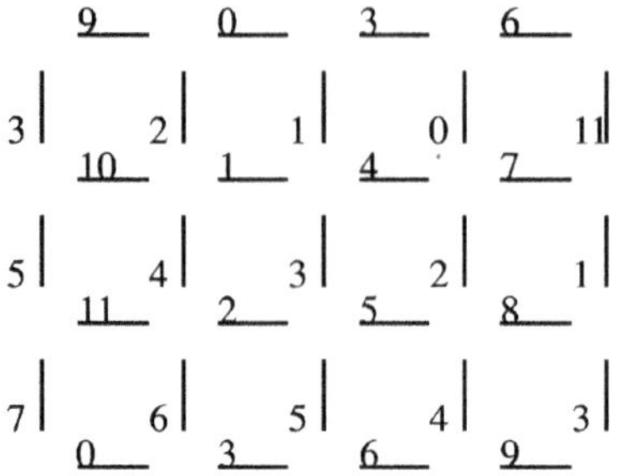


**Fig. 15.** *Fully* dynamic T-net

pitch structures such as *tonnetze*, *affinity spaces*, Alban Berg's "master array" of interval-cycles, and several types of transpositional networks, here referred to as homogeneous, progressive and dynamic T-nets. While the paper engages in incremental modifications to the constructive features of the *Tonnetz*, additional steps could be taken in the exploration of further deformations. For instance, Fig. 15 presents a dynamic T-net[15], which further deforms the changes of transpositional values while retaining a consistent structure (14).

$$\Delta dx|x = 3, \Delta dy|y = 10, \Delta dx|y = \Delta dy|x = 11. \tag{14}$$

## Acknowledgments

I would like to thank the three anonymous reviewers of MCM 2011 for their helpful commentaries as well as Akiyuki Anzai, Ruben Moreno Bote, Dmitri Tymoczko, and particularly C. Douglas Haessig for their valuable input.

## References

1. Cohn, R.: Introduction to Neo-Riemannian Theory: A Survey and a Historical Perspective. Journal of Music Theory 42(2), 167–180 (1998)
2. Journal of Music Theory 42(2), 167–348 (1998)
3. Cohn, R.: Neo-Riemannian Operations, Parsimonious Trichords and Their Tonnetz Representations. Journal of Music Theory 41(1), 1–66 (1997)
4. Douthett, J., Steinbach, P.: Parsimonious Graphs: A Study in Parsimony, Contextual Transformations, and Modes of Limited Transposition. Journal of Music Theory 42(2), 241–263 (1998)
5. Lewin, D.: Notes on the Opening of the F# Minor Fugue from WTCI. Journal of Music Theory 42(2), 235–239 (1998)

[15] This T-net could be referred to as *fully dynamic* as opposed to the previous *partially dynamic* where $\Delta dx|x = \Delta dy|y$; both dynamic T-nets, however, have the following constraint: $\Delta dx|y = \Delta dy|x$.

6. Martins, J.O.: Dasian, Guidonian, and Affinity Spaces in Twentieth-century Music. Ph.D. diss, U Chicago (2006)
7. Pesce, D.: The Affinities and Medieval Transposition. Indiana University Press, Bloomington (1987)
8. Gollin, E.: Multi-Aggregate Cycles and Multi-Aggregate Serial Techniques in the Music of Bla Bartk. Music Theory Spectrum 29(2), 143–176 (2007)
9. Lambert, P.: Interval Cycles as Compositional Resources in the Music of Charles Ives. Music Theory Spectrum 12(1), 43–82 (1990)
10. Carey, N., Clampitt, D.: Regions: A Theory of Tonal Spaces in Early Medieval Treatises. Journal of Music Theory 40(1), 113–147 (1996)
11. Martins, J.O.: Affinity Spaces and Their Host Set Classes. In: Klouche, T., Noll, T. (eds.) MCM 2007. Communications in Computer and Information Science, vol. 37, pp. 499–511. Springer, Heidelberg (2009)
12. Perle, G.: Berg's Master Array of the Interval Cycles. The Musical Quarterly 63(1), 1–30 (1977)
13. Headlam, D.: George Perle: An Appreciation. Perspectives of New Music 47(2), 59–195 (2009)
14. Hoffmann, P.: Die Kakerlake sucht den Weg zum Licht: Zum Streichquartett op. 1 von Gyrgy Kurtg. Die Musikforschung 44, 32–48 (1991)
15. Martins, J.O.: Bartók's Vocabulary and Kurtág's Syntax in the First Movement of the 'Quartetto per Archi', op. 1. Presented at the symposium: Bartók's String Quartets: Tradition and Legacy, University of Victoria (2008)

# Two-Dimensional Visual Inspection of Pitch-Space, Many Time-Scales and Tonal Uncertainty over Time

Agustín Martorell and Emilia Gómez

Music Technology Group, Universitat Pompeu Fabra
{agustin.martorell,emilia.gomez}@upf.edu

**Abstract.** This work explores the representational limitations of toroidal pitch-spaces, when multiple temporal resolutions, tone center ambiguity, and the time dimension are considered for visualization of music pieces. The algorithm estimates key from chroma features, over time at many time-scales, using the key-profile correlation method. All these estimations are projected as tonal centroids within Krumhansl and Kessler's toroidal space of inter-key distances. These centroids, belonging to a toroidal surface, are then mapped to colours by 3-dimensional geometric inscription of the whole pitch-space in the CIELAB colourspace. This mapping provides a visual correlate of pitch-space's double circularity, approximates perceptual uniformity of colours throughout near regions, and allows for representing key ambiguity. We adapt Sapp's keyscapes to summarize tonal centroids in pitch-space at many time-scales over time, in a two-dimensional coloured image. Keyscapes are linked with higher-dimensional tonal representations in a user interface, in order to combine their informative benefits for interactive analysis. By visualizing some specific music examples, we question the potential of continuous toroidal pitch-spaces in supporting long term analytical conclusions and tonal ambiguity description, when assisted by time vs. time-scale representations.

**Keywords:** Pitch-space, Toroidal, Multi-scale, Ambiguity, Visualization.

## 1 Introduction

Pitch-spaces, as theoretical formalizations of tonal concepts and relationships, facilitate mathematical and visual analysis of tonality from specific music examples. Introducing the time dimension into such analyses often requires understanding pitch-spaces as fixed structures, in which different musical *states* are activated at different spatial locations. This leads to a movement-in-space metaphor to model processes, since many of these spaces are fully informative just for *static* situations.

C. Agon et al. (Eds.): MCM 2011, LNAI 6726, pp. 140–150, 2011.

In this work[1], we address the representational limitations of toroidal pitch-spaces when analysis over time, at multiple temporal resolutions, and ambiguity of description are jointly pursued. We combine spatial models of tonality and summarization techniques to fit multi-scale representations over time, and we propose a computer-aided interactive framework to assist tonal interpretations of long segments of music.

## 2 Background

### 2.1 Tonal Cognition and Toroidal Spaces

Krumhansl and Kessler's (henceforth, K-K) psychological experiments lead to several toroidal pitch-space organizations [1]. One of such models, proposed as a space of inter-key distances, is derived from correlation between ring-shifted key profiles. Multidimensional scaling (MDS) of these correlations yields a reasonable four-dimensional solution of Euclidian distances between any pair of tone centers. In this 4-D space, two circular structures become evident: the double circle of fifths appears in the first two dimensions, while the last two dimensions reveal a circular arrangement of the relative and parallel relationships (Fig. 1(a)).

Given the topological homeomorphism between a ring torus and the cartesian product of two circles, solution points can be thought as belonging to a 3-D toroidal surface (Fig. 1(b)). This surface can be parameterized in two dimensions by the angular information, as long as double circularity is provided by *glueing* left-right and top-bottom borders (Fig. 1(c)).

Although the model of inter-key distances is only accurate as a 4-D space, representations in 3-D and 2-D are convenient for visualization and geometric manipulation. Similar toroidal distributions can be derived from Schoenberg's *charts of regions* [2], as well as from certain boundary conditions of Chew's *spiral array* [3]. Explicit toroidal spaces have been proposed in [4] and [5], among others.

### 2.2 Multidimensional Unfolding and Tonal Centroids

Being K-K's torus a space of inter-key distances, it can be used to represent the induced key of any input *pitch-class profile* by multidimensional unfolding [6]. The method discussed by Krumhansl and Toiviainen [7] finds, for a given 12-dimensional input vector, a point in the solution space (Fig. 1(c)) that optimizes distances from all key profiles. Although problematic for quantitative interpretation, this method helps representation of tonal ambiguity, since these *centroids* can be proportionally close to several keys.

[1] Given the relevance of colour, readers are encouraged to consult the electronic version. Figures are available at
`http://www.dtic.upf.edu/~amartorell/publications/`

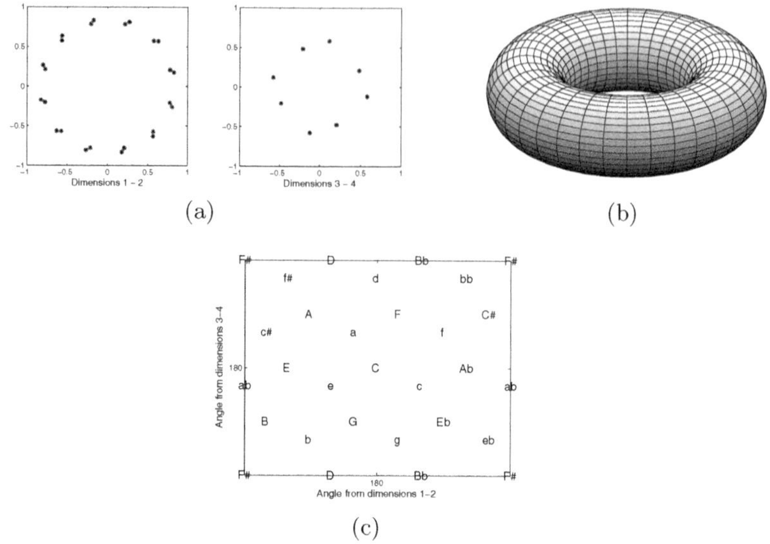


**Fig. 1.** (a) K-K's 4-D inter-key space. (b) 3-D toroidal surface, homeomorphic to both circles. (c) Angle-based 2-D unfolding of K-K's space.

### 2.3 The SOM Approach

Centroids in K-K's toroidal surface are not good summaries in some cases, in which proper solutions might exist in the 4-D solution, but fall *outside* the 3-D projection. For instance, centroids from symmetric inputs[2] cannot remain in the torus' surface while being proportionally close to all their closest candidates. This is a consequence of the asymmetrical hierarchies of K-K's (major and minor) key profiles, which characterize the solution space.

In [7], a toroidal self-organized map (SOM) is trained with K-K's key profiles, and the complete space is activated according to its similarity with the input vector. The higher-dimensionality of SOM allows for representing and discriminating ambiguous cases, and provides clear information about key strength.

### 2.4 Representation over Time and Temporal Multi-scale

These space-based representations are, however, clearly informative only on a *frame-based* approach. When it comes to represent musical processes, movement-in-space is a convenient metaphor, but visual problems start to arise for leaving traces of past frames. However, it is often of high interest to analyze long segments of music to observe far-reaching relationships along time.

Additionally, interpretation of tonality is particularly sensitive to temporal scale. At a given time point in music, it is often legitimate to understand different

[2] For example, pitch-class profiles from Messiaen's *modes of limited transpositions.*

keys when considering shorter or longer contexts. Depending on tonal discourse, some properties of key cannot be captured below or above a certain time-scale, as it seems counterintuitive to experience, in a general case, to grasp a sense of key under a fixed time-scale and hop-size basis.

Activation of pitch-space over time has been visualized as single images in SOM [8], and geometrical representations of different pitch hierarchies, involving two time-scales, have been animated on *Spiral Array* [9]. Both works present critical visual concerns. Spatial representation of concepts in SOM exhausts screens' dimensional constraints, making the time dimension barely recognizable, while Chew's moving visual information quickly overloads the short-time memory of observer. This reduces the chances for meaningful long-term visualization or for further exploitation of multi-scale.

Sapp's *keyscapes* [10] approach the other side of the problem, and represent key estimations over time at many time-scales in a 2-D image. For each frame at each scale, a pixel is coloured according to the strongest key candidate, reducing key representation to a single dimension.

## 3 Proposed Model

We propose a method for exploring the analytical potential of combining key estimations projected into continuous toroidal pitch-spaces with time vs. time-scale representations. We aim to preserve or to complement information across different dimensional reductions, taking both analytical and perceptual implications into consideration. The following discussed elements have been prototyped as an interactive tonal analysis application, built using the MIDI Toolbox for Matlab [11].

### 3.1 Signal Segmentation and Key Estimation

First, a chroma feature time-series is extracted from audio or MIDI files, and is segmented following a multi-scale sliding window policy. This is achieved by aligning many conventional sliding windows of sizes ranging from fractions of second to the whole duration of the piece. Following a basic key-finding method (MIDI Toolbox's implementation), a pitch-class profile is computed from each segment (henceforth, *frame*), and these 12-dimensional vectors are correlated with ring-shifted K-K's key profiles. The resulting 24-dimensional estimations, for all frames and time-scales, are projected to a K-K's pitch-space[3].

### 3.2 Mapping Key Estimations into Pitch-Space

Two complementary mapping methods are used for projecting the estimates, which respectively maximize dimensionality reduction and frame informativeness. The first method, virtually identical to that in [7], finds a tonal centroid in an angle-based parameterization of K-K's torus. The number of key candidates

[3] Coordinates of the 4-dimensional MDS solution are taken from [1], p.42.

to be considered and an *ambiguity factor*, used to promote or relax the strongest candidate, are introduced as parameters for unfolding. The second projection is provided by MIDI Toolbox's SOM visualization for key analysis.

### 3.3 Time vs. Time-Scale Representation

We propose an adaptation of Sapp's keyscape to visualize centroids, for all frames and time-scales, in a 2-D image. Instead of mapping categorical keys to colours, our continuous pitch-space requires some considerations for colouring: a) Centroid locations aim for perceptual closeness to keys. b) Ambiguity allows centroids to be anywhere in space. c) Toroidal pitch-spaces feature double spatial circularity. Given that, it is desirable for the colouring policy to account for perceptual distance and continuity or colour blending, and to keep these properties across the space's double circularity.

### 3.4 Geometrical Colourspace

CIE 1976 L*a*b* (known as CIELAB) colourspace, defined by the Commission Internationale de l'Éclairage as conversion standard [12], can approximate our needs. It uses absolute scales[4], it is device independent[5], and it is compressed to approximate perceptual uniformity. CIELAB is a 3-dimensional geometric space in which most of the human visible gamut is covered by a spherical sub-space. Three parameters define colour: $L*$ for luminance, $a*$ for green-magenta colour-opponent axis, and $b*$ for blue-yellow axis. Other CIELAB's parameterization, $LCh$, uses cylindrical coordinates: $L$ for luminance, $C$ for chroma saturation, and $h$ for hue angle.

### 3.5 Colouring Strategy

For assigning colours to centroids, the whole 3-D toroid (volume) is inscribed into CIELAB's *sphere*. This gives a unique colour for each point in pitch-space (toroidal surface) and gradual colour transitions along any direction, it approximates perceptual correlation with distance, and it guarantees double circularity. We match hue angle with the circle in first 2 dimensions of K-K's 4-D solution to provide maximum hue differences along both circles of fifths. The last two dimensions of the solution are parameterized as a circle in L and C axes. Geometrical manipulation allows for rotating the torus to match any colour to any point, which can be used for local compensation of the projection's distortion[6] respecting the 4-D space. The colouring method is schematized in Fig. 2.

The final keyscape is then plotted by aligning all coloured pixels, representing centroids in pitch-space, in a time vs. time-scale image.

4 Scales are relative to CIE's *standard illuminant* $D50$ white point.

5 Final result depends on the device's gamut and mapping equations. We use Matlab's LAB to sRGB conversion, based on CIE's *perceptual intent* recommendation.

6 Comprehensive discussion respecting inter-key distances in [1].

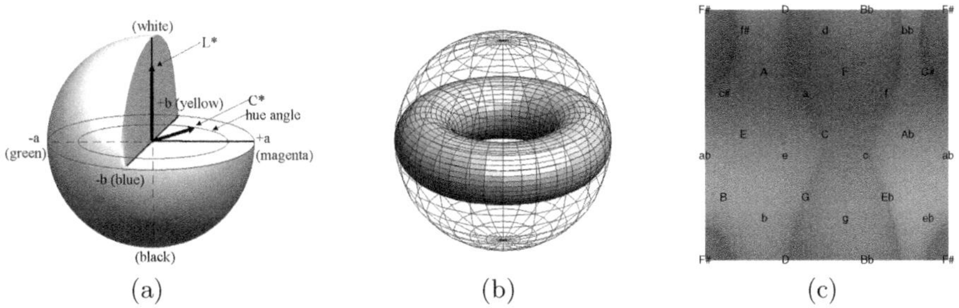


**Fig. 2.** Colouring process. (a) Spherical sub-space of CIELAB colourspace. (b) Geometrical inscription of toroidal pitch-space in spherical colourspace. (c) Unfolded torus' surface, coloured after its embedding in colourspace.

## 4 Discussion

The Finale of Haydn's String Quartet op.74 n.3 "Rider", in G minor, will serve to show the combined representational capabilities of the proposed model. The keyscape in Fig. 3(a) represents tonal estimations in time (x-axis) vs. time-scale (y-axis). The higher the position in the keyscape, the larger the time-scale used for key estimation. Unfolding has been parameterized to promote the first key candidate, so centroids are close to single keys. In the image, homogeneous colours covering large areas reveal stable tonal sections. Both coloured pitch-spaces in Fig. 3(b) and Fig. 3(c) serve as map legend between the keyscape and the pitch-space.

### 4.1 Formal Analysis

The piece begins with a first theme in Gm, followed by a longer second theme group in Bb. Then, a development section wanders around 3 and 4-flat key signatures (each one as different blue tone). Recapitulation takes the first theme in Gm, and a dominant pedal leads to the second theme group and conclusion in G major[7].

The complete piece is estimated as Gm (very top of the keyscape), a somewhat surprising result given the relative short duration in this key. The proposed connection of keyscape with pitch-space helps to clarify this analytical aspect at structural level. This is shown in Fig. 3(b) by projecting all the centroids (black dots) for the highest time-scales (above horizontal dashed line in the keyscape) and observing the relative *weight* reached at each location (since centroids fall very close to keys, their accumulation is approximated by the circles' sizes). The exposition and recapitulation sections have been summarized as Bb and G respectively, given the longer duration of their second theme groups. They are almost balanced in duration, and Gm falls approximately half-way between them in pitch-space. The shorter development section, summarized at Cm, gently

[7] Analysis by first author.

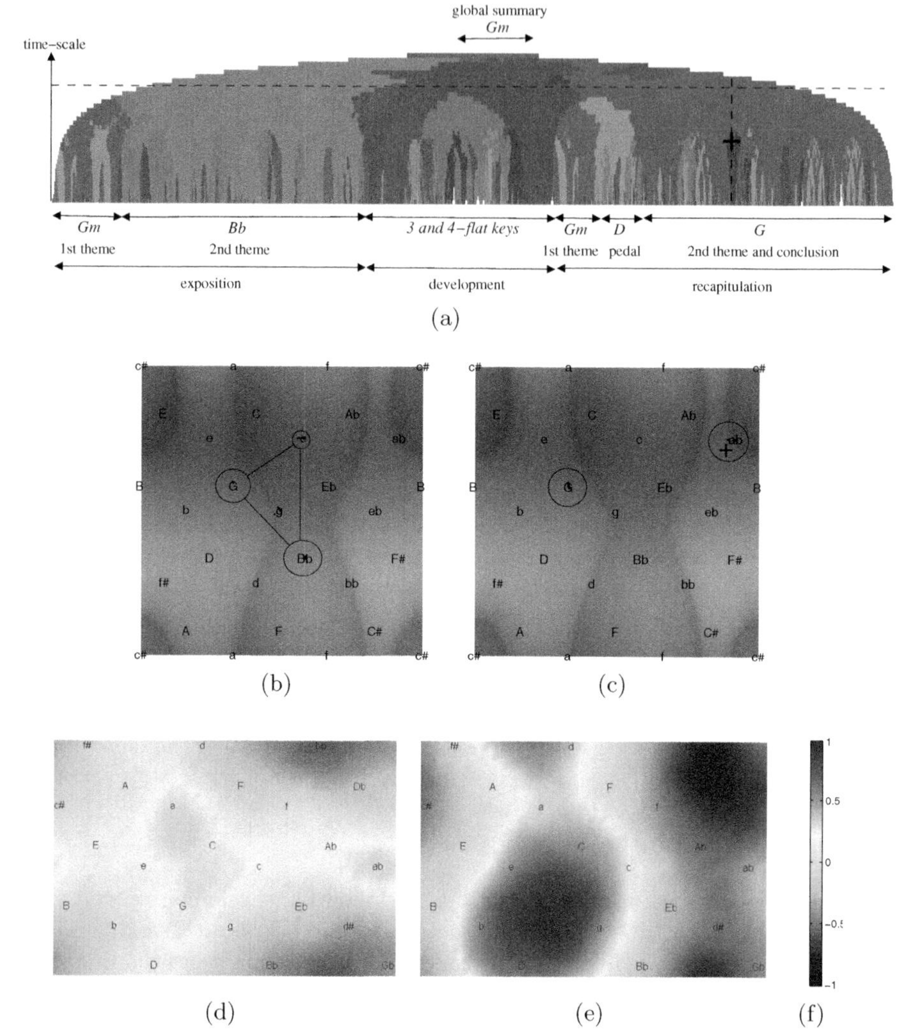


**Fig. 3.** Finale of Haydn's "Rider" String Quartet. (a) Keyscape. (b) Spatial-structural relationships at the highest time-scales. (c) Clustered centroids for the selected time-frame. (d) SOM activation for the selected event. (e) SOM activation for the selected event's context. (f) SOM key strength colourbar.

pulls the whole-piece centroid towards the correct key of the piece. This spatial interpretation of *classical symmetry* is a consequence of the topology of pitch-space, and two aspects linking the key estimation method with some of the piece's *aesthetical* concerns. First, the hierarchy implied in K-K's profiles is well suited to pitch-class statistical analysis of classical works. Second, our duration-based computation of pitch-class profiles matches the classical style's ideals of temporal balance. It's worth mentioning that the given interpretation has not been computed as a *systemic* reduction: all the estimates are computed taking into account the overall pitch-class content in their segments.

### 4.2 Tonal Mining in More Dimensions

The keyscape provides a big picture tonal description of the piece, although quite summarized. We propose to use keyscapes in interaction, in order to uncover the missing information. A point in the keyscape has been selected (+ sign) in Fig. 3(a) for a more detailed inspection of the corresponding segment. All the centroids aligned in time with the selected frame are represented in Fig. 3(c), showing a *vertical reading* at that time point (vertical dashed line in the keyscape). The tonal meaning arises: clustered centroids (surrounded by circles in the pitch-space) identify an Ab minor *event* within a G context.

Centroids are strong simplifications of the actual estimates. The correlation stage of our key-finding model ouputs 24-dimensional vectors, which are *forced* to become single points in the torus by the unfolding stage. A more explanatory view of the selected segment is given by the SOM in Fig. 3(d), where a notable key uncertainty appears as weak activations throughout and no clustered isolated region. This ambiguity contrasts with the clarity and strength of the activation produced by its context in G major, just few time-scales above, as shown in Fig. 3(e).

### 4.3 Other Aspects of Tonal Ambiguity

Chopin's Prelude in E major will serve to illustrate two different unfolding parameterizations (Fig. 4). This piece can be structured in three phrases, each of them beginning and ending in E. The last two phrases take distant journeys in pitch-space, and considerable analytical freedom respecting tone center induction is allowed.

The keyscape in Fig. 4(a) maps centroids in pitch-space by promoting the strongest key, showing homogeneous colours and neat boundaries between regions. Typical analyses of modulation and tonicizations for this piece are found for time-scales between the dashed lines. Compared with the tonal stability of the first phrase, the keyscape shows important key shifting activity for the last two sections. The second phrase visits G-C-Fm-Ab-G♯m, and the third section *crosses* Am-F-Dm-Gm-G-Em. A comparable multi-scale tonal analysis through pitch-space for this piece is given in [5]. The model estimates Am in the third phrase, which is suggested by Lerdahl to fit the *shortest path* rule from E to F. Similarly, Dm is estimated as mediating between F and Gm. The algorithm fails to estimate the first phrase in E, in favour of B, as well as the global key of the piece.

Centroid unfolding for the second keyscape, in Fig. 4(b), considers the two strongest candidates and their weights. The image looks similar but it's significantly *fuzzier*, being informative in a different (not necessarily better) way. Most areas represent spatial locations between two keys, by proportional *blending* of both candidates' colours. The continuous colourspace aims to facilitate

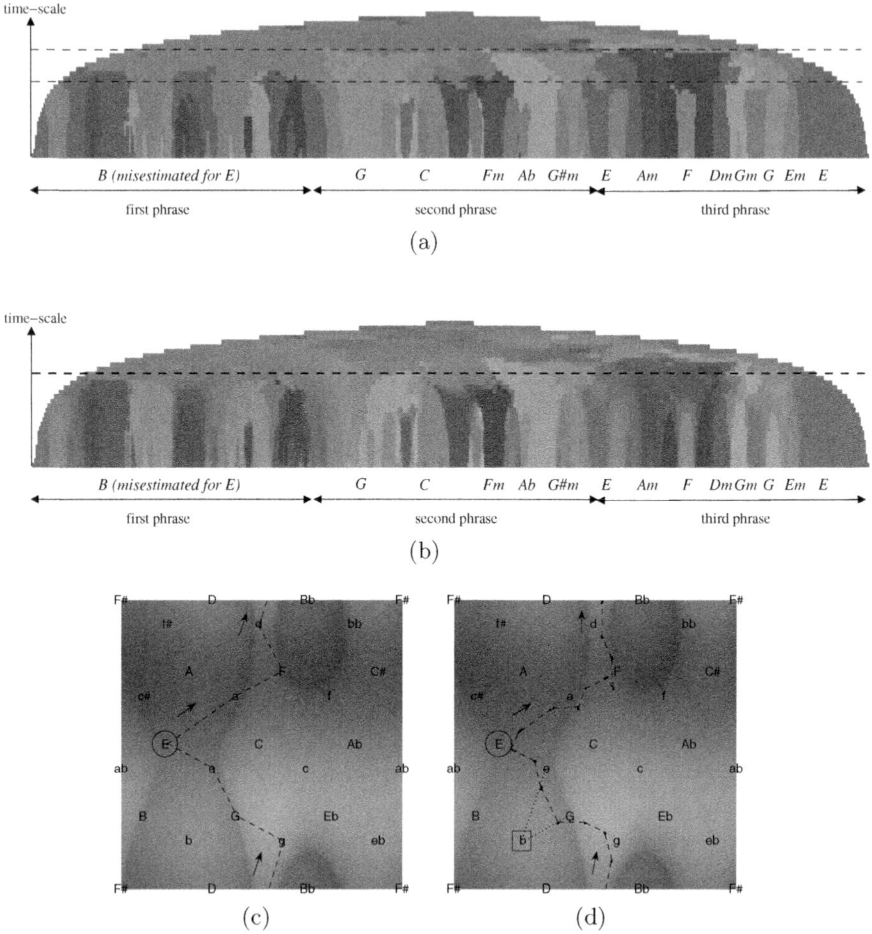


**Fig. 4.** Chopin's Prelude in E. (a) Keyscape promoting strongest candidate. (b) Keyscape from the first two candidates. (c) Path followed by the third phrase under categorical unfolding. (d) Path of the third phrase under ambiguous unfolding.

the distinction between movement towards neighbour keys (soft mixture of both colours) and far jumps (sudden strong contrast). As a difference with other colouring schemes, like that in [10], the proposed geometrical mapping approximates gradual colour transitions in any direction, so all neighbour tonal relationships would get benefited from the perceptual closeness.

Fig. 4(c) shows the path followed by the centroids for the third phrase under categorical unfolding (black dots fall right in the key centers). The same phrase is shown in Fig. 4(d) for the ambiguous case. The time-scale used in both cases is shown in the ambiguous keyscape (horizontal dashed line). Instead of jumping to categorical keys, some intermediate positions soften the way and the corresponding colours in the ambiguous path. By observing the centroids' deviations from their closest keys it is possible to grasp which keys were taken as second candidates, as it's illustrated in the ambiguous path during the transition between G and E minor. It's worth noting that the most critical aspect for achieving gradual colour shifting is the capability of the key estimation and centroid unfolding methods to provide such gradual movement in pitch-space, and this necessarily depends on musical discourse as well.

## 5 Conclusions and Future Work

We have proposed a representation of multi-scale key estimation over time, where pixels represent uncertain centroids in a continuous pitch-space. This mapping is achieved by introducing a geometrical continuous colourspace. Big picture keyscapes are linked with higher-dimensional representations, which can account for extended tonal description. We have pointed to the potential of continuous pitch-spaces in supporting long term analytical conclusions, when assisted by time vs. time-scale representations. The method also facilitates the inspection of different cases of tonal ambiguity.

Our prototype proposes an interactive selection of segments and trajectories in the keyscape, which acts as the entry point for analysis. This information is complemented by the higher dimensional models and contrasted by audition, so it could be useful for refined evaluation of key modeling algorithms.

We expect that interacting with these combined representations can contribute to better understanding of the analytical relevance of time-scale in tonality. The exploitation of human vision abilities for reaching conclusions from ambiguous evidence might play an important role in such interaction.

## Acknowledgements

This work has been supported by the projects Classical Planet: TSI-070100-2009-407 (MITYC) and DRIMS: TIN2009-14247-C02-01 (MICINN).

## References

1. Krumhansl, C.L.: Cognitive Foundations of Musical Pitch. Oxford University Press, New York (1990)
2. Schoenberg, A.: Structural Functions of Harmony (rev. ed.), Norton, New York (1969) (Originally published, 1954)
3. Chew, E.: Towards a Mathematical Model of Tonality. Ph.D. dissertation, Massachussets Institute of Technology (2000)
4. Werts, D.: A Theory of Scale References. Ph. D. dissertation, Princeton University (1983)
5. Lerdahl, F.: Tonal Pitch Space. Oxford University Press, New York (2001)
6. Coombs, C.H.: A Theory of Data. Wiley and sons, New York (1964)

7. Krumhansl, C.L., Toiviainen, P.: Tonal Cognition. In: Peretz, I., Zatorre, R. (eds.) The Cognitive Neuroscience of Music, pp. 95–108. Oxford University Press, New York (2003)
8. Janata, P.: Navigating Tonal Space. In: Hewlett, W.B., Selfridge-Field, E., Correia Jr., E. (eds.) Tonal theory for the digital age. Computing in Musicology, vol. 15, pp. 39–50. CCARH, Stanford (2007)
9. Chew, E., François, A.R.J.: Interactive Multi-scale Visualizations of Tonal Evolution in MuSA.RT Opus 2. Computers in Entertainment 3(4), 1–16 (2005)
10. Sapp, C.S.: Visual Hierarchical Key Analysis. Computers in Entertainment 3(4), 1–19 (2005)
11. Eerola, T., Toiviainen, P.: MIDI Toolbox: MATLAB Tools for Music Research. University of Jyväskylä, Kopijyvä, Jyväskylä, Finland (2004), Electronic version `http://www.jyu.fi/musica/miditoolbox/`
12. CIE 015: 2004. Colorimetry, 3rd ed. Commission Internationale de L'Éclairage (2004)

# Musical Composition and Gestural Diagrams

Guerino Mazzola and Florian Thalmann

School of Music, University of Minnesota
{mazzola,thalm007}@umn.edu

**Abstract.** By an adjoint functor argument, we reinterpret categorical gestures as being "continuous diagrams" with values in topological categories, which we therefore call "gestural diagrams". This allows to view traditional transformational diagrams as canonical restrictions of gestural diagrams and to reinterpret musical gesture theory in a natural way as a topological extension of transformational theory. We apply these tools to extend the concept of a musical score to a "processual diagrammatic score". Such a score not only captures the result of a compositional effort but also the poietic process and its underlying gestures. These conceptual extensions can be modeled on the level of denotators and forms so that an implementation for the Rubato Composer software becomes feasible. Recent developments in this software enable the definition of affine transformations using finger gesture input on trackpads. Once such gestures are abstracted in a transformational processual diagram we introduce a Bruhat decomposition argument for $SL_2(\mathbb{Z})$ to reconstruct canonical gestural diagrams. Based on this model, we suggest new ways of graphical software interaction that facilitate dynamic navigation and intervention in the composition's history.

**Keywords:** Gesture, Transformational Theory, Poiesis, Musical Composition, Bruhat Decomposition, Rubato Composer, Big Bang Rubette, Diagrammatic Score.

## 1 Introduction

This paper takes some essential steps towards the formal, i.e. mathematical, and eventually computational, integration of different ontological layers of music as they have been described in [1, 24.6]. This so-called *oniontology* distinguishes three layers that relate to the degree of facticity of musical phenomena, see Fig. 1. These are the layers of facts, processes, and gestures, and they differentiate between the results, the schemes of accessing these results, and the innermost gestural activities that give life to those schemes. We believe that it is important to understand not only the resulting fact of musical activities, be they compositional or performative, but also their genesis, the making of music from its gestural germs through the processes which cast gestural dynamics, and finally the processes' output, the facts that define the endpoints of musical endeavors.

In terms of oniontological levels the traditional notion of a musical composition is merely associated with the factual level. Its score only contains the final

C. Agon et al. (Eds.): MCM 2011, LNAI 6726, pp. 151–164, 2011.

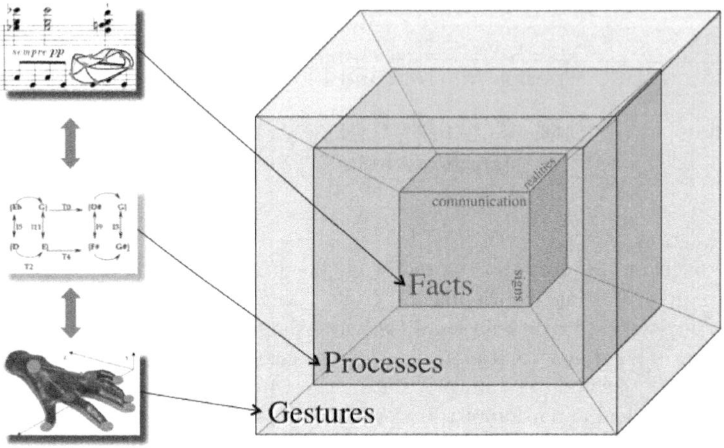


**Fig. 1.** The three layers of the oniontological topography of music

result of the poietical process of composition. The entire logic of its construction (forming the processual level) is lost after its completion. Evidently, for both the process of composition and the reception of a musical work, the processual logic is of great importance. In turn, the reification of such processual logic is based on the gestural level of communication.

It is remarkable that the history of musical scores is above all one of an incessant refinement of performative expression, while the poietical roots are virtually never traced in the score. It might be that this asymmetry is due to the composers' cultivation of mystery, their refusal of explicit declarations about how they unfold their compositions. But it is certainly also due to the mere technical difficulty to formally capture poietical work. In this paper we propose such a formalism. We do not only believe that it may grasp a good number of compositional strategies, but we also hope that it may make transparent many technical strategies, and thereby also enable composers to work with more subtlety on their creative landscapes.

This more philosophical insight is however not easy to make explicit, precise and eventually operational. In this paper we embark on the mathematical conceptualization of these layered realities, and more precisely on the transitions between the processual and gestural oniontological levels. We do this in the general conceptual architecture of forms and denotators described in [2] and set into action in the Rubato Composer music software [3]. This has two advantages: first, it requires a disciplined shaping of all concepts and second, it opens the door to a computational implementation of abstract mathematical constructions, and provides a new way of interacting with musical compositions, not only with the result itself (factual level), but with its historical evolution (processual level). Such interaction can be operationalized using gestures, e.g. using a computer mouse or a trackpad (gestural level).

Our approach consists in two basic steps: First, we develop the theory of categorical gestures as introduced in [4] in order to be able to deal with compositional activities as a dancing movement as dreamed of by David Lewin [5], and second, we step from single objects (notes, motives, chords, etc.) to objects which embrace their historical trace, the path from original shapes to the actual expression.

We have to express the trace of poietical labor as an integral part of our musical objects.

## 2 Preliminaries on Gesture Categories

Before we embark on the discussion of compositional issues, we have to introduce a number of mathematical constructions related to categorical gestures and necessary to describe our formal discourse.

Recall from [4] that a gesture is a digraph morphism $\gamma : \Gamma \to \overrightarrow{X}$ from a directed graph $\Gamma$ (the gesture's skeleton), to the spatial digraph $\overrightarrow{X}$ of a topological category $X$ (the latter being called the gesture's body). The digraph $\overrightarrow{X}$ has the objects of $X$ as its vertices, while the arrows are the continuous functors $c : \nabla \to X$ on the simplex category $\nabla$ associated with the unit interval $I$. Recall from [4] that $\nabla$'s morphism set is $\nabla = \{(x,y)|x,y \in I \text{ and } x \leq y\}$, $d(x,y) = (x,x)$, $c(x,y) = (y,y)$, the composition of morphisms is obvious, and the topology on $\nabla$ is the relative topology inherited from the usual product topology on $I \times I \subset \mathbb{R} \times \mathbb{R}$. In [6], we described another view of gestures in terms of the natural gesture $g(\Gamma)$ associated with digraph $\Gamma$. Let us now adapt this construction to the categorical gesture context.

To this end, we endow the topological space $|\Gamma|$ associated with $\Gamma$ in [6, section 5] with the structure of a topological category[1], the *continuous path category of* $\Gamma$, denoted by $CPath(\Gamma)$, as follows.

To do so, and generalizing the construction of $\nabla$ from the unit interval $I$, we define the following set of morphisms. Given a natural number $n = 0, 1, 2, \ldots$, the morphisms of length $n$ are these: for $n = 0$, the morphisms are the identities (the objects of the category), namely either the limiting points of one of the curves $c_a$ corresponding to arrows $a$ of $\Gamma$, or else the pairs $(c_a, \mu)$, where $0 < \mu < 1$, i.e. the internal points.

The morphisms of positive length $n = 1$ are the triples $(c_a, \mu, \nu)$, where $0 \leq \mu < \nu \leq 1$ are points on the curve $c_a$. Observe that even if both points with $\mu = 0, \nu = 1$ coincide for a loop $c_a$, the morphism is not the identity on that point! For morphisms of length $n > 1$, we start with a path $p = a_n a_{n-1} \ldots a_1$ of length $n$ in the path category $Path(\Gamma)$ of $\Gamma$ and then pick a point $\mu < 1$ on $c_1$ and a point $0 < \nu$ on $c_n$. In other words, we consider a "continuous path" from $\mu < 1$ on $c_1$ to $0 < \nu$ on $c_n$, represented by the triple $(p, \mu, \nu)$, passing through all intermediate curves $c_i, i = 2, \ldots n-1$ in the direction indicated by the arrows $a_i$ they represent, see Fig. 2 for an example.

[1] i.e. of a category internal to the category **Top** of topological spaces. We denote the category of topological categories by **TopCat**.

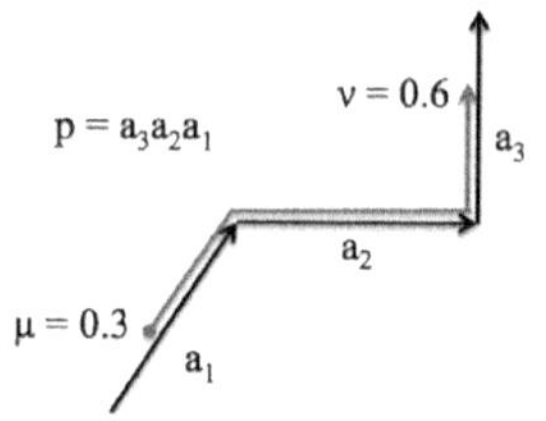


**Fig. 2.** A morphism of length 3 in the path category $Path(\Gamma)$

Call this category the *continuous path category of* $\Gamma$, denote it by $CPath(\Gamma)$. Clearly, this defines a functor $CPath : Digraph \to$ **TopCat**. If we embed the category **Cat** of (small) categories in **TopCat** by the indiscrete topology, we have a natural transformation $C : Path \to CPath$, which identifies the paths $p = a_n a_{n-1} \dots a_1$ on a digraph $\Gamma$ with the continuous paths $(p, 0, 1)$ reaching from initial to final points on arrow curves. Conversely, the discrete gesture construction $\searrow(g) : \nabla \to K$ for any morphism $g$ in a category $K$ (endowed with the indiscrete topology) gives rise to a continuous functor $\searrow(\Gamma) : CPath(\Gamma) \to Path(\Gamma)$ ($Path(\Gamma)$ with the indiscrete topology), which then combines to an embedding plus left-inverse projection

$$Path(\Gamma) \xrightarrow{C(\Gamma)} CPath(\Gamma) \xrightarrow{\searrow(\Gamma)} Path(\Gamma).$$

This is the evident extension of the classical path category construction to "continuous arrows", and it can also be completed to linear functors

$$RPath(\Gamma) \xrightarrow{RC(\Gamma)} RCPath(\Gamma) \xrightarrow{R\searrow(\Gamma)} RPath(\Gamma)$$

on linearized categories for given commutative rings $R$.

With these continuous path categories, we now have an evident adjoint functor isomorphism

$$CP : Digraph(\Gamma, \overrightarrow{X}) \xrightarrow{\sim} \mathbf{TopCat}(CPath(\Gamma), X)$$

for digraphs $\Gamma$ and topological categories $X$. We call the continuous functor $CP(\gamma) : CPath(\Gamma) \to X$ associated with gesture $\gamma$ its *gestural diagram* since it is a continuous extension of a classical diagram in $X$, the latter being given here by the restriction $CP(\gamma)|Path(\Gamma) : Path(\Gamma) \to X$, or, equivalently, by the digraph morphism $CP(\gamma)|\Gamma : \Gamma \to Dig(X)$ into the digraph $Dig(X)$ underlying the category $X$. In order to keep terminology clear, we shall sometimes call *processual diagrams* the classical diagrams (defined by a path category), as opposed to gestural diagrams (defined by a continuous path category).

### 2.1 Form Diagrams

In what follows, we shall deal with processual and gestural diagrams associated with the category **Forms** of forms, as they are used in the concept architecture of denotators described in [2, G.5.3]. Recall that a form $F$ has a name denotator[2] $N(F)$ and a contravariant functor $fun(F)$ : **Mod** $\to$ **Sets**, which is uniquely determined by the form's name, $fun(F) = fun(N(F))$.

A diagram of forms is a diagram $d : D \to$ **Forms**, or, equivalently, a diagram $d : D \to$ **FormNames** of form names. To define paths, we use the linear digraph $[n]$ with $n+1$ vertices $0, 1, \ldots n$ and $n$ arrows $a_i = (i, i+1), i = 0, 1, \ldots n$. Whenever we have a path $p : [n] \to D$ of length $n$ in the digraph $D$ of a form diagram, this induces a path of forms $dp : [n] \to$ **Forms**. For any such path $dp$, if $x = DN : A@F_0(\xi)$ is a denotator of form(name) $F_0 = dp(0)$ and value $\xi \in A@fun(F_0)$, $x$ generates a path of denotators $dp(x)$, whose value $x_i = dp(x)(i)$ at vertex $i$ is the evaluation of the corresponding concatenation of the form morphisms of path $dp$ at the initial denotator's value $\xi$ until form $F_i = dp(i)$.

If we want to apply gesture theory to forms, it is necessary to consider topological categories. We may therefore consider the situation where form diagrams are situated in subcategories $X \subset$ **Forms** which have been given a topology, in short: with topological subcategories. In musical situations many such categories may be constructed. We shall only need a minimum of such subcategories in what follows, but we stress that gesture theory on denotators will enforce adequate choices of topological subcategories of forms. A first example of such a subcategory is given by the transformational context on the standard space $\mathbb{R}^5$ of five real parameters onset, pitch, loudness, duration, and voice. One may then consider the simple form with name $Note$ and functor $fun(Note) = @\mathbb{R}^5 = Hom(-, \mathbb{R}^5)$. Our category $X$ then has as morphisms all invertible affine morphisms $@T^t f, t \in \mathbb{R}^5, f \in \mathrm{GL}_5(\mathbb{R})$ on $\mathbb{R}^5$. It is just the topological category of the general affine group $\overrightarrow{\mathrm{GL}_5}(\mathbb{R})$. More generally, we may consider a number of different simple spaces $F_i, i = 1, 2, \ldots k$ with associated modules $\mathbb{R}^{n_i}$ and morphisms $f : F_i \to F_j$ defined by matrices $M_f \in \mathbb{M}_{i \times j}(\mathbb{R})$, the topology being the canonical topology of the disjoint union of the real topology matrix spaces associated with the morphism spaces $F_i@F_j$.

In what follows, we shall also use form morphisms which are not defined by affine morphisms. We shall typically work on the category of functors that are not necessarily representable. Take for example the form $ZAN$ ($ZAN$ standing for "zero addressed notes") that takes the values $fun(ZAN)(A) = \{f : A \to \mathbb{R}^5 | f \text{ factorizes through } 0\}$ for a module $A$. Then consider as morphisms all continuous maps $c : \mathbb{R}^5 \to \mathbb{R}^5$ with the maps they induce on $fun(ZAN)(A)$, and take the compact-open topology, which defines a topological category since $\mathbb{R}^5$ is locally compact Hausdorff. This applies to situations, where musical transformations are not defined by affine morphisms and also not defined for arbitrary addresses, but only for the zero address module. Let us give an example of such a situation.

*Example 1.* In performance theory [7], one typically considers two forms: first, the above form $Note$, whose denotators represent symbolic notes as typically

[2] The name denotator is a denotator whose form is simple, of module $\mathbb{Z}\langle UNICODE\rangle$ e.g. the monoid $\mathbb{Z}$-algebra over the set $UNICODE$ of Unicode symbols.

described in scores. Performance then maps such note denotators to sound event denotators in a second simple form *Sounds*, which replaces the symbolic parameters by corresponding physical ones. Symbolic "Onset", typically measured in quarter note units, is replaced by physical "onset", measured by seconds, "Pitch", typically measured by semitone steps, is replaced by physical "pitch", measured in logarithms of frequency, and so on. The *Sound* form module is again $\mathbb{R}^5$ and the performance map $\wp$ is a differentiable map $\wp : \mathbb{R}^5 \to \mathbb{R}^5$, representing the performance transformation on zero-addressed note denotators.

It may also happen that we do not assume such maps to be continuous in a standard topology. Let us look at the following example:

*Example 2.* For the topological operation of alteration on the *Note* form, one generates a "force field" $A$ that moves zero-addressed note denotators $n = N : 0@Note(\nu)$ in certain directions and amounts to: $\nu \mapsto A(\nu)$, see [8, 16.3 and 17.4.3] for details. The force field is defined by attraction forces induced by a set $\{n_1, \ldots n_k\}$ of attractor notes. In general, this is not a continuous map on $A : \mathbb{R}^5 \to \mathbb{R}^5$ for the usual topology, since neighboring notes may be attracted to different attractor notes according to their relative position with respect to nearest attractors. This situation occurs in particular for the well-known operations of parameter quantization, which is a very special type of alteration. In order to include this case in our topological approach, one would select any locally compact Hausdorff topology on $\mathbb{R}^5$ for which these alteration maps are continuous, and proceed as above with the compact-open topology.

### 2.2 Gesturally Defined Form Diagrams

In general, the definition of a diagram of forms $D : \Gamma \to X \subset$ **Forms** on a topological subcategory $X \subset$ **Forms** can be given by arbitrarily abstract procedures, such as matrices and similar formulaic procedures. But in practice, there are concrete methods for the diagram construction, which simplify considerably the necessary procedures. Here is a method, which is also used in the Big Bang Rubette of the Rubato Composer software [8, 18.5.2]. It works for the construction of two-dimensional affine transformations, but since any $n$-dimensional affine transformation is the product of two-dimensional factors (leaving fixed all but two dimensions), this is a quite generic method. To clarify, we are now dealing with the transition from gestural diagrams to processual diagrams (see Fig. 1). The inverse of this transition will be discussed in the next section.

As discussed in [9], any two-dimensional affine transformation can be defined by a standardized three-finger multitouch gesture. For three starting points $p_s^i = (x_s^i, y_s^i)$ on a two-dimensional plane and three intermediate or ending points $p_e^i = (x_e^i, y_e^i)$, $i = 1, 2, 3$, corresponding to the starting and current locations of three fingers, we define the two vectors $v_j = p_j^2 - p_j^1$ for $j = s, e$. Furthermore, we define the distances $d^i = p_e^i - p_s^i, i = 1, 2, 3$, and the normalized vectors $\hat{v}_j = \frac{v_j}{|v_j|}$. Consequently we define the following five components of an affine transformation:

- the difference $d^1$ defines the translation component,
- $\delta = \frac{|v_e|}{|v_s|}$ forms the dilation component,
- the angle $\theta = arccos(\hat{v}_e \cdot \hat{v}_s)$ forms the rotation component,
- the projection length $\sigma = |(d^3 \cdot \hat{v}_e)\hat{v}_e|$ determines the shearing component, and
- the projection length of $\rho = |(d^3 \cdot \hat{u}_e)\hat{u}_e|$, where $u_e$ is a vector perpendicular to $v_e$, defines the reflection component.

From these numerical components, exemplified in Fig. 3, we can finally define the unique corresponding affine transformation. The Big Bang Rubette's graphical interface enables the use of these standard gestures to quickly apply sequences of morphisms to selections of *Note* denotators. These sequences of morphisms are in fact instances of such Form diagrams.

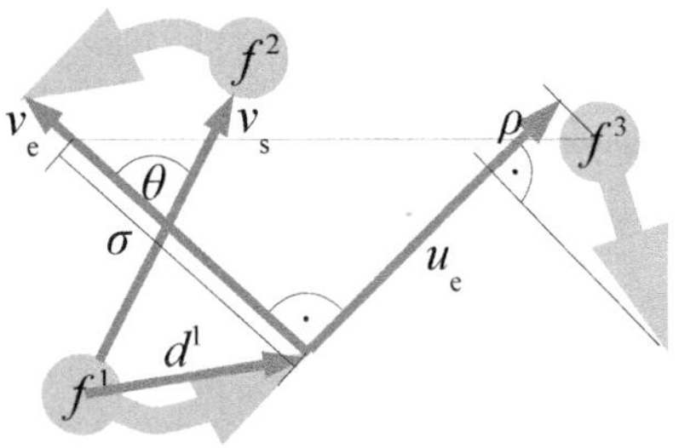


**Fig. 3.** A schematic three-finger multitouch gesture showing all the defined numerical components, except for $\delta$, which would be $\frac{|v_e|}{|v_s|}$

## 2.3 Bruhat Decomposition for Gestural Diagrams

After having discussed the transition from a gestural diagram to a processual diagram, we now propose a procedure to reconstruct gestural diagrams from processual diagrams.

Often one is not given a gestural diagram, but just a (process) diagram of morphisms $D : \Gamma \to Dig(X)$, and there is no canonical gestural diagram $cp : CPath(\Gamma) \to X$ which gives rise to $D$ by canonical restriction. We want to describe the construction of such a gestural extension of $D$ in the case of a simple, but very useful topological category, namely the category $X(N)$ associated with one form $N$, whose functor is $fun(N) = \mathbb{R}^5$, and whose set of morphisms is the set $\mathbf{Aff}_5(\mathbb{R})$ of all affine maps on $\mathbb{R}^5$. Clearly, this set is in bijection with $M = \mathbb{R}^5 \times \mathbb{M}_{5\times5}(\mathbb{R})$ (translation vector plus matrix for the linear part). The problem is this: if we are given an affine map $T^t f$, then how can we write it as the endpoint of a continuous curve in $M$? Of course, one can just draw a straight line from $T^0 Id$ to $T^t f$ and move on this line as the curve parameter grows from 0 to 1. But the meaning of this curve in terms of intuitively understandable transformations is not guaranteed. No rotation, reflection, or other standard transformation will be likely to appear in this trajectory.

A more intuitive solution of this problem is provided by the Bruhat decomposition of the special linear group $\mathrm{SL}_2(\mathbb{R})$ [10, XI.2]. Here is the decomposition we propose for an intuitive gestural curve from $T^0 Id$ to $T^t f$. To begin with, the translation can be parametrized in an intuitive way by the curve $c(\lambda) = T^{\lambda.t}$. As to the linear part $f$, one point is the control of $Ker(f)$. If we take a canonical decomposition $\mathbb{R}^n = Ker(f) \oplus \mathbb{R}^k, k = n - dim(Ker(f))$, then $f$ is essentially described by the induced general linear map $g : \mathbb{R}^k \to \mathbb{R}^k$. We may then concentrate on this part and suppose that $f \in \mathrm{GL}_n(\mathbb{R})$. Next, we want to write $f$ as a product of factors that only affect two of the $n$ coordinates. This is a standard procedure in computational linear algebra, and we may step over to the critical case of $n = 2$. Moreover, the determinant $d = det(f)$ can be eliminated by writing

$$f = \begin{pmatrix} 1 + \lambda(d-1) & 0 \\ 0 & 1 \end{pmatrix} f_0,$$

$f_0 \in \mathrm{SL}_2(\mathbb{R})$, we can focus on $f \in \mathrm{SL}_2(\mathbb{R})$, while the determinant is generated from the identity by a line that has at most one point of singularity if $d < 0$ and none else. If $f \in \mathrm{SL}_2(\mathbb{R})$, we have the unique Bruhat decomposition with two possible cases. Either

$$f = u(a)s(b), u(a) = \begin{pmatrix} 1 & a \\ 0 & 1 \end{pmatrix}, s(b) = \begin{pmatrix} b & 0 \\ 0 & 1/b \end{pmatrix},$$

or

$$f = u(a)s(b)wu(c), w = \begin{pmatrix} 0 & 1 \\ -1 & 0 \end{pmatrix}$$

with the unique coefficients $a, b, c$. Here $u(a)$ is a shearing, $s(b)$ a dilation in both coordinates, and $w$ the clock-wise rotation by $90^0$. We may then replace $a, c$ by curves $\lambda a, \lambda c, 0 \leq \lambda \leq 1$. The rotation $w$ can be replaced by a rotation curve $w(\lambda)$ starting at $w(0) = Id$ and ending at $w(1) = w$. If $0 < b$, this can be replaced by the curve $1 + \lambda(b - 1)$. If $0 > b$, we have to add a rotation matrix curve $\rho(\lambda)$ which is the identity for $\lambda = 0$ and a rotation by $180^0$ for $\lambda = 1$. Then consider the product

$$\rho(\lambda) \begin{pmatrix} 1 + \lambda(|b| - 1) & a \\ 0 & 1/(1 + \lambda(|b| - 1)) \end{pmatrix}$$

which connects the identity to the final matrix by intuitive factors without going through singularities. So we have a canonical embedding of our transformations as being extremal points in continuous gestural curves of transformations. We can do this for every transformation representing an arrow of our diagram. This generates a diagram of gestures $cp : CPath(\Gamma) \to X(N)$ that extends the given "abstract" transformational diagram $D : \Gamma \to X(N)$.

## 3 A New Perspective on Musical Composition

Now let us turn to the discussion of a way we might use the idea of gestural and processual diagrams to look at a musical composition. We will first describe the extension of the traditional notion of a composition (a finished score) to a gesturally informed diachronic one. Then we will consider several practical aspects of such a view. Finally we will describe the benefits of a potential implementation of the principle, especially with regard to the Rubato Composer software and its module for gestural music composition Big Bang Rubette.

### 3.1 The Process Diagram of a Score

In terms of forms and denotators, we traditionally view a musical composition as a *Score*, a powerset of *Notes*. In our notion of the process of music composition, a composer typically defines a number of *Notes* and transforms and multiplies them using a number of morphisms. Thereby we always see the latest transformed *Score* denotator as the respective composition. In this paper, we propose to see a composition as the initial set of *Notes* and associate it with a diagram of morphisms containing all performed transformations. Figure 4 visualizes the two different notions. Since a transformation is usually merely applied to a subset of the *Score*, we further associate with each note a path in the diagram. If one wants to refer to paths on form diagrams, it is necessary to create a denotator representation of such paths in order to integrate these objects in musical contexts, where denotators apply. To begin with, suppose we have constructed a form diagram, i.e. a diagram $D : \Gamma \to \mathbf{Forms}$. We first want to normalize the digraph $\Gamma$. Without loss of generality, one may suppose that its vertices are all denotators of form names. One may also use synonyms of form names in order to create distinct vertices if this is required. Since form names are just denotators of the name form $NF : Id.\mathbf{Simple}(\mathbb{Z}\langle UNICODE\rangle)$ (see [11]), the digraph can be formally described as a set of arrows $\Gamma : 0@Arrow(\gamma)$ in the limit form $Arrow : Id.\mathbf{Limit}(NF, NF, AI)$, with the arrow index form

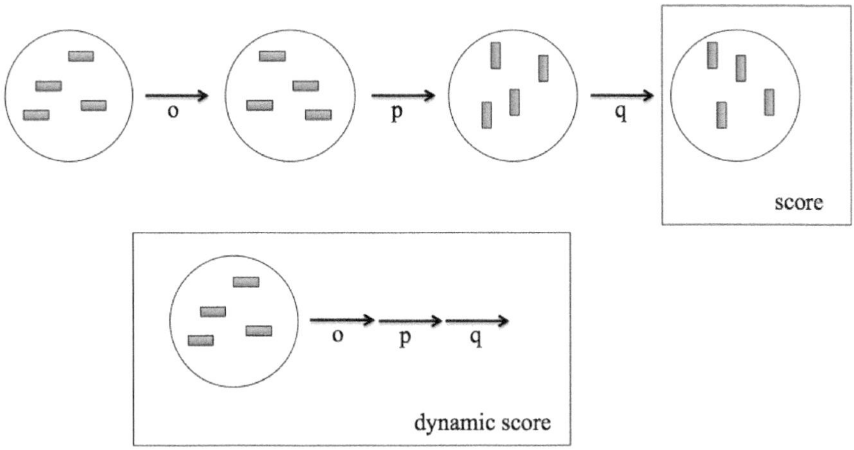


**Fig. 4.** Two ways of seeing a composition. The big rectangles represent the respective definition.

$AI : Id.\mathbf{Simple}(\mathbb{Z})$. With this information, the form diagram $D$ must associate a form with each vertex, and we take the form whose name stands in $\Gamma$, whereas one has to give a form morphism for each digraph arrow $i : FN_1 \to FN_2$. With this information, a path $p$ in $\Gamma$ of length $k$ is a denotator in the limit form $P_k : Id.\mathbf{Limit}(Arrow \xrightarrow{pr_2} NF \xleftarrow{pr_1} Arrow \xrightarrow{pr_2} NF \xleftarrow{pr_1} \ldots Arrow)$. Then a path $p$ of length $k$ in our diagram is just a denotator in $P_k$ with the condition that its arrows are all in $\Gamma$. If we want to integrate any path length, the colimit $Path : Id.\mathbf{Colimit}(P_0, P_1, \ldots)$ of all path forms can be taken. One may now define notes that contain their generative history as follows:

$$DynamicNote : Id.\mathbf{Limit}(Onset, Pitch, Loudness, Duration, Voice, Path)$$

And then scores of dynamic notes are defined as usual by

$$DynamicScore : Id.\mathbf{Power}(DynamicNote)$$

### 3.2 Integrating Satellite Structures with Dynamic Notes

Let us briefly look at the compatibility of this model with the more advanced score forms used in recent publications. In order to handle hierarchies among *Notes*, we defined the circular *MacroScore* form:

$$\begin{aligned} MacroScore &: Id.\mathbf{Power}(Node) \\ Node &: Id.\mathbf{Limit}(Note, MacroScore) \end{aligned}$$

The point of this structure is to be able to express dependencies between *Notes* on a higher hierarchical level and their satellites (see [3] for more information). If a so-called anchor *Note* is transformed, all its satellites keep their relative position to it. We call the operations for moving *Notes* down and up in their hierarchy *satellite building* and *flattening* operations. We now show that in the new model, both operations can be represented by simple translations.

The introduction of dynamic notes has a deep impact on this satellite construction since we now not only have to bind notes to anchor notes as their satellites, but also have to take care of the notes' generative history. In the *MacroScore* form a satellite was just a note (with its own satellite baggage), the note's coordinates were set to express the relative position with regard to the anchor note. In order to include the satellite's generative history, it is now represented by a dynamic note $dn : A@DynamicNote(x, pth)$, where the first five coordinates $x = (o, p, l, d, v)$ describe the $A$-addressed normal note $x$, while $pth$ is the path that defines the generative history of the note, and where the denotators generated successively through the $pth$ end up defining the present state denotator $pth(x)(k)$, where $k$ is the length of the path. Declaring $dn$ as being the dynamic satellite of an dynamic anchor note $da : A@DynamicNote(y, pt)$ now not only means adding $dn$ to the set of satellites and replacing its coordinates by the relative coordinates with respect to the anchor, as was the case for *MacroScore*, but we now have to add the relative position as a new transformation in the satellite's history path $pth$.

More precisely (see also Fig. 5), if $\Delta$ is the translation mapping the coordinates $y$ of the anchor to the coordinates $x$ of the satellite, then we have to add the inverse $-\Delta$ of this transformation $-\Delta : Note \to Note$ to the satellite's path, generating the path $-\Delta.pth$. This path however only testifies that any future transformations of the anchor note will be acting on the satellite, too. So, if the anchor presently has path $pt$, $-\Delta.pth$ targets at the endpoint of $pt$. But when future transformations of the anchor add to $pt$ to give a longer path $pt'.pt$, the satellite will have to move with the added transformations. This means that when the anchor's transformations are accomplished, we shall have to add the inverse $\Delta$ to obtain the final position of the satellite. So the final path of the satellite is the "conjugate" path $\Delta.pt'. - \Delta.pth$. If the satellite is disconnected from its anchor, future transformations of the anchor will not influence the path of the former satellite anymore.

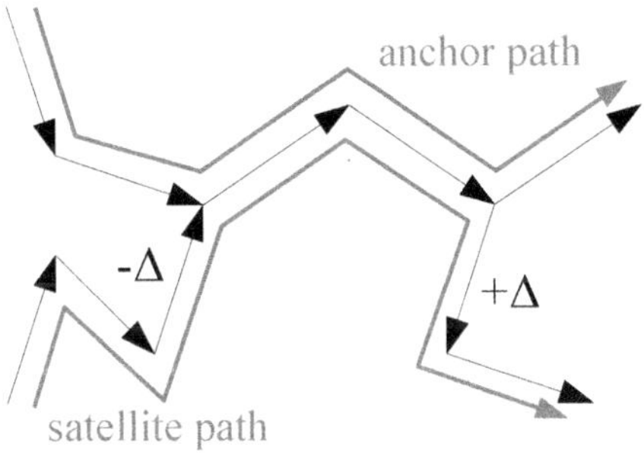


**Fig. 5.** The *MacroScore* operation introducing a dynamic satellite note as being attached to a given dynamic anchor note in a transformational way

In another recent publication [12], the *MacroScore* form was extended to a *SoundScore* in order to represent timbre. In addition to having a set of satellites, each *Note* can have a set of regular notes modulating its frequency. For the time being, when transformed, modulators move with their carrier in an identical fashion as satellites move with their anchor. Thus, the new transformational model also enables a simplification of these structures.

### 3.3 General Score Operations in Process Diagrams

There are several basic and intuitive score operations that should be available to composers using any system. Here is a list of some of well-known examples:

- The first one is *adding notes* to the composition, however these may be defined. Usually notes are added to the present composition, by simply adding them to the latest set of notes (the rightmost set in Fig. 4 above, for instance). In our transformational model, however, this is a slightly more complex yet much more powerful matter. We only have one set of notes (see Fig. 4 below) and this is where we add notes. Nevertheless, if we define notes after we already have defined a diagram, we will have to define a path in that diagram for each of them. The most straightforward thing is to just add them with trivial paths which corresponds to adding them to the latest composition in the traditional case. But the additional notes can each be assigned a specific path in the diagram which virtually adds them to an earlier state of the composition. If a composition, for instance, is generated by an initial motif that is gradually transformed and multiplied, we are now able to change or add notes to the initial motive after the generating diagram has already been made.

- Another frequently used operation is the *copy and transform* operation as described in [13]. Although in this case notes are added as well, these are not individual notes but copies of notes that are already present in the original score. These copies are then given the same transformational paths as their originals and finally transformed, i.e. a morphism is added to their path. The diagram is thus extended by a forking piece whereas with the previously described operation it does not change.
- Another operation that is easily integrated with our dynamic model is the *shaping* operation described in [12]. There, notes are immediately given a new value for a certain coordinate. With a dynamic composition, we would simply add this change as a translation, thereby keeping all the notes' individual history.
- For a more complex example, we look at two of the more advanced and compound operations used in connection with denotators and morphisms. The first one, *wallpapers*, has been discussed in [14]. It can be seen as a special construction of "regular" form diagrams. To define a wallpaper, a sequence of $n$ morphisms on the score form is used to generate a regular grid structure as shown in Fig. 6. For each of the vertices of the generated diagram we obtain a transformed version of the initial wallpaper motif.

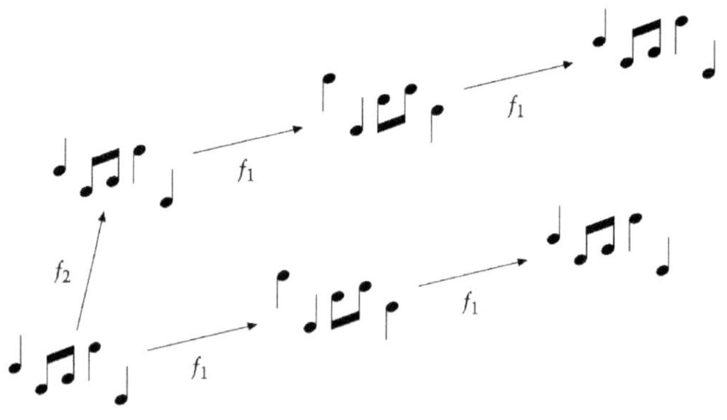


**Fig. 6.** The wallpaper construction (here from two morphisms $f_1, f_2$) is a special case of a dynamic score construction. It uses a grid diagram of morphisms and an input motif to build a motivic "wallpaper".

The composition's paths from the grid origin to defined vertices are generated based on a predefined linear order among the morphisms since the diagram need not be commutative. We can now understand general form diagrams used in dynamic scores as being more irregular types of wallpapers, which may be defined in an intuitive and gestural way.
- The second more advanced operation was already briefly discussed above: *alterations* (example 2). They are an instance of non-affine transformations commonly used to distort a composition. In a form diagram an alteration would merely be featured as a single arrow. But larger diagrams of alterations with changing force fields can be envisaged.

### 3.4 Outlook: Navigating and Changing the History of a Composition

Considering our new definition of a dynamic score from a more practical perspective we might imagine for it a dedicated graphical and gestural computer interface, such as the Big Bang Rubette. While the composer is manipulating the score it keeps track of the performed transformations through form processual diagrams. The software could then allow a browsing of the underlying diagram, letting the user view and hear previous states of the composition (similar to version control systems used in software development) and even enabling to modify previous transformations by finger-driven gestures. The effects of these changes could then be visualized just as with the Big Bang Rubette, in a similar live and gestural fashion, so that the user can directly follow their effects. In a way, we would obtain an advanced dynamic *undo/redo* functionality letting the composer, not only undo the past, but gesturally modify it.

Another significant strength of the use of form diagrams in dynamic scores is that, instead of undoing transformations by performing their inversion, we can now simply remove them from the diagram. Thus undoing even noninvertible transformations becomes easy.

Finally, using methods akin to the previously described Bruhat decompositions, such a software could generate a movie of the evolving composition from the initial set of notes, thereby reenacting all executed transformations in a standardized way. The composer could virtually observe the big bang of his own composition.

## References


1. Mazzola, G.: La vérité du beau dans la musique. Delatour, Paris (2007)
2. Mazzola, G.: The Topos of Music. Birkhäuser, Basel (2002)
3. Mazzola, G., Milmeister, G., Morsy, K.: Functors for Music: The Rubato Composer System. In: Randy Adams, S., Gibson, S. (eds.) Transdisciplinary Digital Art: Sound, Vision and the New Screen. CCIS. Springer, Heidelberg (2008)
4. Mazzola, G.: Categorical Gestures, the Diamond Conjecture, Lewin's Question, and the Hammerklavier Sonata. Journal of Mathematics and Music 3(1), 31–58 (2009)

5. Lewin, D.: Generalized Musical Intervals and Transformations. Oxford University Press, Oxford (2007)
6. Mazzola, G., Andreatta, M.: Formulas, Diagrams, and Gestures in Music. Journal of Mathematics and Music 1(1), 21–32 (2007)
7. Mazzola, G.: Musical Performance. Springer, Heidelberg (2011)
8. Milmeister, G.: The Rubato Composer Music Software. Springer, Heidelberg (2009)
9. Thalmann, F., Mazzola, G.: Affine Musical Transformations Using Multi-touch Gestures. To appear in Ninad-Journal of ITC SRA
10. Lang, S.: $SL_2(\mathbb{R})$. Addison-Wesely, Reading (1975)
11. Mazzola, G.: Towards a Galois Theory of Concepts. In: Mazzola, G., Noll, T. (eds.) Perspectives in Mathematical Music Theory. EpOs, Osnabrück (2004)
12. Thalmann, F., Mazzola, G.: Gestural Shaping and Transformation in a Universal Space of Structure and Sound. In: Proceedings of the ICMC 2010. ICMA, Ann Arbor (2010)
13. Thalmann, F., Mazzola, G.: The BigBang Rubette: Gestural Music Composition With Rubato Composer. In: Proceedings of the ICMC 2008. ICMA, Ann Arbor (2008)
14. Mazzola, G., Thalmann, F.: Grid Diagrams for Ornaments and Morphing. In: Klouche, T., Noll, T. (eds.) MCM 2007. Communications in Computer and Information Science, vol. 37. Springer, Heidelberg (2009)

# Tonal Scales and Minimal Simple Pitch Class Cycles

David Meredith

Aalborg University
dave@create.aau.dk

**Abstract.** Numerous studies have explored the special mathematical properties of the diatonic set. However, much less attention has been paid to the sets associated with the other scales that play an important rôle in Western tonal music, such as the harmonic minor scale and ascending melodic minor scale. This paper focuses on the special properties of the class, $\mathcal{T}$, of sets associated with the major *and minor* scales (including the harmonic major scale). It is observed that $\mathcal{T}$ is the set of pitch class sets associated with the shortest simple pitch class cycles in which every interval between consecutive pitch classes is either a major or a minor third, and at least one of each type of third appears in the cycle. Employing Rothenberg's definition of stability and propriety, $\mathcal{T}$ is also the union of the three most stable inversional equivalence classes of proper 7-note sets. Following Clough and Douthett's concept of maximal evenness, a method of measuring the evenness of a set is proposed and it is shown that $\mathcal{T}$ is also the union of the three most even 7-note inversional equivalence classes.

**Keywords:** Diatonic set, Pitch class sets, Pitch class cycles, Scales, Minor scales.

## 1 Introduction

Piston [1] (pp. 1-2) begins his classic textbook on harmony by introducing the chromatic scale and what he calls the "fundamental diatonic scales", that is, the major scale and the natural, harmonic and melodic forms of the minor scale. Each of these scale types is listed in Table 1, together with its associated translational and inversional equivalence classes of pitch class set[1]. Let $\mathcal{F}$ denote the translational equivalence class of pitch class sets associated with Piston's "fundamental diatonic scales". Table 1 shows that $\mathcal{F} = \mathcal{D} \cup \mathcal{H}_{\min} \cup \mathcal{M}$, where $\mathcal{D}$, $\mathcal{H}_{\min}$ and $\mathcal{M}$ are as defined in this table. In this paper, the sets $\mathcal{D}$, $\mathcal{H}_{\min}$ and $\mathcal{M}$ will be called, respectively, the *diatonic sets*, the *harmonic minor scale sets* and the *ascending melodic minor scale sets*.

Numerous studies have explored the special mathematical properties of the diatonic sets (i.e, $\mathcal{D}$) (see [2] for a review). However, as pointed out by Cross

[1] $[S]_{\mathrm{T}}$ and $[S]_{\mathrm{I}}$ (as used in Table 1) denote, respectively, the translational and inversional equivalence class to which pitch class set $S$ belongs.

C. Agon et al. (Eds.): MCM 2011, LNAI 6726, pp. 165–179, 2011.

**Table 1.** Piston's "fundamental diatonic scales" along with their associated translational and inversional equivalence classes of pitch class set

| *Common name* | *Example* | *Pitch class set* $(P)$ | $[P]_{\mathrm{T}}$ [a] | $[P]_{\mathrm{I}}$ [a] |
|---|---|---|---|---|
| Major | C D E F G A B C | $\{0,2,4,5,7,9,11\}$ | $\mathcal{D}$ | $\mathcal{D}$ |
| Natural minor | A B C D E F G A | $\{9,11,0,2,4,5,7\}$ | $\mathcal{D}$ | $\mathcal{D}$ |
| Harmonic minor | C D E♭ F G A♭ B C | $\{0,2,3,5,7,8,11\}$ | $\mathcal{H}_{\mathrm{min}}$ | $\mathcal{H}$ |
| Melodic minor ascending | C D E♭ F G A B C | $\{0,2,3,5,7,9,11\}$ | $\mathcal{M}$ | $\mathcal{M}$ |
| Melodic minor descending | C B♭ A♭ G F E♭ D C | $\{0,2,3,5,7,8,10\}$ | $\mathcal{D}$ | $\mathcal{D}$ |

[a] $\mathcal{D} = [\{0,1,3,5,6,8,10\}]_{\mathrm{T}} = [\{0,1,3,5,6,8,10\}]_{\mathrm{I}}$
$\mathcal{H}_{\mathrm{min}} = [\{0,1,3,4,6,8,9\}]_{\mathrm{T}}$
$\mathcal{H} = [\{0,1,3,4,6,8,9\}]_{\mathrm{I}}$
$\mathcal{M} = [\{0,1,3,4,6,8,10\}]_{\mathrm{T}} = [\{0,1,3,4,6,8,10\}]_{\mathrm{I}}$

*et al.* [3] (p. 233), many of these properties are not possessed by the harmonic and ascending melodic forms of the minor scale which clearly "play an extremely significant role in tonal music". In this paper, the focus is on structural properties possessed by all and only those pitch class sets associated with the major *and minor* scales in all their usual forms.

The sets $\mathcal{D}$ and $\mathcal{M}$ are each closed under inversion. However, the inversion of a harmonic minor scale set is in the transpositional equivalence class $[\{0,1,3,5,6,8,9\}]_{\mathrm{T}}$, denoted here by $\mathcal{H}_{\mathrm{maj}}$. This is the set class associated with what Rahn [4] (p. 41) calls the "major harmonic scales", an example of which is C-D-E-F-G-A♭-B-C. This scale type has been recognized within tonal theory since at least the mid-19th century. For example, Hauptmann [5] (p. 39) called it the "Moll-Dur-Tonart" (see also McCune [6], pp. 10-11) and Schenker [7] (p. 89) called it the "second series" in his six "products of combination". A cadential progression that ends with a Picardy third often uses all and only the pitch classes in a harmonic major scale—see Fig. 1.

Let $\mathcal{H} = \mathcal{H}_{\mathrm{min}} \cup \mathcal{H}_{\mathrm{maj}}$. $\mathcal{H}$ will be called the *set of harmonic scale sets*. The set $\mathcal{F}$, associated with Piston's "fundamental diatonic scales" is closed under transposition, but not under inversion. The smallest inversionally closed superset

**Fig. 1.** Example of a Picardy third from the end of J. S. Bach's chorale, *Zeuch ein zu deinen Toren* (BWV 28/6). This cadential figure employs all and only the pitch classes in a harmonic major scale set.

of $\mathcal{F}$ is $\mathcal{F} \cup \mathcal{H}_{\text{maj}}$ which will be denoted by $\mathcal{T}$ and called the *set of tonal scale sets*. $\mathcal{T}$ is therefore given by the following equation.

$$\mathcal{T} = \mathcal{F} \cup \mathcal{H}_{\text{maj}} = \mathcal{D} \cup \mathcal{M} \cup \mathcal{H}_{\text{min}} \cup H_{\text{maj}} = \mathcal{D} \cup \mathcal{M} \cup \mathcal{H} . \tag{1}$$

Schenker [7] (Chapter 2) describes how six different scales or "series" can be generated from the major and natural minor scales on a given tonic by flattening one or more of the third, sixth and seventh degrees of the major scale (see Table 2). These six "products of combination" along with the major and natural minor scales from which they are generated, transposed to all possible keys, give a set of scales that I'll call *Schenker's combination scales*. The set of tonal scale sets, $\mathcal{T}$, is the set of pitch class sets associated with Schenker's combination scales (see Table 2). It follows that $\mathcal{T}$ is also the set of pitch class sets associated with scales that can be constructed by combining tonic, dominant and subdominant triads where each can be either major or minor (see Table 2). More tightly, $\mathcal{T}$ is the set of pitch class sets associated with scales that can be constructed by combining a major dominant triad with tonic and subdominant triads that can each be either major or minor (see emboldened lines in Table 2). In other words, $\mathcal{T}$ is the set of sets associated with the major scales and scales constructed by flattening the third and/or sixth degrees in these major scales.

**Table 2.** Scales with tonic C, generated by combining tonic, dominant and subdominant triads that can be either major or minor. $\mathcal{T}$ is generated by all 8 combinations as well as by only those combinations containing V (in bold).

| *Triads*[a] | *Scale with C as tonic* | *Pitch class set (P)* | $[P]_\text{T}$ | $[P]_\text{I}$ | *Mode* | *Series*[b] |
|---|---|---|---|---|---|---|
| **I IV V** | **C D E F G A B C** | $\mathbf{\{0,2,4,5,7,9,11\}}$ | $\mathcal{D}$ | $\mathcal{D}$ | **Ionian** | — |
| I IV v | C D E F G A B♭ C | $\{0,2,4,5,7,9,10\}$ | $\mathcal{D}$ | $\mathcal{D}$ | Mixolydian | Third |
| **I iv V** | **C D E F G A♭ B C** | $\mathbf{\{0,2,4,5,7,8,11\}}$ | $\mathcal{H}_{\text{maj}}$ | $\mathcal{H}$ | — | **Second** |
| I iv v | C D E F G A♭ B♭ C | $\{0,2,4,5,7,8,10\}$ | $\mathcal{M}$ | $\mathcal{M}$ | — | Sixth |
| **i IV V** | **C D E♭ F G A B C** | $\mathbf{\{0,2,3,5,7,9,11\}}$ | $\mathcal{M}$ | $\mathcal{M}$ | — | **First** |
| i IV v | C D E♭ F G A B♭ C | $\{0,2,3,5,7,9,10\}$ | $\mathcal{D}$ | $\mathcal{D}$ | Dorian | Fifth |
| **i iv V** | **C D E♭ F G A♭ B C** | $\mathbf{\{0,2,3,5,7,8,11\}}$ | $\mathcal{H}_{\text{min}}$ | $\mathcal{H}$ | — | **Fourth** |
| i iv v | C D E♭ F G A♭ B♭ C | $\{0,2,3,5,7,8,10\}$ | $\mathcal{D}$ | $\mathcal{D}$ | Aeolian | — |

[a] Upper case denotes a major triad, lower case denotes a minor triad.
[b] Schenker's "series" [7] (pp. 87–93).

## 2 Minimal Simple Pitch Class Cycles

A *pitch class sequence* is an ordered set of pitch classes. For example, the pitch class sequence $\langle 0,2,4,5,7,9,11,0 \rangle$ represents one octave of an ascending C major scale. Let $\mathbf{S}$ be a pitch class sequence. $|\mathbf{S}|$ denotes the *length* of $\mathbf{S}$ which is the number of (not necessarily distinct) elements in $\mathbf{S}$. For example $|\langle 0,2,4,2,0 \rangle| = 5$. $\mathbf{S}[i]$ denotes the $(i+1)$th element in $\mathbf{S}$. For example, $\langle 0,2,4,0 \rangle[2] = 4$. $\mathbf{S}[a,b]$ denotes the *subsequence* of $\mathbf{S}$ that contains, in order, the $(a+1)$th to the $b$th elements of $\mathbf{S}$, inclusive. For example, $\langle 0,1,2,3,4,5\rangle[1,4] = \langle 1,2,3\rangle$. If $a = 0$ in the expression $\mathbf{S}[a,b]$, then the subsequence is a *prefix* of $S$. If $b = |S|$, then $\mathbf{S}[a,b]$ is a *suffix* of $S$. If $\mathbf{A}$ and $\mathbf{B}$ are any two sequences, such that $\mathbf{A} = \langle a_1, a_2, \ldots a_m\rangle$ and $\mathbf{B} = \langle b_1, b_2, \ldots b_n\rangle$, then $\mathbf{A} \oplus \mathbf{B} = \langle a_1, a_2, \ldots a_m, b_1, b_2, \ldots b_n\rangle$. If $\mathbf{S}$ is a sequence, then $\rho(\mathbf{S})$ denotes the *reverse* of $\mathbf{S}$. That is, $\rho(\mathbf{S}) = \langle \mathbf{S}[|\mathbf{S}|-1], \mathbf{S}[|\mathbf{S}|-2], \ldots \mathbf{S}[0]\rangle$. $\mathrm{P}(\mathbf{S})$ denotes the *associated pitch class set* of $\mathbf{S}$ which contains all and only those pitch classes in $\mathbf{S}$. That is, $\mathrm{P}(\mathbf{S}) = \bigcup_{k=0}^{|\mathbf{S}|-1}\{\mathbf{S}[k]\}$. $\mathrm{I}(\mathbf{S})$ denotes the *interval set* of the pitch class sequence $\mathbf{S}$, which is the set of pitch class intervals that occur between *consecutive* elements in $\mathbf{S}$. For example, $\mathrm{I}(\langle 0,2,4,0\rangle) = \{2,8\}$ and $\mathrm{I}(\langle 0,4,7,11,2,5,9,0\rangle) = \{3,4\}$. $\mathbf{J}(\mathbf{S})$ denotes the *associated interval sequence* of the pitch class sequence $\mathbf{S}$ and it is defined as follows:

$$\mathbf{J}(\mathbf{S}) = \langle \mathrm{i}(\mathbf{S}[0], \mathbf{S}[1]), \mathrm{i}(\mathbf{S}[1], \mathbf{S}[2]), \ldots \mathrm{i}(\mathbf{S}[|\mathbf{S}|-2], \mathbf{S}[|\mathbf{S}|-1])\rangle \tag{2}$$

where $\mathrm{i}(p_1, p_2)$ is the pitch class interval from pitch class $p_1$ to pitch class $p_2$. That is,

$$\mathrm{i}(p_1, p_2) = (p_2 - p_1) \bmod \mu \tag{3}$$

where $\mu$ is the *chromatic cardinality* ([8], p. 94). In this paper, $\mu = 12$ unless otherwise stated. From the definitions above, it follows that

$$\mathrm{I}(\mathbf{S}) = \bigcup_{k=0}^{|\mathbf{J}(\mathbf{S})|-1} \{\mathbf{J}(\mathbf{S})[k]\}\ . \tag{4}$$

If $\mathbf{S}_1$ and $\mathbf{S}_2$ are two pitch class sequences, then $\mathbf{S}_1$ and $\mathbf{S}_2$ are defined to be *transpositionally equivalent* if and only if $|\mathbf{S}_1| = |\mathbf{S}_2|$ and there exists a single pitch class interval, $i$, such that $\mathrm{i}(\mathbf{S}_1[k], \mathbf{S}_2[k]) = i$ for all $0 \leq k < |\mathbf{S}_1|$.

A *pitch class cycle* is a pitch class sequence that begins and ends on the same pitch class. If $\mathbf{S}$ is a pitch class cycle and $n$ is an integer, then the function $\mathrm{CYC}(\mathbf{S}, n)$ cycles $\mathbf{S}$ by $n$ steps. That is

$$\mathrm{CYC}(\mathbf{S}, n) = \mathbf{S}[n \bmod (|\mathbf{S}|-1), |\mathbf{S}|] \oplus \mathbf{S}[1, n \bmod (|\mathbf{S}|-1)+1]\ . \tag{5}$$

Two pitch class cycles, $\mathbf{S}_1$ and $\mathbf{S}_2$, are *cyclo-transpositionally equivalent* if and only if there exists an integer $n$ such that $\mathbf{S}_1$ is transpositionally equivalent to $\mathrm{CYC}(\mathbf{S}_2, n)$. A *simple pitch class cycle* is a pitch class cycle that contains no element more than once except the first element which occurs only at the beginning and at the end. For example $\langle 0,2,4,2,0\rangle$ is a pitch class cycle but it is not simple because 2 occurs twice, and not only at the beginning and the end. $\langle 0,2,4,0\rangle$ is an example of a simple pitch class cycle. A simple pitch class cycle is *minimal* if and only if there exists no shorter pitch class cycle that has the same interval set. For example, $\langle 0,4,7,11,2,5,9,0\rangle$ is minimal because there exists no shorter pitch class cycle that has the interval set $\{3,4\}$. $\Gamma(I)$ denotes the *set of minimal simple cycles* that have the interval set $I$. $\mathcal{C}(I)$, the *set of minimal simple cycle sets* for interval set $I$, is defined to contain all and only those pitch class sets associated with cycles in $\Gamma(I)$.

## 3 Minimal Simple Pitch Class Cycles and the Tonal Scale Sets

It can be shown that $\mathcal{T}$ is the set of minimal simple cycle sets for the interval set $\{3,4\}$, that is,

$$\mathcal{T} = \mathcal{C}(\{3,4\}) \,. \tag{6}$$

Eq. 6 is important because it expresses a simple mathematical property that is shared by all *and only* the major *and minor* scales recognized in traditional tonal theory. It contrasts with those properties of the diatonic set on which recent mathematical work has focused that are not possessed by the common forms of the minor scales. Moreover, Eq. 6 shows how the normal major and minor scales are simply generated by the consonant major and minor third intervals on which tonal harmony is based.

Eq. 6 can be proved by showing separately that $\mathcal{T} \subseteq \mathcal{C}(\{3,4\})$ and that $\mathcal{T} \supseteq \mathcal{C}(\{3,4\})$. Suppose $P \in \mathcal{C}(\{3,4\})$. There therefore exists a minimal simple cycle, $\mathbf{S}$, such that

$$P = \mathrm{P}(\mathbf{S}) \,, \tag{7}$$
$$\mathbf{S}[0] = \mathbf{S}[|\mathbf{S}|-1] \,, \tag{8}$$
$$|\mathbf{S}| = |P| + 1 \,, \tag{9}$$
$$\mathrm{I}(\mathbf{S}) = \{3,4\} \,. \tag{10}$$

Moreover, there is no shorter sequence than $\mathbf{S}$ that satisfies Eqs. 7–10. Eq. 10 implies that $\mathbf{J}(\mathbf{S})$ contains $x$ intervals of size 3 and $y$ intervals of size 4 where $x, y \geq 1$. Eq. 9 implies that

$$x + y = |P| \,. \tag{11}$$

Eq. 8 implies that the sum of the intervals in $\mathbf{J}(\mathbf{S})$ is $12n$ where $n > 0$ and $n \in \mathbb{Z}$. Therefore

$$3x + 4y = 12n \,, \text{ where } n > 0, n \in \mathbb{Z} \,. \tag{12}$$

$\mathbf{S}$ is minimal, so $x + y$ must be a minimum. $n = 2$ is the smallest value of $n$ that satisfies Eqs. 11 and 12. Therefore $y = 3$ and $x = 4$. The 3 intervals of size 4 cannot be consecutive in $\mathbf{J}(\mathbf{S})$ as this would either cause a pitch class other than the initial one to be repeated or it would cause the initial pitch class to occur three times. If $\mathbf{J}(\mathbf{S})$ has two consecutive 4s separated from the third 4 by one interval of size 3, then $\mathbf{S}$ will be cyclo-transpositionally equivalent to $\langle 0,4,8,11,3,6,9,0 \rangle$, implying that $P \in \mathcal{H}_{\mathrm{maj}}$. If the two consecutive 4s are separated from the third 4 by two 3s, then $\mathbf{S}$ will be cyclo-transpositionally equivalent to $\langle 0,4,8,11,2,6,9,0 \rangle$, implying that $P \in \mathcal{M}$. If the two consecutive 4s are separated from the third 4 by three 3s, then $\mathbf{S}$ will be cyclo-transpositionally equivalent to $\langle 0,4,8,11,2,5,9,0 \rangle$, implying that $P \in \mathcal{H}_{\mathrm{min}}$. If there are no consecutive intervals of size 4, then $\mathbf{S}$ will be cyclo-transpositionally equivalent to $\langle 0,4,7,11,2,6,9,0 \rangle$, in which case $P \in \mathcal{D}$. This argument is summarised in Table 3. This accounts for all the cyclo-transpositionally distinct possibilities.

**Table 3.** Column headed $\mathbf{J}(\mathbf{S})$ shows the five cyclically distinct ways of arranging 3 intervals of size 4 and 4 intervals of size 3. $[\mathrm{P}(\mathbf{S})]_{\mathrm{T}}$ and $[\mathrm{P}(\mathbf{S})]_{\mathrm{I}}$ give the translational and inversional equivalence classes, respectively, associated with the given pitch class cycle.

| $\mathbf{J}(\mathbf{S})$ | $[\mathrm{P}(\mathbf{S})]_{\mathrm{T}}$ | $[\mathrm{P}(\mathbf{S})]_{\mathrm{I}}$ | *Comment* |
|---|---|---|---|
| $\langle 4,4,4,3,3,3,3\rangle$ | $[\langle 0,3,4,6,8,9\rangle]_{\mathrm{T}}$ | $[\langle 0,3,4,6,8,9\rangle]_{\mathrm{I}}$ | Not a simple pitch class cycle. |
| $\langle 4,4,3,4,3,3,3\rangle$ | $\mathcal{H}_{\mathrm{maj}}$ | $\mathcal{H}$ | |
| $\langle 4,4,3,3,4,3,3\rangle$ | $\mathcal{M}$ | $\mathcal{M}$ | |
| $\langle 4,4,3,3,3,4,3\rangle$ | $\mathcal{H}_{\mathrm{min}}$ | $\mathcal{H}$ | |
| $\langle 4,3,4,3,4,3,3\rangle$ | $\mathcal{D}$ | $\mathcal{D}$ | |

Therefore $P \in \mathcal{C}(\{3,4\}) \Rightarrow P \in \mathcal{H}_{\mathrm{maj}} \cup \mathcal{M} \cup \mathcal{H}_{\mathrm{min}} \cup \mathcal{D}$. Therefore (from Eq. 1) $P \in \mathcal{C}(\{3,4\}) \Rightarrow P \in \mathcal{T}$. Therefore

$$\mathcal{T} \supseteq \mathcal{C}(\{3,4\}) . \tag{13}$$

We now need to show that $\mathcal{T} \subseteq \mathcal{C}(\{3,4\})$. Let $H_1 \in \mathcal{H}_{\mathrm{maj}}$, $M \in \mathcal{M}$, $H_2 \in \mathcal{H}_{\mathrm{min}}$ and $D \in \mathcal{D}$. There therefore exist simple cycles $\mathbf{H_1}$, $\mathbf{M}$, $\mathbf{H_2}$ and $\mathbf{D}$ that are transpositionally equivalent to, respectively, $\langle 0,4,8,11,3,6,9,0\rangle$, $\langle 0,4,8,11,2,6,9,0\rangle$, $\langle 0,4,8,11,2,5,9,0\rangle$ and $\langle 0,4,7,11,2,6,9,0\rangle$, such that $H_1 = \mathrm{P}(\mathbf{H_1})$, $M = \mathrm{P}(\mathbf{M})$, $H_2 = \mathrm{P}(\mathbf{H_2})$ and $D = \mathrm{P}(\mathbf{D})$. $\mathbf{H_1}$, $\mathbf{M}$, $\mathbf{H_2}$ and $\mathbf{D}$ each contain 3 intervals of size 4 and 4 intervals of size 3, therefore $H_1, M, H_2, D \in \mathcal{C}(\{3,4\})$. Therefore $\mathcal{H}_{\mathrm{maj}} \cup \mathcal{M} \cup \mathcal{H}_{\mathrm{min}} \cup \mathcal{D} \subseteq \mathcal{C}(\{3,4\})$ and so (from Eq. 1) $\mathcal{T} \subseteq \mathcal{C}(\{3,4\})$. This, in conjunction with Eq. 13, implies Eq. 6.

## 4 Minimal Simple Cycle Sets and Interval Set Inversion

If $i$ is a pitch class interval, then the *inversion* of $i$, denoted $\mathrm{INV}(i)$, is given by $\mathrm{INV}(i) = (-i) \bmod \mu$ where $\mu$ is the chromatic cardinality. If $I$ is a pitch class interval set, then the *inversion* of $I$ is given by

$$\mathrm{INV}(I) = \bigcup_{i \in I} \{\mathrm{INV}(i)\} . \tag{14}$$

**Lemma 1.** *If* $\mathbf{S}$ *is a pitch class sequence, then* $\mathrm{I}(\rho(\mathbf{S})) = \mathrm{INV}(\mathrm{I}(\mathbf{S}))$.

***Proof.*** From Eq. 4, $\mathrm{INV}(\mathrm{I}(\mathbf{S})) = \mathrm{INV}(\bigcup_{k=0}^{|\mathbf{S}|-2}\{\mathrm{i}(\mathbf{S}[k], \mathbf{S}[k+1])\})$. This and Eq. 14 together imply that $\mathrm{INV}(\mathrm{I}(\mathbf{S})) = \bigcup_{k=0}^{|\mathbf{S}|-2}\{\mathrm{INV}(\mathrm{i}(\mathbf{S}[k], \mathbf{S}[k+1]))\}$. Since $\mathrm{INV}(\mathrm{i}(p_1, p_2)) = \mathrm{i}(p_2, p_1)$, it follows that $\mathrm{INV}(\mathrm{I}(\mathbf{S})) = \bigcup_{k=0}^{|\mathbf{S}|-2}\{\mathrm{i}(\mathbf{S}[k+1], \mathbf{S}[k])\}$ which implies that $\mathrm{INV}(\mathrm{I}(\mathbf{S})) = \mathrm{I}(\rho(\mathbf{S}))$. ■

**Theorem 1.** *If* $I$ *is a pitch class interval set, then* $\mathcal{C}(I) = \mathcal{C}(\mathrm{INV}(I))$.

***Proof.*** Let $P \in \mathcal{C}(I)$. There therefore exists a minimal simple cycle, $\mathbf{S}$, such that $P = \mathrm{P}(\mathbf{S})$ and $\mathrm{I}(\mathbf{S}) = I$. Let $\mathbf{R} = \rho(\mathbf{S})$. From Lemma 1 it follows that $\mathrm{I}(\mathbf{R}) = \mathrm{INV}(\mathrm{I}(\mathbf{S})) = \mathrm{INV}(I)$. $\mathbf{S}$ is a simple cycle, therefore so is $\mathbf{R}$. $\mathbf{R}$ is therefore a minimal simple cycle for $\text{INV}(I)$ provided there is no simple cycle shorter than $\mathbf{R}$ whose interval set is $\text{INV}(I)$. We'll suppose $\mathbf{Q} \in \Gamma(\text{INV}(I))$ and that $|\mathbf{Q}| < |\mathbf{R}|$ and show that this leads to a contradiction. $\text{I}(\mathbf{Q}) = \text{INV}(I)$, therefore $\text{I}(\rho(\mathbf{Q})) = I$ (from Lemma 1). $\rho(\mathbf{Q})$ is therefore a simple cycle shorter than $\mathbf{R}$ whose interval set is $I$. But $\mathbf{R}$ is the same length as $\mathbf{S}$ so this would mean that $\mathbf{Q}$ is shorter than $\mathbf{S}$, which is impossible because $\mathbf{S}$ is a minimal simple cycle for $I$. Therefore $\mathbf{Q}$ cannot exist and $\mathbf{R} \in \Gamma(\text{INV}(I))$. But $P = \text{P}(\mathbf{S}) = \text{P}(\mathbf{R})$, therefore $P \in \mathcal{C}(\text{INV}(I))$ and so $\mathcal{C}(I) \subseteq \mathcal{C}(\text{INV}(I))$. If we now let $J = \text{INV}(I)$, then $\mathcal{C}(J) \subseteq \mathcal{C}(\text{INV}(J))$ and so $\mathcal{C}(\text{INV}(I)) \subseteq \mathcal{C}(I)$. Since $\mathcal{C}(\text{INV}(I)) \subseteq \mathcal{C}(I)$ and $\mathcal{C}(I) \subseteq \mathcal{C}(\text{INV}(I))$, it follows that $\mathcal{C}(I) = \mathcal{C}(\text{INV}(I))$. ■

Theorem 1 and Eq. 6 together imply that $\mathcal{T} = \mathcal{C}(\{8, 9\})$.

## 5 Minimal Simple Cycle Sets and the Diatonic Spectra

A pitch class sequence is *simple* if and only if it contains no pitch class more than once. If $P$ is a pitch class set, then the *associated ascending simple sequence* of $P$, denoted $\mathbf{S}(P)$, is the simple pitch class sequence that contains every pitch class in $P$, exactly once, sorted into ascending order. For example, $\mathbf{S}(\{3, 2, 7, 10, 8, 0, 5\}) = \langle 0, 2, 3, 5, 7, 8, 10\rangle$. Following Clough and Myerson [9] (p. 262) and Clough and Douthett [8] (p. 95), let the *k-spectrum*, $e_k(P)$, of a pitch class set, $P$, be defined as follows:

$$e_k(P) = \{\text{i}(\mathbf{S}(P)[j], \mathbf{S}(P)[(j + k) \bmod |P|]) \mid 0 \leq j < |P|\} , \tag{15}$$

where $k \in \mathbb{Z}$ and $1 \leq k < |P|$. For example, the 1-spectrum of $\{0, 2, 4, 5, 7, 9, 11\}$ is $\{1, 2\}$ and the 2-spectrum of the same set (a diatonic set) is $\{3, 4\}$.

If $p$ is a pitch class, $i$ is a pitch class interval and $P$ is a pitch class set, then the *transposition of $p$ by $i$* is given by $\text{TRAN}(p, i) = (p + i) \bmod \mu$ where $\mu$ is the chromatic cardinality and the *transposition of $P$ by $i$* is given by $\text{TRAN}(P, i) = \bigcup_{p \in P}\{\text{TRAN}(p, i)\}$. The following theorem is easy to prove.

**Theorem 2.** *If $P_1$ and $P_2$ are transpositionally equivalent pitch class sets, then $e_k(P_1) = e_k(P_2)$ for any given value of $k$.*

If $S \subseteq [P]_\text{T}$ for some pitch class set $P$, then Theorem 2 allows us to define $e_k(S)$ to be the $k$-spectrum common to every pitch class set in $S$ (for some specified value of $k$). We can therefore state that $e_2(\mathcal{D}) = \{3, 4\}$ and that $e_5(\mathcal{D}) = \{8, 9\}$. In general, $e_k(\mathcal{D})$ is the *diatonic k-spectrum*. Therefore, from Eq. 6 and Theorem 1,

$$\mathcal{T} = \mathcal{C}(e_2(\mathcal{D})) = \mathcal{C}(e_5(\mathcal{D})) . \tag{16}$$

If $S \subseteq [P]_\text{T}$ for a pitch class set, $P$, then the *spectrum set* of $S$, denoted by $\text{E}(S)$, is defined to be equal to $\{e_k(S) \mid 1 \leq k < n\}$ where $n$ is the cardinality of a pitch class set in $S$. $\text{E}(\mathcal{D})$ is therefore the set of *diatonic spectra*. Table 4 shows the set of minimal simple cycle sets for each diatonic spectrum.

As can be seen in Table 4, $\mathcal{C}(e_1(\mathcal{D})) = \mathcal{C}(e_6(\mathcal{D})) = \mathcal{D} \cup \mathcal{M} \cup \mathcal{W}$, where $\mathcal{W} = [\{0, 1, 2, 4, 6, 8, 10\}]_\text{I}$. $\mathcal{W}$ is the inversional equivalence class of pitch class sets

**Table 4.** The set of minimal simple cycle sets for each diatonic spectrum, $\mathcal{C}(e_k(\mathcal{D}))$

| $k$ | $e_k(\mathcal{D})$ | $\mathcal{C}(e_k(\mathcal{D}))$[a] |
|---|---|---|
| 1 | $\{1,2\}$ | $\mathcal{D} \cup \mathcal{M} \cup \mathcal{W}$ |
| 2 | $\{3,4\}$ | $\mathcal{D} \cup \mathcal{M} \cup \mathcal{H}$ |
| 3 | $\{5,6\}$ | $\mathcal{D}$ |
| 4 | $\{6,7\}$ | $\mathcal{D}$ |
| 5 | $\{8,9\}$ | $\mathcal{D} \cup \mathcal{M} \cup \mathcal{H}$ |
| 6 | $\{10,11\}$ | $\mathcal{D} \cup \mathcal{M} \cup \mathcal{W}$ |

[a] $\mathcal{D} = [\{0,1,3,5,6,8,10\}]_{\mathrm{I}}$
$\mathcal{M} = [\{0,1,3,4,6,8,10\}]_{\mathrm{I}}$
$\mathcal{H} = [\{0,1,3,4,6,8,9\}]_{\mathrm{I}}$
$\mathcal{W} = [\{0,1,2,4,6,8,10\}]_{\mathrm{I}}$

associated with what Rahn [4] (p. 37) calls the "neapolitan ascending melodic minor scale" or "whole tone scale with filler tone". Table 4 also illustrates the following lemma.

**Lemma 2.** *If $S \subseteq [P]_{\mathrm{T}}$ for some specified pitch class set, $P$, then $e_k(S) = \mathrm{INV}(e_{n-k}(S))$ where $1 \leq k < n$ and $n = |P|$.*

***Proof*** By definition, $\mathrm{INV}(e_{n-k}(S)) = \mathrm{INV}(\{\mathrm{i}(\mathbf{S}(P)[m], \mathbf{S}(P)[(m+n-k) \bmod n]) \mid 0 \leq m < n\})$. Therefore $\mathrm{INV}(e_{n-k}(S)) = \mathrm{INV}(\{\mathrm{i}(\mathbf{S}(P)[m], \mathbf{S}(P)[(m-k) \bmod n]) \mid 0 \leq m < n\}) = \{\mathrm{i}(\mathbf{S}(P)[(m-k) \bmod n], \mathbf{S}(P)[m]) \mid 0 \leq m < n\}$ where $1 \leq k < n$. Let $j = (m-k) \bmod n$ so that $(j+k) \bmod n = m$. It follows that $\mathrm{INV}(e_{n-k}(S)) = \{\mathrm{i}(\mathbf{S}(P)[j], \mathbf{S}(P)[(j+k) \bmod n]) \mid 0 \leq j < n\} = e_k(S)$ where $1 \leq k < n$. ■

We can then infer the following theorem, also illustrated in Table 4.

**Theorem 3.** *If $S \subseteq [P]_{\mathrm{T}}$ for some specified pitch class set, $P$, then $\mathcal{C}(e_k(S)) = \mathcal{C}(e_{n-k}(S))$ where $1 \leq k < n$ and $n = |P|$.*

***Proof*** From Theorem 1 we know that $\mathcal{C}(e_k(S)) = \mathcal{C}(\mathrm{INV}(e_k(S)))$. But Lemma 2 tells us that $e_k(S) = \mathrm{INV}(e_{n-k}(S))$. Therefore $\mathcal{C}(e_k(S)) = \mathcal{C}(\mathrm{INV}(\mathrm{INV}(e_{n-k}(S))))$. Therefore $\mathcal{C}(e_k(S)) = \mathcal{C}(e_{n-k}(S))$. ■

If $S \subseteq [P]_{\mathrm{T}}$ for some specified pitch class set, $P$, then let $\mathcal{C}(S) = \bigcup_{e \in \mathrm{E}(S)} \mathcal{C}(e)$. As can be seen in Table 4,

$$\mathcal{C}(\mathcal{D}) = \mathcal{D} \cup \mathcal{M} \cup \mathcal{H} \cup \mathcal{W} = \mathcal{T} \cup \mathcal{W} . \tag{17}$$

In other words, the set of minimal simple cycle sets associated with the diatonic spectra is equal to the union of the set of tonal scale sets, $\mathcal{T}$, and the set of sets associated with what Rahn [4] (p. 37) calls the "whole-tone scale with filler tone", $\mathcal{W}$.

## 6 $\mathcal{C}(\mathcal{D})$ and Contradictions

The concept of a *contradiction* discussed by Rahn [4] (p. 36) and Clough and Douthett [8] (p. 125) can be defined as follows.

**Definition 1.** *If* $P$ *is a pitch class set and there exist integers,* $j$, $k$, $\ell$ *and* $m$ *such that* $\mathrm{i}(\mathbf{S}(P)[\ell], \mathbf{S}(P)[m]) > \mathrm{i}(\mathbf{S}(P)[j], \mathbf{S}(P)[k])$ *and* $(m - \ell) \bmod |P| < (k - j) \bmod |P|$, *then this is a case of contradiction.*

Let a *non-contradictory* set be one that has no contradictions. Let $\mathcal{R}_{\mu,n}$ denote the set of non-contradictory sets of cardinality $n$ when the chromatic cardinality is $\mu$. Rahn [4] (p. 37) observes that $\mathcal{R}_{12,7} = \mathcal{D} \cup \mathcal{M} \cup \mathcal{H} \cup \mathcal{W}$. From Eq. 17 above, it follows that $\mathcal{C}(\mathcal{D}) = \mathcal{R}_{12,7}$. That is, a set is a minimal simple cycle set for a diatonic spectrum if and only if it is a non-contradictory set of cardinality 7.

## 7 The Anhemitonic Pentatonic Scale and the Major Dominant Ninth Chord

An interval set $I$ is *contiguous* if and only if there exist $j$ and $k$ such that $I = \{i | j \leq i < k\}$. Each diatonic spectrum is a contiguous set of the form $\{i, i+1\}$ where $i \in \{1, 3, 5, 6, 8, 10\}$ (see Table 4). Each of these spectra has an interesting set of minimal simple cycle sets associated with it, which makes one wonder what the minimal simple cycle sets might be for interval sets of the form $\{i, i+1\}$ where $i \notin \{1, 3, 5, 6, 8, 10\}$—that is, where $i \in \{2, 4, 7, 9\}$. This set of contiguous-pair interval sets is equal to the spectrum set of the usual, "black note", anhemitonic pentatonic scale, {F♯,G♯,A♯,C♯,D♯}, that is, $\mathrm{E}(\mathcal{P}) = \{\{2,3\}, \{4,5\}, \{7,8\}, \{9,10\},\}$ where $\mathcal{P} = [\{0,2,4,7,9\}]_{\mathrm{T}} = [\{0,2,4,7,9\}]_{\mathrm{I}}$. Let us call $\mathrm{E}(\mathcal{P})$ the *set of pentatonic spectra* and $e_k(\mathcal{P})$ the *pentatonic k-spectrum.* It can readily be shown that $\mathcal{C}(e_2(\mathcal{P})) = \mathcal{C}(e_3(\mathcal{P})) = \mathcal{P}$ and that $\mathcal{C}(e_1(\mathcal{P})) = \mathcal{C}(e_4(\mathcal{P})) = \mathcal{P} \cup \mathcal{N}$ where $\mathcal{N} = [\{0,2,4,6,9\}]_{\mathrm{T}} = [\{0,2,4,6,9\}]_{\mathrm{I}}$. $\mathcal{N}$ is the set of sets associated with the complete major dominant ninth chords (see, e.g., Piston [1], pp. 334–346). It therefore follows that $\mathcal{C}(\mathcal{P}) = \mathcal{P} \cup \mathcal{N}$.

There are intriguing parallels between the sets of minimal simple cycle sets for the diatonic spectra and those for the pentatonic spectra. First, for $k$-spectra where $k \in \{\lfloor |P|/2 \rfloor, \lceil |P|/2 \rceil\}$, the minimal simple cycle sets are diatonic if $P$ is diatonic and pentatonic if $P$ is pentatonic. For the other spectra, the minimal simple cycle sets include other scale or chord types as well as the diatonic /pentatonic sets: for the other diatonic spectra, the other scale types generated are the harmonic major and minor, the ascending melodic minor and the "whole tone plus filler tone" scales; for the other pentatonic spectra, the other set class that emerges is that associated with the complete major dominant ninth chord. From the perspective of minimal simple cycle sets, the relationship between the major dominant ninth set and the pentatonic scale set is analogous to that between the diatonic set and the minor scales. Moreover, just as the set complement of a diatonic set ($\mathcal{D}$) is a pentatonic scale set ($\mathcal{P}$), so the set complement of an ascending melodic minor scale set ($\mathcal{M}$) is a major dominant ninth set ($\mathcal{N}$).

# 8 $\mathcal{C}(\mathcal{D})$, $\mathcal{C}(\mathcal{P})$ and Rothenberg's Concept of Propriety

Let $x$, $y$, $z$, $w$, $k$ and $\ell$ be any six tones on a *continuous* frequency scale, $S$, and let $(hj)$ denote the undirected interval between tones $h$ and $j$. Rothenberg [10] (pp. 206–207) defines a function $F(hj)$ that maps all musical intervals, $(hj) \in S \times S$, onto a set of discrete points called the *code*. Such a mapping is defined to be *proper* if and only if

$$(F(xy) = F(zw)) \wedge ((xy) < (kl) < (zw)) \Rightarrow (F(kl) = F(xy)) \,. \tag{18}$$

Rothenberg defines a *contradiction* to be any violation of this condition (i.e., $F(kl) > F(zw)$ or $F(kl) < F(xy)$). If $S$ is a discrete set of pitches (e.g., the ones that can be produced on a normal piano) and we define a *specific* interval to be one between pitches in this discrete $S$ and a *generic* interval to be the code onto which each specific interval is mapped, then a contradiction in Rothenberg's sense occurs when the ordering of two specific intervals is the opposite of the ordering of their corresponding generic intervals. The concept of contradiction defined in the more recent papers of Rahn [4] (p. 36) and Clough and Douthett [8] (p. 125) is therefore identical to Rothenberg's concept in the special case where $S$ is a discrete set of pitch points.

Rothenberg [10] (pp. 206-7) defines a mapping $F$ to be *strictly proper* if and only if it satisfies Eq. 18 and

$$(xy) \sim (zw) \Rightarrow F(xy) = F(zw) \tag{19}$$

where $(xy) \sim (zw)$ means that the intervals $(xy)$ and $(zw)$ sound the same. Thus, for a mapping to be strictly proper, the range of intervals that maps onto a given code must be contiguous and must not overlap with the range of intervals that map onto some other code. If Eq. 19 is not satisfied (i.e., two intervals, that sound the same, map onto different codes), then Rothenberg calls this an *ambiguity*. In the special case where $S$ is discrete, Rothenberg's concept of ambiguity is therefore identical to that of Rahn [4] (p. 36) and Clough and Douthett [8] (p. 125), despite the fact that neither of these more recent papers attributes the concept to Rothenberg.

Rothenberg [10] (p. 207) defines an *ambiguous interval* to be one that is mapped by $F$ onto more than one code. If $P$ is the range of $F$, then the *stability* of $P$ is the ratio of the number of non-ambiguous undirected intervals in $P$ to the total number of undirected intervals [11] (p. 354). For example, the stability of the diatonic set is $\frac{20}{21}$. If $P$ and $Q$ are pitch class sets, then (following Rothenberg [11], p. 369) $Q$ is a *sufficient set* for $P$ if and only if $P$ is the only element of $[P]_{\mathrm{T}}$ that contains $Q$; and $Q$ is a *minimal sufficient set* for $P$ if $Q$ is a sufficient set for $P$ and does not contain any proper subset which is also a sufficient set for $P$. Let $\mathbf{S}_i$ be any pitch class sequence that contains every element of $P$ exactly once. There are therefore $(|P|)!$ distinct sequences $\mathbf{S}_i$. Now let $s_i$ be the length of the shortest prefix of $\mathbf{S}_i$ whose associated pitch class set is a sufficient set for $P$. The average value of $s_i$ over all sequences $\mathbf{S}_i$ for a pitch class set $P$ is given by $\sigma(P) = (\sum_i s_i)/(|P|!)$ and Rothenberg [11] (p. 370) defines the *efficiency* of

**Table 5.** The inversional equivalence classes that contain proper 5- and 7-note sets for $\mu = 12$, together with the stability and efficiency of the sets in each class. (From Rothenberg [11], p. 367).

| $P$ | $[P]_{\mathrm{I}}$ | Stability | Efficiency |
|---|---|---|---|
| $\{0,1,4,6,8\}$ | $[\{0,1,4,6,8\}]_{\mathrm{I}}$ | 0.6000 | 0.5800 |
| $\{0,2,4,6,8\}$ | $[\{0,2,4,6,8\}]_{\mathrm{I}}$ | 0.4000 | 1.0000 |
| $\{0,2,4,7,9\}$ | $\mathcal{P} = [\{0,2,4,7,9\}]_{\mathrm{I}}$ | 1.0000 | 0.8000 |
| $\{0,1,4,6,9\}$ | $[\{0,1,4,6,9\}]_{\mathrm{I}}$ | 0.9000 | 0.6000 |
| $\{0,2,4,6,9\}$ | $\mathcal{N} = [\{0,2,4,6,9\}]_{\mathrm{I}}$ | 0.9000 | 0.6400 |
| $\{0,1,3,6,9\}$ | $[\{0,1,3,6,9\}]_{\mathrm{I}}$ | 0.4000 | 0.6400 |
| $\{0,1,3,4,6,8,9\}$ | $\mathcal{H} = [\{0,1,3,4,6,8,9\}]_{\mathrm{I}}$ | 0.4762 | 0.6259 |
| $\{0,1,3,5,6,8,10\}$ | $\mathcal{D} = [\{0,1,3,5,6,8,10\}]_{\mathrm{I}}$ | 0.9542 | 0.7687 |
| $\{0,1,3,4,6,8,10\}$ | $\mathcal{M} = [\{0,1,3,4,6,8,10\}]_{\mathrm{I}}$ | 0.7143 | 0.6299 |
| $\{0,1,2,4,6,8,10\}$ | $\mathcal{W} = [\{0,1,2,4,6,8,10\}]_{\mathrm{I}}$ | 0.2857 | 0.6327 |

$P$ to be $\varepsilon(P) = \sigma(P)/|P|$. The efficiency of a set $P$ is therefore the average size of a sufficient set for $P$ as a proportion of the size of $P$. It indicates how much of a set needs to be presented on average in order for it to be distinguished from its transpositions.

Table 5 lists every proper inversional equivalence class of 5- and 7-note sets for a chromatic modulus of 12, together with its stability and efficiency. Note that a proper set is strictly proper if and only if its stability is 1. This table reveals a number of interesting facts. First, if we rank the inversional equivalence classes of proper 5-note sets into decreasing order by, first, stability and then efficiency, then the union of the two highest-ranked classes is equal to $\mathcal{C}(\mathcal{P}) = \mathcal{P} \cup \mathcal{N}$. Next, the set of proper 7-note sets is equal to $\mathcal{C}(\mathcal{D}) = \mathcal{R}_{12,7}$. Finally, the union of the three most stable inversional equivalence classes of proper 7-note sets is equal to $\mathcal{T}$, the set of tonal scale sets. We can speculate that the rather low stability of $\mathcal{W}$ may be related to the fact that it seems to have been used much less commonly by composers than the other minimal simple cycle sets for the diatonic spectra.

## 9 $\mathcal{T}$, $\mathcal{C}(\mathcal{P})$ and Carey's Coherence Quotient

Like Balzano [12] (p. 71), Carey [13] (p. 6) defines a set to be *coherent* only if it contains neither contradictions nor ambiguities (in the sense of Rothenberg [10], p. 207, discussed above). Let $S$ be a pitch class set of cardinality $N$. Carey denotes by $AF(S)$ the total number of coherence failures in $S$—that is, the total number of contradictions and ambiguities in $S$ [13] (p. 19). Carey also denotes by $SF(N)$ the maximum number of simultaneous failures for a set of cardinality $N$ [13] (p. 14). He then defines the *coherence quotient* of $S$, denoted by $CQ(S)$, to be $1-(AF(S)/SF(N))$ and proposes that sets with higher coherence quotients will be more "scale-like" [13] (p. 19). Carey shows that the three, 7-note, inversional equivalence classes with the highest coherence quotients are (in decreasing order) $[\{0,1,3,5,6,8,10\}]_{\mathrm{I}}$ (i.e., $\mathcal{D}$), $[\{0,1,3,4,6,8,10\}]_{\mathrm{I}}$ (i.e., $\mathcal{M}$)

and $[\{0,1,3,4,6,8,9\}]_\mathrm{I}$ (i.e., $\mathcal{H}$) [13] (p. 36). In other words, the set of tonal scale sets, $\mathcal{T}$, is equal to the union of the three, 7-note, inversional equivalence classes with the highest coherence quotients. When we consider the most coherent 5-note sets, we find one inversional equivalence class, $\mathcal{P} = [\{0,2,4,7,9\}]_\mathrm{I}$, with a maximal coherence quotient of 1 and then two classes, $\mathcal{N} = [\{0,2,4,6,9\}]_\mathrm{I}$ and $[\{0,1,4,6,9\}]_\mathrm{I}$ with a coherence quotient of 0.96 [13] (p. 42). The set $\mathcal{C}(\mathcal{P})$ is therefore a subset of the union of the 5-note classes with the two highest coherence quotients.

## 10 Measuring the Evenness of a Pitch Class Set

Clough and Douthett [8] (p. 96) define a set to be *maximally even* if and only if every spectrum of the set is either a single integer or a pair of consecutive integers. However, they do not suggest how the evenness of a set might be *measured.*

Figure 2 represents the diatonic set $\{0,1,3,5,6,8,10\}$. Each pitch class in the set is connected to the centre of the diagram by a radial line segment. The arcs represent springs that start and end on these radial line segments. Suppose all the springs actually connect to the radial line segments at the same distance from the centre (despite the way the diagram is drawn). Suppose also that every spring is constrained in some frictionless way to lie on the circumference of a circle whose centre is at the centre of the diagram. Assume each spring obeys Hooke's law and that they all have the same spring constant $k$. The potential energy stored in a spring is $u(\Delta x) = \frac{k}{2}(\Delta x)^2$, where $\Delta x$ is the difference between the length of the spring and its resting length. A spring is defined to be at its resting length when the angle between the two pitch classes that it connects is the angle that would be between them if all the pitch classes in the set were perfectly evenly distributed around the circle.

If $p_1$ and $p_2$ are two pitch classes in a pitch class set $P$, then let the *diatonic interval*, $d(p_1,p_2,P)$, be defined to be $(\mathrm{POS}(p_2,\mathbf{S}(P)) - \mathrm{POS}(p_1,\mathbf{S}(P))) \bmod |P|$ where $\mathrm{POS}(x,\mathbf{S})$ is the index of the first occurrence of $x$ in $\mathbf{S}$ (or null if $x$ is not in $\mathbf{S}$). For a pitch class set $P$, the resting length of the spring connecting $p_1$ and $p_2$ is $2\pi rd/|P|$ where $d$ is the diatonic interval associated with the lesser of $\mathrm{i}(p_1,p_2)$ and $\mathrm{i}(p_2,p_1)$ and $r$ is the distance from the centre to the spring. If we suppose that $r = 1/(2\pi)$, then the resting spring length reduces to $d/|P|$ and the actual spring length is equal to $c/\mu$ where $c$ is the smaller of $\mathrm{i}(p_1,p_2)$ and $\mathrm{i}(p_2,p_1)$. $\Delta x$ for a spring is therefore $\frac{c}{\mu} - \frac{d}{|P|}$ and the potential energy stored in the spring is $k\left(\frac{c}{\mu} - \frac{d}{|P|}\right)^2/2$, which reduces to $((c/\mu)-(d/|P|))^2$ if we set $k=2$. We can use this potential energy as a measure of the "unevenness" within $P$ of the interval between the two pitch classes connected by the spring. We can then sum the potential energies of all the springs to get a measure of the unevenness of $P$. Formally, if $P$ is a pitch class set and $p_1, p_2 \in P$, then the *unevenness* associated with the pair of pitch classes $p_1$ and $p_2$ is

$$u(p_1,p_2,P) = ((\gamma(p_1,p_2)/\mu) - (\delta(p_1,p_2,P)/|P|))^2$$

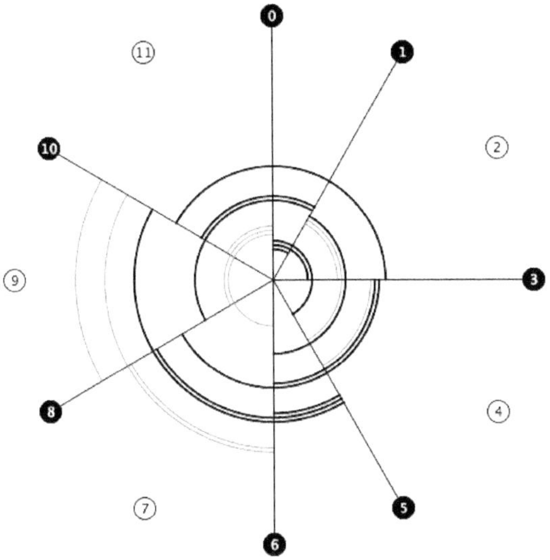


**Fig. 2.** Representation of the diatonic set illustrating a proposed method of measuring evenness. Each arc represents a spring connecting the pitch classes whose radial line segments intersect the endpoints of the arc. Stretched springs are thin and grey, compressed springs are thick and black.

where $\gamma(p_1, p_2) = \min(\mathrm{i}(p_1, p_2), \mathrm{i}(p_2, p_1))$ and

$$\delta(p_1, p_2, P) = \begin{cases} d(p_1, p_2, P) & \text{if } \mathrm{i}(p_1, p_2) < \mathrm{i}(p_2, p_1), \\ d(p_2, p_1, P) & \text{if } \mathrm{i}(p_1, p_2) > \mathrm{i}(p_2, p_1), \\ \min(d(p_1, p_2, P), d(p_2, p_1, P)) & \text{if } \mathrm{i}(p_1, p_2) = \mathrm{i}(p_2, p_1). \end{cases} \quad (20)$$

The *unevenness* of the set $P$ is then given by

$$U(P) = \sum_{i=0}^{|P|-2} \left( \sum_{j=i+1}^{|P|-1} u(\mathbf{S}(P)[i], \mathbf{S}(P)[j], P) \right) . \quad (21)$$

When the unevenness of a set is measured in this way, it turns out that: (1) the set of non-singleton, perfectly even sets (i.e., the ones that have an unevenness of 0 and a cardinality greater than 1) is equal to the set of minimal simple cycle sets for interval sets of cardinality 1; (2) $\mathcal{T}$ is equal to the union of the three most even 7-note inversional equivalence classes; and (3) $\mathcal{C}(\mathcal{P})$ is equal to the union of the two most even 5-note inversional equivalence classes.

Note that $\mathcal{T}$ is *not* the union of the three most even 7-note inversional equivalence classes when certain other measures that have been proposed in the literature are used to measure evenness. Table 6 shows the 7-note set classes that are judged to be most even using Eq. 21 along with their ranks using Block and

**Table 6.** Most even 7-note set classes when measured using Eq. 21 with ranks generated by Block and Douthett's [14] (pp. 32–35) measure for comparison

| *Rank using Eq. 21* | $U(P)$ | $[P]_{\mathrm{I}}$ | *Block and Douthett's rank* |
|---|---|---|---|
| 1 | 0.0278 | $\mathcal{D} = [\{0, 1, 3, 5, 6, 8, 10\}]_{\mathrm{I}}$ | 1 |
| 2 | 0.0417 | $\mathcal{M} = [\{0, 1, 3, 4, 6, 8, 10\}]_{\mathrm{I}}$ | 2 |
| 3 | 0.0556 | $\mathcal{H} = [\{0, 1, 3, 4, 6, 8, 9\}]_{\mathrm{I}}$ | 4 |
| 4= | 0.0694 | $\mathcal{W} = [\{0, 1, 2, 4, 6, 8, 10\}]_{\mathrm{I}}$ | 3 |
| 4= | 0.0694 | $[\{0, 1, 2, 4, 6, 8, 9\}]_{\mathrm{I}}$ | ? |

Douthett's measure [14] (pp. 32–35). Note that $\mathcal{C}(\mathcal{D})$ is a subset of the union of the set classes with the highest 4 ranks when evenness is calculated with either measure. However, $\mathcal{T}$ is no longer the union of the three most even 7-note classes when Block and Douthett's measure is used. This may suggest that the measure defined in Eq. 21 is a better model of intuition than that proposed by Block and Douthett, a hypothesis that could be investigated further by carrying out appropriate psychological experiments.

## 11 Summary of Main Results

The set $\mathcal{F} = \mathcal{D} \cup \mathcal{M} \cup \mathcal{H}_{\mathrm{min}}$ is the set of pitch class sets associated with what Piston [1] (pp. 1-2) calls the "fundamental diatonic scales". The smallest inversionally closed superset of $\mathcal{F}$ is the set $\mathcal{T} = \mathcal{F} \cup \mathcal{H}_{\mathrm{maj}}$ where $\mathcal{H}_{\mathrm{maj}}$ is the set of sets associated with the harmonic major scales. In this paper, $\mathcal{T}$ is called the set of tonal scale sets as it contains all and only those sets associated with the usual major and minor scales. It has been shown that the set $\mathcal{T}$ is equal to:

1. the smallest inversionally closed superset of $\mathcal{F}$;
2. the set of pitch class sets associated with Schenker's combination scales;
3. the set of pitch class sets generated by combining a major dominant triad with a major or minor tonic triad and a major or minor subdominant triad;
4. the set of pitch class sets generated by combining primary triads that can each be either major or minor;
5. the set of minimal simple cycle sets for the diatonic 2-spectrum, $\{3, 4\}$;
6. the union of the three most stable inversional equivalence classes of proper 7-note sets (using Rothenberg's [10,11] definitions of stability and propriety);
7. the union of the three, 7-note inversional equivalence classes with the highest coherence quotient (as defined by Carey [13]); and
8. the union of the three most even 7-note inversional equivalence classes (using the spring-based method of measuring evenness proposed in Sect. 10).

It was also shown that the set of non-contradictory 7-note sets, $\mathcal{R}_{12,7}$ is equal to the set $\mathcal{C}(\mathcal{D})$, that is, the union of the sets of minimal simple cycle sets for the diatonic spectra.

Finally, it was shown that the union of the sets of minimal simple cycle sets for the pentatonic spectra, $\mathcal{C}(\mathcal{P})$, is equal to $\mathcal{P} \cup \mathcal{N}$ where $\mathcal{N}$ is the class of pitch class sets associated with the complete major dominant ninth chord. The relationship between $\mathcal{N}$ and $\mathcal{P}$ is therefore analogous to that between $\mathcal{D}$ and the other tonal scales. $\mathcal{C}(\mathcal{P})$ is also:

1. the union of the two most highly-ranked inversional equivalence classes of proper 5-note sets when these are sorted by stability and then efficiency;
2. a subset of the union of the 5-note inversional equivalence classes with the two highest coherence quotients (as defined by Carey [13]); and
3. the union of the two most even 5-note inversional equivalence classes (when evenness is measured using the spring-based method proposed in Sect. 10).

## References


1. Piston, W.: Harmony. Gollancz, London (1991)
2. Clough, J., Engebretsen, N., Kochavi, J.: Scales, Sets and Interval cycles: A Taxonomy. Music Theory Spectrum 21(1), 74–104 (1999)
3. Cross, I., West, R., Howell, P.: Cognitive Correlates of Tonality. In: Howell, P., West, R., Cross, I. (eds.) Representing Musical Structure, pp. 201–243. Academic Press, London (1991)
4. Rahn, J.: Coordination of Interval Sizes in Seven-Tone Collections. Journal of Music Theory 35(1/2), 33–60 (1991)
5. Hauptmann, M.: Die Natur der Harmonik und der Metrik zur Theorie der Musik. Breitkopf und Härtel, Leipzig (1853)
6. McCune, M.: Moritz Hauptmann: Ein haupt Mann in Nineteenth Century Music Theory. Indiana Theory Review 7(2), 1–28 (1986)
7. Schenker, H.: Harmony. University of Chicago Press, Chicago (1954)
8. Clough, J., Douthett, J.: Maximally Even Sets. Journal of Music Theory 35(1/2), 93–173 (1991)
9. Clough, J., Myerson, G.: Variety and Multiplicity in Diatonic Systems. Journal of Music Theory 29(2), 249–270 (1985)
10. Rothenberg, D.: A Model for Pattern Perception with Musical Applications. Part I: Pitch Structures as Order-Preserving Maps. Mathematical Systems Theory 11, 199–234 (1978a)
11. Rothenberg, D.: A Model for Pattern Perception with Musical Applications. Part II: The Information Content of Pitch Structures. Mathematical Systems Theory 11, 353–372 (1978b)
12. Balzano, G.J.: The Group-Theoretic Description of 12-Fold and Microtonal Pitch Systems. Computer Music Journal 4(4), 66–84 (1980)
13. Carey, N.: On Coherence and Sameness, and the Evaluation of Scale Candidacy Claims. Journal of Music Theory 46(1/2), 1–56 (2002)
14. Block, S., Douthett, J.: Vector Products and Intervallic Weighting. Journal of Music Theory 38(1), 21–41 (1994)

# Scratching the Scale Labyrinth

Andrew J. Milne[1], Martin Carlé[2], William A. Sethares[3], Thomas Noll[4], and Simon Holland[1]

[1] The Open University, Milton Keynes, UK
{a.j.milne,s.holland}@open.ac.uk
[2] Humboldt-Universität zu Berlin, Germany
martin.carle@culture.hu-berlin.de
[3] University of Wisconsin-Madison, USA
sethares@ece.wisc.edu
[4] Escola Superior de Música de Catalunya, Barcelona, Spain
noll@cs.tu-berlin.de

**Abstract.** In this paper, we introduce a new approach to computer-aided microtonal improvisation by combining methods for (1) interactive scale navigation, (2) real-time manipulation of musical patterns and (3) dynamical timbre adaption in solidarity with the respective scales. On the basis of the theory of well-formed scales we offer a visualization of the underlying combinatorial ramifications in terms of a *scale labyrinth.* This involves the selection of generic well-formed scales on a binary tree (based on the Stern-Brocot tree) as well as the choice of specific tunings through the specification of the sizes of a period (pseudo-octave) and a generator (pseudo-fifth), whose limits are constrained by the actual position on the tree. We also introduce a method to enable transformations among the modes of a chosen scale (generalized and refined "diatonic" and "chromatic" transpositions). To actually explore the scales and modes through the shaping and transformation of rhythmically and melodically interesting tone patterns, we propose a playing technique called Fourier Scratching. It is based on the manipulation of the "spectra" (DFT) of playing gestures on a sphere. The coordinates of these gestures affect score and performance parameters such as scale degree, loudness, and timbre. Finally, we discuss a technique to dynamically match the timbre to the selected scale tuning.

**Keywords:** MOS Scales, Well-Formed Scales, Diatonic, Chromatic, Stern-Brocot Tree, Farey Sequence, Fourier Scratching.

## 1 Introduction

The scale labyrinth is a visualization of a widely studied class of musical scales, which form a deeply structured and interconnected scale universe. The terminology applied in this paper originates from two separate strands of music-theoretical discourse. The term *moment of symmetry scale* (MOS) is rooted in the field of investigations into musical tunings, while the term *well-formed* (WF) scales is common in algebraic scale theory. The present paper attempts to

C. Agon et al. (Eds.): MCM 2011, LNAI 6726, pp. 180–195, 2011.

merge these two traditions within an experimental paradigm. For now, however, the terms are used interchangeably.

Well-formed scales are generated by a single interval and contain exactly two different step sizes (large and small) that are maximally evenly distributed. The case where these step sizes coincide is included as a degenerate instance. Each scale has a *valid tuning range* over which it maintains a structural identity but over which the sizes of the two steps co-vary. At any given tuning, each scale is embedded within a unique family of scales with successively larger numbers of tones (and successively smaller valid tuning ranges). Carey [1] denotes the class of all well-formed $N$-note scales as $WF(N, g)$ where $g$ is the factor that converts generator order into scale step order (mod $N$). For example, the diatonic scale and its inverse belong to $WF(7, 2)$ while the chromatic scale and its inverse belong to $WF(12, 7)$. These relationships are represented visually in Sect. 4.

Visualizing the structure of these scales is useful from an analytic point of view, and it can also function as a graphical user interface (GUI) object for use in musical applications. An example of an analytical application is given in Figure 5 in [2], where a scale labyrinth is used to indicate MOS scale tunings that provide good approximations of just intonation. A concrete example of use in a GUI is given in Sec. 5, where the scale labyrinth allows a musician to choose, simultaneously, a scale structure (number of small and large steps) and its tuning (the sizes of its period and generator).

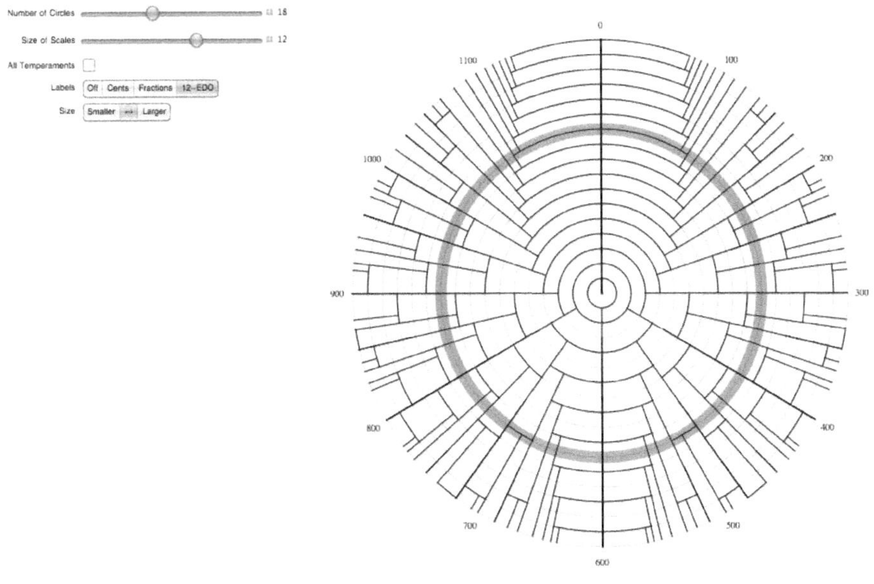


**Fig. 1.** This scale labyrinth, generated interactively in Mathematica, can be adjusted in several ways to emphasize certain aspects of the tunings as described in Sect. 4. The program can be downloaded from
`http://homepages.cae.wisc.edu/~sethares/MOSLabyrinth.nbp`

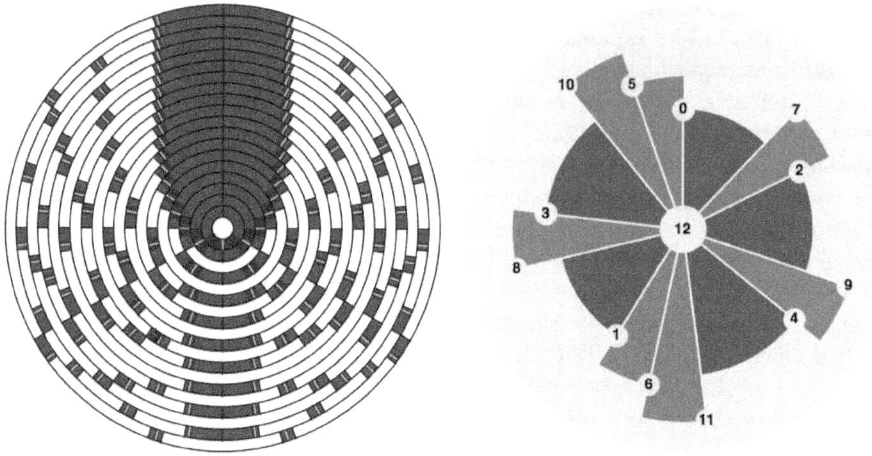


**Fig. 2.** This scale labyrinth can be used directly to control the tuning of an interactive algorithmic music application, as described in Sect. 5. See also `http://www.youtube.com/watch?v=0BUUcjLbCIk` for a short movie

**Fig. 3.** The scale labyrinths of Figures 1 and 2 have a remarkable symmetry and are reminiscent of labyrinths of Gothic architecture, such as the famous Chartres cathedral labyrinth shown here (original photograph by Mich De Mey `http://www.flickr.com/photos/dumbo/2555996059/`)

The scale labyrinth is related to a Stern-Brocot tree (Fig. 4), which is a systematic enumeration of the rational numbers. In the case of the labyrinth, the tree has been bent into a circular shape so that 0/1 and 1/1 lie at the same location. In the scale labyrinth, each nested circle corresponds to a fraction with denominator equal to the radius of the circle; in the Stern-Brocot tree, each

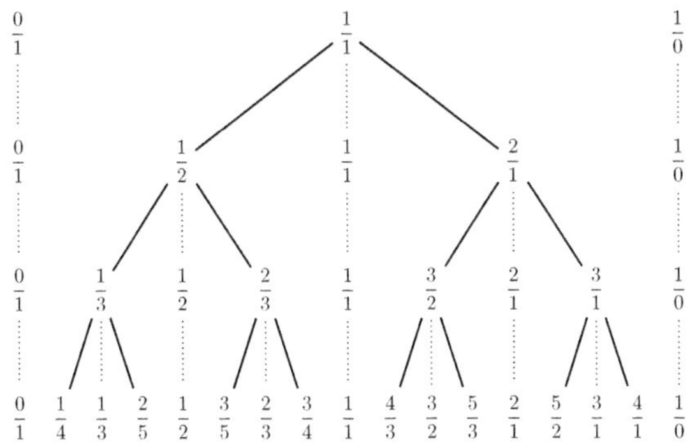

**Fig. 4.** The Stern-Brocot tree

row contains fractions with different denominators. The Stern-Brocot tree has been used by Erv Wilson to illustrate the structure of MOS scales using scale tree diagrams [3]. More recently, Holmes [4] produced a version, rotated by 90 degrees, where the left-right position of each fraction was determined by its denominator. Inspired by this diagram, Milne [2] produced a circular version to visually emphasize the period (octave) equivalence so that rotation by 360 degrees corresponds to the period.

## 2 Some Properties of MOS/Well-Formed Scales

There are several alternative but equivalent definitions: an MOS/well-formed scale is a *generated scale* containing exactly two step sizes that are distributed with maximal evenness[1]. A generated scale is produced by repeatedly adding a *generator* interval (typically a perfect fifth) and then reducing all such intervals by repeatedly subtracting a *period* interval (typically the octave) so all intervals are smaller than the period. The number of times the generator can be stacked so as to produce just two evenly distributed step sizes depends on the ratio of the generator and period. For example, if the generator is 702 cents and the period is 1200 cents (a generator/period ratio of 0.585), the two-note scale C-G belongs to scale class $WF(2,1)$, the 3-tone scale C-D-G belongs to $WF(3,2)$, the 5-tone scale C-D-E-G-A belongs to $WF(5,2)$, the 7-tone scale C-D-E-F♯-G-A-B belongs to $WF(7,2)$, the 12-tone chromatic scale belongs to $WF(12,7)$, $WF(17,12)$ is a 17-tone scale, and so forth [5]. Different generator/period ratios require different numbers of tones to produce well-formed scales. For example, when the generator is 316 cents (a just intonation "minor third") and the period is 1200 cents, the following numbers of tones produce well-formed scales: 2, 3, 4, 7, 11, 15, 19, and so forth. Whenever the size of the period is not explicitly mentioned, it is assumed to be 1200 cents.

[1] That is, the distribution of the two steps forms a Christoffel word.

Each MOS scale can be characterized by its number of large (L) and small (s) steps, so the familiar (anhemitonic) pentatonic scale can be characterized with the signature 2L, 3s, the familiar diatonic scale by the signature 5L, 2s. For any given generator/period ratio, the number of tones required to form an MOS scale and the range of tunings over which any such scale maintains its identity (i.e., its number of large and small steps is invariant), is given by the Stern-Brocot tree (as discussed in the next section).

Well-formed scales have a number of properties that are thought to give them aesthetic value. For example: every scale span (generic interval size) occurs in exactly two interval sizes (Myhill's property [6]). The two scale step sizes are evenly distributed throughout the period. Within the period, every scale degree has a unique pattern of intervals surrounding it [7], which helps support tonal functionality. When transposed by the generator, the resulting scale shares all but one tone, facilitating modulation [7]. Collectively, these features suggest a good compromise between the excessive simplicity of equal step scales and the complexity of completely irregular scales [8].

Western theory recognizes the first five fifth-generated well-formed scales: authentic division of the octave, tetractys, the pentatonic, diatonic and chromatic[2]. But there are a number of MOS scales that, due to their microtonal intervals, are unfamiliar and may be difficult to play on standard instruments. Interestingly, a number of such well-formed scales contain numerous intervals and chords that approximate consonant just intonation intervals and chords as effectively as the familiar diatonic scale [9]. And, as discussed in Sect. 6, it is possible to use synthesizers capable of spectral retuning to minimize sensory dissonance for any MOS scale at any tuning [10]. It seems, therefore, that with novel musical controllers and synthesizers, the musical possibilities of MOS structures may become more accessible to musicians and composers. This paper discusses one such application.

Some intriguing features of scales with the MOS/well-formed structure are:

*Co-prime step numbers:* In every MOS, the number of small steps and number of large steps (in each period) is always co-prime. For example, the pentatonic scale is 2L, 3s; the diatonic is 5L, 2s. There is no MOS scale with, for example, 2 large steps and 4 small steps (within the period).

*Inverse scales:* Every MOS scale has an "inverse" form, where the number of large and small steps swaps. For example, the diatonic scale (5L, 2s) has an inverse, the *anti-diatonic* scale, which is 2L, 5s.

*Landmark equal tunings:* As the tuning of generator changes, the sizes of the small and large steps co-vary. For example, when the generator is 700 cents, the diatonic scale's large steps (major seconds) are 200 cents, its small steps (minor seconds) are 100 cents; when the generator is 710 cents, the large steps are 220 cents, the small are 50 cents (in all cases, $5 \times$ large step size $+ 2 \times$ small step size $=$ period size).

[2] Strictly speaking, the chromatic scale in 12-TET is a *degenerate* well-formed scale because its two step sizes are identical [5].

The co-varying step sizes produce three "landmark" tunings: a) the tuning where the small and large steps become the same size (this tuning marks the transition between an MOS scale and its inverse), b) the tuning at which the small steps of the MOS shrink to zero, c) the tuning at which the small steps of the inverse MOS scale shrink to zero. All three landmark tunings are equal temperaments: the cardinality of a) is the number of L steps plus the number of s steps, the cardinality of b) is the number of L steps, and the cardinality of c) is the number of s steps. For example, the diatonic 5L, 2s scale meets its inverse (the anti-diatonic scale) at 7-TET (685.714 cents), where the large and small steps become identically sized; the diatonic is also bounded at 5-TET (720 cents), where its two small steps shrink to zero size; the anti-diatonic is also bounded at 2-TET (600 cents), where its five small steps shrink to zero.

*Embeddings:* Every MOS scale with signature $p$L, $q$s is embedded in a family of MOS scales. The lowest cardinality embedding scale has $2p + q$ steps. For example, the lowest cardinality MOS scale that embeds the 5L, 2s diatonic scale has $2 \times 5 + 2 = 12$ tones; the lowest cardinality scale that embeds the 2L, 5s anti-diatonic has $2 \times 2 + 5 = 9$ tones. A method to determine the tuning of this embedding scale using the Stern-Brocot tree is given in Sect. 3.

*Coherence within a well-defined tuning range:* A scale is *coherent* [11] or *proper* [12] if there is a monotonic relationship between that scale's generic interval sizes and its specific interval sizes. This requires, for example, that every fifth be larger than every fourth, which are larger than every third, which are larger than every second, which in turn are larger than the unison. Well-formed scales are coherent over the tuning range within which the ratio of L to s (Blackwood's $R$ [13]) is less than 2. This is precisely the tuning at which the lowest cardinality embedding scale is equally tuned. For instance, the 5L, 2s diatonic scale is coherent between 4/7 and 7/12. The anti-diatonic scale 2L, 5s is coherent between 5/9 and 4/7.

## 3 The Stern-Brocot Tree

The Stern-Brocot tree [14,15] is a systematic enumeration of the rational numbers independently discovered in the 19th century by the mathematician Moritz Stern and the watchmaker Achille Brocot. Stern's focus was mathematical whereas Brocot's focus was the specification of gear ratios for clock design. The tree provides a method to iteratively generate all rational numbers, in reduced form, exactly once. In Fig. 4, note that the top row consists of the three rationals 0/1, 1/1, and 1/0; the next row down consists of the mediants of each adjacent pair of the above row, which gives the rationals 1/2 and 2/1 (the *mediant* of $a/b$ and $c/d$ is $(a+c)/(b+d)$, where all fractions are reduced). The iterative process of populating each new row with the mediants of the adjacent pairs above produces the Stern-Brocot tree, and all fractions are ordered, by size, from left to right. The following paragraphs show the correspondences between the Stern-Brocot tree and the scale properties described in the previous section.

In the Stern-Brocot tree, each fraction can be thought of as representing a generator/period tuning ratio. Because these are co-prime ratios $n/d$, they

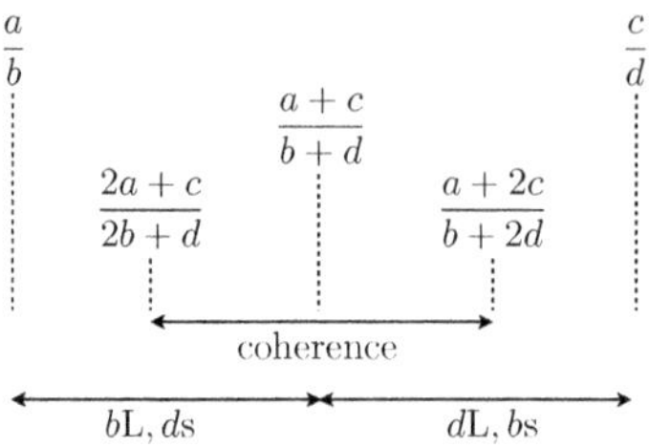


**Fig. 5.** A well-formed scale's landmark tunings, its range of tuning coherence, and its lowest cardinality embedding can be read directly off the Stern-Brocot tree

generate an equal division of the period of cardinality $d$. For instance, 4/7 can represent a generator of 685.714 cents and a period of 1200 cents, and thus generates 7-TET. With this in mind, the tree can be used to provide the precise landmark tunings, range of coherence, and embeddings of any well-formed scale. Fig. 5 provides an illustration of the methods described below.

Consider two fractions in the same row of the tree, $\frac{a}{b}$ and $\frac{c}{d}$, and their mediant $\frac{a+c}{b+d}$. The interval between $\frac{a}{b}$ and $\frac{a+c}{b+d}$ corresponds to the valid (generator/period) tuning range of an MOS scale with $b$ large steps and $d$ small. There must be $d$ small steps because the boundary tuning at $\frac{a}{b}$ has $b$ tones, which implies that $d$ small steps have shrunk to zero size. Conversely, the interval between $\frac{a+c}{b+d}$ and $\frac{c}{d}$ corresponds to the valid tuning range of the inverse MOS scale, which has $d$ large steps and $b$ small—the boundary tuning at $\frac{c}{d}$ has $d$ tones, so $b$ small steps have shrunk to zero size. The tuning at precisely $\frac{a+c}{b+d}$ is $(b+d)$-TET, so this is the tuning at which the "large" and "small" steps become equivalent in size and the MOS meets its inverse.

For example, the fractions $\frac{1}{2}$ and $\frac{3}{5}$ have a mediant $\frac{4}{7}$, so the interval between $\frac{1}{2}$ and $\frac{4}{7}$ is the valid tuning range of an MOS with 2 large and 5 small tones (the anti-diatonic), while the interval between $\frac{4}{7}$ and $\frac{3}{5}$ is the valid tuning range of its inverse—the diatonic 5L, 2s. At precisely $\frac{4}{7}$, the large and small steps are equally sized and the two scales meet. Each triple of fractions made from two adjacent fractions and the fraction between them on the row below corresponds, therefore, to the three landmark tunings: The central value gives the generator/period ratio where the MOS scale meets its inverse, the outer values give the tunings at which the small steps of the MOS, or its inverse shrink to zero.

The lowest cardinality scales within which any MOS is embedded can be identified by the mediant of its two boundary tunings. As above, the boundary tunings of $b$L, $d$s are $\frac{a}{b}$ and $\frac{a+c}{b+d}$, and their mediant is $\frac{2a+c}{2b+d}$. This means the embedding scale contains $2b+d$ tones, with either $b+d$ large steps and $b$ small, or $b$ large and $b+d$ small, depending on the tuning. For example, the diatonic scale (whose boundaries are 4/7 and 3/5) is embedded within their mediant, which is 7/12 (which represents the scale 7L, 5s or its inverse 5L, 7s). Similarly the anti-diatonic whose boundary tunings are 1/2 and 4/7 is embedded within their mediant, which is 5/9 (7L, 2s, or 2L, 7s).

The tuning range over which $b$L, $d$s, and its inverse $d$L, $b$s, are coherent is bounded by the tunings at which their respective lowest cardinality embedding scales are equally tuned. That is, at $\frac{2a+c}{2b+d}$ and $\frac{a+2c}{b+2d}$. For instance, the diatonic has boundary tunings of 4/7 and 3/5, and their mediant (embedding scale) is 7/12, so the range over which the diatonic scale is coherent is 4/7 to 7/12; similarly, the anti-diatonic has boundary tunings at 1/2 and 4/7 with a mediant (embedding scale) of 5/9, so the range over which it is coherent is 5/9 to 4/7, which can be gleaned from [5].

## 4 Reading the Interactive Labyrinths

Depending on its purpose, different visualizations of the scale labyrinth may be preferred. Figure 1 displays several kinds of information that may be useful in a detailed analysis while the simpler Fig. 2 may be more appropriate as an interface element for the purpose of choosing a scale and tuning. In Fig. 1, the angle indicates the ratio between the generator and the period—the top of the circle represents a generator/period ratio of zero, the bottom of the circle a ratio of 1/2, the 11 o'clock position, a ratio of 11/12. For example, the familiar 12-tone equal tempered diatonic scale, which requires a generator of 700 cents (7 semitones) and a period of 1200 cents (12 semitones), can be found at the location $700/1200 = 7/12 = 0.583$. Figure 1 is labeled in cents (assuming a period of 1200) but, in the interactive version, fractions can be displayed instead.

Note that the structure is left-right symmetric, because a scale created using any generator is identical to the scale generated by the complement of the generator within the period (e.g. precisely the same scale is produced by generators of 697 cents and $1200 - 697 = 503$ cents). Each ring corresponds to the set of MOS scales that contain the numbers of notes indicated by its integer label (the dot at the center of the circle is ring 1). Each radial line (spoke) extending inwards from the edge of the circle represents an equal temperament scale. For each spoke, there is some circle that the spoke touches but does not cross. This circle gives the number of notes in the corresponding equal step scale. The angle of the spoke gives the tuning of its generator relative to the period. For instance, at 700 cents, there is a spoke that extends from the edge of the circle to the 12th ring, which indicates that this tuning produces 12-TET. Similarly, at 685.714 cents ($685.714/1200 = 4/7$) there is a spoke extending to the 7th ring, illustrating that this tuning produces 7-TET.

As described in Sect. 2, as the generator/period tuning ratio changes, the small and large steps co-vary across three landmark tunings, which are visually prominent in the labyrinth. Using the 7-note diatonic and anti-diatonic scales as an example, focus on the seventh ring as in Fig. 6, which zooms in about the 700 cent region. There are two lines which cross (rather than merely touch) this part of the 7-ring, those at 600 and 720 cents. These delimit an arc which corresponds to the possible tunings of a 7-note MOS scale, (in this case the diatonic) and its inverse (in this case the anti-diatonic). The 7-TET spoke at 685.7 cents is the only spoke which meets (but does not cross) this arc. This spoke represents the

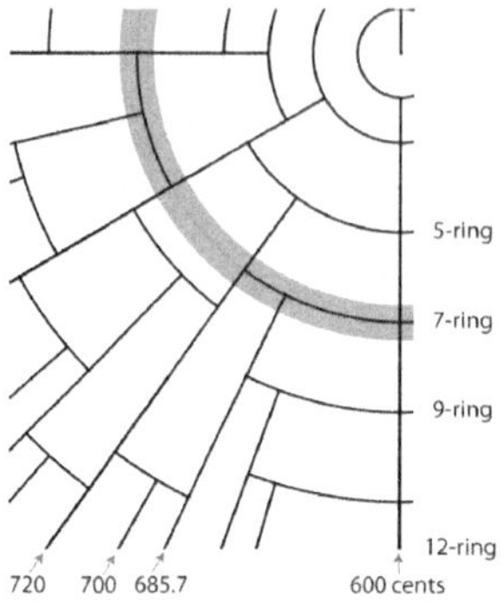


**Fig. 6.** This is a zoom into the scale labyrinth of Figure 1. Extra annotations have been added to help clarify the discussion in this section.

point at which the sizes of the small and large steps equalize before reversing roles, and so marks the tuning at which the diatonic and anti-diatonic meet.

At 720 cents, the spoke marking the left edge of the arc extends inwards to touch the 5th ring. This tuning marks 5-TET, the point at which the 2 small scale steps shrink to zero. Similarly, the spoke at the right edge of the arc (at 600 cents) extends inwards to the 2nd ring. This marks 2-TET, the point at which 5 small scale steps shrink to zero. Thus, this MOS scale has 5 large and 2 small steps (5L, 2s), with the numbers reversed for the inverse scale. This simple procedure of following the spokes inward gives the structure of the MOS sale and its inverse. Conversely, following outward the spokes that delimit the scale and its inverse shows which MOS scales 7-TET is embedded inside. In this case, 7-TET is embedded in MOS scales of size 9 and 12. Similarly, any arc and its associated three spokes correspond to an MOS scale and its three landmarks. Thus the labyrinth can be used to investigate visually the structure, inverse, embedding scales, associated equal tunings, and valid tuning range of any MOS scale.

Figure 2 uses a slightly different visualization. In this version, each scale's valid tuning range is indicated with a thick colored band; this enables the scales to be quickly spotted and easily clicked upon when used in a GUI. The landmark tunings are now indicated by the boundaries of each scale arc (as before), and by the radial segment inside it, which marks the location at which the inverses meet. The thicker arcs also allow for the tuning range of coherence to be clearly indicated with a darker shading.

The scale labyrinth in Fig. 1 is drawn interactively in a Mathematica program that allows the user to control how the information is presented. The basic size of the labyrinth (how many concentric circles it contains) is controlled by the top slider, while the second slider moves the golden ring to highlight scales of the specified size. The circumference of the circle can be labelled in cents or fractions, and in absolute terms or with respect to the chosen scale size. While Fig. 1 shows scales up to 18 tones, the interactive application can display scales of any size.

A large variety of tunings that provide approximations to (temperaments of) just intonation—such as meantone, srutal, magic, hanson, etc.—are built in and detailed information about them is shown by clicking on the names. When displaying individual temperaments, red lines are superimposed whose angles show their 5-limit TOP (Tenney optimal) tunings (such tunings minimize the maximum error of all possible 5-limit just intonation intervals [9]). The number of the ring they extend to indicates the lowest cardinality scale for which every scale degree is a member of a major or minor triad. For instance, in the pentatonic scale C, D, E, G, A, there is no major or minor triad that contains tone D. Adding another fifth (to make the Guidonian hexachord) gives C, D, E, G, A, B, and now every tone is a member of at least one major or minor triad. This scale, however, is not well-formed. The lowest cardinality well-formed scale, all of whose tones are a member of a major or minor triad is, in this case, the seven-tone diatonic scale. So, in general, it may be musically useful to choose an MOS close to a location where a red line passes through its arc.

The concept of *scale*, as we use it, presupposes periodicity. Under this perspective all rotations of a scale are equivalent. We use the concept of *mode* in order to grasp all the combinatorial refinements, which emerge from the variation of a fundamental domain for this period.

The modes of a well-formed scale have a remarkable property of parsimony, by which one mode can be transformed into another by the replacement of a single tone either by a pseudo-octave (= period) or by the augmented prime (= difference interval between large and small step). These transformations are

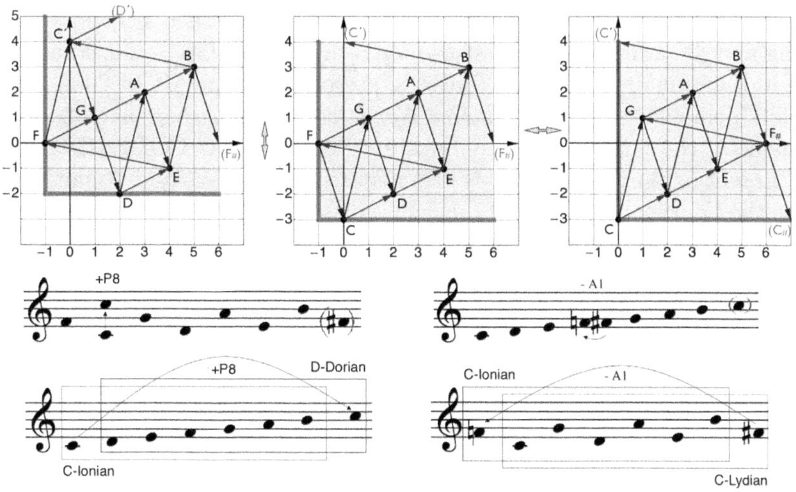


**Fig. 7.** Two types of mode transformation correspond to vertical and horizontal shifts of a fundamental frame in a generic height and width coordinate system. The figure shows the transformation of a C-Ionian mode into the common origin mode D-Dorian (left) and the transformation of a C-Ionian mode into the common finalis mode C-Lydian (right), and vice versa.

modal refinements of diatonic and chromatic transposition. Every well-formed scale has an associated universe of modes which is freely generated from two basic and commuting transformations and which is therefore isomorphic to the free commutative group $\mathbb{Z}^2$ of rank 2. One transformation preserves the origin of a mode and yields a new finalis (lowest tone) a second higher. The other transformation preserves the finalis of a mode and yields a new origin a pseudo-fifth (generator) sharp-wards. The details of the interactive navigation through the modes of a well-formed scale are based on a mathematical approach given in [22,23]. Every mode occupies a fundamental frame in $\mathbb{Z}^2$ whose borders correspond to the augmented prime (horizontal dimension) and the pseudo-octave (vertical dimension). The zig-zag trajectories within each frame represent the step interval patterns and the folding patterns. Minimal changes of the mode can be described as vertical or horizontal shifts of the fundamental domain. Any modal change can be decomposed into such minimal changes. The height-width representation therefore offers an effective navigation method with the modal universe of a well-formed scale.

## 5 Fourier Scratching as a Performance Technique

The Fourier Transform has been successfully applied in many different ways to the analysis and manipulation of musical sound. It has also been fruitfully applied to non-acoustic problems in mathematical music theory such as those in [16,17] and [18]. The *Fourier Scratching* project attempts to transfer the established analysis - manipulation - resynthesis paradigm from the domain of sound processing to the domain of macroscopic musical events such as melody, rhythm, tuning and dynamics. It was first proposed in [19] and later realized as a prototype in the domain of rhythm (initially presented at the MCM 2009 [20], later publicly demonstrated at the *Lange Nacht der Wissenschaften in Leipzig*, and discussed at the SuperCollider symposium { SOUNDING CODE } in 2010 [21]). The present extension to scales is also inspired by Quinn's and Amiot's findings about the Fourier properties of musical scales [17,18].

Traditionally, playing microtonal music in different scales requires developing multiple instruments with suitable interfaces, which one then needs to learn to play. For example, think of the Cembalo Cromatico with 19 keys per octave. To build and master keyboards like this is quite a challenge, both technologically and pianistically. The present project began with the idea of improvising over the tunings of a scalar labyrinth, something that is infeasible without the help of a computer. Even with suitable software, which makes it easy to play any pattern in any scale at any speed, it is still necessary to specify every pattern in advance. But how can one improvise in any pattern in any scale at any speed?

The idea is to actuate a (virtual) playing robot in a musically sensible way. The basic behavior of the robot is to perform keystrokes with variable strength on variable positions of a keyboard with its $n$ fingers one after the other at a given pulse rate. The finger movements occur with an unperturbed orderliness: finger 1 followed by finger 2 etc. and eventually finger $n$ followed again by finger 1.

The pulse and the number $n$ of fingers can be changed but for the moment consider them fixed. The main paradigm of playing with the help of this robot is that the improviser can change the strength and the position of any finger at any time. The (virtual) keyboard is designed as a continuous circle, i.e. any point on this circle may have a (potentially) different sound. The keyboard layout for a finite scale can have specific key widths which are proportional to the sizes of the step intervals above each tone (see Fig. 8). In these examples the number of robot fingers coincides with the number of scale tones, but this is not mandatory.

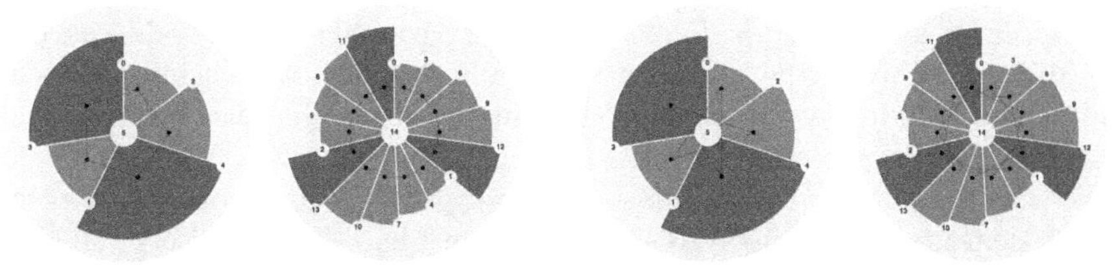

**Fig. 8.** The black polygons represent elementary play states with a scale being distributed on the circular keyboard (5 tones and 14 tones, respectively). In these cases the playing robot has as many fingers as there are tones in the scale and their distribution is regular. The coherence of these scale examples therefore guarantees that each tone is being played exactly once. In the Fourier picture these play states may be regarded as pure partials. In the left two examples the first Fourier coefficient satisfies $a_1 = 1$ and all others vanish. In the right two examples, $a_3 = 1$ and all other coefficients vanish.

The idea behind the Fourier Scratching technique and interface is the interactive control of musical parameters through the visualization and gestural manipulation of their Fourier coefficients. This presupposes that the musical parameters in question are encoded as complex vectors $f = (f_0, ..., f_{n-1}) \in \mathbb{C}^n$. For moderately small values of the dimension $n$ there is no need to use the FFT to obtain realtime control, and thus the values of $n$ are not restricted to powers of 2. The slower discrete Fourier transform $\hat{f} = DFT(f) = (a_0, ..., a_{n-1})$ can be computed quickly enough and the inverse Fourier transform can be applied after the gestural manipulation of a Fourier coefficient. The didactical incentive for the Fourier Scratching of Rhythms in the above-mentioned previous implementation is a musically faithful realization of the complex coefficients $f_k = r_k e^{it_k}$. The magnitudes $r_k$ of the coefficients $f_k$ encode loudness values and their phases $t_k$ encode sound colors—given by realtime adjustable FM-sounds—audibly describing a circle. The vector $f \in \mathbb{C}^n$ represents a rhythmical loop $\mathbb{Z}_n \to \mathbb{C}$. Thereby the finite cyclic group $f : \mathbb{Z}_n$ of order $n$ denotes a cyclic pulse of length $n$, which in performance triggers the values $f_k = r_k e^{it_k}$ one after another; that is, it plays percussive FM-sounds of loudness $r_k$ and of sound color with phase-index $t_k$. The vector $f$ is subject to continuous change through gestural manipulation of $\hat{f}$ by the performer and can be best described as a "traveling rhythm" in a circle of sound.

In the present application the play state of the robot with $n$ fingers is also encoded by a complex vector $f = (f_0, ..., f_{n-1}) \in \mathbb{C}^n$. With respect to polar coordinates $f_k = r_k e^{it_k}$, $r_k$ can be interpreted as the strength (loudness) and $t_k$ as the play position of the finger with index $k$. The DFT of the play states in Fig. 8 are thus (from left to right) $(0, 1, 0, 0, 0)$, $(0, 1, 0, 0, 0, 0, 0, 0, 0, 0, 0, 0, 0)$, $(0, 0, 0, 1, 0)$, and $(0, 0, 0, 1, 0, 0, 0, 0, 0, 0, 0, 0, 0)$, respectively.

From a combinatorial point of view the regular distributions of Fig. 8 are special cases. In general, the position and strength of each robot finger on the circular playtable can be freely chosen at any moment. The naive control mode would be to change these parameters individually for every finger, but this requires a large set of controls. In contrast, the Fourier Scratching technique offers the ability to change the play states globally and smoothly using only a few parameters. While we cannot offer empirical evidence yet that this particular technique is musically more effective than other alternatives, it is useful to observe that the partials (as the most elementary play states) correspond to musically elementary patterns[3]. As exemplified in Fig. 8 it is precisely the family of coherent well-formed scales which will be played in generic scalar order by the first partial play state. Higher partials with indices $k$ coprime to $n$ generate complete generic interval cycles. Early experiments with this system give the impression that play states which are closely related in their Fourier coefficients are sensibly related by the musical ear. Navigation along the scalar hierarchy in real time can suitably be accompanied by a rising or reduction of the dimension of the play state (e.g. by zero-padding the DFT of the current play state or by deleting the Fourier coefficients with minimal energy).

The play states and their Fourier Transforms can be visualized using Riemann number spheres with the vectors $f$ and $\hat{f}$ being displayed as closed polygons with small colored balls at their points as in Fig. 9 (or refer to the demonstration at `http://www.youtube.com/watch?v=-qo09XTtFMA&feature=related`). The small balls are always played in a loop, as indicated by the polygon. The equator represents the continuous circular keyboard where every finger of the robot can hit at any longitude (between 0 and $2\pi$). Gestural control of the locations where the fingers "hit" the spherical surface of the "keyboard", selection of the "keys" (small balls), of the spheres, and activation of sonification, is currently enabled through a standard 4-axis 12-button game controller. Points which differ only in latitude from some point on the equator have the same sound and the same pitch but are played louder (north) or softer (south) than their projection to the equator. Both pitch and timbre are mapped to longitude, but pitch moves in discrete steps (as set by the choice of scale—see the sectors illustrated in Fig. 7), while timbre is continuously variable. This implies that two different pitches never have the same timbre, and that the same pitch may have more than one timbre.

The paradigm of Scratching the ascending pattern of a scale can be extended to the paradigm of Scratching a circular musical score (a loop). In addition to the play state, the score can be visualized on the surface of the sphere as a landscape

[3] The $N$th *partial* is defined as $p_{N,k}(t) = e^{\frac{2\pi i t k}{N}}$.

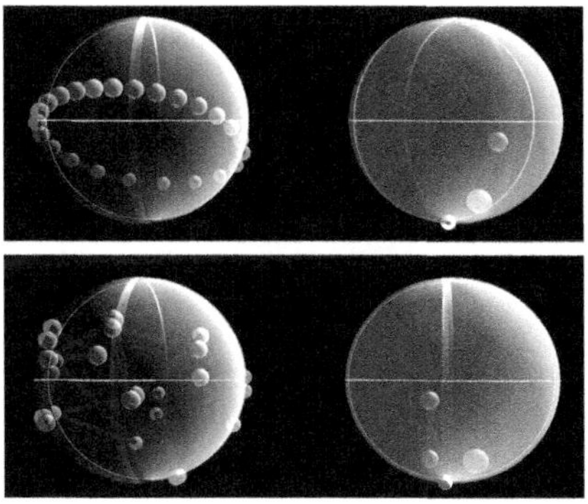

**Fig. 9.** The screenshots display two 24-dimensional play states with relatively simple Fourier Transforms. In each screenshot (top and bottom) there are two spheres. The left ones (passive) show the play states, while the right ones (active and selected) show the associated Fourier transforms. The screenshot on top represents a play sate with only two non-vanishing Fourier coefficients: $a_0$ and $a_1$. Here $a_0$ is selected (and enlarged). The screenshot at the bottom shows, how this simple play state is being "scratched" by a manipulation of the coefficient $a_3$, being selected (and enlarged).

and the robot finger with index $k$ in a certain play state $(..., r_k e^{t_k}, ...)$ plays the notes (or generalized musical events) which are positioned at the score position specified by its phase $t_k$ and with the loudness specified by its magnitude. In this paradigm the composer of the score defines the elementary patterns.

## 6 Automatic Adaption of Sound Color

For tones with harmonic spectra (such as most Western instruments and the human voice), intervals close to low integer frequency ratios (e.g. the octave and perfect fifth) are typically considered to be harmonically consonant and have high melodic affinity. However, the intervals found within dynamically tuned well-formed scales can take any size and so may be quite unfamiliar and dissonant in character, or sound out-of-tune.

The inharmonic timbres produced by FM-synthesis are effective at ameliorating these issues, but there is an alternative technique—Dynamic Tonality [10]—that can be used to minimize the sensory dissonance [24] and maximize the melodic affinity and in-tuneness [25] of microtonal intervals. This is achieved by "matching" the tunings of the tones' partials (overtones or harmonics) to the underlying tuning of the scale (matching means that when a typical scale interval is played, many of the partials in one tone have the same pitches as partials in the other tone). This technique can be applied using any form of synthesis in which the tunings of partials can be precisely controlled; appropriate methods of sound synthesis include analysis-resynthesis, additive, and modal. Software synthesizers utilizing these techniques can be downloaded from `http://www.dynamictonality.com`. The pitch (relative to the fundamental) of each partial is mapped to a linear combination of the pitch heights of the period and generator of the underlying scale. This enables some of the intervals between the partials to correspond to some of the intervals found in the MOS scale, even as the underlying tuning is changed dynamically.

## 7 Discussion

In speaking about and proposing a technique for playing music, there is an æsthetic aspect that inevitably arises. We see affinities between the Fourier Scratching techniques suggested here and minimal music, streamlined loops in DJ Culture, and even with some ideas of early serialism. Our initial motivation was to depart from the traditional dichotomy between instrument and virtuoso musician so as to allow a novel interaction between mathematical composition techniques, real-time computation, and improvised performance. Likewise, our first experiences from the more didactically oriented realization of the Fourier Scratching concept ([20,21]) transformed into a music-æsthetic and media-theoretic challenge. It appears that the "immediate" interaction of theoretical thought and musical experience calls for a suitable environment of reflection and action, where music, mathematics and computation are "naturally" entangled. Looking towards the future, perhaps the ultimate performance interface for this playing technique would be a large interactive sphere with multitouch and display.

## 8 Conclusion

In this paper, we have discussed the scale labyrinth—a visualization of a universe of interconnected well-formed scales that enables any generic well-formed scale, and its specific tuning, to be easily selected. We have shown how the selected scale can be played by the technique of Fourier Scratching, which operates on spectra of playing gestures, thereby patterning the order in which scale degrees are played, as well as their timbres and their loudnesses. We also discuss how mode transformations of the scale can be effected, and a method to ensure the timbre used is optimally matched to the underlying scale so as to maximize tonal affinity. We hope this novel system provides a set of useful constraints and parameters for performing improvisations across the valid tuning range of a given well-formed scale, across "diatonic" and "chromatic" transpositions of such a scale, and across the interconnected universe of different well-formed scales that is pictured in the labyrinth.

## References

1. Carey, N.: Distribution Modulo 1 and Musical Scales. Ph.D Dissertation, University of Rochester (1998)
2. Milne, A.J., Sethares, W.A., Plamondon, J.: Tuning Continua and Keyboard Layouts. J. of Math. Music 2, 1–19 (2008)

3. Wilson, E.: Scale-tree, Peirce Sequence (1994), `http://www.anaphoria.com/sctree.PDF`
4. Holmes, R.: 7-limit Scales (2006), `http://doctroid.wordpress.com/pages/music/xenharmonic-scales/7-limit-scales/`
5. Carey, N., Clampitt, D.: Aspects of Well-formed Scales. Music Theory Spectrum 11, 187–206 (1989)
6. Clough, J., Engebretsen, N., Kochavi, J.: Scales, Sets, and Interval Cycles: A Taxonomy. Music Theory Spectrum 21, 74–104 (1999)
7. Balzano, G.J.: The Pitch Set as a Level of Description for Studying Musical Perception. In: Clynes, N. (ed.) Music, Mind, and Brain. Plenum Press, New York (1982)
8. Carey, N.: Coherence and Sameness in Well-formed and Pairwise Well-formed Scales. J. of Math. Music 2, 79–98 (2007)
9. Erlich, P.: A Middle Path Between Just Intonation and the Equal Temperaments, Part 1. Xenharmonikôn 18, 159–199 (2006)
10. Sethares, W.A., Milne, A.J., Tiedje, S., Prechtl, A., Plamondon, J.: Spectral Tools for Dynamic Tonality and Audio Morphing. Comput. Music J. 33, 71–84 (2009)
11. Balzano, G.J.: The Group-theoretic Description of 12-fold and Microtonal Pitch Systems. Comput. Music J. 4, 66–84 (1980)
12. Rothenberg, D.: A Model for Pattern Perception with Musical Applications. Math. Systems Theory 11, 199–234 (1978)
13. Blackwood, E.: Structure of Recognizable Diatonic Tunings. Princeton University Press, Princeton (1985)
14. Stern, M.A.: Über eine Zahlentheoretische. Funktion 55, 193–220 (1858)
15. Brocot, A.: Calcul des Rouages par Approximation Nouvelle Methode. Revue Chronometrique 6, 186–194 (1860)
16. Lewin, D.: Re: Intervallic Relations between Two Collections of Notes. J. of Music Theory 3, 298–301 (1959)
17. Quinn, I.: A Unified Theory of Chord Quality in Equal Temperaments. Ph.D. Dissertation, Eastman School of Music (2004)
18. Amiot, E.: David Lewin and Maximally Even Sets. J. of Math. Music 2, 157–172 (2007)
19. Amiot, E., Noll, T., Agon, C., Andreatta, M.: Fourier Oracles for Computer-Aided Improvisation. In: Proceedings of the ICMC: Computer Music Conference. Tulane University, New Orleans (2006)
20. Carlé, M., Hahn, S., Matern, M., Noll, T.: Demonstration of Fourier Scratching. Presented at MCM 2009, New Haven, US (2009)
21. Carlé, M., Noll, T.: Fourier Scratching: SOUNDING CODE. In: SuperCollider Symposium 2010, Berlin (2010)
22. Noll, T.: Ionian Theorem. J. of Math. Music 3, 137–151 (2009)
23. Clampitt, D., Noll, T.: Modes, the Height-Width Duality, and Handschin's Tone Character. Music Theory Online 17(1) (2011)
24. Sethares, W.A.: Tuning, Timbre, Spectrum, Scale. Springer, London (2005)
25. Milne, A.J., Prechtl, A., Laney, R., Sharp, D.B.: Spectral Pitch Distance & Microtonal Melodies. Presented at ICMPC11, Seattle (2010)

# $N^{th}$ Roots of Pitch-Class Inversion

Robert W. Peck

School of Music, Louisiana State University
rpeck@lsu.edu

**Abstract.** In this study, we investigate the square, cubic, and other *n*th roots of inversion in discrete pitch-class spaces. We examine the group-theoretical structures that they inhabit, as well as various multi-dimensional regular polytopes whose symmetries model those structures. Moreover, we determine which *n*th roots of inversion occur in pitch-class spaces of various sizes, and their multiplicities. Because of their relevance to the majority of music in the Western canon, as well as to the transformational theories that engage this repertoire, we focus largely on inversions and their *n*th roots in mod-7 diatonic space and in mod-12 chromatic space. Our objective is to further the understanding of pitch-class inversion as a gesture, through an exploration of its *n*th roots in discrete transformational music theory.

**Keywords:** Transformation theory, Pitch-class set theory, Group theory, Hypercube symmetries, Hyperoctahedral group, *n*th roots.

## 1 Introduction

In transformational music theory—particularly in the tradition of Lewin [1], Klumpenhouwer [2], and Cohn [3]—the operation of pitch-class inversion is of fundamental importance. The analytical significance of inversion is also evident in earlier musical theories, including Morris's [4] and Forte's [5] work with pitch-class-sets, Babbitt's writings on twelve-tone serialism [6], and Oettingen's [7] and Riemann's [8] notion of harmonic dualism. Indeed, theories of inversion have distant and obscure origins, dating at least to Guido d'Arezzo, who, in the eleventh century, likened melodic inversion to the reflection of a face in a well [9].

Like Guido, the above theories generally regard the operation of inversion as a reflection. Such inversions are typically modeled on a line, in the case of infinite pitch space (Fig. 1(a)) (or discretely on a set of equidistant points on a line), or on a circle, in the case of modular pitch-class space (Fig. 1(b)) (or discretely on the vertices or edges of a regular polygon inscribed in a circle). More recently, Mazzola [10] describes inversion in terms of a gesture that is instead a 180° rotation in the complex plane. Accordingly, such a gesture passes through an imaginary dimension. Peck [11] extends Mazzola's concept to additional imaginary dimensions, using discrete music-theoretical techniques in the Lewinian transformational tradition.

In this study, we investigate the square, cubic, and other *n*th roots of inversion in discrete pitch-class spaces. We examine the group-theoretical structures that they inhabit, as well as various multi-dimensional regular polytopes whose symmetries

C. Agon et al. (Eds.): MCM 2011, LNAI 6726, pp. 196–206, 2011.

model those structures. Specifically, we model *n*th roots of inversion as symmetries of hypercubes. Whereas roots of permutations have been studied previously in the mathematical literature, such as in [12], and hypercube symmetries have existing applications in music theory, including [13—15], their integration in this study is novel. Because of their relevance to the majority of music in the Western canon, as well as to the transformational theories that engage this repertoire, we focus largely on inversions and their *n*th roots in mod-7 diatonic space and in mod-12 chromatic space.

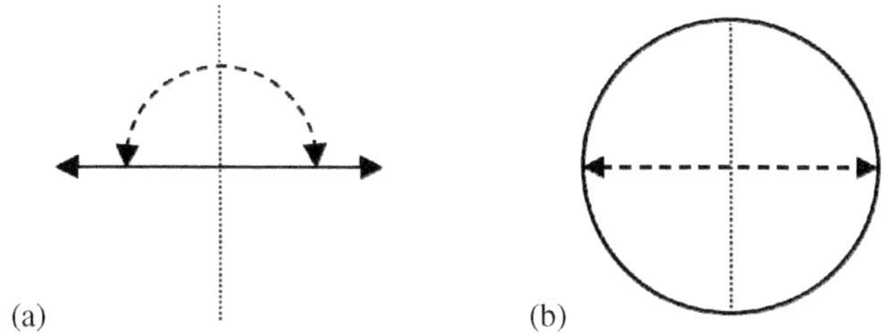


**Fig. 1.** Inversion as a reflection through an axis (a) on a line and (b) on a circle

Our objective is to further the understanding of pitch-class inversion, through an exploration of its *n*th roots in discrete transformational music theory. Our results demonstrate that in certain pitch-class spaces, models of inversion that are also inclusive of its *n*th roots conform to reflections; whereas in others, they correspond to rotations. We choose to situate this investigation in discrete terms to emphasize the inherent limitations of these models. In particular, not all *n*th roots occur in spaces of particular sizes. For instance, no square roots of inversion exist in mod-7 diatonic space. The types of symmetries that model these inversions, as well as the resulting limitations in the discrete case, contribute to our overall interpretation of inversion as a gesture, particularly within the restrictions of pitch-class space.

## 2 *n*th-Root Inclusive Models of Inversion in Pitch-Class Spaces

In this section, we construct representations of pitch-class inversion that accommodate its various *n*th roots. First, we consider modular spaces of even order, and then odd-ordered spaces. In the former, we note an essential distinction between the types of gestures that model *n*th roots of inversion in spaces of orders $2m$ where $m$ itself is even, and spaces of orders $2m$ where $m$ is odd. In spaces of odd order, a similar division exists between sizes $2m+1$ where $m$ is even and where it is odd.

### 2.1 Inversion in Pitch-Class Spaces of Order 2*m*

Let $\Gamma$ be a discrete, modular pitch-class space of order $2m$, which we model using $\mathbb{Z}_{2m}$. Put $x \in \mathbb{Z}_{12}$. We define an inversion operation $i_x$ on $\Gamma$, and its various nth roots, as follows.

**Definition 1.** $i_x$ consists of the product of all pairwise disjoint transpositions $(a,b)$ in $\Gamma$, such that $a+b = x \bmod 2m$.

(a) Call $i_x$ even if $x$ is even.
(b) Call $i_x$ odd if $x$ is odd.

**Definition 2.** An $n$th root of $i_x$, $\sqrt[n]{i_x}$, agrees with some $g \in \mathrm{Sym}(\Gamma)$, where $g^n = i_x$. There may be more (or less) than one $n$th root for any $i_x$.

We observe two general categories of inversions: those for which $x$ is even, and those for which $x$ is odd, leading to our first result.

$$\text{If } 2 | x \text{ (as an integer), then } fix(i_x) = \{x/2, (x/2)+m\}; \text{ else, } fix(i_x) = \emptyset . \tag{1}$$

If $i_x$ is even, then $i_x$ fixes two pitch-class points in $\Gamma$; if $i_x$ is odd, it is fixed-point free. For reasons of simplicity, we limit ourselves henceforth to studying $n$th roots of odd inversions, as the lack of stabilized pitch-classes produces a more straightforward model. The interested reader can extend our results to even inversions with some additional effort and consultation of the references.

Given an odd $i_x$, we have the following significant structure.

**Definition 3.** Let $D = \langle (a_1,b_1)\rangle \times \ldots \times \langle (a_m,b_m)\rangle$ be the direct product of 2-groups generated by transpositions in $i_x$.

We note that $D$ is elementary abelian of order $2^m$. The following observations are also of continuing importance.

$$i_x \sqrt[n]{i_x} = \sqrt[n]{i_x}\, i_x \tag{2}$$

$$\sqrt[n]{i_x} \in C_{\mathrm{Sym}(\Gamma)}\langle i_x \rangle \tag{3}$$

$$Z(C_{\mathrm{Sym}(\Gamma)}\langle i_x \rangle) = \langle i_x \rangle \tag{4}$$

Essentially, (2) observes that $i_x$ commutes with any of its $n$th roots; hence, all $n$th roots of $i_x$ are members of the centralizer in the symmetric group on $\Gamma$ of the group generated by $i_x$ (3). Then, (4), which states that the center of the centralizer is $\langle i_x \rangle$ itself, follows from a standard result in the theory of orthogonal groups [16]. We also note the following points.

$$\text{The action of } \langle i_x \rangle \text{ on } \Gamma \text{ is semiregular (or free).} \tag{5}$$

$$\langle i_x \rangle \text{ is cyclic of order 2.} \tag{6}$$

$$\langle i_x \rangle \text{ has } m \text{ orbits in } \Gamma. \tag{7}$$

We may therefore determine the structure of $C_{Sym(\Gamma)}\langle i_x \rangle$ for any odd $i_x$ [17].

$$C_{\mathrm{Sym}(\Gamma)}\langle i_x \rangle \cong S_2 \wr S_m . \tag{8}$$

As a wreath product of order $2^m \cdot m!$, $C_{\text{Sym}(\Gamma)}\langle i_x \rangle$ is a semidirect product of its normal subgroup $D$ (from Definition 3) by a subgroup whose action on $D$ under conjugation is isomorphic to $S_m$.

$$C_{\text{Sym}(\Gamma)}\langle i_x \rangle \cong S_2^m \rtimes S_m \ . \tag{9}$$

Via results in representation theory [18], $S_2^m \rtimes S_m$ is isomorphic to the Weyl group of the root system $B_m$ generated by reflections, which is also the group $Q_m$ of symmetries of the $m$-dimensional hypercube (i.e., the $m$-dimensional hyperoctahedral group) [19].

$$C_{\text{Sym}(\Gamma)}\langle i_x \rangle \cong B_m = Q_m \ .$$

Because $Q_m$ contains both reflections and rotations, it is in this particular connection that we situate $i_x$ as either a reflection or a rotation.

An $m$-dimensional hypercube ($m$-cube) possesses various sets of elements in dimensions less than $m$, ranging from vertices (0-cubes) to "faces" (($m-1$)-cubes). Specifically, it contains $2^m$ vertices and $2m$ faces. Then, the $m$-dimensional hyperoctahedron is dual to the $m$-cube, containing $2m$ vertices and $2^m$ faces. For example, a 3-cube (i.e., a cube) has $2^3 = 8$ vertices and $2 \cdot 3 = 6$ faces (2-cubes, or squares). It is dual to the octahedron, which has six vertices and eight faces (Fig. 2).

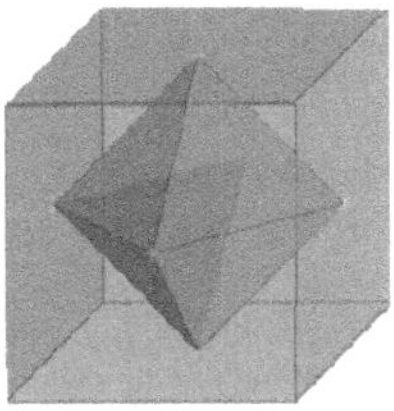

**Fig. 2.** The 3-cube and its dual, the octahedron

Because we are concerned in this section with spaces of size $2m$, we map the pitch-classes in our models to the $2m$ faces of an $m$-cube (or, equivalently, to the $2m$ vertices of a hyperoctahedron). For instance, to model $n$th roots of inversion in the usual twelve pitch-classes requires $12/2 = 6$ dimensions: we do so on the twelve 5-cubes that comprise the faces of a 6-cube (Fig. 3).

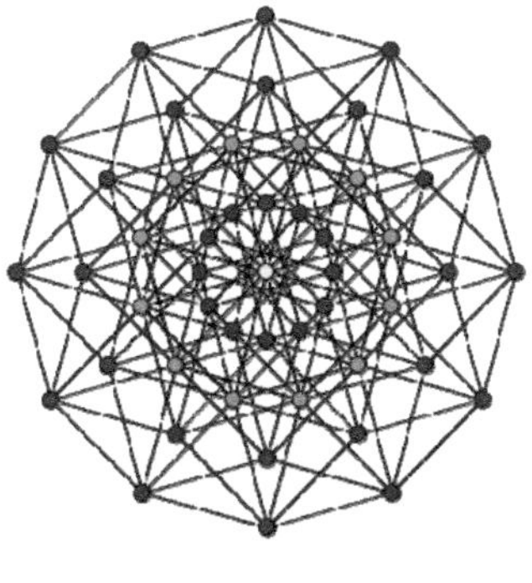

**Fig. 3.** Orthogonal projection of a 6-cube

We observe that certain symmetries of the $m$-cube are also members of the alternating group on the set of faces, which we call even rotations and even reflections; symmetries that are not in this alternating group, we call odd rotations and odd reflections.

**Definition 4.** Symmetries in $Q_m$:

- the set $R+$ of $|Q_m|/4$ even rotations,
- the set $R-$ of $|Q_m|/4$ odd rotations,
- the set $F+$ of $|Q_m|/4$ even reflections, and
- the set $F-$ of $|Q_m|/4$ odd reflections.

The index-4 subgroup $R+$ of even rotations is isomorphic to the following semidirect product.

$$R+ \cong S_2^{m-1} \rtimes A_m \,. \tag{11}$$

$R+$ is normal in $Q_m$; therefore, we may give the following quotient group.

$$Q_m/R+ \cong V_4 \,. \tag{12}$$

Therein, $R+$ functions as the kernel of the map from $Q_m$ to the Klein four-group $V_4$, as demonstrated by Table 1. (See also [19].)

**Table 1.** Cayley table of $Q_m/R+$

| | $R+$ | $R-$ | $F+$ | $F-$ |
|---|---|---|---|---|
| $R+$ | $R+$ | $R-$ | $F+$ | $F-$ |
| $R-$ | $R-$ | $R+$ | $F-$ | $F+$ |
| $F+$ | $F+$ | $F-$ | $R+$ | $R-$ |
| $F-$ | $F-$ | $F+$ | $R-$ | $R+$ |

The table exhibits two binary sets of qualities: rotations vs. reflections, and even vs. odd symmetries. The product of two rotations or two reflections is a rotation, whereas the product of a rotation and a reflection (in either order) is a reflection. Similarly, the product of two even symmetries or two odd symmetries is even, whereas the product of two symmetries of the opposite types is odd. Hence, the square (or any even power) of any symmetry is an even rotation, and lives in $R+$.

These cosets combine in various ways to form three index-2 subgroups in $Q_m$. Because they are all of index 2, they are also maximal normal subgroups.

**Definition 5.** Maximal normal subgroups of $Q_m$:

(a) $R+ \cup F+$ of even rotations and even reflections,
(b) $R+ \cup F-$ of even rotations and odd reflections, and
(c) $R+ \cup R-$ of even and odd rotations.

The structures of the first two subgroups may be given as follows.

$$R{+} \cup F{+} \cong \mathrm{S}_2^{m\text{-}1} \rtimes \mathrm{S}_m \,. \tag{13}$$

$$R{+} \cup F{-} \cong \mathrm{S}_2^{m} \rtimes \mathrm{A}_m \,. \tag{14}$$

The first of these subgroups, $R{+} \cup F{+}$, is also a subgroup of the alternating group on the set of the $2m$ faces of the $m$-cube. As such, it is the set of symmetries of an $m$-dimensional demicube (with alternate vertices in $Q_m$), and is isomorphic to the Weyl group $\mathrm{D}_m$ ([20], which uses the prefix "hemi-").

Following results in cohomology [21], the structure of the rotation subgroup, $R{+} \cup R{-}$, varies in odd and even dimensions. In odd dimensions, it is isomorphic to $R{+} \cup F{+}$.

$$R{+} \cup R{-} \cong \mathrm{S}_2^{m\text{-}1} \rtimes \mathrm{S}_m \,,\ m \text{ is odd.} \tag{15}$$

In even dimensions, however, its structure is more complex. It is also a semidirect product, but of $R{+}$ (itself a semidirect product) by an order-2 subgroup generated by an odd rotation.

$$R{+} \cup R{-} \cong (\mathrm{S}_2^{m\text{-}1} \rtimes \mathrm{A}_m) \rtimes \mathrm{S}_2 \,,\ m \text{ is even.} \tag{16}$$

(The one exception to this general rule exists in the case with $m = 2$. Here, no odd rotations of order 2 exist. Rather, the rotation subgroup is merely cyclic, generated by an odd rotation of order 4.)

We are now ready to situate $i_x$ as either a rotation or a reflection. Our determination is informed first by the parity of $m$. $i_x$ corresponds to the product of exchanges of opposite faces of the $m$-cube. Therefore, if $m$ is even, then $i_x$ must be either an even rotation or an even reflection; if $m$ is odd, then $i_x$ must be an odd rotation or an odd reflection. The centers of the various maximal subgroups, per (13)-(16) above, indicate whether these symmetries are rotations or reflections.

$$|Z(R{+} \cup F{+}) = Z(\mathrm{R}{+} \cup \mathrm{R}{-})| = 1;\ |Z(\mathrm{R}{+} \cup \mathrm{F}{-})| = 2 \,,\ m \text{ is odd.} \tag{17}$$

$$|Z(R{+} \cup F{+}) = Z(R{+} \cup F{-}) = Z(R{+} \cup R{-})| = 2 \,,\ m \text{ is even.} \tag{18}$$

We recall from (4) that $i_x$ commutes with every element of $Q_m$. Hence, it is a member of the center of any subgroup in which it is contained. Then, as $R{+} \cup F{-}$ is the only subgroup with a nontrivial center in odd dimensions, we must conclude that $i_x$ is an odd reflection in odd dimensions. In contrast, in even dimensions, the order-2 center in each of the subgroups is the same subgroup; therefore, $i_x$ must be an even rotation. This distinction between parities corresponds to a more general result in the theory of orthogonal groups, which states that reflection through the origin (to which our $i_x$ corresponds) is orientation-reversing in odd dimensions, but orientation-preserving in even dimensions [16].

### 2.2 Inversion in Pitch-Class Spaces of Order $2m+1$

In pitch-class spaces of orders $2m+1$ (i.e., odd orders), we use the same representations that we do for odd inversions in spaces of order $2m$. That is, we model $i_x$ and its $n$th roots in the set of symmetries of an $m$-cube. Any inversion in an odd-ordered pitch-class space fixes one and only one pitch-class; consequently, no difference between even and odd inversions exists in these spaces that corresponds to the distinction made in (1) above. Further, because the permutation representation of a point stabilizer restricted to the trivial orbit is merely the identity, the centralizer in Sym(Γ) of the 2-group generated by any such inversion is isomorphic to the direct product of 1 by $S_2 \wr S_m$, where the wreath product is the structure of the centralizer of the orbit restriction of the group to its non-trivial orbits [17].

$$C_{\mathrm{Sym}(\Gamma)}\langle i_x\rangle \cong S_2 \wr S_m . \tag{19}$$

For instance, in mod-7 diatonic space, we model inversion and its various $n$th roots in the set of symmetries of a 3-cube (i.e., $7 = (2 \cdot 3) + 1$).

Fig. 4 presents one such example. It models the inversion $i_6 = (0,6)(1,5)(2,4)$ in mod-7 diatonic space, as an odd reflection in the figure's origin (i.e., odd because the reflection transposes the respective members of the figure's three pairs of opposite faces). In this case, $i_6$ fixes the pitch class 3; it is therefore not represented in the faces of the cube. Rather, we could situate it at the origin, which remains invariant under any symmetry of the cube.

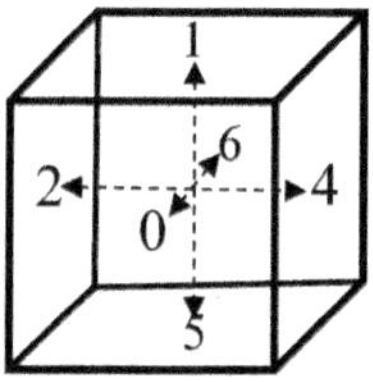


**Fig. 4.** $i_6$ (mod 7) modeled as an odd reflection in the origin of a cube

## 3 *n*th Roots of Inversion in the Models

In discrete pitch-class spaces of sizes $2m$ or $2m+1$, only particular $n$th roots of the above inversions occur. In this section, we determine which $n$th roots are possible in these spaces, and how many instances exist for each $n$th root. For example, in the 3-dimensional model of diatonic space (mod 7), no square roots of inversion occur, but we find eight cubic roots. In contrast, the 6-dimensional model of chromatic space (mod 12) allows for 120 square roots, 800 cubic roots, 2304 fifth roots, and 3840 sixth roots of any odd inversion, but no fourth roots.

### 3.1 The Existence of *n*th Roots in Moduli 2*m* and 2*m*+1

A particular *n*th root of an inversion exists in an *m*-dimensional model of pitch-class space if and only if there exists some *k*-tuple $\Phi = [\varphi_i \in 1, \ldots, m \mid \varphi_i \leq \varphi_{i+1}]$ for which the following three criteria hold.

$$\textstyle\sum_{i=1}^{k} \varphi_i = m \,. \tag{20}$$

$$\mathrm{LCM}(\varphi_1, \ldots, \varphi_k) = n \,. \tag{21}$$

$$\varphi_i = n \bmod 2\varphi_i, \text{ for } i = 1, \ldots, k \,. \tag{22}$$

$\Phi$ associates with certain permutations of $\Gamma$, where the lengths of the respective disjoint cycles in the permutation are given by $2\varphi_i$, for all $\varphi_i \in \Phi$. Therefore, because the inversion involves *m* transpositions, (20) associates $\Phi$ potentially with *n*th roots that are fixed-point-free in spaces of order 2*m*, or which have a single fixed point in spaces of order 2*m*+1. Next, as the length of any permutation is the least common multiple of the lengths of its constituent cycles, (21) ensures that the *n*th root has an appropriate length (i.e., an overall length of 2*n*, where 2 is the length of the inversion itself). Finally, (22) ensures that all subsequent powers of the *n*th root, up to its *n*th power (the inversion itself), are respectively fixed-point-free or have only a single fixed point, as we demonstrate in the examples below.

Accordingly, inversion in mod-7 diatonic space possesses at least one cubic root, but no square roots. We recall that we model *n*th roots of inversion in diatonic space in the symmetries of a 3-cube. For $m = 3$, we may define three *k*-tuples in the form of $\Phi$ that meet condition (20): [1,1,1], [1,2], and [3]. However, only [1,1,1] for $n = 1$, per condition (21), which is the inversion itself; and [3] for $n = 3$ (per 21), a cubic root, also meet condition (22). For example, a cubic root of the inversion $i_6 :=$ (0,6)(1,5)(2,4) that associates with $\Phi = [3]$ is the operator (0,1,2,6,5,4). Because $m =$ 3 is odd in this example, we model $i_x$ as an odd reflection. Then, as $n = 3$ is also odd, any $\sqrt[3]{i_x}$ is an odd reflection: the product of an odd number of reflections is a reflection. The only possibility of a square root, [1,2] for $n = 2$, fails to meet condition (22); whereas $2 = 2 \bmod 4$, $1 \neq 2 \bmod 2$. Consequently, a permutation that associates with [1,2]—such as (0,6)(1,2,5,4)—is not fixed-point free for all subsequent powers up to the *n*th power: e.g., its square fixes pitch classes 0 and 6.

In mod-12 chromatic space, $m = 6$; hence, we have more possibilities. The forms of $\Phi$ that are meet condition (20) are the following: [1,1,1,1,1,1], [1,1,1,1,2], [1,1,2,2], [2,2,2], [1,1,1,3], [1,2,3], [3,3], [1,1,4], [2,4], [1,5], and [6]. Among these, [1,1,1,1,1,1] for $n = 1$, [2,2,2] for $n = 2$, both [1,1,1,3] and [3,3] for $n = 3$, [1,5] for $n =$ 5, and [6] for $n = 6$ also satisfy condition (22). Therefore, no possibilities exist for fourth roots. Interestingly, we find two subclasses of cubic roots. For the inversion $i_{11} :=$ (0,11)(1,10)(2,9)(3,8)(4,7)(5,6), a cubic root that corresponds to [1,1,1,3] is the operation (0,1,2,11,10,9)(3,8)(4,7)(5,6); one that corresponds to [3,3] is (0,1,2,11,10,9)(3,4,5,8,7,6). Both subclasses are modeled as even rotations. First, [1,1,1,3] and [3,3] both represent even permutations. Second, for an even *m*, $i_x$ is an even rotation. Finally, as $n = 3$ is odd, a cubic root in this context cannot be a reflection: the product of an odd number of rotations is a rotation.

Figure 5 presents one example of a cubic root for the inversion $i_{11}$. This example uses the mod-12 operation $x := (0,10,6,11,1,5)(2,3,4,9,8,7)$, a member of the subclass [3,3] of cubic roots of $i_{11}$. Specifically, $x^3 = i_{11}$. The figure demonstrates the inversion in terms of gesture that carries a D major triad to a D minor triad (i.e., {2,6,9} to {2,5,9} as unordered pitch-class sets, where the pitch class C = 0). Under $x$, the D major triad maps to a G# minor triad, which in turn maps to a C# diminished triad, which maps subsequently under $x$ to the D minor triad.

**Fig. 5.** Inversional gesture under $x := (0,10,6,11,1,5)(2,3,4,9,8,7)$, a cubic root of $i_{11}$

### 3.2 Multiplicities of the *n*th Roots

All the symmetries of an $m$-cube that satisfy conditions (20)-(22) for some particular $n$ and $m$ form a class of $n$th roots. This class may divide further—as in the case of cubic roots for $m = 6$ above—into subclasses that satisfy the two conditions, but which are unique. These (sub)classes, then, correspond to conjugacy classes in $Q_m$. To establish the number of $n$th roots for any inversion, it is necessary to calculate the sizes of the respective conjugacy classes.

For a group $G$ with an element $g$, the size of the conjugacy class of $g$, Cl($g$), is given by the index in $G$ of the centralizer of the group generated by $g$. Let $\sqrt[n]{i_x}$ represent a member of one of the (sub)classes for $n$. The following formula gives the number of operations in $Q_m$ that are conjugate to $\sqrt[n]{i_x}$ (hence, the number of $n$th roots in the same subclass).

$$|\mathrm{Cl}(\sqrt[n]{i_x})| = \frac{|Q_m|}{|\mathrm{C}_{Q_m}(\langle\sqrt[n]{i_x}\rangle)|}\,. \qquad (23)$$

To determine the size of the appropriate centralizer, we need to know both the number of elements that exist in the underlying set for $\Phi$ (i.e., without their multiplicities), as well as their multiplicities in $\Phi$. Therefore, let $\bar{\Phi}$ be the underlying set for $\Phi$, and let $k_{\varphi_i}$ represent the number of occurrences of the element $\varphi_i$ in $\Phi$. The order of the centralizer is given by the following formula [17].

$$|\mathrm{C}_{Q_m}(\langle\sqrt[n]{i_x}\rangle)| = \prod_{i=1}^{|\bar{\Phi}|}(2\varphi_i)^{k_{\varphi_i}} \cdot k_{\varphi_i}!\,. \qquad (24)$$

Then, the total number of $n$th roots is the sum of the sizes of the respective conjugacy classes for operations that associate with all sets in the form of $\Phi$, satisfying a particular $n$ for condition (21).

In terms of the example of cubic roots in diatonic space, the size of the conjugacy class of an inversion that corresponds to the $k$-tuple $\Phi = [3]$ is determined as follows. $\bar{\Phi}$ has only one element, 3, and its multiplicity in $\Phi$ is 1. Therefore, (24) gives us $|\mathrm{C}_{Q_m}(\langle\sqrt[n]{i_x}\rangle)| = 6$. As $|Q_3| = 48$, we have $|\mathrm{Cl}(\sqrt[3]{i_x})| = 48/6 = 8$, per (23). Accordingly, $i_6$ (mod 7) has eight cubic roots.

The example in chromatic space gave two subclasses of cubic roots: [1,1,1,3] and [3,3]. The sizes of conjugacy classes of the cubic roots that these respective multisets represent are different, and are determined as follows. The underlying set for $\Phi = [1,1,1,3]$ has two elements, 1 and 3, as the multiplicity in $\Phi$ of 1 is 3, and of 3 is 1. By (24), the size in $Q_6$ of the centralizer of a cubic root in this subclass is accordingly $((2 \cdot 3)^1 \cdot 1!) \cdot ((2 \cdot 1)^3 \cdot 3!) = 288$. Therefore, for $|Q_6| = 46080$, the size of the subclass is 46080/288 = 160. The underlying set of $\Phi$ in the other subclass, $\Phi = [3,3]$, has only one element, 3, but its multiplicity in $\Phi$ is 2. The order of a corresponding centralizer in $Q_6$ of such a cubic root is $((2 \cdot 3)^2 \cdot 2!) = 72$. Hence, the size of the corresponding conjugacy class is 46080/72 = 640. Altogether, then, the number of cubic roots of $i_{11}$ (mod 12) is 160 + 640 = 800.

## 4 Conclusions and Future Work

The discussion above presents a preliminary theory of *n*th roots of inversion in discrete pitch-class spaces. Specifically, it situates inversion and its *n*th roots in a model of hypercube symmetries, and it investigates which *n*th roots occur in spaces of various sizes and their multiplicities. Since applications to analysis of musical examples are outside the scope of our present discussion, we offer some suggestions here to potential uses in musical contexts. *n*th roots of inversion are particularly applicable to situations in which inversion obtains not instantaneously, but through a process, or gesture. One might describe such a gesture in terms of the orbit of a cyclic group generated by an *n*th root as it acts on a set of pitch-class sets. Another application is to Klumpenhouwer networks (2). Here, *n*th roots of inversion might function as hyper-operators that act on a K-net via conjugation, arriving at a negative isography as the result of an iterative process. In these ways, *n*th roots of inversion might also inform compositional processes, and so on.

The theory presented here deals exclusively with finite and discrete pitch-class spaces. As a result, it bears certain limitations: in even modular spaces, a significant distinction exists between *n*th roots of even and odd inversions. This feature does not exist in continuous modular spaces [22]. Likewise, only particular *n*th roots are possible in a discrete space of some given size, which again is not the case in the continuous model. This study highlights some of these essential differences between the two theoretical approaches.

## References

1. Lewin, D.: Generalized Musical Intervals and Transformations. Yale University Press, New Haven (1987)
2. Klumpenhouwer, H.: A Generalized Model of Voice-Leading for Atonal Music. Ph.D. diss., Harvard University (1991)
3. Cohn, R.: Neo-Riemannian operations, parsimonious trichords, and their tonnetz representations. J. Mus. Theory 41(1), 1–66 (1997)
4. Morris, R.: Composition With Pitch-Classes: A Theory of Compositional Design. Yale University Press, New Haven (1987)
5. Forte, A.: The Structure of Atonal Music. Yale University Press, New Haven (1973)

6. Babbitt, M.: The Function of Set Structure in the 12-tone system. Ph.D. diss., Princeton University (1992) [written 1946]
7. von Oettingen, A.: Harmoniesystem in dualer Entwicklung: Studien zur Theorie der Musik. Verlag W-Glaser, Dorpat (1866)
8. Riemann, H.: Musik-Lexikon, 7th edn. Mar Hesses Verlag, Leipzig (1909)
9. d'Arezzo, G.: Micrologus. Trans.: Warren Babb. In: Palisca, C. (ed.) Hucbald, Guido, and John: Three Medieval Treatises. Yale University Press, New Haven (1978)
10. Mazzola, G.: Categorical gestures, the diamond conjecture, Lewin's question, and the Hammerklavier Sonata. J. Math Music 3, 31–58 (2009)
11. Peck, R.: Imaginary transformations. J. Math Music 4, 157–171 (2010)
12. Čupona, G., Samardžiski, A., Celakoski, N.: Roots of permutations. In: Proceedings of the Symposium $n$-ary Structures. Makedonska Akad. Nauk, Umetnost., Skopje (1982)
13. Douthett, J., Peck, R.: Wreath products and n-cube symmetry: a music-theoretical application. Presentation to the Joint Mathematics Meeting. Special Session on Mathematical Techniques in Musical Analysis, New Orleans, LA (January 2011)
14. Lombardi, P.: Serial $n$-cubes. Presentation to the Thirty-third Annual Meeting of the Society for Music Theory, Indianapolis, IN (November 2010)
15. Tymoczko, D.: A Geometry of Music: Harmony and Counterpoint in the Extended Common Practice. Oxford University Press, Oxford (2011)
16. Taylor, D.: The Geometry of the Classical Groups 9. Heldermann Verlag, Berlin (1992)
17. Peck, R.: Generalized Commuting Groups. J. Music Theory 54(2), 143–177 (2010)
18. Humphreys, J.: Introduction to Lie Algebras and Representation Theory. Graduate Texts in Mathematics, vol. 9. Springer, Heidelberg (1972)
19. Kerber, A.: Representations of permutation groups I. Lecture Notes in Mathematics, vol. 240. Springer, Heidelberg (1971)
20. Conway, J., Burgiel, H., Goodman-Strauss, C.: The Symmetries of Things. A K Peters, Ltd, Wellesley (2008)
21. Hatcher, A.: Algebraic Topology. Cambridge University Press, Cambridge (2001)
22. Callender, C., Quinn, I., Tymoczko, D.: Generalized Voice Leading Spaces. Science 320, 346–348 (2008)

# Cardinality Transformations in Diatonic Space

Richard Plotkin

University at Buffalo, The State University of New York
richardp@buffalo.edu

**Abstract.** This paper introduces a system in which parsimonious and continuous transformations occur seamlessly between triads and tetrachords. Such fluidity is abundant in common practice music, but unprecedented in theoretical literature, largely because there has been no consistent way to approach transformations independent of cardinality. Neo-Riemannian theory elegantly unites harmonic change and voice-leading efficiency, but deals exclusively with set class [037] in a 12-gamut pcset space. Attempts to extend the neo-Riemannian approach to tetrachords in 12-gamut space often fall short; the elegant characteristics of the triadic theory do not carry over. However, when a scalar context arbitrates the parsimoniousness of transformations, triads and tetrachords can be treated in a consistent manner. Within this consistently modeled space, cardinality itself can be transformed. In this paper, we see that filtered point-symmetry is an essential tool for working through the iterated maximally even sets that establish scalar contexts. To understand cardinality transformations, we also extend filtered point-symmetry to model partially symmetric distributions and relatively even sets.

**Keywords:** Filtered point-symmetry, Maximally even sets, Partially symmetric distribution, Relatively even sets, neo-Riemannian transformations, Tetrachord transformations, Scalar contexts.

## 1 Introduction

This paper introduces a system in which parsimonious and continuous transformations occur seamlessly between triads and tetrachords. Such fluidity is abundant in common practice music, but unprecedented in theoretical literature, largely because there has been no consistent way to approach transformations independent of cardinality. By identifying the distributional characteristics of some familiar pitch spaces as maximally even, we can modify relevant equations to push beyond the boundaries typically established between chords of different cardinalities.

This paper is divided into three parts. In the first part, we review filtered point-symmetry (FiPS), a geometric tool useful for evaluating and manipulating maximally even sets. In the second part, we use FiPS to examine neo-Riemannian triadic transformations, and to create an identical system for dealing with tetrachordal transformations. In the final part, we revise our notion of maximal evenness to allow for distributions involving fractional cardinalities. We then see how

C. Agon et al. (Eds.): MCM 2011, LNAI 6726, pp. 207–219, 2011.

fractional cardinality allows for the smooth transformation between triads and tetrachords in a consistent transformational space.

## 2 Filtered Point-Symmetry

As previously stated, filtered point-symmetry is a geometric tool useful for evaluating and manipulating maximally even (MaxE) sets [1]. The process of geometric calculation follows:

1. Start with some number of consecutively embedded rings, each having a specific number of symmetrically distributed holes. A ring with holes is called a *filter*.
2. The innermost filter is called the *beacon*, and emits a beam from each hole. Each beam is perpendicular to the tangent of the circle at the location of the hole it emanates from (or passes through).
3. When the beams encounter the next filter, each passes through a hole on that filter by satisfying one of two conditions:
   (a) If a beam aligns with a hole, it will pass through that hole. This is the case for all beams passing through *Hole #6* in Fig. 1.
   (b) If a beam does not align with a hole, it will travel counter-clockwise until it encounters a hole to pass through.
4. A beam will continue its journey until it passes through a hole on the outermost filter. Every time a beam passes through a hole, its angle is altered as in step two.

FiPS diagrams are geometric analogs to the $J$ function, an algebraic method for calculating maximal evenness [2]:

$$J_{c,d}^{m}(k) = \left\lfloor \frac{ck+m}{d} \right\rfloor_{\text{mod } c} . \tag{1}$$

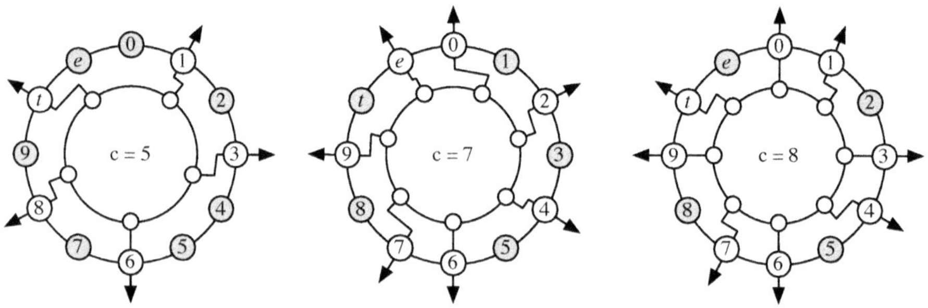


**Fig. 1.** Filtered point-symmetry is used to calculate three familiar maximally even distributions. The value $c$, at the center of each drawing, indicates the cardinality of the beacon and, consequently, the cardinality of the overall set. The *drawing on the left* represents a pentatonic set as a MaxE distribution of $5 \rightarrow 12$. The *drawing in the center* represents a diatonic set as a MaxE distribution of $7 \rightarrow 12$. The *drawing on the right* represents an octatonic set as a MaxE distribution of $8 \rightarrow 12$.

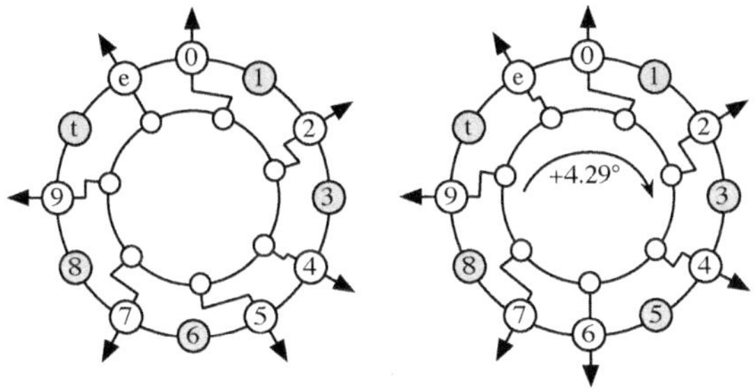


**Fig. 2.** A MaxE distribution of 7 → 12. The *diagram on the left* shows a C major collection. The *diagram on the right* shows a G major collection. The transformation from one collection to the other involves a clockwise rotation of the interior filter, equivalent to continuously adjusting $m$ to the desired offset.

To determine a MaxE distribution of a number of elements $d$ over a number of slots $c$, one takes the aggregate result for all $k$ in $d$, where $k$ is the index of the particular element in $d$ one is curious about. For example, in a maximally even distribution of 7 → 12, the fourth element of the derived seven-note set would be 5, as shown below. Note that the fourth element has an index of 3, since the first element is at index 0.

$$J^{m=0}_{c=12,d=7}\,(k=3) = \left\lfloor \frac{12 \times 3 + 0}{7} \right\rfloor_{\text{mod } 12} = \left\lfloor \frac{36}{7} \right\rfloor = \lfloor 5.143 \rfloor = 5\,. \qquad (2)$$

The oddly-shaped brackets represent a floor function, which removes any decimal value. The counter-clockwise movement of a beam to get to a hole is the geometric representation of the floor function.

The $J$ function also includes a variable $m$, which alters the *mode index*. When we calculate the distribution $J_{12,7}$ with $m = 0$, our result is {013568t}, a C♯ major collection. The other eleven diatonic collections are available at different

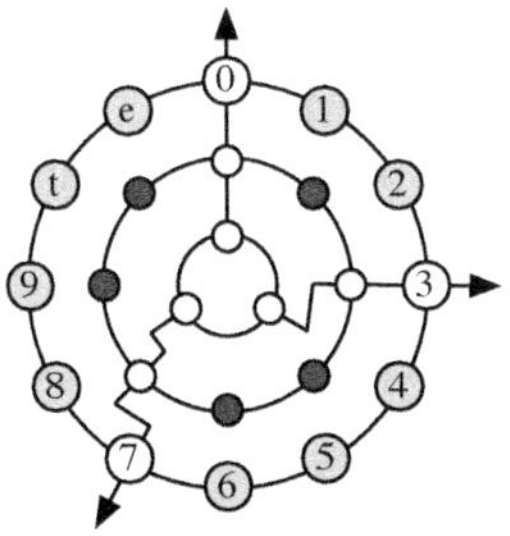


**Fig. 3.** A MaxE distribution of 8 → 12 ($J^0_{12,8}$) gives the octatonic set {0134679*t*}. A MaxE distribution of 3 → 8 ($J^0_{8,3}$) gives $\{025\}_{mod8}$, shown here as {037} in the iterated distribution $J^{0,0}_{12,8,3}$, since the first (0), third (2), and fifth (5) elements of the octatonic set are 0, 3, and 7.

values of $m$. In FiPS, $m$ becomes a continuous function, represented by the rotation of a filter (Fig. 2).

Finally, the $J$ function can be iterated; Clough, Cuciurean, and Douthett [3] formally define an $n^{th}$-order (iterated) maximally even set as

$$J_{d_0,d_1,d_2,\ldots,d_n}^{m_1,m_2,\ldots,m_n}(k) = J_{d_0,d_1}^{m_1}\left(J_{d_1,d_2}^{m_2}\left(J_{d_2,d_3}^{m_3}\left(\ldots J_{d_{n-1},d_n}^{m_n}(k)\right)\right)\right) . \quad (3)$$

In FiPS, geometric iteration is carried out by increasing the number of filters. Figure 3 gives an iterated MaxE distribution of $3 \to 8 \to 12$—the MaxE distribution of three elements over an octatonic set.

## 3 Graphic Transformations with Scalar Contexts

Neo-Riemannian theory emphasizes parsimonious transformations between major and minor triads. There are three such transformations—*Leittonwechsel* (L), Parallel (P), and Relative (R)—and each transformation abides by the neo-Riemannian rules for triadic parsimony, summarized in Table 1.

**Table 1.** Rules of parsimony in neo-Riemannian triadic transformations

| Rule | Description |
|---|---|
| 1) Note preservation | A triad will preserve two of its three notes. |
| 2) Neighbor motion | A changed note may only change to one of its nearest neighbors, by a voice-leading distance of a semitone or a whole tone. |
| 3) Type preservation | The changed note of the triad must change in a way that preserves the $T_n/T_nI$ type. |

These maximally smooth transformations are generally conceived of in chromatic space, beyond the reign of any scalar or tonal context. Such a property is exceptional; voice-leading efficiency and harmonic transformation are effortlessly united in an unrestricted, 12-gamut universe. Theories of tetrachord transformations struggle to achieve the same sort of exceptional union. Adrian Childs speculates on transformations conceptually analogous to the neo-Riemannian model for triadic relationships; his study is self-limiting with its focus on type preservation, and cannot achieve transformations beyond two $T_n$ types of tetrachord (nor does it set out to do so) [4]. Douthett and Steinbach [5] adopt a different approach that uses single-note displacement transformations to link four tetrachord types. In their model, set class is not preserved, but voice-leading efficiency is maximal (e.g. D°7 → Dm7 → F7 → C°7).

I propose an alternate approach: By generally adopting a scalar context in which to view transformations, we retain the union of harmonic transformation and voice-leading efficiency that is so desirable in neo-Riemannian theory, while gaining the ability to consistently transform chords of any cardinality within

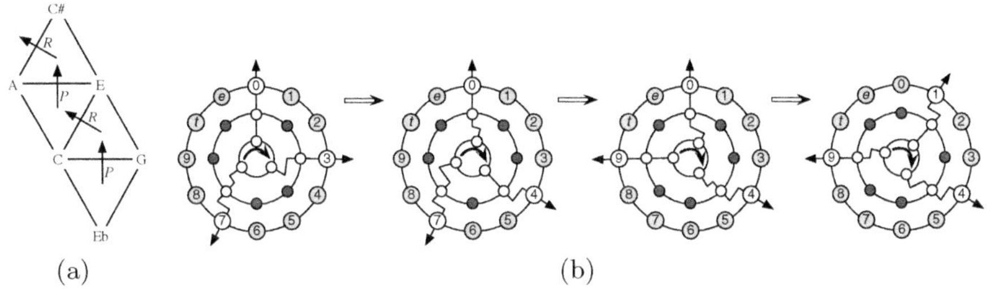


**Fig. 4.** (a) A *PR* cycle on the neo-Riemannian Tonnetz. (b) A *PR* cycle from a $3 \rightarrow 8 \rightarrow 12$ configuration in FiPS, with the interior ring rotating dynamically in 15° increments. Both drawings show the start of the same cycle from C− → C+ → A− → A+.

those contexts. To demonstrate, let's shift from thinking of neo-Riemannian transformations as chromatically contextualized, to envisioning those transformations in an octatonic context. A *PR* cycle becomes the smooth change of a maximally even trichord within a single octatonic set (Fig. 4). A *PL* cycle involves a smooth change of the trichord, as well as an alternating change in the active octatonic set (Fig. 5).

Though the transformational output remains intact, the entries in Table 1 become results instead of rules. Such a change undoes the compulsion for type preservation and chromatic motion, deferring to the overall plan of maximally smooth voice leading according to the chosen distributions. We can further refine our approach by adopting any analytically viable scalar context. To resolve the discrepancy in the treatment of triads and tetrachords, we use a diatonic scalar context that will consistently arbitrate the parsimonious change of and between

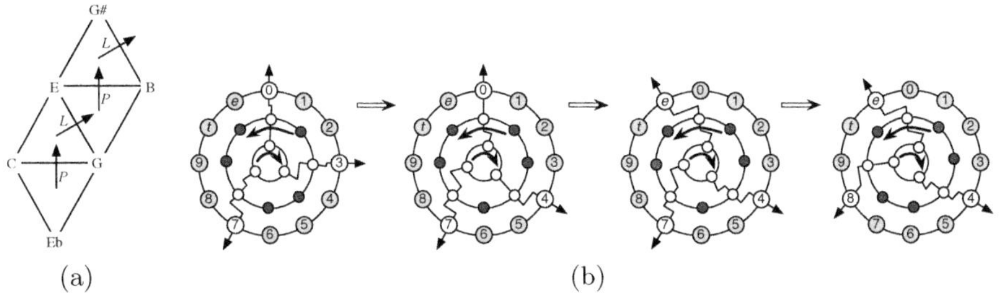


**Fig. 5.** (a) A *PL* cycle on the neo-Riemannian Tonnetz. (b) A PL cycle from a $3 \rightarrow 8 \rightarrow 12$ configuration in FiPS, with the interior ring rotating dynamically in 7.5° increments, and the middle ring rotating in the opposite direction, in 2.14° increments. Both drawings show the start of the same cycle from C− → C+ → E− → E+. Each ring alternates as the instigator of harmonic change; it is as though a *P* transformation takes place within an octatonic collection, and then a change in the active octatonic collection leads to the *L* transformation.

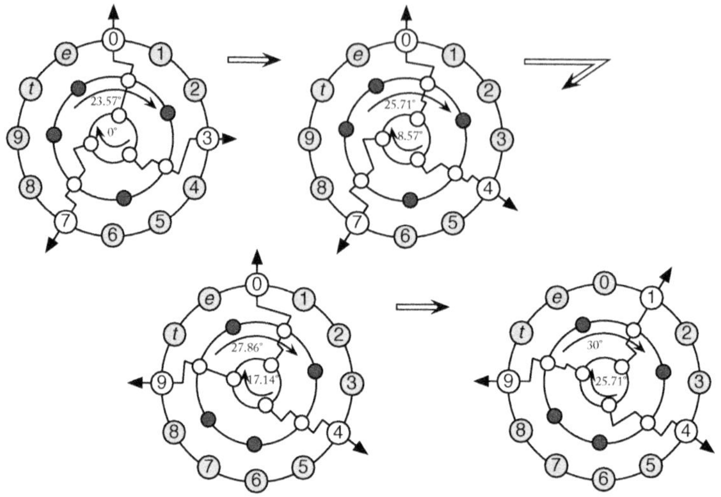


**Fig. 6.** Filtered point-symmetry showing the first four transformations of a $PR$ cycle within a diatonic context. Note that the diagram seems to indicate total independence of each ring; in actuality, interior rings rotate *with* all exterior rings, and the change shown in these diagrams is in addition to the sympathetic rotations.

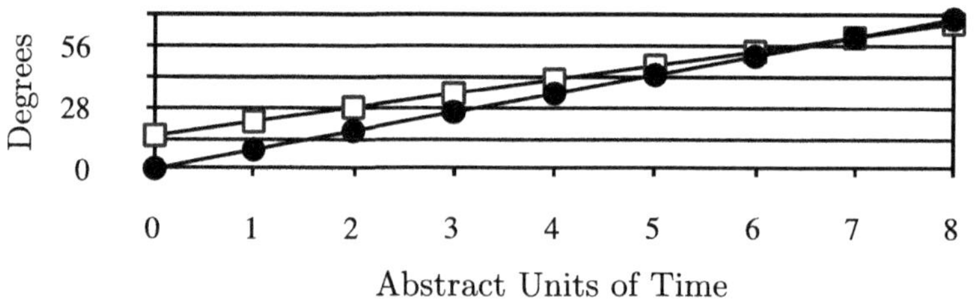


**Fig. 7.** A graph of the ring motion shown in Fig. 6, continued through a full $PR$ cycle from C− → C+ → A− → ... → C−. The *x axis* represents abstract units of time, and the *y axis* represents degrees of rotation of a ring. The *filled, circular points* represent measures for the 3-hole beacon; the *hollow, square points* represent measures for the 7-hole filter. The rotation of the beacon is *relative* to the 7-hole filter. As indicated by each line's positive slope, the rings are rotating at different rates in the same direction. The cause of harmonic transformation alternates between both rings.

chords of either cardinality. Figure 6 shows the neo-Riemannian $PR$ cycle within a diatonic context. Figure 7 moves the contents of Fig. 6 into graph notation, where the simultaneous motion of the filters is captured. Figure 8 shows a graph of FiPS transformations applied to diatonic tetrachords. As detailed in the figure, the tetrachordal cycle is related to the triadic $PR$ cycle.

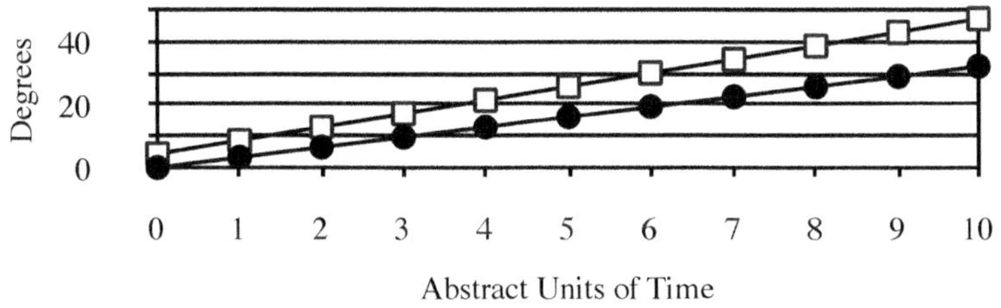


**Fig. 8.** A graph of ring motion in a 4 → 7 → 12 FiPS configuration, carried out in a manner similar to Fig. 7, in the sense that both rings are rotating at different rates in the same direction, and the cause of harmonic transformation alternates between rings. The cycle consists of {0158} → {0258} → {0259} → {4590} → {4690} → {4691} → {8914} → {8$t$14} → {8$t$15} → {0158}. As before, the $x$ *axis* represents abstract units of time, and the $y$ *axis* represents degrees of rotation of a ring. The *filled, circular points* represent measures for the 4-hole beacon; the *hollow, square points* represent measures for the 7-hole filter. The rotation of the beacon is relative to the 7-hole filter.

# 4 Cardinality Transformations

The previous section creates a consistent transformational approach for both triads and tetrachords. With this correlation, we can now consider transformations *between* triads and tetrachords. To do so, we will step outside the normal bounds of set theory in music, and imagine fractional cardinalities.

## 4.1 Partially Symmetric Distribution

In FiPS, we envision an integer-based cardinality as a point-symmetric distribution of holes around a ring. Figure 9 shows two such distributions, one of cardinality 3, and one of cardinality 4.

I propose that we think of fractional cardinality as a fuzzy abstraction. Pitch-class sets do not have fractional numbers of elements; however, there are sets that can share properties common to two different cardinalities. This abstraction is most easily understood if we invoke a visualization suggested by FiPS: If one imagines that either distribution could change fluidly into the other, then

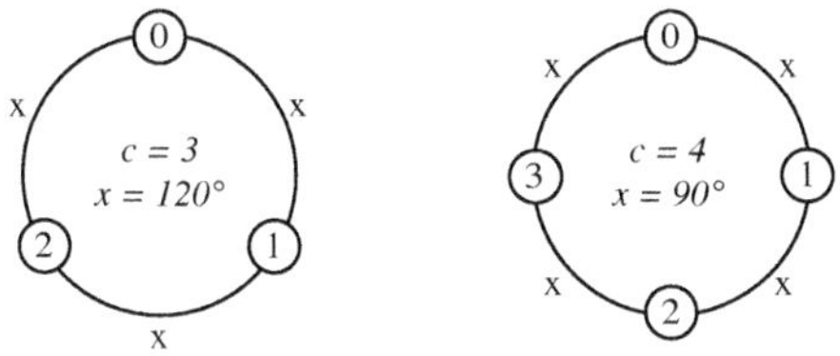


**Fig. 9.** Point-symmetric distributions representing integer-based cardinalities. The *diagram on the left* has a cardinality of 3; the *diagram on the right* has a cardinality of 4. The value of $x$ is the distance between each hole.

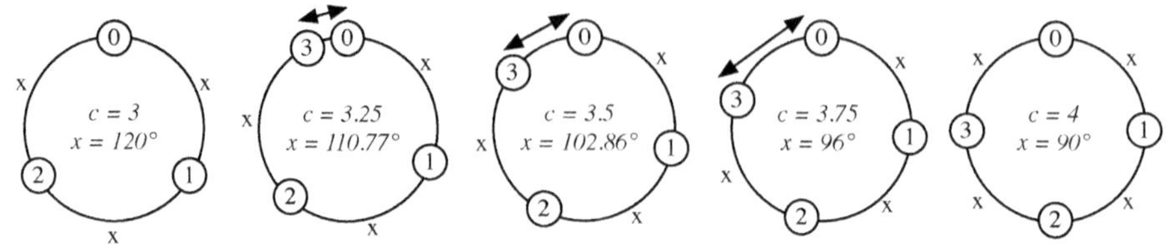


**Fig. 10.** Transformation from cardinality 3 to 4, using point-symmetric and partially symmetric distributions. The value of $x$ is the distance between evenly separated holes.

any distribution that occurs during that change will be "partially symmetric[1]". Figure 10 shows a smooth transformation between the cardinalities 3 and 4, in increments of 0.25.

To bring about this fluid change, a hole is added – in this case *Hole #3* – on top of the first hole. The number of holes for cardinality $c$:

$$\lceil c \rceil = 4, \text{where } 3 < c \leq 4 \,. \tag{4}$$

Except for the newly added hole, every hole will be spaced $360°/c$ degrees apart. For instance, in the point-symmetric distribution of 3 holes, each hole is 120° apart. In the partially symmetric distribution where $c = 3.25$, the first three holes are each separated by

$$^{360°}/_{3.25} = 110.769° \,. \tag{5}$$

The distance between the fourth and first hole is the remaining distance, or

$$360° - 3 \times \left(^{360°}/_{3.25}\right) = 27.692° \,. \tag{6}$$

### 4.2 Relatively Even Sets

By replacing a point-symmetric filter with a partially symmetric filter in FiPS, we create the possibility of a *relatively* even set—one that is as as even as possible given the fractional cardinality involved. The $J$ function (1) can be generalized to accommodate fractional cardinality, in the manner given below. Note that relatively even sets can be computed with the simpler, non-generalized $J$ function when only the interior ring is of fractional cardinality. This means, perhaps counterintuitively, that despite the extensive appearance of the outer ring cardinality $c$ in the generalized function, $d$ can be, and at some point in an iterated function *will* be, the fractional value in use.

Treat the generalized $J$ function as a computation based on the variance caused by a non-integer cardinality. To do this, replace $c$ with $\lceil c \rceil$, multiply mode

[1] Partial symmetry reduces the full dihedral symmetry of a fully symmetric distribution to one reflection and the identity.

index $m$ by $\lceil c \rceil / c$, and subtract some variance $V$, thereby giving our generalized function a new basic form of

$$J_{c,d}^{m}(k) = \left\lfloor \frac{\lceil c \rceil k + \frac{\lceil c \rceil}{c} m - V}{d} \right\rfloor_{\bmod \lceil c \rceil} . \tag{7}$$

To determine the variance $V$, let $DEG_{final}$ be the distance in degrees between the final two holes of an exterior ring of cardinality $c$, such that

$$DEG_{final} = 360 \left(1 - \frac{\lfloor c \rfloor}{c}\right) . \tag{8}$$

Where $c < \lceil c \rceil$, the distance between holes on a ring of cardinality $c$ will be greater than the distance between holes on a ring of cardinality $\lceil c \rceil$. Let $\Delta_{hole}$ represent the difference in degrees between two consecutive, non-final holes on a ring of cardinality $c$ and two consecutive holes on a ring of cardinality $\lceil c \rceil$, such that

$$\Delta_{hole} = \frac{DEG_{final}}{\lceil c \rceil} = \frac{360}{\lceil c \rceil} \left(1 - \frac{\lfloor c \rfloor}{c}\right) . \tag{9}$$

$\Delta_{hole}$ can be expressed as an offset value $O_{hole}$ in terms of $c$ and $d$, as

$$O_{hole} = (dc \times \Delta_{hole}) / 360 = \left(\frac{360}{\lceil c \rceil}\right) \left(\frac{dc}{360}\right) \left(1 - \frac{\lfloor c \rfloor}{c}\right) = \frac{d}{\lceil c \rceil} (c - \lfloor c \rfloor) . \tag{10}$$

For each hole on $c$, apply the offset $O_{hole}$, by multiplying the current hole's index by the value of $O_{hole}$ . Because it is impossible to know how many offsets are necessary until determining the result of the $J$ function, the $J$ function itself must be recursively embedded with this multiplier. Thus

$$V = \left\lfloor \frac{\lceil c \rceil k + \frac{\lceil c \rceil}{c} m}{d} \right\rfloor_{\bmod \lceil c \rceil} \times \frac{d}{\lceil c \rceil} (c - \lfloor c \rfloor) . \tag{11}$$

and

$$J_{c,d}^{m}(k) = \left\lfloor \frac{\left(\lceil c \rceil k + \frac{\lceil c \rceil}{c} m\right) - \left\lfloor \frac{\lceil c \rceil k + \frac{\lceil c \rceil}{c} m}{d} \right\rfloor_{\bmod \lceil c \rceil} \times \frac{d}{\lceil c \rceil} (c - \lfloor c \rfloor)}{d} \right\rfloor_{\bmod \lceil c \rceil} . \tag{12}$$

The geometric evaluation of the more general $J$ function remains the same. Figure 11 gives two possible results for $3.25 \to 7 \to 12$.

In this figure, we see both a MaxE triad, as we would see in $3 \to 7 \to 12$ (with a doubled tone), and a MaxE seventh chord, as we would see in $4 \to 7 \to 12$. In fact, when $c \in (3, 3.5)$, we encounter all possible diatonic trichords *and* tetrachords as the filters rotate.

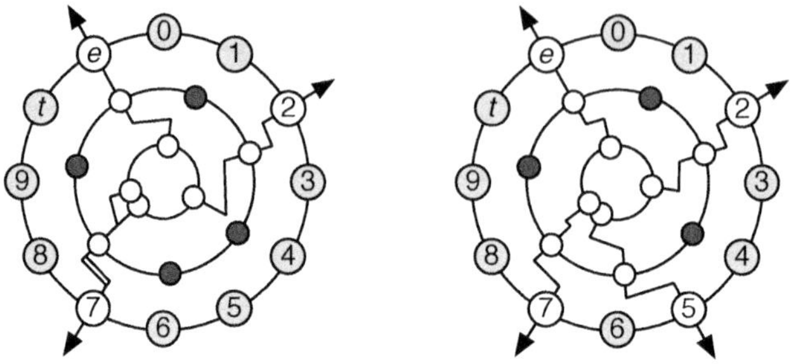


**Fig. 11.** Relatively even distributions of 3.25 → 7 → 12. The *drawing on the left* shows a G+ triad with a doubled root; the *drawing on the right* shows a G7 dominant seventh chord, with the beacon at a slightly different offset.

### 4.3 Harmonic Relationships

In integer-based distributions of 3 → 7 and 4 → 7, there are repeating families of transpositional set types and repeating patterns of set families.

For example, 120° of beacon rotation in a FiPS distribution of 3 → 7 produces (mod 7): $< \{024\}, \{025\}, \{035\}, \{135\}, \{136\}, \{146\}, \{246\}, \{240\} >$. Every third set is $\mathrm{T}_1$-related, and the original set recurs after seven sets. The 3 → 7 and 4 → 7 distributions, and their iterated distributions through 12 (given as part of a C major collection), are provided in Tables 2 and 3, respectively.

Our experience of iterating 3.25 → 7 is slightly different. Although every fourth set is $T_1$-related, the original set does not reappear until a full 360° rotation of the beacon, when all possible sets are exhausted (though many occur more than once). Table 4 gives the first 16 sets in this cycle. A pattern is readily apparent: we insert three parsimoniously related triads between each seventh chord of Table 3. In all cases, the triad immediately preceding the seventh chord is the seventh chord without its root, and the triad immediately following is the seventh chord without its seventh.

**Table 2.** Family of sets in 3 → 7 and 3 → 7 → 12

| Set (mod7) | Iteration through a C major set | Beacon Offset |
|---|---|---|
| {024} | C | 0° |
| {025} | Am | 17.14° |
| {035} | F | 34.29° |
| {135} | Dm | 51.43° |
| {136} | *B*° | 68.57° |
| {146} | G | 85.71° |
| {246} | Em | 102.86° |
| {240}*(repeat) | C | 120° |

**Table 3.** Family of sets in $4 \to 7$ and $4 \to 7 \to 12$

| Set (mod7) | Iteration through a C major set | Beacon Offset |
|---|---|---|
| {0135} | Dm7 | 0° |
| {0235} | Fmaj7 | 12.86° |
| {0245} | Am7 | 25.71° |
| {0246} | Cmaj7 | 38.57° |
| {1246} | Em7 | 51.43° |
| {1346} | G7 | 64.29° |
| {1356} | B°7 | 77.14° |
| {1350}*(repeat) | Dm7 | 90° |

**Table 4.** Family of sets in $3.25 \to 7$ and $3.25 \to 7 \to 12$

| Set (mod7) | Iteration through a C major set | Beacon Offset |
|---|---|---|
| {0135} | Dm7 | 0° |
| {1135} | Dm | 3.96° |
| {1136} | B° | 7.91° |
| {1146} | G | 11.87° |
| {1246} | Em7 | 15.82° |
| {2246} | Em | 19.78° |
| {2240} | C | 23.74° |
| {2250} | Am | 27.69° |
| {2350} | Fmaj7 | 31.65° |
| {3350} | F | 35.60° |
| {3351} | Dm | 39.56° |
| {3361} | B° | 43.52° |
| {3461} | G7 | 47.47° |
| {4461} | G | 51.43° |
| {4462} | Em | 55.38° |
| {4402} | C | 59.34° |

### 4.4 Cardinality Transformations

Table 4 demonstrates that triads and tetrachords with three notes in common are neighbors in a fractional distribution of $3.25 \to 7 \to 12$. Table 4 also shows all of the triads as multisets. If there are no doublings in the music being modeled, then multisets may be undesirable. We can avoid multisets by allowing a filter to simultaneously change position and cardinality. Figure 12 shows a transformation from $Em \to G \to G7$, without multisets. Figure 13 shows a graph of these transformations. The graph remains consistent whether it is representing triads or tetrachords.

Our next step would be to make use of cardinality transformations in analysis. To do so, we would need to broaden the scope of Table 4 with the different

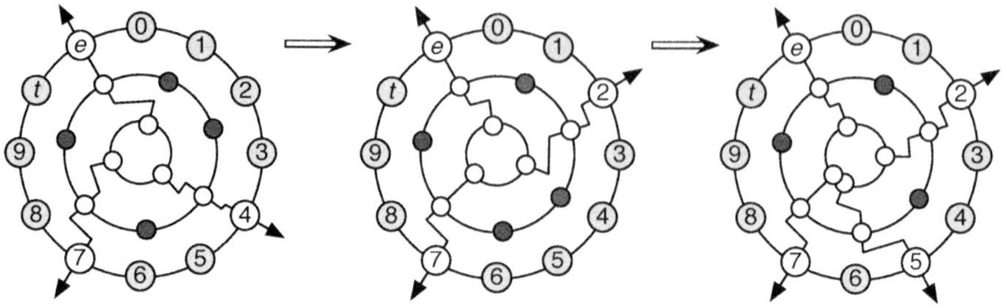


**Fig. 12.** A continuous transformation from $Em \to G \to G7$ in the key of C major. Between the *left* and *middle diagrams*, the beacon rotates counter-clockwise. Between the *middle* and *final diagrams*, the beacon's cardinality transforms from 3 to 3.25.

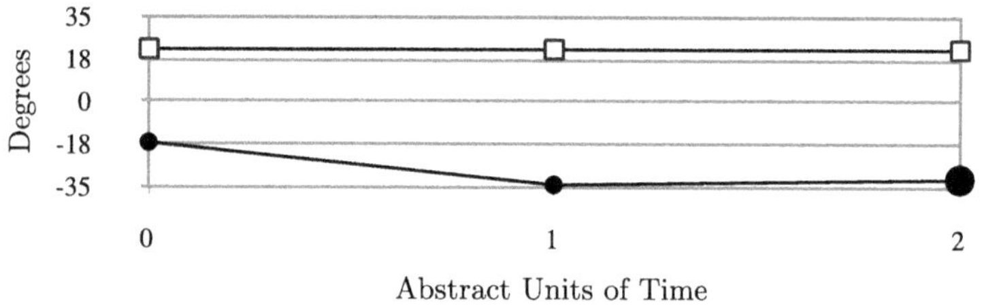


**Fig. 13.** A graph of the transformation shown in Fig. 12. The $x$ axis represents abstract units of time, and the $y$ axis represents degrees of rotation of a ring. The *filled, circular points* represent measures for the beacon; the *hollow, square points* represent measures for the 7-hole filter. The *size of the circular points* indicates the current cardinality: 3 for the *smaller circles*, 3.25 for the *larger circle*. For a longer time span, a three-dimensional plot may be used to convey cardinality; for consistency, a two-dimensional plot is used here.

but equally viable relationships expressed when the middle ring rotates, as in Fig. 5 and Fig. 6. We would also need to add two more rings to our model, one duplicating the beacon, and one duplicating the middle filter [6]. For the purposes of the current discussion, think of the duplication as a means of allowing us to expand and contract at will from any point on a point-symmetric ring. If, for example, we were analyzing $Dm7 \to Dm \to B^\circ \to G7 \to G$, we would transform through the first three elements of Table 4, reduce our beacon to cardinality 3, and then re-expand the beacon from the twelfth element in the table (the subsequent $B^\circ$), bringing us through $B^\circ$ to $G7$ and $G$.

## 5 Conclusion

Filtered point-symmetry allows us to apply a scalar context to transformations, and brings consistency to our treatment of transformations with any cardinality. Incorporating relatively even sets into our conception of musical transformation, we are able to transform between sets of different cardinality. This capability is significant: without being reductive, we can analyze parsimonious and continuous transformations in music that make use of both triads and tetrachords.

## References

1. Douthett, J.: Filtered Point-Symmetry and Dynamical Voice-Leading. In: Douthett, J., Hyde, M.M., Smith, C.J. (eds.) Music Theory and Mathematics: Chords, Collections, and Transformations. University of Rochester Press, NY (2008)
2. Clough, J., Douthett, J.: Maximally Even Sets. Journal of Music Theory 35(1/2), 93–173 (1991)
3. Clough, J., Cuciurean, J., Douthett, J.: Hyperscales and the Generalized Tetrachord. Journal of Music Theory 41(1), 67–100 (1997)
4. Childs, A.P.: Moving Beyond Neo-Riemannian Triads: Exploring a Transformational Model for Seventh Chords. Journal of Music Theory 42(2), 181–193 (1998)
5. Douthett, J., Steinbach, P.: Parsimonious Graphs: A Study in Parsimony, Contextual Transformations, and Modes of Limited Transposition. Journal of Music Theory 42(2), 241–263 (1998) (Note: See also Waller [7])
6. Plotkin, R.: Transforming Transformational Analysis: Applications of Filtered Point-Symmetry. PhD Thesis, University of Chicago (2010)
7. Waller, D.A.: Some Combinatorial Aspects of the Musical Chords. The Mathematical Gazette 62(419), 12–15 (1978)

# Indeterminate Music and Probability Spaces: The Case of John Cage's Number Pieces

Alexandre Popoff

al.popoff@free.fr

**Abstract.** Indeterminate music is characterized by the use of random outputs, either during the compositional process or during its performance. John Cage's Number Pieces are works indeterminate in their realization in which the performer, through a framework of "time-brackets", has control over the temporal limits of fixed sounds. In this paper we analyze John Cage's temporal system of time-brackets using a statistical approach. It is shown that for a single time-bracket a probability space can be defined concerning the choice of the temporal limits of a sound. The performer's attitude toward choice is modelled through different probability distributions over the sample space and the audible quantities (in particular, length) of the sound contained within a time-bracket are calculated. We show how time-brackets can be considered as flexible structures ensuring complex outputs from simple assumptions. The limits of our statistical model as compared to real human behavior are discussed, and perspectives are given concerning the study of complete sets of time-brackets.

**Keywords:** John Cage, Time-bracket, Probability Space, Probability Distributions, Point processes.

## 1 Introduction

Indeterminate music, which has risen prominently during the second half of the 20th century mainly due to the works of John Cage, relies on aleatoric choice over the sensible parameters of sound, such as pitch, duration, rhythm, order, and so on. Aleatoric choice can be introduced by the composer while writing the piece: in that case, a fixed score is produced from random outcomes. John Cage's *Music of Changes* is an example of such a score: random outcomes from the Yi-King were used to determine piano notes and the performer has no choice but to follow the final score. On the other hand, aleatoric choice can be introduced by the performer following instructions from the composer: the music is thus said to be indeterminate in its realization. Morton Feldman's *Projections* series, for example, relies on the performer to choose the pitch of the notes according to a particular grid-like framework. Another example is Stockhausen's *Klavierstück XI* in which the performer can choose the order of fixed sequences. In this paper, we will focus on the so-called "Number Pieces" of John Cage. In these works the position in time of fixed sounds or notes is left to the performer : it is therefore an example of indeterminacy in realization.

C. Agon et al. (Eds.): MCM 2011, LNAI 6726, pp. 220–229, 2011.

The structure of a piece of music is determined by the collection of sounds contained within and by the relations between those sounds. Speaking of the structure of a piece which is indeterminate in its realization has no sense: by definition every performance is different. Instead, one can speak of the structure of a realization, for in this case the parameters of the sounds are known. Thus, an important question arises : given that realizations stems from aleatoric choices by the performer(s), is the resulting structure (or the perceptual qualities of the realization) absolutely random ? Answering this question is mathematically equivalent to studying the statistical properties of many realizations in order to determine the "degree of randomness". Therefore, one can speak of a structure of structures, or "meta-structure". This meta-structure encompasses all possible realizations and is the proper approach for the structure of a piece indeterminate in its realization.

Statistical properties of indeterminate music have already been studied in the past, although few examples are available, as shown by the work of Thomas de Lio [1] on John Cage's *Variations II*. In this piece, the performer(s) places points and lines on a surface: the distances between them determines parameters such as pitch, amplitude, etc. De Lio commented on the underlying meta-structure of *Variations II* as "one large comprehensive system which itself represents the total accumulation of its many constituents realization", and described it as a collection of several inter-correlated statistical distributions.

In this paper, the late Number Pieces of John Cage will be discussed from a statistical point of view. The purpose of this contribution is to examine the meta-structure (or equivalently the nature of the stochastic process) given to the Number Pieces by the system of time-brackets and their rules of execution. In particular, we will show that the system of "time-brackets" leads naturally to the definition of a probability space over which a probability measure can be defined. Statistical properties of the sounds in terms of length in a single time-bracket will be analyzed. Finally, perspectives will be given concerning successions of time-brackets in a piece.

## 2 Introduction and Model

In 1987, John Cage began to write a series of scores, the titles of which refer to the number of performers involved, hence the name "Number Pieces". In these pieces, a superscript in the title indicates the rank of the piece among those with the same number of performers. Thus *Four* is the first Number Piece written for four performers, whereas *Four*$^3$ is the third one. From 1987 to Cage's death in 1992 forty-seven such pieces were written.

In almost all the Number Pieces, John Cage used a particular time-structure for ordering the sounds which was named "time-bracket". These time-brackets already appeared in earlier works such as *Thirty Pieces for Five Orchestras* and *Music for ...* However, in the Number Pieces, Cage simplified the contents of the time-brackets, most of them containing only a single tone or sound, especially in his late works.

A time-bracket is basically made of three parts: a fragment of one or many staves, lying under two time intervals, one on the left and one on the right. Time intervals themselves consist of two real-time values separated by a two-way arrow. The staves contain one or more sound events without any duration indications. The time-bracket is performed as follows : the performer decides to start playing the written sounds anytime inside the first time interval on the left, and chooses to end them anytime inside the second one. These parameters are thus left free to the performer, provided he respects the time-bracket structure. Successive time-brackets occurs in a Number Piece score with possible overlap between each other, meaning that the ending time interval of one time-bracket may overlap the beginning interval of the next one.

It is important to notice that time in the Number Pieces is similar to physicists' time, i.e a continuous real-valued quantity, as opposed to the discrete time of classical music measured in beats. Besides, Cage often referred to the use of a stopwatch for musicians to perform the piece. It should be noted that the notation of classical music defines the duration as well as the starting time (which often coincides with the ending time of the previous sound) of a sound *a priori*, its ending time being known *a posteriori*. On the opposite, in the framework of a time-bracket, a sound is defined by its starting and ending time and all the perceptual qualities of the sound such as duration or repartition in time are defined *a posteriori*.

## 2.1 Definitions

We now introduce a formalized definition of the temporal structure of a time-bracket. From now on, time will be expressed in seconds. For the sake of clarity it is assumed, without loss of generality, that a single time-bracket starts at 0 seconds.

**Definition 1.** A time-bracket is a set TB={ST;ET} of two closed intervals over $\mathbb{R}$, referred to as the Starting Time interval (ST) and Ending Time interval (ET). ST is written as $[0;T_2]$ and ET as $[T_1;T_3]$ with $T_1,T_2,T_3 \in \mathbb{R}$, and $0 < T_1 < T_2 < T_3$.

**Definition 2.** The internal overlap of a time-bracket is defined as $ST \cap ET=[T_1;T_2]$. The external overlap of two successive time-brackets TB1={ST1;ET1} and TB2={ST2;ET2} is defined as $ET1 \cap ST2$.

**Definition 3.** A realization of a time-bracket is a set $\{t_s;t_e\}$ of two elements of $\mathbb{R}$, with $t_s \in$ ST, $t_e \in$ ET, and $t_s < t_e$.

**Definition 4.** The length of the sound is defined as $L=t_e-t_s$.

Studying the statistical properties of a time-bracket involves analyzing all possible outcomes. Therefore, the definition of the *sample space* of a time-bracket arises naturally:

**Definition 5.** The sample space $\Omega$ of a single time-bracket is the set of all possible realizations $\{t_s;t_e\}$ according to Definition 3. It is therefore a subset of $\mathbb{R}^2$.

In the rest of this analysis, we choose the first time-bracket in the Number Piece "Five" which is defined by $T_1$=30, $T_2$=45 and $T_3$=75, as shown in Fig. 1a. We will assume this time-bracket contains a single sound only. The representation of the sample space of this time-bracket is shown in Fig. 1b.

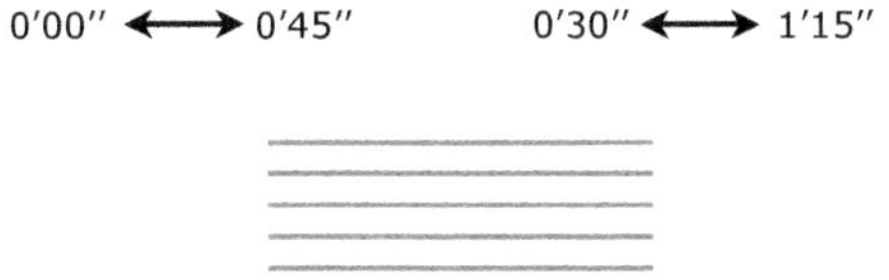


(a)

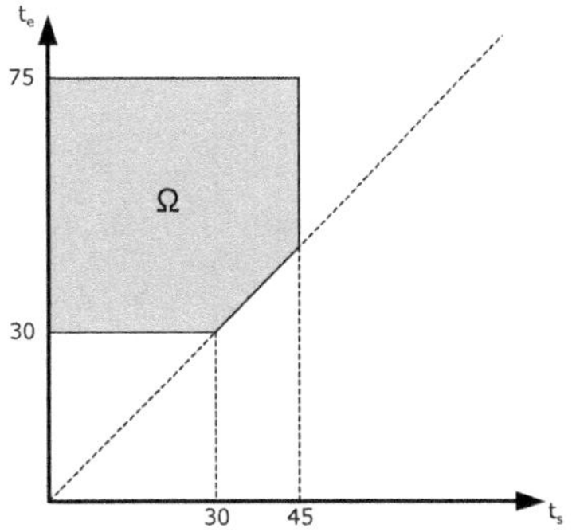


(b)

**Fig. 1.** (a) Temporal structure of the first time-bracket of the Number Piece *Five*. (b) The sample space $\Omega$ associated with this time-bracket.

## 2.2 The Probability Space of a Single Time-Bracket

In order to study a time-bracket statistically, we need to quantify the choice of a set $\{t_s;t_e\}$. Mathematically, it means turning the sample space $\Omega$ into a probability space by defining a probability distribution $\mu(t_s, t_e)$ over it. Variables $t_s$, $t_e$ and L thus become random variables, and the quantity $\mu(t_s, t_e) \times dt_s \times dt_e$ expresses the probability of finding a realization (the outcome of the above random variables) within the small interval $[t_s; t_s + dt_s] \times [t_e; t_e + dt_e]$.

Given that a performer makes choices about time-brackets in chronological order, the probability measure can therefore be expressed as:

$$\mu(t_s, t_e) = \mu_s(t_s)\mu_e(t_e|t_s) \tag{1}$$

where $\mu_e(t_e|t_s)$ is the conditional probability distribution of choosing $t_e$ after $t_s$.

The choice of a probability measure that accounts for human decisions is difficult. This stems from the fact that humans are poor random generators [2], usually biased by phonological or visuo-spatial activities [3]. Keeping this fact in mind, our analysis will use three different distributions:

- The first one assumes unbiased random choice, which will be accounted for by uniform distributions. For a time-bracket with a non-null internal overlap, we thus have:

$$\mu_s(t_s) = \frac{1}{T_2} \tag{2}$$

and

$$\begin{aligned} \mu_e(t_e|t_s) &= \tfrac{1}{T_3-T_1} \text{ if } t_s < T_1 \\ &= \tfrac{1}{T_3-t_s} \text{ if } t_s > T_1 \end{aligned} \tag{3}$$

– The second one assumes classical gaussian distributions in the form of :

$$\mu_s(t_s) = e^{-\frac{(t_s-T_2/2)^2}{\sigma^2}} \tag{4}$$

and

$$\begin{aligned} \mu_e(t_e|t_s) &= e^{-\frac{(t_e-(T_1+T_3)/2)^2}{\sigma^2}} \text{ if } t_s < T_1 \\ &= e^{-\frac{(t_e-(t_s+T_3)/2)^2}{\sigma^2}} \text{ if } t_s > T_1 \end{aligned} \tag{5}$$

where $\sigma$ will be chosen here as roughly equal to $T_2/4$.

– The third one assumes quadratic distributions in the form of :

$$\mu_s(t_s) = (t_s - T_2/2)^2 \tag{6}$$

and

$$\begin{aligned} \mu_e(t_e|t_s) &= (t_s - (T_1 + T_3)/2)^2 \text{ if } t_s < T_1 \\ &= (t_e - (t_s + T_3)/2)^2 \text{ if } t_s > T_1 \end{aligned} \tag{7}$$

This unusual distribution is introduced here in order to depart from classical distributions, which are probably too simple to model human choices. However we do not claim that this particular distribution reflects human behavior, which is known to be much more complicated.

The corresponding plots of $\mu(t_s, t_e)$ over $\Omega$ according to the above equations are shown in Figure 2.

## 3 Sound Lengths in a Single Time-Bracket

As mentionned in the introduction, the peculiarity of the time-bracket system lies in the fact that the audible random variables, such as the length of a sound, are *not* the random variables which are chosen by the performer. Indeed, only when the sound is finished, i.e when both $t_s$ and $t_e$ have been chosen, can we appreciate its length or its position in the piece. Having defined the probability distributions $\mu(t_s, t_e)$, the purpose of this section is therefore to analyze the statistical structure induced by $\mu(t_s, t_e)$ over audible quantities, focusing in particular on the length L.

In most cases, the length distribution is too complicated to be expressed analytically (the uniform distribution case has been carried in a previous analysis

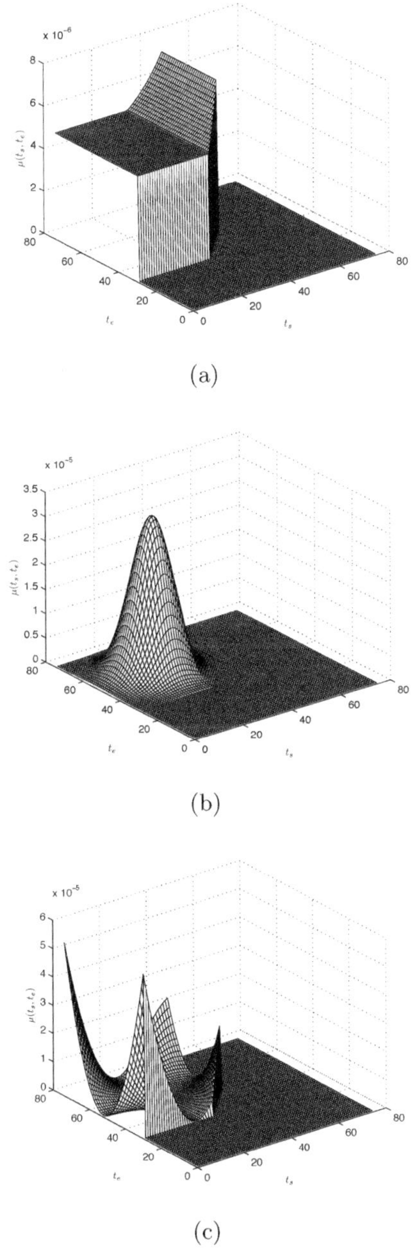


**Fig. 2.** Plot of $\mu(t_s, t_e)$ in the sample space $\Omega$ according to the (a) uniform distribution assumption, (b) gaussian distribution assumption and the (c) quadratic distribution assumption

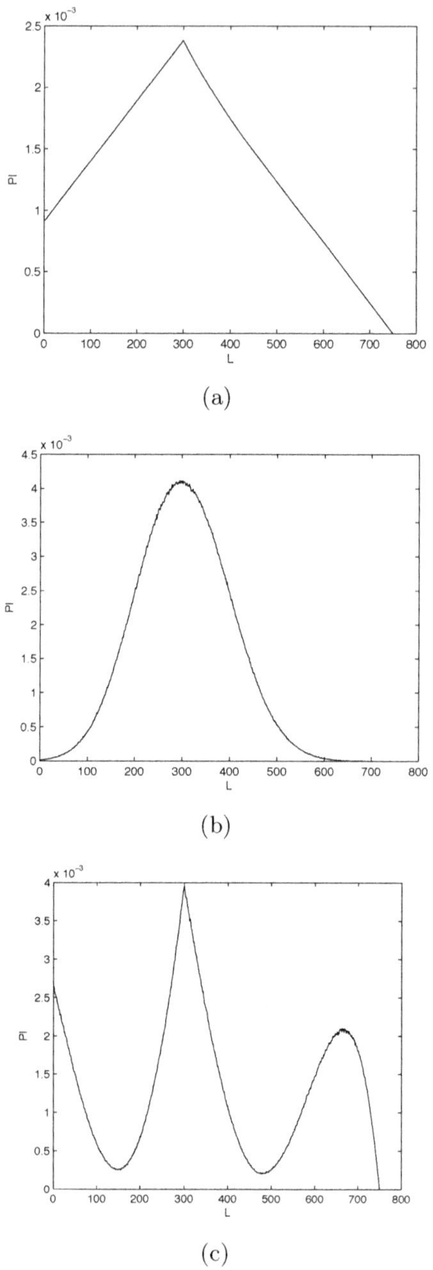


**Fig. 3.** Plot of the length distribution of sound in a single time-bracket according to the (a) uniform distribution assumption, (b) gaussian distribution assumption and the (c) quadratic distribution assumption. Length is expressed in tenth of seconds due to numerical discretization.

[4]). We thus used a computer program to sample the space $\Omega$ through Monte-Carlo techniques and obtain numerical values for the length distribution. The plots of this distributions are shown in Figure 3.

From these plots, we see that although the $\mu(t_s, t_e)$ distributions are quite simple in nature, they lead to complicated distributions of lengths, especially in the case of the uniform or quadratic assumptions (As could have been expected from the law of large numbers, the gaussian assumption leads to a gaussian distribution of lengths). It is therefore a strength of the time-bracket system that it induces complex behavior from simple selection rules.

Another interesting aspect of these distributions comes from the fact that they all have a maximum at L=30 seconds. This stems from the fact that the time-bracket used in this study is symetrical around t=37.5 seconds and has $T_1$ equals to 30 seconds. It is therefore possible to start a 30-seconds sound anywhere between 0 and $T_1$: longer sounds would fall outside the $T_3$ limit, whereas shorter ones would not end inside the $[T_1;T_3]$ interval. We thus see another strength of the time-bracket system: while allowing flexibility in its realization, it guarantees an overall regular behavior (incidentally, it was shown in [4] that the sound is also localized in terms of position inside the time-bracket). In Cage's words [5]: "It is not entirely structural, but it is at the same time not entirely free of parts", or similarly from Benedict Weisser's Ph.D [6] : "What time-brackets are capable of in this regard is in ensuring predictiveness on a very remote and distant scale global in its proportions".

It should be noted that all three distributions show null probability of finding a 75-seconds sound. From our model, this would require to exactly pick $t_s$=0 and $t_e$=$T_3$, an event which has a null probability and thus is mathematically impossible, although acceptable for a performer. This apparent paradox was discussed in [4] and points to the limits of our statistical model as applied to human random generation behavior.

## 4 Multiple Time-Brackets and Point Processes

In this section we wish to open perspectives concerning the statistical analysis of successive time-brackets, as encountered in a complete Number Piece. Following the same approach, a 2N-dimensional sample space can be defined from the set of N time-brackets, as well as a corresponding probability distribution $\mu(t_s(1), t_e(1), ..., t_s(N), t_e(N))$.

In this multi-dimensional space, the succession of time-brackets is better viewed as a marked point process (MPP) on the real line. A point process is a mapping from a probability space to a collection of points on a set (see [7] for reference on point processes). A marked point process furthermore assigns a mark to each point. In our cases, the point values are simply the values $(t_s(i), t_e(i))$ together with a two-valued mark indicating if the point is a starting- or ending-time. A graphical representation of this process is shown on Figure 4.

In many Number Pieces, as in *Five*, time-brackets admit the same structure and the external overlap between successive time-brackets is equal to the internal

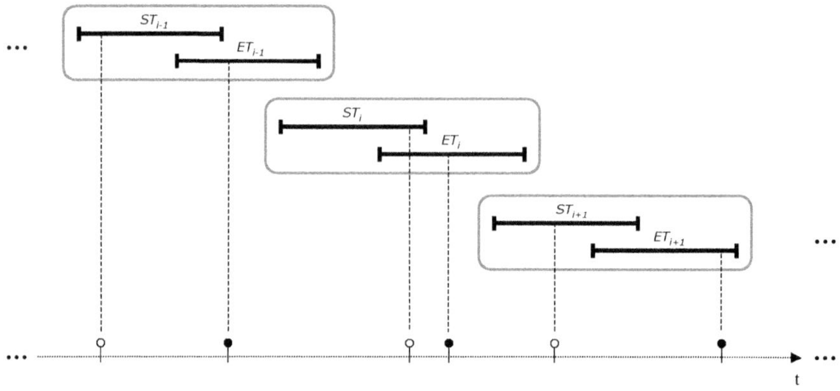


**Fig. 4.** A succession of time-brackets viewed as a marked point process (MPP) on the real line

overlap. Thus, by considerations of symmetry, the mark becomes useless and the MPP simplifies to a simple point process (SPP). A SPP is equivalently defined by the time values of the points or by the sequence of the inter-events times. Therefore, in the particular case of a symmetric succession of time-brackets, the SPP is entirely defined by the knowledge of the length distribution. Furthermore, given that inter-events times are independent and equally distributed, the SPP thus obtained is called a *renewal process*.

As stated in the previous section, modelling human behavior through fixed distributions is most probably not appropriate. It would therefore be interesting in future works to analyze time-brackets in the framework of point processes with more complicated models. In particular, Markov chains models could prove useful given their ability to take into account past states and outcomes. Indeed, it was shown in the above mentioned references that short-term memory effects might play a role in the departure from randomness in human random generation tasks. At the same time, comparison with real-world performances and recordings is needed to validate the models. Analysis of recordings could be achieved by identifying each part, in a spectrogram for example, provided the number of performers is small and the parts are clearly differentiated. On the other hand, data could be collected from live performances by recording the status of the performers. This could be done either by the musicians themselves, through an appropriate triggering mechanism, or by external observers.

## 5 Conclusions

In this paper, a statistical framework was introduced in order to analyze the properties of John Cage's time-bracket system. It was shown that a probability space can be defined for a single time-bracket which leads to complex behavior in terms of perceptual qualities of the sound, distinct from complete randomness.

While providing flexibility in small time scales, the time-bracket system also ensures uniformity over the large time scale of the Number Piece. The limits of our statistical model were shown as compared to human random generation behavior. In this line, perspectives concerning the use of point processes theory were given. A more detailed study is thus called for in order to analyze the real-world musicians' attitude towards choice during the performance of the Number Pieces.

## References

1. De Lio, T.: John Cage's Variations II: The Morphology of a Global Structure. Perspectives of New Music 19(1-2), 351–371 (1980)
2. Wagenaar, W.A.: Generation of Random Sequences by Human Subjects: A Critical Survey of Literature. Psychological Bulletin 77(1), 65–72 (1972)
3. Vandierendonck, A.: Is Judgment of Random Time Intervals Biased and Capacity-Limited? Psychological Research 63, 199–209 (2000)
4. Popoff, A.: John Cage's Number Pieces: The Meta-Structure of Time-Brackets and the Notion of Time. Perspectives of New Music 48(1), 65–82 (2010)
5. Cage, J.: I-VI, p. 442. Harvard University Press, Cambridge (1990)
6. Weisser, B.J.: Notational Practice in Contemporary Music: a Critique of Three Compositional Models (L. Berio, J. Cage, B. Ferneyrough). University of New York, Ph.D Dissertation (1998)
7. Daley, D.J., Vere-Jones, D.: An Introduction to the Theory of Point Processes, vol. I-II. Springer, Heidelberg (2004)

# Voice-Leading Prototypes and Harmonic Function in Two Chorale Corpora

Ian Quinn[1] and Panayotis Mavromatis[2]

[1] Yale University
ian.quinn@yale.edu
[2] New York University
panos.mavromatis@nyu.edu

**Abstract.** We describe a data representation for voice leading between two sonorities in a chorale texture, and a similarity measure for these voice leadings. These tools are used in an empirical study of the relationship between voice leading and harmonic function in a corpus of Bach chorales and a corpus of Lutheran chorales from a hundred years earlier. Common voice-leading types in the corpora are subjected to a cluster analysis that is readily interpreted in terms of harmonic functional syntax. We are thus able not only to read a theory of harmony directly out of a corpus, but to do so without building in a priori notions of chord structure, rootedness, or even key. The cluster analysis also clarifies important syntactic differences between the pre-tonal (modal) corpus and the Bach (tonal) corpus.

**Keywords:** Corpus Analysis, Voice Leading, Tonality, Modality, Harmonic Function, Clustering.

## 1 Introduction

In 1722, Jean-Philippe Rameau [1] proposed an early version of the concept that has come to be known as *harmonic function*. For Rameau, harmonic function is a product of both chord structure and voice leading. Since that time, there has been broad agreement among music theorists that these two aspects of pitch structure play a central role in the syntax of tonal music. There is, however, no consensus as to how chord structure and voice leading can be formalized, or precisely how harmonic function emerges from their interaction.

In recent years, a number of researchers have sought to provide an empirical basis to the theory of tonal syntax by undertaking computer-assisted analyses of Bach chorales. Dmitri Tymoczko [2] isolated pairs of successive tertian sonorities (triads and seventh chords) in 186 chorales and analyzed the root motions between them, finding that descending fifths, descending thirds, and ascending steps were more common than ascending fifths, ascending thirds, and descending steps. Martin Rohrmeier [3] examined every simultaneous pitch-class set in 386 chorales after transposing them to C major or C minor, clustering the most frequent sets by the degree to which they tend to appear in similar harmonic successions. He noted with surprise that the IV chord and V chord end up in the same cluster. These results,

C. Agon et al. (Eds.): MCM 2011, LNAI 6726, pp. 230–240, 2011.

however, are easily explained: (a) both IV and V frequently follow I, and (b) IV can go to I in plagal-type formations, and V typically goes to I. Tymoczko's approach can distinguish between IV–I (ascending fifth) and V–I (descending fifth) progressions, but cannot accommodate sonorities without a determinate root. Rohrmeier's approach is general enough to accommodate non-tertian sonorities, but is limited to assigning each chord to a single functional category. The two approaches differ in another important respect: Tymoczko's method, which considers only the relationship between adjacent sonorities, disregards the relationship of the sonorities to the prevailing tonic. Rohrmeier's method, on the other hand, identifies sonorities based on their relationship to the overall tonic of each chorale, disregarding modulations within chorales.

Darrell Conklin [4] proposed a representation that shares with Rohrmeier's method the advantage of not requiring sonorities to have roots, and with Tymoczko's method the advantage of focusing on the relationship between sonorities rather than on the sonorities themselves. His representation takes the form of an ordered 4-tuple of integers modulo 12 representing the pitch-class interval traversed by each of the four voices (bass up to soprano) in the voice leading between the two sonorities. This representation has certain problems, however. First, we view it to be insufficiently abstract insofar as it hard-codes the ordering of the four voices. A chorale that is transformed by swapping the tenor and alto voices, for example, would receive a completely different representation. Most contemporaneous theories of harmony (e.g. [1]) do not encourage such a rigid differentiation between contrapuntal voices, holding instead that, in principle, voices are permutable. Second, Conklin's representation is too abstract in the following sense: imagine a Bach chorale transformed so that the entire bass line is transposed down a semitone, the soprano is transposed up a semitone, and the inner voices are left where they are. This operation would clearly have a devastating effect on the tonal structure of the chorale, though it would not affect Conklin's representation. Seeking a happy medium, Ian Quinn [5] developed a representation inspired by 17th-century figured-bass theory that takes the form of an ordered triple ($S_1$, $S_2$, $I$), where $S_1$ and $S_2$ are sets of intervals (in semitones modulo the octave) above the bass in two adjacent sonorities, and $I$ is the pitch-class interval from the first bass note to the second. This representation is key-independent and captures the voice-leading of the bass, as well as the pitch-class set identity of the two sonorities (up to transposition).

Rameau proposed that the contrapuntal voices in a voice-leading between two chords can be permuted without changing the functional relationship between the chords [1]. Inspired by this observation, we refined Quinn's figured-bass representation to capture more details of voice leading without privileging any one voice. In this report, we describe this representation, which we call the voice-leading type (VLT). We use this representation to further explore the question whether harmonic syntax can be learned from the chorale corpus, and in particular, whether function can be read out of voice leading.

Our exploration of the relationship between harmonic function and voice leading must address an important problem: whereas most theories of harmony involve a very small number of functions, the Bach chorale corpus involves thousands of distinct VLTs. This very large number of combinatorial possibilities must ultimately be reduced down to a much smaller number of functional categories. Moreover, as we have observed above, the general consensus is that harmonic function is determined by both chord structure and voice leading. If this is so, then it is reasonable to suppose that members of the same functional category will be similar in both of these respects. On this basis, we set out to investigate the following hypothesis: *a cluster analysis that reduces the number of VLTs to a manageable number of clusters, and that does so on the basis of a reasonable measure of voice-leading similarity, will afford an interpretation in terms of harmonic function.*

In order to investigate the differences between Bach's tonal style and its historical precedents, we also supplement the Bach corpus with a corpus of chorales from the pre-tonal era.

## 2 Methods

We examined two corpora of Lutheran chorales [6-10]. This genre is well-suited to the empirical study of harmony and voice-leading since many chorales are set in a four-voice texture that is strictly homophonic (all four voices sound together all the time, without staggered entrances). The **Modal corpus** consists of 404 four-voice homophonic chorales found in four collections published between 1586 and 1627 [6-9]. The earliest of these collections was the first in which the composer set the chorale melody in the soprano rather than the tenor. The **Bach corpus** consists of 353 chorales set by J.S. Bach and compiled by Riemenschneider [10]. Duplicates in the published sources were not duplicated in the corpora, and chorales not set in a strict four-voice texture were not included in the corpora. Each chorale in the corpora was encoded as a MIDI file.

We treated each chorale as a series of four-note chords (pitch-class sets with the possibility of doubling), identifying a new chord each time a note-on message occurred in the MIDI file. We did not restrict the identification of chords to those with roots, as did Tymoczko [2], nor did we discriminate between consonant and dissonant chords or pay attention to the metrical position of chords, as did Rohrmeier [3].

We addressed voice leading by representing each pair of chords as an unordered set of four ordered pairs of pitch classes. Each of these ordered pairs represents the voice-leading motion of an individual voice. These four ordered pairs were then transposed simultaneously so as to put the first chord in prime form [11]. The voice-leading type (VLT) was thus represented as an unordered set of four ordered pairs of pitch classes, one for each voice. For example, the two most frequently attested resolutions of a dominant seventh chord, regardless of the actual key, are represented as {(0,1), (3,1), (6,5), (8,8)} and {(0,8), (3,1), (6,5), (8,1)}. In this way, the VLT treats two voice-leadings as equivalent up to permutation of voices or uniform transposition.

The combined corpus contains 51,418 transitions between simultaneities, representing 4748 distinct VLTs. The frequencies of occurrence $f(i)$ of these VLTs follow a Zipfian distribution, which is characteristic of word frequencies in natural-language corpora [12]. In this type of distribution, the frequency with which each token occurs in the corpus is approximately inversely proportional to its rank on a frequency table of all tokens. Zipfian distributions have long and heavy tails—in our combined corpus, about half of all voice-leading types are *hapax legomena*—so an ideal cluster analysis of this sort of data would be able to give greater weight to the fewer high-frequency types than to the many low-frequency types. Since no generally accepted weighting scheme for cluster analysis exists, we roughly approximated its effect by simply trimming the low-frequency tail of the distribution.We set a cutoff of $f(i) \geq 15$, which accounts for just 11.9% of voice-leading types (a total of 567 out of 4748), but 77.7% of all transitions between simultaneities (39,940 of 51,418).

In order to make a cluster analysis on the space of VLTs frequently attested in the corpus, we needed a measure of similarity for VLTs. We preferred to have a similarity measure that introduced as few assumptions as possible about the nature of harmonic function: we did not want to assume any particular tonal (scale-degree) interpretation of the chords, nor did we want to make hierarchical distinctions between different types of chord tones (roots, thirds, fifths, sevenths) or between chord and non-chord tones. We settled on a very simple definition of similarity between two VLTs, calculated by fixing one VLT and taking the other through all twelve possible transpositions in order to find a transposition level that maximizes the number of ordered pairs shared by both VLTs. At the optimum level of relative transposition, two non-identical VLTs may have 0, 1, 2, or 3 common ordered pairs. We normalize this value by dividing by four and subtracting from 1, to obtain a dissimilarity value between 0 and 1. For example, the first two black-note chords of Fig. 2 overlap in three out of four voices, producing a dissimilarity value of 0.25.

We used this similarity measure to create a complete dissimilarity matrix for the 567 most frequent VLTs in our combined corpus. This matrix served as input for the *Partitioning Around Medoids* (PAM) clustering algorithm, a non-hierarchical technique for partitioning a set of $n$ items into $k < n$ clusters, implemented in the R language.

## 3 Results

The value for $k$ was selected using the *silhouette width* statistic, which summarizes the quality of the clustering. For each data point $i$, the silhouette $s(i)$ measures how well that particular point has been classified by quantifying how close that point is to other cluster members and how far it is from the nearest neighbor cluster. A well-classified point that is near to its fellow cluster members and far from the members of the nearest cluster has a silhouette of 1; a poorly-classified point that is as close to the members of its cluster as to the members of the nearest cluster has a silhouette of –1. The silhouette width of a cluster is the average of $s(i)$ over all points in the cluster. The overall quality of a clustering is estimated by averaging the silhouette widths of all $k$ clusters. We used PAM to perform $k$-medoids clustering of our 567 VLTs, with $2 \leq k \leq 100$. The graph of the silhouette widths plotted against $k$ is shown in Fig. 1. From this graph, two values of $k$ stand out: the first is $k$=34, determined by a local maximum near the shoulder of the curve as it approaches its peak value among higher values of $k$; the second value is $k$=5, representing the other clustering that achieves comparable silhouette width while maintaining enough clusters to afford a reasonable level of discrimination.

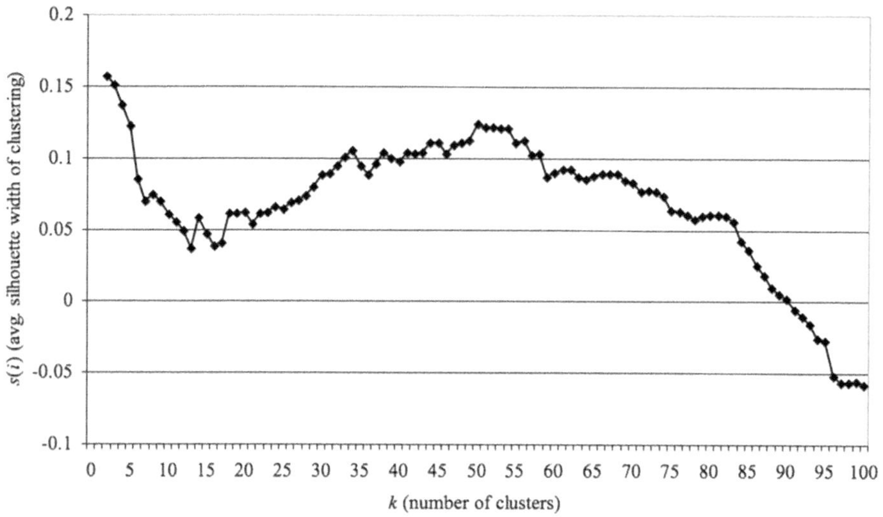


**Fig. 1.** Quality of $k$-medoids clustering of voice-leading types (VLTs) for $2 \leq k \leq 100$, as measured by average silhouette width of clusters

### 3.1 Five Clusters

Figure 2 displays a summary of the PAM clustering where $k = 5$. Each staff corresponds to a cluster, with representative member VLTs shown in black noteheads. The VLTs are realized using transpositions and registrations that maximize common pitch sequences across each staff. The white-note *voice-leading prototypes* at left informally characterize each cluster by showing the two or three most common pitch sequences in each cluster. To a great extent, the voice leadings in each cluster can be viewed as variations or elaborations of the prototypes shown. The rest of each line consists of three groups of five voice-leadings. First are the five voice-leadings with the widest silhouette (largest value of $s(i)$) in the cluster; these are interpreted as being the most *prototypical* of the cluster, in the sense that, on average, they are the closest to all other cluster members and the furthest away from non-members. Next are the five voice-leadings that are most frequently represented in the corpus, excluding the five most prototypical. Since each voice-leading is input only once into the clustering algorithm regardless of how many times it appears in the corpus, it is possible for an infrequent VLT to be highly prototypical (e.g. 4(*b*) in Fig. 2), or for a common VLT to be far from the center of a cluster (e.g. 3(*f*)–(*j*) in Fig. 2). At right are five additional VLTs that are representative of the cluster by virtue of being both high frequency and highly prototypical, without being in the top five of either category.

This method for classifying voice-leadings is simple, even simplistic, yet it can captures important aspects of the nuanced relationship between voice-leading and harmony. Annotations below the top staff of Fig. 2 show the root motion underlying each of the VLTs that represent Cluster 1. A plurality of these ($n$=7) correspond to descending-fifth motion, with motions by ascending whole- or half-step ($n$=5)

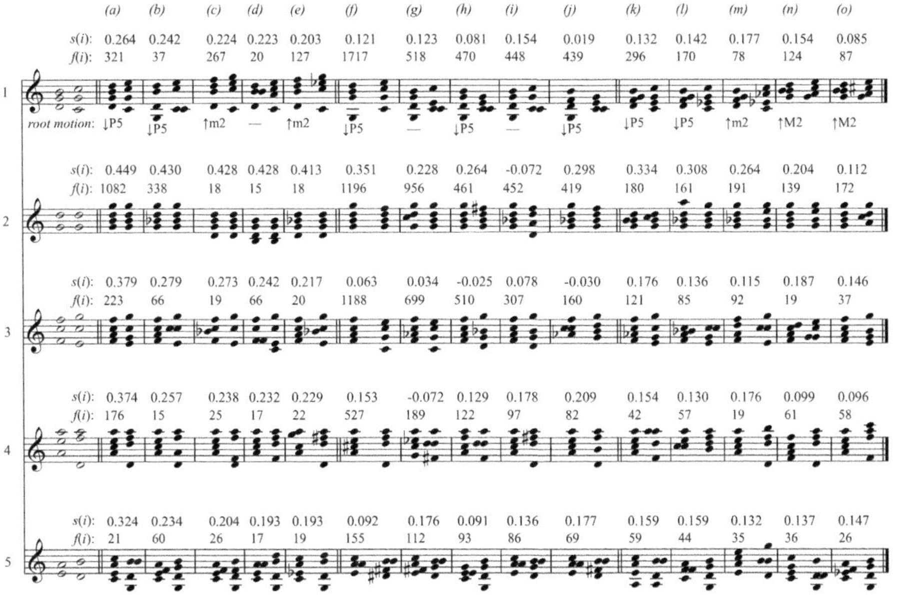


**Fig. 2.** VLTs characteristic of the clustering with $k = 5$. (*a*)–(*f*) are highly prototypical, (*f*)–(*j*) are frequently attested in the corpus, and (*k*)–(*o*) are other noteworthy VLTs. Root motions are indicated as appropriate below the staff. Silhouette width, $s(i)$, and frequency, $f(i)$, are indicated above the staff. White-note prototypes at left show common elements of VLTs in each cluster.

forming the next most populous category. The staff is filled out with three voice-leadings involving root motion that, due to the presence of incomplete chords, is indeterminate. The descending-fifth progressions as realized here can all be interpreted as V–I progressions in the key of C, and many can be interpreted as I–IV progressions in the key of G. While V–I and I–IV are clearly quite different progressions, they both feature syntactically prograde motion on the functional circle tonic (T) — subdominant (S) — dominant (D) — tonic (T), a type of motion Nicolas Meeùs [13] terms *dominant*, as opposed to the syntactically retrograde motions he calls *subdominant*. (To avoid confusion, we use *prograde* and *retrograde* to refer to motion between harmonic functions, and T, D, and S to refer to the functions themselves.) Progressions by ascending step are also in this category. Among those in Cluster 1 we find three D–T progressions: 1(c) and 1(e) resolve a leading-tone chord to the tonic and 1(m) resolves a dominant-seventh chord to the submediant in a minor-mode deceptive cadence. We also find a T–S progression at 1(n) and a S–D progression at 1(o).

We have seen that Cluster 1, despite a variety of root motions and chord functions, seems to bring together progressions that are highly typical of the functionally prograde category. Such progressions are also found in Cluster 4, though these often have a specific intervallic quality thanks to the ascending semitone realized as E–F in the voice-leading prototype. Table 1 summarizes the root motions found in each

**Table 1.** Root motions associated with voice-leading types in the clustering with $k$=5

| root motion | cluster | | | | | total |
|---|---|---|---|---|---|---|
| | 1 | 2 | 3 | 4 | 5 | |
| unison | | 42 | | | | 42 |
| $\uparrow$m2 | 19 | | | 1 | 2 | 22 |
| $\uparrow$M2 | 26 | 1 | 16 | 6 | 14 | 63 |
| $\uparrow$m3 | | 16 | | | 1 | 17 |
| $\uparrow$M3 | | 6 | 3 | | | 9 |
| $\downarrow$P5 | 55 | 6 | | 20 | 1 | 82 |
| tritone | 4 | 4 | | | | 8 |
| $\uparrow$P5 | 6 | 6 | 23 | | 11 | 46 |
| $\downarrow$M3 | 3 | 6 | | 5 | | 14 |
| $\downarrow$m3 | 1 | 15 | | 6 | | 22 |
| $\downarrow$M2 | 14 | | 3 | | 7 | 24 |
| $\downarrow$m2 | 2 | | 3 | | 1 | 6 |
| — | 41 | 133 | 26 | 4 | 8 | 212 |
| total | 171 | 235 | 74 | 42 | 45 | 567 |

cluster. Clusters 1 and 4 both feature predominantly prograde root progressions, though they have two major differences in their root-motion profiles. First, progressions by ascending step are much more likely to be classified in Cluster 1 than Cluster 4. Second, the descending-step motions found in Cluster 1 (retrograde motions in a cluster of mostly prograde motions) are not found in Cluster 4. Both of these differences can be traced to the characteristic ascending semitone of Cluster 4.

Just as Clusters 1 and 4 are characterized by prograde harmonic motion, Clusters 3 and 5 are characterized by retrograde motions: ascending fifths and descending steps, though there are also a number of ascending-step progressions in both clusters. Cluster 2 contains most of the VLTs in the corpus with a root progression of an ascending or descending third and all of the VLTs connecting two chords with the same root. The cluster is characterized by harmonic motion that is neither prograde nor retrograde, thanks to an abundance of common tones [14]. Three common voice-leading patterns in Cluster 2 are 8–7 motions that turn a triad into a seventh chord, as at 2(f), 2(h), and 2(j); 5–6 motions that turn a triad into another triad with a root a third lower, as at 2(m) and 2(n); or the trivial voice-leading between repeated chords, as at 2(a)-(e) in Fig. 2. Much of the remainder of Cluster 2 consists of miscellaneous VLTs that arise from passing tones and neighbor tones.

### 3.2 Thirty-Four Clusters

While the five-fold clustering generally reflected a distinction between VLTs that are harmonically prograde, retrograde, and static, some anomalies arose. Clusters 1 and 4, despite a general tendency to include harmonically prograde VLTs, also included retrograde root motions by descending step; Clusters 3 and 5 included most of the retrograde VLTs but also a significant number of prograde ascending-step progressions.

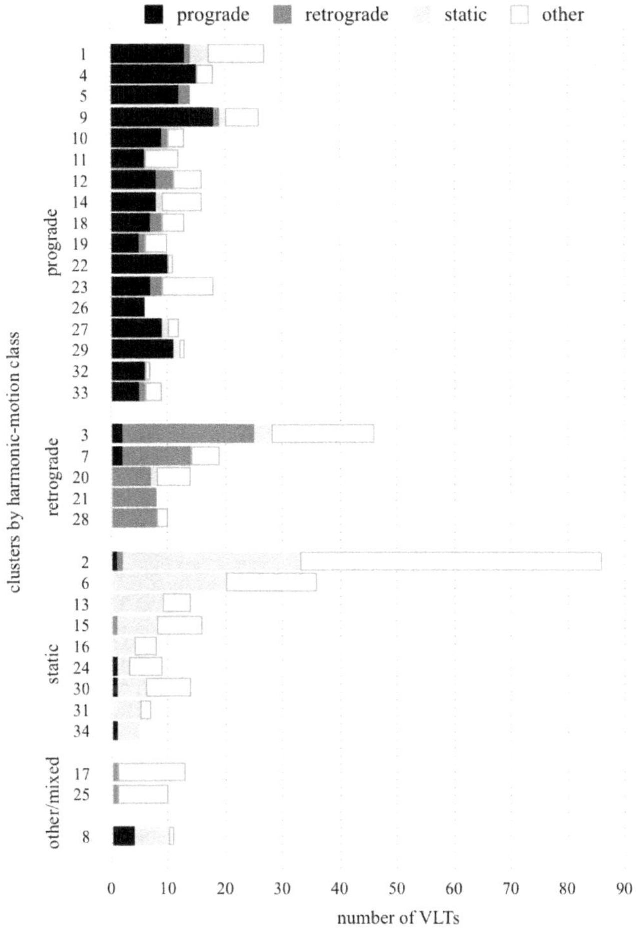


**Fig. 3.** Distribution of VLTs over clusters ($k$=34), by harmonic-motion class. Most clusters are dominated by a single class of VLT.

**Table 2.** Coverage of the corpora by each cluster ($k$=5). The log ratio of the two coverage figures indicates whether the cluster is more characteristic of one corpus than the other (positive = M, negative = B).

| cluster | $cov_M$ | $cov_B$ | $\log(cov_M/cov_B)$ |
|---|---|---|---|
| 3 | 0.1955 | 0.0937 | 0.32 |
| 4 | 0.0697 | 0.0417 | 0.22 |
| 5 | 0.0481 | 0.0337 | 0.15 |
| 1 | 0.3660 | 0.3108 | 0.07 |
| 2 | 0.3207 | 0.5201 | -0.21 |
| total | 1.0000 | 1.0000 | |

These conflicts are resolved in the 34-fold clustering. Figure 3 summarizes the distribution of the 567 VLTs under consideration over the 34 clusters, classifying each VLT as *prograde* (root motion by descending fifth or ascending step), *retrograde* (ascending fifth or descending step), *static* (unison or third), or *other* (root motion by tritone, or VLTs where one or both of the simultaneities has no determinate root). Confining ourselves to the first three classes, we observe that only one cluster contains a substantial number of representatives of more than one class, and the rest have a clear functional interpretation: cluster 1 is dominated by prograde VLTs,

**Table 3.** Coverage of the corpora by each cluster ($k$=34). The log ratio of the two coverage figures indicates whether the cluster is more characteristic of one corpus than the other (positive = M, negative = B).

| cluster | class | $cov_M$ | $cov_B$ | $\log(cov_M/cov_B)$ |
|---|---|---|---|---|
| 34 | **S** | 0.0091 | 0.0003 | 1.49 |
| 21 | **R** | 0.0141 | 0.0012 | 1.06 |
| 26 | **P** | 0.0090 | 0.0010 | 0.94 |
| 7 | **R** | 0.0636 | 0.0109 | 0.76 |
| 8 | -- | 0.0317 | 0.0064 | 0.70 |
| 29 | **P** | 0.0163 | 0.0067 | 0.39 |
| 31 | **S** | 0.0071 | 0.0030 | 0.37 |
| 10 | **P** | 0.0386 | 0.0164 | 0.37 |
| 3 | **R** | 0.1598 | 0.0696 | 0.36 |
| 4 | **P** | 0.0734 | 0.0375 | 0.29 |
| 1 | **P** | 0.1295 | 0.0675 | 0.28 |
| 28 | **R** | 0.0097 | 0.0053 | 0.26 |
| 5 | **P** | 0.0333 | 0.0188 | 0.25 |
| 13 | **S** | 0.0251 | 0.0155 | 0.21 |
| 16 | **S** | 0.0149 | 0.0118 | 0.10 |
| 22 | **P** | 0.0096 | 0.0119 | -0.09 |
| 2 | **S** | 0.1666 | 0.2470 | -0.17 |
| 18 | **P** | 0.0137 | 0.0203 | -0.17 |
| 23 | **P** | 0.0112 | 0.0195 | -0.24 |
| 20 | **R** | 0.0116 | 0.0207 | -0.25 |
| 14 | **P** | 0.0154 | 0.0287 | -0.27 |
| 12 | **P** | 0.0189 | 0.0359 | -0.28 |
| 32 | **P** | 0.0034 | 0.0066 | -0.29 |
| 27 | **P** | 0.0076 | 0.0157 | -0.32 |
| 33 | **P** | 0.0040 | 0.0086 | -0.33 |
| 15 | **S** | 0.0136 | 0.0297 | -0.34 |
| 11 | **P** | 0.0123 | 0.0272 | -0.35 |
| 6 | **S** | 0.0301 | 0.0798 | -0.42 |
| 9 | **P** | 0.0263 | 0.0756 | -0.46 |
| 19 | **P** | 0.0044 | 0.0152 | -0.54 |
| 17 | -- | 0.0070 | 0.0243 | -0.54 |
| 30 | **S** | 0.0049 | 0.0270 | -0.74 |
| 25 | -- | 0.0035 | 0.0192 | -0.74 |
| 24 | -- | 0.0005 | 0.0153 | -1.48 |
| total | | 1.0000 | 1.0000 | |

cluster 2 by static VLTs, cluster 3 by retrograde VLTs, and so on. (Cluster 8, with 4 prograde VLTs and 6 static VLTs, is the sole exception).

### 3.3 Comparison of the Corpora

The two corpora used in this study have a crucial difference: Bach's chorales were written a century later than the chorales in the modal corpus, and this intervening century is generally thought to have witnessed the widespread consolidation of functional harmonic tonality in Europe. Since we claim that the VLT clusters we have discovered are bound together by principles related to harmonic function, we should expect to see differences in the way these VLT clusters are represented in the two corpora.

To test this, we determined each cluster's *coverage* of each corpus: i.e. the number of times VLTs in each cluster were used in each corpus. We then normalized the coverage value, dividing it by the total number of times any of the 567 most frequent VLTs was instantiated in each corpus. We symbolize the normalized coverage values for the modal and Bach corpora as $cov_M$ and $cov_B$, respectively. Finally, we estimated the degree to which a cluster is more characteristic of one corpus than the other by taking the logarithm of $cov_M/cov_B$. A positive value means the cluster's VLTs are more characteristic of the modal corpus; likewise for negative values and the Bach corpus.

Our findings are summarized in Tables 2 and 3. Consistent with our expectations, we found a great deal of variance in the log ratio of the coverage values, suggesting that the VLT clusters are unevenly represented in the two corpora. In particular, 4 of 5 clusters associated with retrograde harmonic motion are skewed toward the modal corpus, while 11 of 17 clusters associated with prograde harmonic motion are skewed toward the Bach corpus.

## 4 Conclusions

The VLT representation, in conjunction with a cluster analysis based on a simple similarity relation, has allowed us to recover from our chorale corpora a threefold classification of harmonic motions (prograde, retrograde, static) without building into the model any assumptions about harmonic function. We have also seen evidence that the 34-fold classification of voice leadings produced by our cluster analysis may prove useful in the investigation of syntactic differences between Bach's tonal chorales and the modal chorales composed a century earlier. Encouraged by these findings, we intend to continue this work by using the clusters described here as a basic alphabet for finite-state models of the syntactic structure of these two corpora.

## Acknowledgments

We thank Chris White and the anonymous referees for their comments on an earlier draft of this paper.

## References


1. Rameau, J.-P.: Traité de l'harmonie réduite à ses principes naturels, Paris (1722)
2. Tymoczko, D.: Progressions fondamentales, functions, degrés, une grammaire de l'harmonie tonale élémentaire. Musurgia 10, 35–64 (2003)
3. Rohrmeier, M., Cross, I.: Statistical Properties of Tonal Harmony in Bach's Chorales. In: Proceedings of the 10th International Conference on Music Perception and Cognition, pp. 619–627. ICMPC, Sapporo (2008)
4. Conklin, D.: Representation and Discovery of Vertical Patterns in Music. In: Anagnostopoulou, C., Ferrand, M., Smaill, A. (eds.) ICMAI 2002. LNCS (LNAI), vol. 2445, pp. 32–42. Springer, Heidelberg (2002)
5. Quinn, I.: Are Pitch-Class Profiles Really Key for Key? Zeitschrift der Gesellschaft der Musiktheorie 7, 151–163 (2010)
6. Osiander, L.: Fünfftzig Geistliche Lieder und Psalmen, Nürnberg (1586)
7. Praetorius, H., et al.: Melodeyen Gesangbuch, Hamburg (1604)
8. Hassler, H.L.: Kirchengesäng, Nürnberg (1608)
9. Schein, J.: Cantional oder Gesangbuch Augspurgischer Confession, Leipzig (1627)
10. Bach, J.S., Riemenschneider, A. (eds.): 371 Harmonized Chorales and 69 Chorale Melodies with Figured Bass. Schirmer, New York (1941)
11. Forte, A.: The Structure of Atonal Music. Yale University Press, New Haven (1973)
12. Zipf, G.K.: The Psycho-biology of Language. Houghton-Mifflin, Boston (1935)
13. Meeùs, N.: Toward a Post-Schoenbergian Grammar of Tonal and Pre-tonal Harmonic Progressions. Music Theory Online 6(1) (2000)
14. Agmon, E.: Functional Harmony Revisited: A Prototype-Theoretic Approach. Music Theory Spectrum 17, 196–214 (1995)

# Emergent Formal Structures of Factor Oracle-Driven Musical Improvisations

Isaac Schankler[1], Jordan B.L. Smith[1], Alexandre R.J. François[2], and Elaine Chew[1]

[1] University of Southern California, Los Angeles, CA, USA
{schankle,jordans,echew}@usc.edu
[2] Harvey Mudd College, Claremont, CA, USA
alex@cs.hmc.edu

**Abstract.** In this article, improvisations created with the factor oracle, a commonly used data structure in machine models of musical improvisation, are shown to exhibit certain formal structures independent of the musical content. We posit that these structures are in fact emergent properties of the behavior of the factor oracle itself. An expert improviser (the first author) performed a series of improvisations with Mimi, a factor oracle-driven multimodal system for human-machine improvisation, and the formal structures of each performance was independently analyzed by the performer and an experienced music structure annotator (the second author). Quantitative assessment of the similarity between the performer's and the listener's analyses was carried out using techniques from the field of automatic structure analysis. Supported by a comparison to baseline analysis approaches, the results suggest a high level of agreement between the two sets of analyses. Drawing upon this foundation of evidence, we discuss these analyses and their relationship to common classical forms, including canon- and rondo-like forms, as well as forms based on the juxtaposition of rhythmic cells.

**Keywords:** Improvisation, Factor oracle, Musical structure, Emergent structure.

## 1 Introduction

When human improvisers perform, they frequently draw on a stock vocabulary of motives, gestures, and phrases. Ideas are recombined, often with very little creation of source material. The factor oracle is an efficient data structure used to model this recombination aspect of musical improvisation using an online construction algorithm and a stochastic traversal algorithm [1]. Assayag and Dubnov were the first to propose and use factor oracles in music improvisation with Omax [2,3].

While there has been considerable development of improvisation engines that make use of the factor oracle, there has as of yet been little study of the products of these engines, i.e. the improvisations themselves, and of whether they actually

C. Agon et al. (Eds.): MCM 2011, LNAI 6726, pp. 241–254, 2011.

resemble improvisations created by human performers. This article examines the large-scale forms created by factor oracle-driven improvisations to determine if any regular or recognizable structural patterns arise. We posit that these structural patterns, if consistently observed, could be emergent properties of the behavior of the factor oracle, or of the particular kind of interaction that exists between the oracle and a performer. Knowing these patterns can lead us to a deeper understanding of the formal structures in musical improvisation and of the nature of human-machine improvisation, and help us advance the development of more sophisticated models of improvisation.

The remainder of this section introduces the Mimi improvisation system and surveys some related research. Section 2 describes how this research was carried out, including the recording and analysis of three improvised performances and the quantitative tools used to compare them. The analyses themselves, the results of their comparison, and the small- and large-scale formal structures oberved in the analyses are reported in Section 3. Section 4, a discussion following the results, concludes the article.

### 1.1 The Mimi System

For this study, the particular factor oracle-driven improvisation engine used was the Mimi (Multimodal Interaction for Musical Improvisation) system [4,5] created by François and collaborators using the Software Architecture for Immersipresence framework for interactive software systems. Mimi was designed for human-machine improvisation on a keyboard or some other MIDI instrument.

While Mimi draws inspiration from the growing family of improvisational systems that use the factor oracle, it differs from its siblings in a few significant ways. In the various incarnations of the OMax improvisation system, the machine learns continually from the human improviser, i.e. the factor oracle continues to grow through the duration of the performance. A human operator at the computer acts as a second performer, managing in real-time the recombination rates (i.e., the frequency with which skips are made while traversing the oracle) and the section of the oracle from which the machine draws musical material.

In contrast, the Mimi systems relegate full operational control to the improvising musician. The interaction modality and format are designed so that the human performer is self-sufficient. They also relegate more direct control to Mimi itself, and by extension the factor oracle algorithm. The human performer, for example, has no control over which section of musical material Mimi will choose to perform. (The human performer does, however, have the option at any moment to permanently clear Mimi's memory of the existing musical material.) This makes Mimi an ideal choice for the study of structure, since the structure of the improvisations will more closely reflect the structure of the factor oracle itself.

The Mimi systems were the first, and currently still the only, factor oracle-based systems to experiment with real-time visualization to aid the performer at the piano keyboard in the act of planning and orchestrating an improvisation. Mimi employs a piano roll-style visual display that gives the user information about the current state of the improvisation engine, a ten-second heads-up on

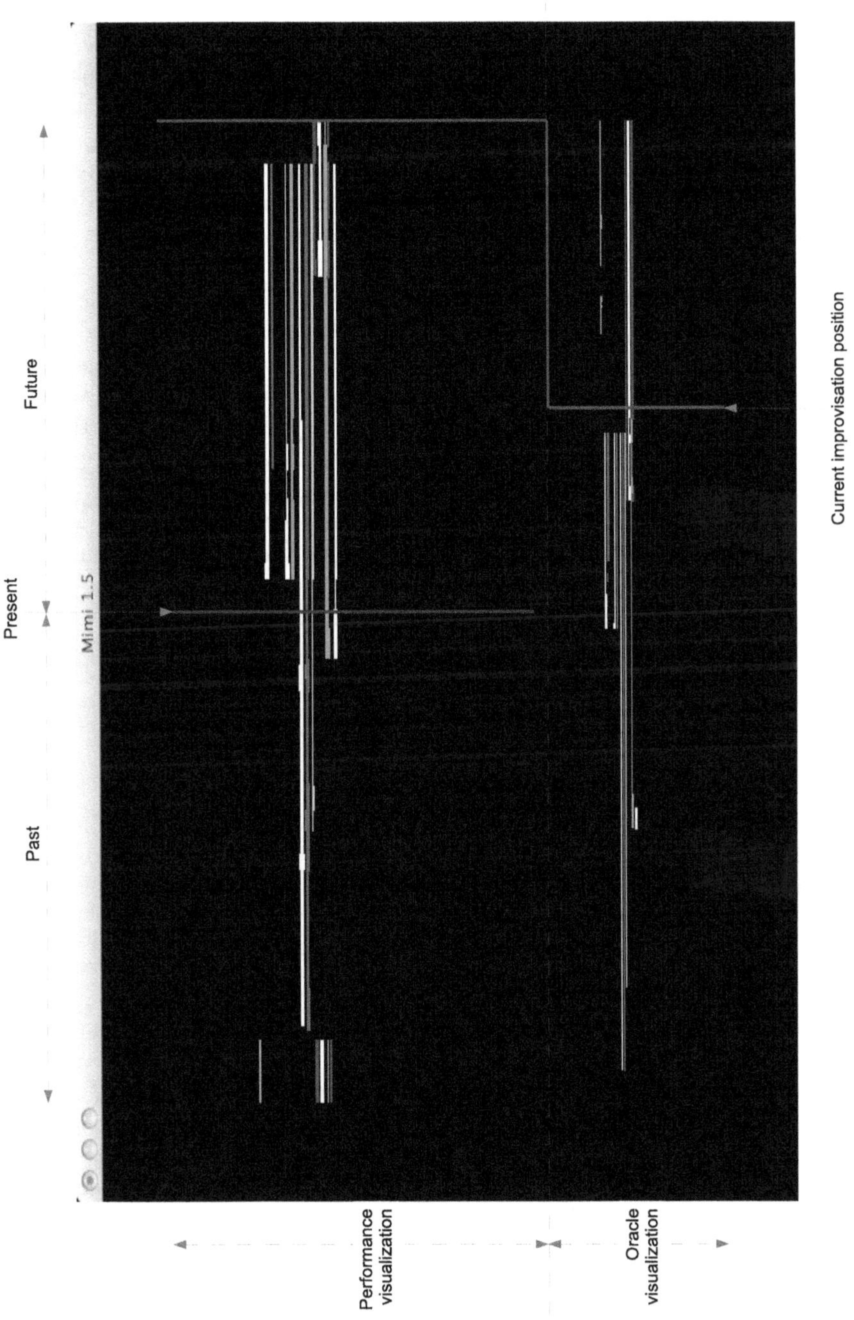


**Fig. 1.** Annotated screenshot of Mimi's visualization during performance

the music the machine will soon play, and ten seconds to review the music that the machine and the user recently played together, as shown in Fig. 1.1.

An alternative visualization scheme for the factor oracle which draws arcs representing suffix links between different sections of material is provided in WoMax[6], another offshoot of OMax. WoMax offers an overview of the connections created by the factor oracle. In the context of an ongoing improvisation, WoMax's visualization is most useful to a second performer (the one at the computer) who is focused on manipulating the audio generated by the oracle. The goal of Mimi's visualization is to provide timely feedback to a single performer focused on coordinating his or her musical ideas with the oracle's.

In the Mimi setup, the human improviser lays down seed material that the machine recombines to generate new music in the style of the seed material. The user, which could be the original human improviser or another person who chooses to use that particular set of seed material, can then improvise over the music that the machine generates. In the Mimi 1.5 system used in the performances described in this article, notes are color coded by pitch class, pedaling is recorded in the seed material, and volume (attack velocity) is indicated through the thickness of the line segments denoting each sounding tone. A MIDI controller provides an intuitive interface for the performer to have quick access to system commands and parameters before and during performance. Through the MIDI controller, the performer can make on-the-fly adjustments to Mimi's learning state (on, off), improvisation state (start, stop), memory (clear), recombination rate and playback volume.

## 2 Procedure

This section provides the details for the experimental and analytical procedures employed, including the recording of the human-machine improvisations, the analyses of these performances, and the quantitative analysis of the resulting annotations. The determination of formal structures, such as rondo or canon forms, was done by inspection based on the structure analyses.

### 2.1 Recording of Human-Machine Improvisations

Over a period of three weeks, an expert keyboard improviser (Isaac Schankler, the first author), produced three separate improvisations, hereafter referred to as Performance 1, 2 and 3, with Mimi. Two recordings of each session were produced: a MIDI recording, with the human improviser's and Mimi's performances saved in separate channels, and a video of Mimi's visual output, recorded using screen-capture software. The two MIDI tracks were used to synthesize a single audio track essentially identical to what the performer heard during the improvisation, and this merged result was the recording consulted during the annotation stage.

It bears mention that the improviser was a practiced Mimi user at the time of the recording of these performances. The performer's experience using Mimi extended several months prior to these recordings, and included several public performances, so that he was already accustomed to working with Mimi's behavior in order to craft coherent improvisations.

### 2.2 Annotation of Performances

To assess how, and the extent to which, the improvisations with Mimi were structured, the formal structure of each piece was independently analyzed and annotated by the performer and an experienced music structure annotator (Jordan Smith, the second author), hereafter referred to as Annotator 1 and Annotator 2, respectively. Annotator 2 has previously performed and studied analysis of large-scale music structure in his Master's research [7] and as part of the Structural Analysis of Large Amounts of Music Information (SALAMI) project[1].

In order to quantitatively compare the structural descriptions, they had to be encoded in a machine-readable format, preferably aligned with the audio. We therefore constrained the format of the analyses to disallow the overlapping of sections, to require that all portions of a piece be assigned a label, and to require that boundaries at a given hierarchical level were respected at smaller-scale (lower) levels. These requirements are akin to Lerdahl and Jackendoff's well-formedness rules for structural grouping analysis [8].

Each piece was analyzed at two hierarchical levels, the large-scale level was annotated with uppercase letters, and the small-scale level with lowercase letters. In each level, identical labels indicate portions of the piece that were judged to contain similar musical material. These analyses were entered by each annotator into the Variations Audio Timeliner [9], a tool for creating easy-to-read diagrams of formal musical structures, where the section boundaries were precisely aligned with the audio file.

### 2.3 Comparison of Annotations

To quantitatively estimate the correspondence between Annotator 1's and Annotator 2's analyses, the two sets of annotations were compared using four metrics that are commonly used to evaluate automatic structure analysis algorithms. The measures are: (1) the boundary $f$-measure; (2) the pairwise $f$-measure; (3) the $K$-measure; and, (4) the Rand index. Brief descriptions of each follow; details on their implementation can be found in Lukashevich [10].

The boundary $f$-measure gives the quality of the match between boundary locations. It is the harmonic mean of *precision*, the percentage of estimated boundaries that are correct, and *recall*, the percentage of annotated boundaries that were retrieved. By convention, either a 1-second or 6-second margin is used to determine whether a matching boundary was found.

The other three quantities measure the similarity between the labels provided by the two descriptions. The pairwise $f$-measure estimates the similarity of two sets of labels. Suppose the two descriptions are divided into very short time windows. Consider the set of all pairs of windows in the annotated description with the same label – as an example, these are indicated by the arcs in Fig. 2.

[1] http://salami.music.mcgill.ca

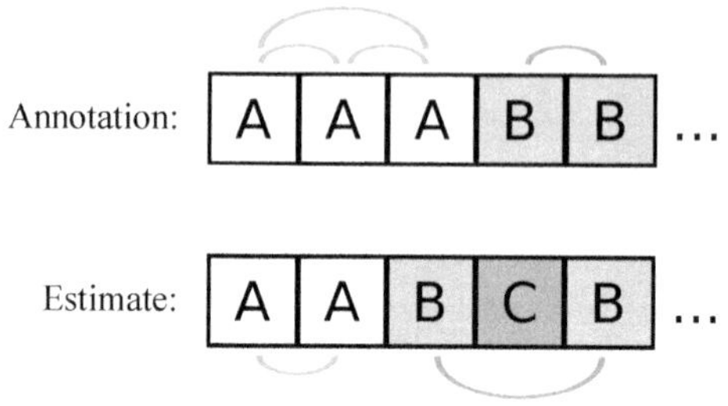


**Fig. 2.** Example of two sequences of windows being compared. Each arc connects two matching time segments. Precision is 50% because 1 of the 2 estimated pairwise matches were correct, and recall is 25% because 1 of the 4 matches in the annotation were retrieved. Their combined $f$-measure is thus the harmonic mean of the two, 33%.

We may then calculate the precision and recall with which this set of similarity relationships has been retrieved.

The $K$-measure follows a similar derivation but is harsher when a labeling error matches one section in a description with many in another, rather than just two or three. The $K$-measure is based on "average cluster purity" and "speaker purity," terms analogous to precision and recall, respectively, and borrowed from a comparable metric used in speech processing research. The Rand index also resembles the pairwise $f$-measure, except that it additionally measures how well the set of dissimilarity relationships was retrieved.

The values output by each metric lie on a scale from 0 to 1, with 1 indicating the best and 0 the worst possible agreement between descriptions. However, the values alone are not very informative without a baseline against which to compare, since previous evaluations have shown that even random descriptions can be quite similar to a given description [7].

Baseline segmentation approaches include placing boundaries every $n$ seconds (for various values of $n$), placing ten evenly-spaced boundaries throughout the piece, placing ten boundaries randomly, and placing no boundaries. For each segmentation, two labeling approaches apply: the segments may all be given the same label, or random labels may be drawn from a set of arbitrary size $m$. To set $m$ fairly, when the baseline was measured against Annotator 1, $m$ was the size of the set of labels in the matching description by Annotator 2, and vice versa. All of the above baseline segmentation and labeling approaches were used to generate descriptions that were compared to each human-generated description.

## 3 Results

### 3.1 Annotation of Performances

Figures 3 through 5 show the Variations Audio Timeliner diagrams created by the annotation method described in Sect. 2.2, in which differently colored "bubbles" represent different sections.

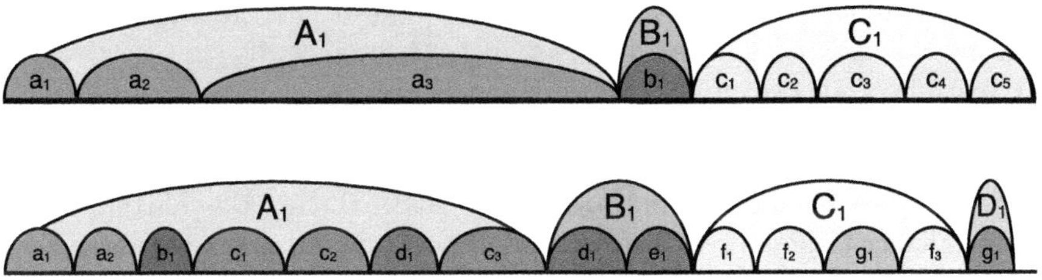


**Fig. 3.** Analysis of Performance 1 (top: Annotator 1; bottom: Annotator 2)

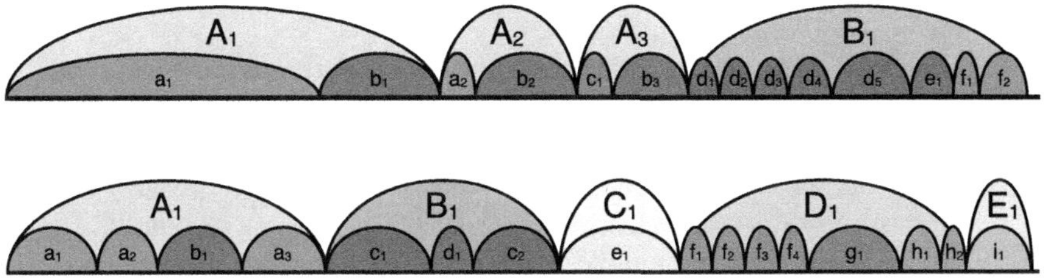


**Fig. 4.** Analysis of Performance 2 (top: Annotator 1; bottom: Annotator 2)

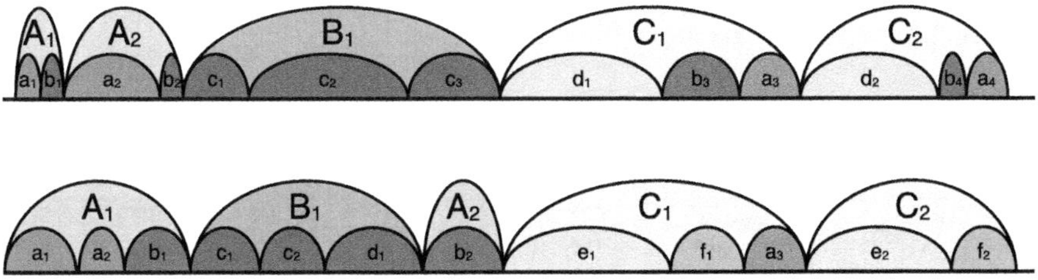


**Fig. 5.** Analysis of Performance 3 (top: Annotator 1; bottom: Annotator 2)

### 3.2 Comparison of Annotations

Although they were produced independently, the structural descriptions generated by the two annotators were quite similar one to another for each piece. Inspection of the three pairs of analyses, pictured in Figs. 3–5, reveals that both Annotator 1 and Annotator 2 perceived similar large-scale structures. While the small-scale structure labels are less clearly alike, both annotators at least appeared to strongly agree on the placement of the boundaries. This is not a trivial result: given that the pieces were improvisations with no exact repetition (recall that Mimi's and the performer's tracks were merged), and given that the annotators did not consult each other except to establish a common format, one might not have expected the analyses to concur significantly.

These observations are reflected in the quantitative measurements as described below. They indicate that the placement of small-scale boundaries was generally quite similar between listeners, with an average $f$-measure of 0.737 using a window of plus or minus three seconds. Labeling was also quite similar, especially at the larger scale: across all performances, the average pairwise $f$-measure, $K$-measure and Rand index were 0.79, 0.80 and 0.85, respectively.

The detailed results are reported in Tables 1–5. The baseline generation method that matched the human annotations best overall was that in which ten boundaries were placed randomly and the labels were randomized. Because this baseline is random, the values reported in Tables 1–5 are the average over 20 trials. The standard deviation of these values ranged from 0.03 to 0.10, with an average of 0.069 and a median of 0.07.

As can be seen from Tables 1–5, in all but one case the two listeners' descriptions were more similar to each other than either was to the baseline description, and usually by a wide margin. These results suggest that the similarity between the listeners' descriptions is not accidental, and support the claim that Figs. 3–5 are a fair representation of the formal structure of the corresponding performances. In the following section, we speculate as to why these structures may have arisen from the interaction between Mimi and the performer.

**Table 1.** Estimated boundary $f$-measure ($\pm$ 1s) between the two annotators' descriptions (Ann1-Ann2), and between the baseline and each annotator's description (baseline-Ann1, baseline-Ann2). Results are reported for the small- and large-scale annotations of each performance. Values range between 0 and 1, with 1 being the optimal value. Highest values are boldfaced.

| Boundary $f$-measure $\pm$ 1 second | Performance 1 | | Performance 2 | | Performance 3 | |
|---|---|---|---|---|---|---|
| Scale: | Small | Large | Small | Large | Small | Large |
| Ann1 - Ann2 | **0.27** | **0.29** | **0.28** | **0.40** | **0.39** | **0.53** |
| baseline - Ann1 | 0.02 | 0.01 | 0.02 | 0.00 | 0.05 | 0.03 |
| baseline - Ann2 | 0.02 | 0.01 | 0.02 | 0.01 | 0.04 | 0.04 |

**Table 2.** Estimated boundary $f$-measure ($\pm$ 6s) between each pair of descriptions. See Table 1 for details.

| Boundary $f$-measure $\pm$ 6 second | Performance 1 | | Performance 2 | | Performance 3 | |
|---|---|---|---|---|---|---|
| Scale: | Small | Large | Small | Large | Small | Large |
| Ann1 - Ann2 | **0.74** | **0.50** | **0.65** | **0.40** | **0.82** | **0.85** |
| baseline - Ann1 | 0.18 | 0.10 | 0.14 | 0.09 | 0.25 | 0.17 |
| baseline - Ann2 | 0.24 | 0.15 | 0.13 | 0.10 | 0.29 | 0.17 |

**Table 3.** Estimated pairwise $f$-measure between each pair of descriptions. See Table 1 for details.

| Pairwise $f$-measure, | Performance 1 | | Performance 2 | | Performance 3 | |
|---|---|---|---|---|---|---|
| Scale: | Small | Large | Small | Large | Small | Large |
| Ann1 - Ann2 | 0.47 | **0.85** | **0.66** | **0.62** | **0.68** | **0.90** |
| baseline - Ann1 | **0.50** | 0.51 | 0.50 | 0.52 | 0.41 | 0.48 |
| baseline - Ann2 | 0.34 | 0.50 | 0.40 | 0.43 | 0.36 | 0.48 |

**Table 4.** Estimated $K$-measure between each pair of descriptions. See Table 1 for details.

| $K$-measure | Performance 1 | | Performance 2 | | Performance 3 | |
|---|---|---|---|---|---|---|
| Scale: | Small | Large | Small | Large | Small | Large |
| Ann1 - Ann2 | **0.60** | **0.84** | **0.70** | **0.70** | **0.68** | **0.87** |
| baseline - Ann1 | 0.56 | 0.53 | 0.55 | 0.56 | 0.44 | 0.51 |
| baseline - Ann2 | 0.43 | 0.54 | 0.50 | 0.50 | 0.42 | 0.52 |

**Table 5.** Estimated Rand index between each pair of descriptions. See Table 1 for details.

| Rand index | Performance 1 | | Performance 2 | | Performance 3 | |
|---|---|---|---|---|---|---|
| Scale: | Small | Large | Small | Large | Small | Large |
| Ann1 - Ann2 | **0.70** | **0.90** | **0.88** | **0.71** | **0.87** | **0.93** |
| baseline - Ann1 | 0.67 | 0.63 | 0.76 | 0.59 | 0.68 | 0.58 |
| baseline - Ann2 | 0.58 | 0.61 | 0.72 | 0.52 | 0.63 | 0.58 |

### 3.3 Formal Structures

This section describes the formal structures deduced from the analyses shown in Figs. 3 through 5, and posits some attributes of Mimi's behavior that may contribute to the creation of these structures.

**The Canon.** The canon is one of the forms most obviously and fundamentally related to Mimi's performance habits. Consider the beginning of an improvisation session with Mimi. By design, Mimi's actions are delayed by 10 seconds to give the performer a visual heads-up on what Mimi is about to perform. If Mimi chooses not to recombine any of the subsequent material—a distinct possibility, especially if the recombination rate is low—Mimi's performance will be an exact copy of the original performance, delayed by 10 seconds, as shown in Fig. 6.

In other words, Mimi will create a canon at the 10-second level with the original performer, regardless of what the performer chooses to play.

| | | | | | | | |
|---|---|---|---|---|---|---|---|
| Human: | A | B | C | D | E | F | etc. |
| Mimi: | | A | B | C | D | E | etc. |

**Fig. 6.** Canon between Mimi and human performer

This pattern was found in Performance 2 (see Fig. 4), where a near-exact but delayed repetition contributed to the sense of regular phrasing (segments $d_1, d_2, d_3 \ldots$ in Annotator 1's analysis, $f_1, f_2, f_3 \ldots$ in Annotator 2's).

Even when this exact sequence does not occur, Mimi's habits create an abundance of canon-like, or heavily imitative, forms, such as that shown in Fig. 7. Imitative forms can be described as those in which fragments of musical material are passed from one voice to another in a polyphonic texture—in the case of Fig. 7, it is passed from the original performer to Mimi. While this and further examples in Figs. 8 through 10 are hypothetical examples, they serve to illustrate patterns observed in the structure analyses.

| | | | | | | | |
|---|---|---|---|---|---|---|---|
| Human: | A | B | C | D | E | F | etc. |
| Mimi: | | A | B | C | A | D | etc. |

**Fig. 7.** Canon-like form created by Mimi and human performer

| | | | | | | | | |
|---|---|---|---|---|---|---|---|---|
| Human: | A | B | C | D | E | F | G | etc. |
| Mimi: | | A | B | A | B | A | C | etc. |

**Fig. 8.** Rondo-like form with Mimi and human performer (through-composed)

| | | | | | | | | |
|---|---|---|---|---|---|---|---|---|
| Human: | A | B | C | A | D | A | E | etc. |
| Mimi: | | A | B | A | C | A | D | etc. |

**Fig. 9.** Rondo-like form with Mimi and human performer (rondo-like)

| | | | | | | | | | | | |
|---|---|---|---|---|---|---|---|---|---|---|---|
| Human: | A | B | C | A | D | E | F | G | E | H | etc. |
| Mimi: | | A | B | A | C | | E | F | E | G | etc. |

**Fig. 10.** Large-scale binary form with nested rondo-like forms

**The Rondo.** As Mimi begins to revisit different sections of musical material, more formal resemblances may emerge. In its conception, Mimi does not privilege any material over any other, but Mimi's design allows it to learn continuously from the performer during the course of an improvisation, if the performer chooses to allow it. In this scenario, material that enters into Mimi's memory earlier will more likely be heard more frequently by virtue of simply having been in memory for a longer period of time. 'A' is more likely to be heard than 'B,' which is more likely to be heard than 'C,' and so on.

As a result, over the course of a typical three- to five-minute improvisation, the very first thing Mimi records is likely to be the most significant piece of musical material. If Mimi revisits this initial idea often enough, it may take on the quality of a refrain, creating a rondo-like form in which the initial idea recurs several times over the course of a performance, interspersed with different, successive episodes.

Performance 3 (see Fig. 5) presented an instance of this design, with both annotators identifying a regular recurrence of material from the opening 'A' section. This will happen quite often even if the human performer proceeds in a through-composed fashion (not revisiting ideas), as shown in Fig. 8. However, a sensitive performer may also choose to respond to Mimi's behavior in a way that brings out these formal divisions, as shown in Fig. 9.

**Formal Divisions.** The performer may also choose to clear Mimi's memory at some point, creating a blank slate to be populated with new material. In this way, the performer is able to create large-scale formal divisions between sections that do not share the same pool of material. Figure 10 gives an example of a large-scale binary form with nested rondo-like forms.

This kind of formal division is the most salient aspect of Performance 1 (see Fig. 3), which both annotators interpreted as consisting of two contrasting sections ('A' and 'C') with a transitionary middle section ('B'). This structure is also apparent in Performance 2 (Fig. 4), in which the canon section ('B' in Annotator 1's analysis, 'D' in Annotator 2's analysis) contrasts strongly with the first half of the piece. This binary structure is apparent despite the annotators' disagreement over the labeling of the first half.

Up until now, much attention has been paid to the large-scale structures that Mimi tends to create, but not much has been said about the small-scale structures they are comprised of. The large-scale section boundaries indicated in the annotations only capture the most broad structures found in the improvisations. A close examination of a short excerpt of Mimi's performance shows that, on a smaller scale, Mimi operates quite differently than one might expect from looking at the large-scale structure.

**Rhythmic Cells.** In Fig. 11, Mimi's opening phrases from the MIDI file of Performance 3 are transcribed and quantized. Repetitions of the same motive are bracketed and indicated with the same letter—this helps to indicate motivic connections that are obscured by the quantization, e.g., motive 'A' in m. 1 and motive 'A' in m. 6, which are identical in performance despite the slightly different notation. Here we see at work what Messiaen called *personnages rythmiques*, or rhythmic cells, individuated by particular sonorities, that can be expanded or contracted [11].

Mimi performs four distinct cells ('A' through 'D') before recombining them in various ways. Contractions of a cell are more common (e.g. the abbreviated form of 'A' at the end of m. 5) but expansions (such as the "stutter" in the middle of the version of 'C' that appears in m. 10) are also possible. Mimi's overwhelming preference for its initial idea can also be observed here. Motive 'A' recurs a total of 7 times, versus 4 iterations of 'B,' 2 of 'C' and 3 of 'D' before moving on to a new idea ('E').

**Fig. 11.** Opening phrases of Mimi's portion of Performance 3, with annotation of rhythmic cells. Note values were quantized to the nearest 16th note, while allowing for simple triplet divisions.

Unlike the canon and the rondo, this particular kind of juxtaposition of rhythmic cells is regarded as a 20th century innovation, and Mimi's juggling of these particular cells is not wholly unlike the jumpy, erratic rhythms found in the pivotal "Danse sacrale" from Stravinsky's *Le sacre du printemps*, which informed Messiaen's conception of *personnes rhythmiques* as well as successive generations of composers who sought a new working theory of rhythm [11,12]. However, since Mimi itself has no theory of rhythm, its improvisations lack the rigorous rhythmic structure and symmetries found in "Danse sacrale" and other composed music.

## 4 Discussion and Conclusion

The fact that large-scale forms can be identified in improvisations with both Mimi and a human performer gives rise to a question: do these forms arise from the human performer's conscious or unconscious intent, Mimi's behavior, or some combination of the above? In the present case, the human performer had improvised extensively with Mimi over several months before performing the improvisations analyzed here. During this time, the performer adapted his usual improvisational technique in order to create successful performances with Mimi (since Mimi is incapable of certain kinds of adaptation, like rhythmic or harmonic coordination with the human performer). In some cases, these adaptations even ran counter to the performer's usual technique (for example, certain kinds of harmonic movement would be restricted by Mimi's inability to follow along). In other words, while the performer had significant control over the improvisation's moment-to-moment *content*, Mimi had much more influence over the *structure* of that content.

We have discussed some of the formal structures observed in the recorded improvisations with the assumption that they emerge, at least in part, from Mimi's inherent behavior. However, whether this is a truly intrinsic aspect of Mimi's behavior or a complex interaction between Mimi and the personal tendencies of the performer is beyond the scope of this paper, and is a definite area for further study involving other performers and more performances.

Addressing rhythmic rigor in Mimi is a compelling avenue for future research. One approach to incorporating rhythmic sensitivity into a factor oracle-based system involves suppressing factor links between contexts with grossly different average rhythmic densities [13]; examination of exactly how this or other approaches might operate in the context of the Mimi system is warranted, as well as further study of the mechanisms by which the observed large-scale structures emerge from these smaller units.

Other recent upgrades to OMax suggest interesting possibilities. Newer versions of OMax traverse the factor oracle using a suffix link tree, improving the contextual continuity during traversal [13]. In addition, the use of an anticipatory model can improve the quality of the suffix links [14]. In this approach, several competing factor oracle models are iteratively improved using reinforcement learning, and the most successful model drives the output. Incorporating these improvements into Mimi may have interesting ramifications for its behavior.

In conclusion, the performances with the Mimi improvisational system appear to be rich in structure, both small-scale and large-scale. Following an investigation of two annotators' analyses of these performances—one being the performer himself and the other an experienced structure analyst—the emergence of several structuring behaviors was observed; these were attributed to Mimi's design and to basic interactions between Mimi and the performer. However, it remains uncertain whether these principles are truly intrinsic to Mimi's programming, or whether the unconscious structuring instincts of the performer were alone sufficient to produce these results. Further investigations, including the analysis of a greater number of performances with other musicians, should address this question.

## Acknowledgments

This work was supported in part by a United States National Science Foundation (NSF) Grant No. 0347988, and by a University of Southern California (USC) Provost's Ph.D. Fellowship. Any opinions, findings, and conclusions or recommendations expressed in this material are those of the authors, and do not necessarily reflect those of the NSF or USC.

## References

1. Allauzen, C., Crochemore, M., Raffinot, M.: Factor oracle: A new structure for pattern matching. In: Bartosek, M., Tel, G., Pavelka, J. (eds.) SOFSEM 1999. LNCS, vol. 1725, pp. 295–310. Springer, Heidelberg (1999)
2. Assayag, G., Bloch, G., Chemillier, M., Cont, A., Dubnov, S.: OMax brothers: A dynamic topology of agents for improvization learning. In: Proceedings of the ACM Workshop on Music and Audio Computing, pp. 125–132 (2006)

3. Assayag, G., Dubnov, S.: Using factor oracles for machine improvisation. Soft Computing 8(9), 604–610
4. François, A.R., Chew, E., Thurmond, D.: Visual feedback in performer-machine interaction for musical improvisation. In: Proceedings of the International Conference on New Interfaces for Musical Expression, pp. 277–280 (2007)
5. François, A.R.: Time and perception in music and computation. In: Assayag, G., Gerzso, A. (eds.) New Computational Paradigms for Computer Music, pp. 125–146. IRCAM, Delatour (2009)
6. Lévy, B.: Visualising OMax. Technical report, Ircam (2009)
7. Smith, J.B.: A comparison and evaluation of approaches to the automatic formal analysis of musical audio. Master's thesis, McGill University (2010)
8. Lerdahl, F., Jackendoff, R.: A Generative Theory of Tonal Music. MIT Press, Cambridge (1983)
9. Yorgason, B., Halliday, J., Colvard, C.: Variations Timeliner [computer software] (2008), `http://variations.sourceforge.net/vat`
10. Lukashevich, H.: Towards quantitative measures of evaluating song segmentationu. In: Proceedings of the International Conference on Music Information Retrieval, pp. 375–380 (2008)
11. Boivin, J.: Musical analysis according to Messiaen: A critical view of a most original approach. In: Dingle, C., Simeone, N. (eds.) Olivier Messiaen: Music, Art and Literature, pp. 137–157. Ashgate Publishing, Ltd, Burlington (2007)
12. Boulez, P.: Stravinsky remains. In: Stocktakings from an Apprenticeship. Trans. Stephen Walsh, pp. 55–110. Clarendon Press, Oxford (1991)
13. Assayag, G., Bloch, G.: Navigating the oracle: A heuristic approach. In: Proceedings of the International Conference on Music Computing, pp. 405–412 (2007)
14. Cont, A., Dubnov, S., Assayag, G.: A framework for anticipatory machine improvisation and style imitation. In: Proceedings of Anticipatory Behavior in Adaptive Learning Systems, ABiALS (2006)

# Open Form and Two Combinatorial Musical Models: The Cases of *Domaines* and *Duel*

Benny Sluchin[1] and Mikhail Malt[2]

[1] EIC, IRCAM
[2] IRCAM, MINT Paris IV
{benny.sluchin,mikhail.malt}@ircam.fr

**Abstract.** Two "open" works, composed within a two-year period by Boulez and Xenakis, could be seen as based on a square matrix of order six and share several properties. Their combinatorial attributes, the theory and the practice of their performances are studied and compared. Our main aim is to establish a relationship between the properties of the mathematical model and its use by Boulez and Xenakis in *Domains* and *Duel*.

**Keywords:** Open Form, Matrix, Combinatory, Permutations, Game Theory, Computational Musicology, Boulez, Xenakis, *Domaines*, *Duel*.

## 1 Introduction

The interest in the "open work" that started in the fifties led to a confusing variety of definitions. To express the different degrees of variance offered by this musical approach (from smallest to largest) to form and to all levels of composition, one could use the terms: "controlled randomness", "mobile forms", "open works" and/or "indetermination". However, we know these terms correspond to different, and sometimes opposed aesthetic concerns.

In this article, we propose to analyze two examples of this very diverse repertoire: *Domaines* by Pierre Boulez, and *Duel* by Iannis Xenakis. These works do not share a common aesthetics, but do have many similarities. We describe the mathematical bases and the compositional preoccupations which led to each of those works being recognized as open forms, thus showing their differences and similarities.

### 1.1 The Common Structure

Let $\mathbb{R}$ be a commutative field and $M_n(\mathbb{R})$ the algebra of a square matrix of order $n$ with coefficients in $\mathbb{R}$. This structure plays a fundamental role in the history of mathematics. We shall study two cases where elements from $M_6(\mathbb{Z})$ allow those works' form to be "mobile" or open. In both cases, we don't have "one work" but a class of works standing for the generic work.

C. Agon et al. (Eds.): MCM 2011, LNAI 6726, pp. 255–269, 2011.

### 1.2 Methodology

The starting point of this work was the pragmatic problems and issues connected with the performance of open works. We mainly focus on the study of the models rather than on the music itself, and try to establish a bridge between the matrix properties and the structure generated by these works.

To study the actual dynamics arising from Boulez and Xenakis' formalisms, we used computer simulation,[1] and specific interfaces for performance. In what follows, we only present the computer interfaces used for simulation[2] and study.

## 2 *Domaines* (1961–1968) by Pierre Boulez

### 2.1 Description

*Domaines* [4] is a work in which the strictness of the composer's writing contrasts with some formal indeterminacy. The ensemble is divided into six groups named A to F as follows:

| | |
|---|---|
| A | Trombone quartet (one alto, two tenors and one bass) |
| B | String sextet (two violins, two altos and two cellos) |
| C | Duo (marimba and double bass) |
| D | Quintet (flute, trumpet, saxophone alto, bassoon and harp) |
| E | Trio (oboe, horn and guitar) |
| F | Bass clarinet |

The six groups are placed on the apexes of a hexagon surrounding the conductor (Fig. 1). The sound sources are thus clearly distinct and the role of space is well defined. We will also notice that the groups are not positioned in alphabetical order (inversion of D and E on the apexes of the hexagon).

In terms of musical performance, *Domaines* is an alternation between a clarinet solo and one of the six instrumental groups. The work consists of two parts: the *Original* and the *Miroir*, each made of six sections in an order to be determined.

For the *Original*, the clarinet player fixes the order of performance of the six cahiers (the clarinet part), titled from A to F, and starts playing in the chosen order. After each clarinet solo, the corresponding instrumental group plays its sequence of *Original*. For example, if the soloist played *cahier* B, the B group (string sextet) will play its own sequence, and so on. At the end of the sixth sequence of *Original*, the process continues with the parts of *Miroir*. For this second part, the conductor fixes the order, and the clarinet player now has to play the corresponding *cahier* after the instrumental groups. The succession of parts is played without interruption. The work starts and finishes with the clarinet solo.

From the standpoint of drama, the work is therefore played by two agents: the clarinet player and the conductor each have to choose a permutation of $S_6$ (the

[1] We used the Max/MSP [1] and OpenMusic [2] environments.

[2] See [3] for a detailed presentation of *Duel* performance interfaces. A computer interface for Boulez's *Domaines* is planned to be used for a performance in May 2011.

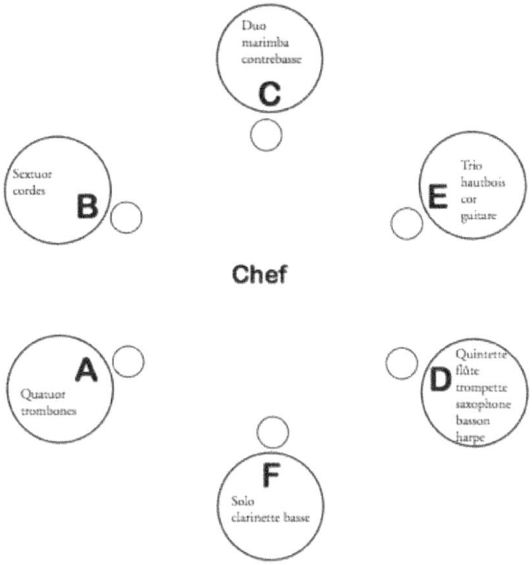


**Fig. 1.** Disposition of instrumental groups in *Domaines*

symmetric group of order 6, subgroup of $M_6(\mathbb{Z})$), which determines the order of the *Original* and the *Miroir* accordingly.

We shall also note that a version of *Domaines* is not only the succession of the parts, but also a physical path of the sound, a spatialization, rendered by the soloist, which must meet the various groups, in agreement with the chosen permutation and the position of each of the 6 instrumental groups. This special feature creates a sound architecture, a motion of sound in space, which results from the combinatorial qualities of the work.

### 2.2 Combinatorics

In this "open form" work, two permutations give birth to $(6!)^2 = 7202 = 518400$ possibilities for *Domaines*. This only accounts for the order of the parts, not any other non-fixed elements. Both agents have to determine the dynamics, the vertical or horizontal reading in the solo parts (see Sect. 2.6), the order of the inserts for the string sextet. The agents play their parts alternately, with only a short superposition for continuity.

### 2.3 Permutations

A permutation of $S_6$ may be represented by a square permutation matrix with elements 0 or 1, only one 1 per line and only one 1 per column, or by a $\sigma$ vector. Thus, a permutation, $\sigma = (1\ 4\ 6\ 5\ 2\ 3)$ is also represented by a matrix $M(\sigma)$ (see Fig. 2).

| | | | | | |
|---|---|---|---|---|---|
| 1 | 0 | 0 | 0 | 0 | 0 |
| 0 | 0 | 0 | 0 | 1 | 0 |
| 0 | 0 | 0 | 0 | 0 | 1 |
| 0 | 1 | 0 | 0 | 0 | 0 |
| 0 | 0 | 0 | 1 | 0 | 0 |
| 0 | 0 | 1 | 0 | 0 | 0 |

**Fig. 2.** Permutation matrix

The square matrices of order $n$ are in bijection with $S_n$ (the permutations of the ensemble 1, 2,, n). One could represent a "version" by two permutations $\sigma$, $\tau$, of $S_6$ which will give the order of the parts, $\sigma$ being the permutation related to the *Original* and $\tau$ being the permutation related to the *Miroir*.

We will use the representation of Fig. 3 to represent the structure of the different sections in a given permutation.

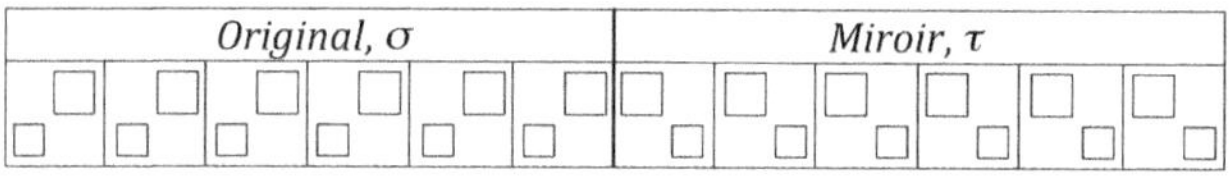


**Fig. 3.** Graphic representation of the basic structure of the instrumental groups and of the clarinet in the *Original* and *Miroir* parts

To each permutation $\sigma$ we will associate two numerical values:

*The density variation* $P_\sigma$*:*

$$P_\sigma = \sum_{i=1}^{5} |P_{i+1} - P_i| \tag{1}$$

where $P_i$ indicates the number of instruments in group playing in section $i$.

*The total angular distance* $T_\sigma$*:*

$$T_\sigma = \sum_{i=1}^{5} |T_{i+1} - T_i| \tag{2}$$

where $T_i$ indicates the angular position (degrees) of the instruments of the group playing in section $i$.

### 2.4 Some Case Studies

It is useful to study a few specific versions of *Domaines*, especially some corresponding to real performances of the piece. A permutation is usually seen as a static process. However, the interest of the analysis of some *Domaines* performances shows that each permutation brings an internal dynamic movement concerning the structure of groups, its density variation and angular position.

In the following figures are visible graphical interfaces inspired from the structure in Fig. 3, where a given permutation ($\sigma$) (that is, one half of the whole piece) is displayed, along with the corresponding $P_i$, $T_i$ and and the resulting values of $P_\sigma$ and $T_\sigma$. Each of the boxes labeled $A$ to $F$ represent a group (above) or the corresponding clarinet *cahier* (below).

The first following examples are special cases of permutation and analysis of the corresponding $P_\sigma$ and $T_\sigma$.

**Identical permutation:** This case corresponds of a diagonal matrix $M(\sigma)$. With this permutation, $P_\sigma = 13$ and $T_\sigma = 420$ (see Fig. 4).

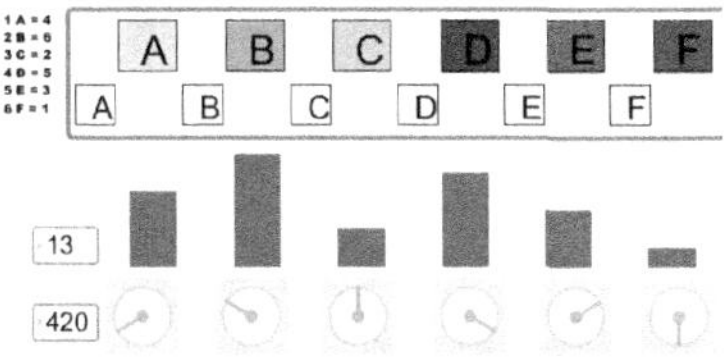


**Fig. 4.** $\sigma = (1\ 2\ 3\ 4\ 5\ 6)$

**Minimal movements path:** Permutations resulting in lowest values of $T_\sigma$. With $\sigma = (1\ 2\ 3\ 5\ 4\ 6)$, $P_\sigma = 13$ and $T_\sigma = 300$ (see Fig. 5(a)). Obviously, circular permutations or inverse paths will be also of the same nature. For instance, with $\sigma = (1\ 6\ 4\ 5\ 3\ 2)$, $P_\sigma = 14$ and $T_\sigma = 300$ (see Fig. 5(b)).

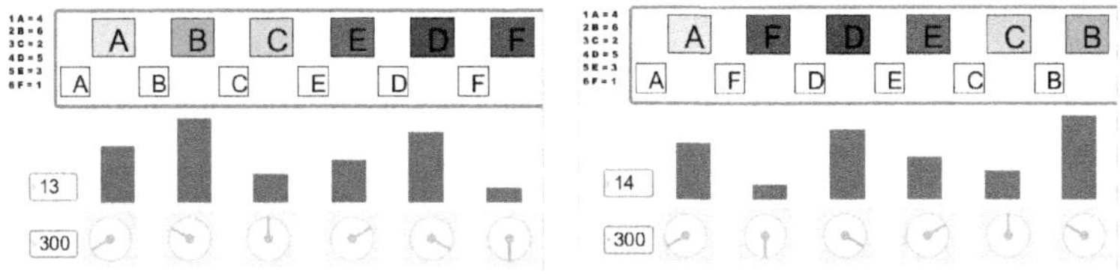


(a) (b)

**Fig. 5.** (a) $\sigma = (1\ 2\ 3\ 4\ 5\ 6)$. (b) $\sigma = (1\ 6\ 4\ 5\ 3\ 2)$

**Maximal movements path:** Permutations resulting in highest values of $T_\sigma$ are obtained with successive 180° transitions. With $\sigma = (1\ 5\ 2\ 4\ 3\ 6)$, $P_\sigma = 9$, $T_\sigma = 780$ (see Fig. 6).

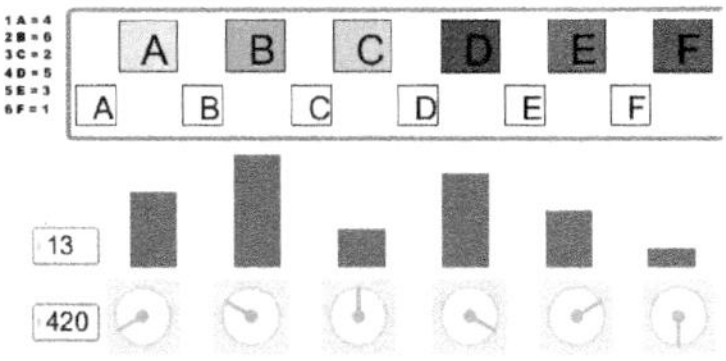


**Fig. 6.** $\sigma = (1\ 5\ 2\ 4\ 3\ 6)$

**Ordered instrumental density:** In both ascending or descending order, $P_\sigma = 5$ (see Fig. 7). Note that in these two cases, the values of spatial movement are high ($T_\sigma$=720). We can suppose that the instrumental groups' positioning has been made to encourage that.

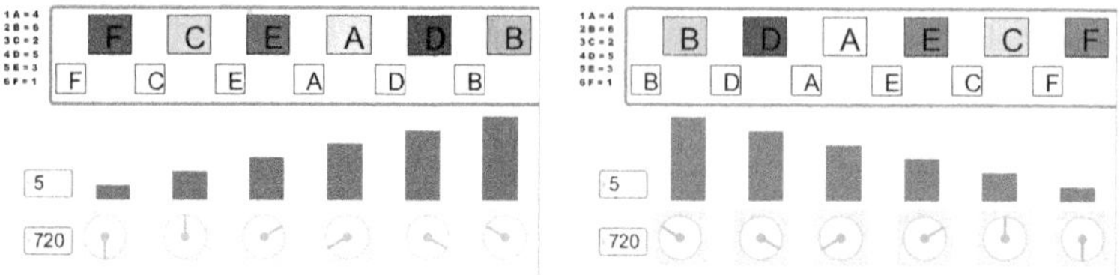


(a) (b)

**Fig. 7.** (a) Instrumental density in ascending order: $\sigma = (6\ 3\ 5\ 1\ 4\ 2)$. (b) Instrumental density in descending order: $\sigma = (2\ 4\ 1\ 5\ 3\ 6)$.

## 2.5 Analyses of Real Performances of *Domaines*

The choice of permutations $\sigma$ and $\tau$ are in principle independent; however, the two agents of *Domaines* are allowed to agree on a path to define their versions. The permutation $\tau$ may therefore be chosen in relation to $\sigma$ and may have mathematical (e.g. inverse or transpose) or musical properties. However, show the following examples, the permutations chosen can influence the dynamics of *Domaines*, giving more or less spatial movement, or creating a density evolution in the sequence of the instrumental groups.

**Boulez/Damiens (Munich, 2005):** This version is a reference, as in this specific case, the composer is also the conductor. According to Boulez: "rehearsal time is lacking to test the possibilities. Some path sequences function better than others."[3] Starting with an homogeneous group, like $A$ (four trombones), introduces the work well, and concluding with $F$ (the bass clarinet—a "solo" group playing with its back to the audience) influences the choice of the two permutations $\sigma$ and $\tau$. Notice that the last element of $\sigma$ ($C$) is also the first of $\tau$.[4]

In this version $P_\sigma = 14$ and $T_\sigma = 480$, for the *Original*, and $P_\tau = 13$ and $T_\tau = 540$, for the *Miroir* (see Fig. 8).

**Masson/Portal (1971):** This version is the only one recorded on a general public media (1971) [5], and has been used in the past for several analyses of the piece [6] (pp. 82–90) [7] (pp. 418–425). Here, $P_\sigma = 13$ and $T_\sigma = 540$ for the *Original*, and $P_\tau = 12$ and $T_\tau = 540$ for the *Miroir* (see Fig. 9). Notice the resemblance with the preceding version (two inversions in the *Original*, and only one in the *Miroir*).

[3] Personal communication, December 2010.

[4] In order to rework the score of group $C$ (marimba, double bass), the $Cm$ part (the mirror part of $C$) has been removed from *Domaines*. A modified and augmented part is now at the Sacher Foundation in Basel: Having only $Co$ (the original part of $C$) to disposal, it has been divided into two parts (which share the same character). In this particular version, the full part $C$ is played as a transition between the *Original* and the *Miroir*.

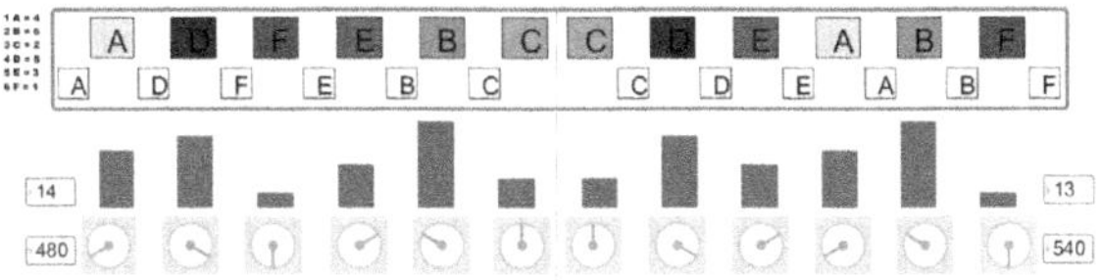


**Fig. 8.** Reconstitution of the performance of *Domaines* in Munich (2005) by P. Boulez and A. Damiens. *Original*: $\sigma = (1\ 4\ 6\ 5\ 2\ 3)$, *Miroir* $\tau = (3\ 4\ 5\ 1\ 2\ 6)$.

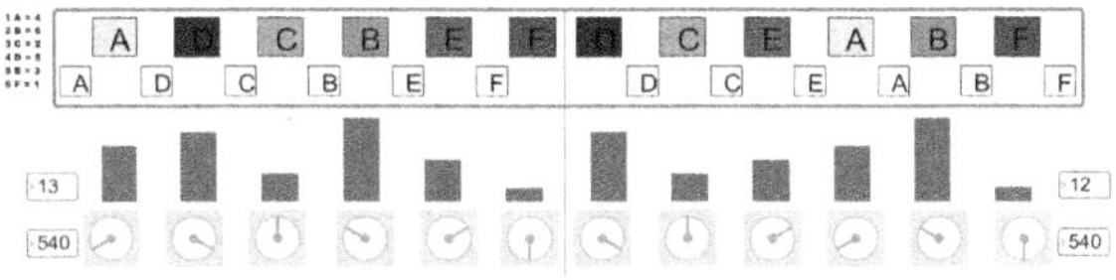


**Fig. 9.** Reconstitution of the performance of *Domaines* in the 1971 recording by D. Masson and M. Portal. *Original*: $\sigma = (1\ 4\ 3\ 2\ 5\ 6)$, *Miroir* $\tau = (4\ 3\ 5\ 1\ 2\ 6)$.

**Eötvös/Damiens (Paris, 1981):** This version (Fig. 10) shows high values for spatial movements ($T_\sigma = 600$, $T_\tau = 720$), as well as for the sum of differences of instrumental density ($P_\sigma = 13$, $P_\tau = 11$). Notice the placement of $B$ at the first position in the *Original* and as last section in the *Miroir*.

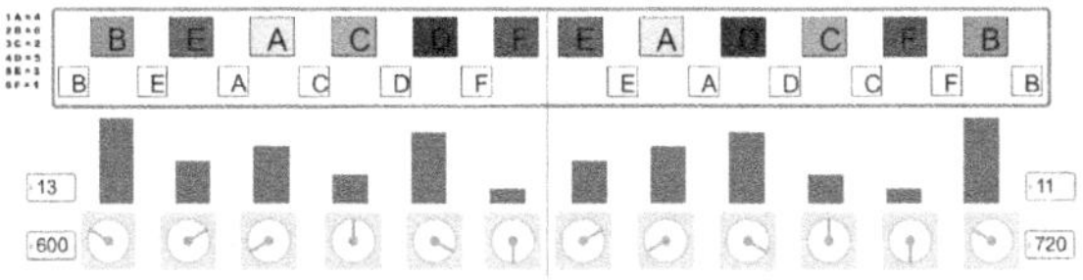


**Fig. 10.** Reconstitution of the performance of *Domaines* in Paris (1981) by P. Eötvös and A. Damiens. *Original*: $\sigma = (2\ 5\ 1\ 3\ 4\ 6)$, *Miroir* $\tau = (5\ 1\ 4\ 3\ 6\ 2)$.

### 2.6 Other Elements of $S_6$ in the Clarinet *Cahiers*

The clarinet sheets show us other permutations, playing a part in the composition and in the performance of *Domaines*. Each of the *cahiers* in the *Original*, from $A$ to $F$, is made of six cells. They are placed on the sheet following the model of Fig. 11.

Each cell gives birth to a mirror form. They are placed on the corresponding sheet according with a permutation $\mu$. So $\mu_A = (6\ 5\ 4\ 3\ 2\ 1)$ because in regard of the *Original (Ao)*, the mirror cells are placed on the *Miroir* part *(Am)* following the model of Fig. 12.

These permutations are characterized by three transpositions (exchange of two elements). We have $\mu_B = (4\ 5\ 6\ 1\ 2\ 3)$, $\mu_C = (2\ 1\ 6\ 5\ 4\ 3)$, and so on. For the performance, the composer uses a horizontal reading mode and a vertical reading mode. For example, the two reading modes for $A$ are shown in Fig. 13.

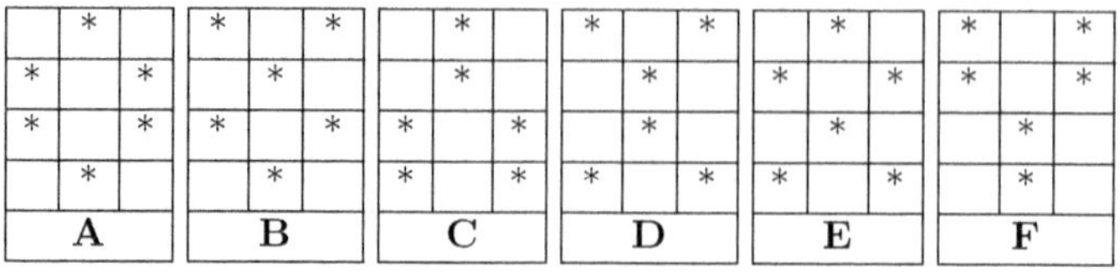


**Fig. 11.** The six dispositions of cells of the clarinet solo

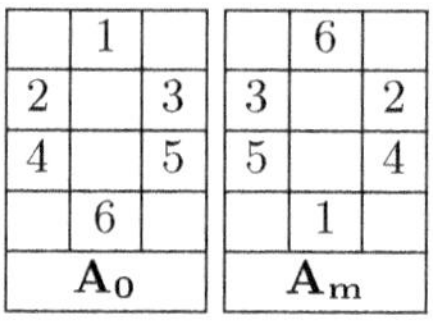


**Fig. 12.** The relationship between cells in the *Original* and in the *Miroir cahiers A*

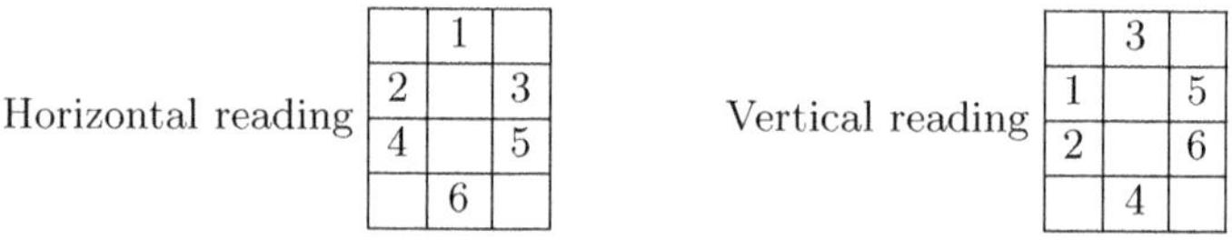


**Fig. 13.** The two reading modes for the clarinet

In Fig. 12 the vertical reading is a permutation $\lambda_A = (2\ 4\ 1\ 6\ 3\ 5)$ of the horizontal one. We have $\lambda_B = (1\ 4\ 3\ 6\ 2\ 5)$, $\lambda_C = (1\ 2\ 3\ 5\ 4\ 6)$, etc. If one chooses the horizontal reading mode for the *Original*, he/she will have to choose the vertical reading mode for the *Miroir*, and vice-versa.[5] Naturally those permutations can be applied to the preceding structures.

### 2.7 Practical Problems of Performance

The "opening" of *Domaines* requires special practice. One can define the path after reflection (solitary or by consensus), but to test it concert conditions and time are needed. One must also be able to swiftly reorganize the scores and to memorize the path. That is why the musicians do not wait for the moment of performance to decide on those matters. The paths are determined beforehand and rehearsed with the soloist and the musicians. The transitions are delicate points to settle.[6]

A computer interface would give the conductor swift ordering, and computer display would make playing this piece, and expounding its combinatorial possibilities, easier. In [8] we have discussed the limits of paper scores and the benefits that could bring computer interfaces in improvisation and open works performances.

[5] See the score explanation pages in [4].

[6] In a personal communication, Peter Eötvös stated that: "Public, ignoring the multitude of possibilities, needs a solid performance where all the transitions and order of parts have been fixed and carefully rehearsed" (March 2011).

## 3 *Duel* (1959) by Iannis Xenakis

### 3.1 Description

*Duel* (1959-60) by Iannis Xenakis is a musical game, for two conductors and two orchestras commissioned by the ORTF (Office de Radiodiffusion-Télévision Française) [9]. It was created in Hilversum in 1971 by Diego Masson and Fernand Tuby. Each orchestra is made of three groups:

1. Winds: 1 piccolo, 1 oboe, 1 E♭ clarinet, 1 B♭ bass clarinet, 1 bassoon, 1 contrabassoon, 2 trumpets, 1 trombone.
2. Percussions: 2 bongos, 3 congas, 1 snare drum, 1 drum.
3. Strings: 6 first violins, 6 second violins, 4 cellos, 2 double basses.

The two orchestras are positioned to the left and to the right of the stage, with the two conductors standing back to back, or on two opposed stages. Each conductor has at his disposal six sound constructs (that we will also call tactics), numbered *I* to *VI* in the score, which are stochastic structures (Fig. 14).[7]

| | |
|---|---|
| *I* | *A cluster of sonic grains* |
| *II* | *Parallel sustained strings with fluctuations* |
| *III* | *Networks of intertwined string glissandi* |
| *IV* | *Stochastic percussion sounds* |
| *V* | *Stochastic wind instrument sounds* |
| *VI* | *Silence* |

**Fig. 14.** The six fundamental tactics

The on-stage disposition is two times three groups (Fig. 15), that is, six groups and the two conductors. The two agents are at variance, acting and creating a sound mixture placed at their disposal by the composer. There are thus six tactics, which each conductor can have his orchestra perform.

### 3.2 Game Theory

Game theory is a formal study of interactions between agents (from cooperation to conflict). A *game* is represented by a matrix giving the gains or losses of the agents depending on their respective choices. Xenakis uses a special case of game called *zero-sum* two-player games [12]. In such game, the gain of an agent is always exactly balanced by the other agent's loss.[8]

[7] In this study consider the discussion in Xenakis' *Musiques Formelles* [10] (p. 9), rather than the score itself [9], where the three first tactics labeled A, B, and C are combinations if tactics I, II, and III: A = strings (pointillism (I) or held (II) or crossed glissandi (III)); B = B, A + percussion or A + winds or percussion and winds; C = A + percussion + winds. However, the game matrix use dis identical in both cases.

[8] The *minmax* theorem (Nash) [13] insures the existence of a unique solution (or *value*) of the game.

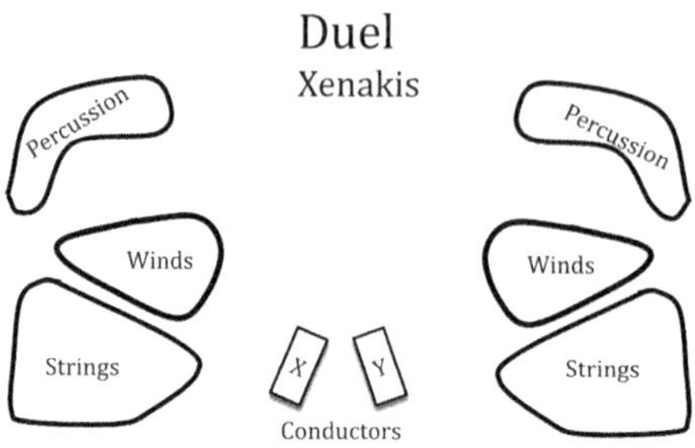


**Fig. 15.** The diagram of the two orchestras on stage (simplified from the score)

The choices of the two conductors are defined by the *game matrix* (in $M_6(\mathbb{Z})$) given by the composer (Fig. 16).[9]

| | I | II | III | IV | V | VI |
|---|---|---|---|---|---|---|
| I | -1 | +1 | +3 | -1 | +1 | -1 |
| II | +1 | -1 | -1 | -1 | +1 | -1 |
| III | +3 | -1 | -3 | +5 | +1 | -3 |
| IV | -1 | +3 | +3 | -1 | -1 | -1 |
| V | +1 | -1 | +1 | +1 | -1 | -1 |
| VI | -1 | -1 | -3 | -1 | -1 | +3 |

**Fig. 16.** The *Duel* game matrix

Each of the 36 elements of this game matrix is a "gain" corresponding to a sound combination. The musical game *Duel* is generated by a choice of a tactic (Fig. 14) by conductor $X$ (playing lines) while conductor $Y$ chooses one of the columns. Positive values are gains for conductor $X$ and negative values are gains for conductor $Y$. The element of the matrix resulting from those choices gives $X$'s gain (a positive or negative integer) equal to the loss of $Y$. The goal of the game is to win a maximum of points at the end of the performance. Winning is not a matter of beauty or good musical choices. The composer itself states that the winner conductor is that follows the matrix rules better than his opponent, and the musical result is the composer's responsibility.[10]

### 3.3 Practical Problems of Performance (I)

The performance of *Duel* is not without difficulties and requires a complex preparation from both agents. Some preliminary decisions are necessary before performance:

[9] In Chapter III of *Musiques Formelles* [10] (or Chapter IV in [11]), Xenakis describes the method to create this matrix from subjective judgements about simultaneous sound event couples.

[10] "Le gagnant a gagné parce que il a simplement mieux suivi les règles du jeu imposés par le compositeur qui, par conséquent, revendique la responsabilité du "beau" ou du "laid" de sa musique." See score instruction notes [9].

- The attribution of lines and columns (by flipping a coin between the two conductors).
- The determination of the duration (by choosing an arbitrary duration $m$ in minutes, the number of moves or the maximum number of points to obtain).
- The decision of who will go first.

Other elements are at the agents' discretion.

- Where in the score to start each tactic.
- Each move's duration.

About the conductors' choices of tactics, Xenakis names two main options:

1. The "degenerate" fashion: Either by the beforehand determination by both agents of a preordained succession, or by arbitrarily following their musical intuition. It is easy to understand that the composer does not encourage this option.
2. Using the matrix and its values, either by choosing on the basis of the potential gain, or by taking balls numbered 1 to 6 from an urn, with proportional quantities in the urn corresponding to the probabilities given by the composer. For this reason *Duel* is a real musical game, and winning or loosing influences the musical result, as is noted by the composer.[11]

### 3.4 Computer Simulations of the Game

To study the musical structure generated by the game model, i.e. the matrix cell sequence (or game *dynamics*), arising from Xenakis' formalism, we used a computer simulation, where each matrix cell is a compound musical texture seen as an element $c \in M_6(T)$ where $T = \{I, II, III, IV, V, VI\}$, the set of six fundamental tactics (Fig. 14), and look for patterns (subsets) and the space covered by the cell sequence $C_s = (c[0], c[1], c[2]...c[k]...)|k \in \mathbb{N}$.

For the purpose of simulation the cells were numbered from 1 to 36. A cell position $c_{x,y}$ is calculated according with the last $X$ ($T_X$) and $Y$ ($T_Y$) moves: $c_{x,y} = 6(T_X - 1) + T_Y | T_X, T_Y \in T$.

Each simulation was composed by 500 moves (250 moves by $X$, and 250 moves by $Y$). For each simulation we used a different playing strategy. Four playing strategies where tested: random choices of each agent, random choices with weights, maximum gain strategy and conventional *minmax* strategy.[12]

**Random choices of each agent:** After 500 moves with random choices, uniformly distributed between the two agents, the game runs through all 36 cells of the game matrix, and the morphology of the paths doesn't seem to form any specific pattern.

---

[11] "La seule [façon] valable, la seule qui apporte quelque chose de nouveau, dans le cas de plusieurs orchestres, est celle qui est sanctionnée par des gains et des pertes, par des victoires et des défaites" [9].

[12] For each simulation, we also approximated the game value, step by step. As the values found were different from Xenakis' calculus (-0.07), this point will not be explained here as it needs further research in the original sketches.

**Random choices of each agent with weighting:** In this second simulation, we used random choices weighted by the probabilities calculated by the composer [10] (p. 148). The game runs through all 36 cells, hence through the 36 proposed sound combinations. The morphology of the paths is dependent on the relative weight of lines and columns and does not show special patterns either.

**Choices of each agent with strategy of maximum gain:** In this third simulation, each agent chooses a line or a column where he hopes to have a maximum gain. This strategy is a simplification of the "minmax" tactic. As a result, the game only runs through 23 cells of the matrix (see Fig. 17). Agent $X$ never chooses tactic $V$, since he always has a better alternative (see Fig. 16). Concerning the trajectory, it is possible to detect a group of irregularly repeating patterns: $\{21 \rightarrow 24 \rightarrow 36 \rightarrow 33 \rightarrow 21\}$, $\{16 \rightarrow 15 \rightarrow 21 \rightarrow 23\}$, $\{3 \rightarrow 1 \rightarrow 13 \rightarrow 18 \rightarrow 36 \rightarrow 33\}$, $\{16 \rightarrow 18 \rightarrow 36 \rightarrow 33 \rightarrow 3 \rightarrow 4\}$ ... (see Fig. 18).

| | I | II | III | IV | V | VI |
|---|---|---|---|---|---|---|
| I | 1 | 2 | 3 | 4 | 5 | 6 |
| II | 7 | 8 | 9 | 10 | 11 | 12 |
| III | 13 | 14 | 15 | 16 | 17 | 18 |
| IV | 19 | 20 | 21 | 22 | 23 | 24 |
| V | 25 | 26 | 27 | 28 | 29 | 30 |
| VI | 31 | 32 | 33 | 34 | 35 | 36 |

**Fig. 17.** Cells run through (in grey) by a simulation with maximum gain

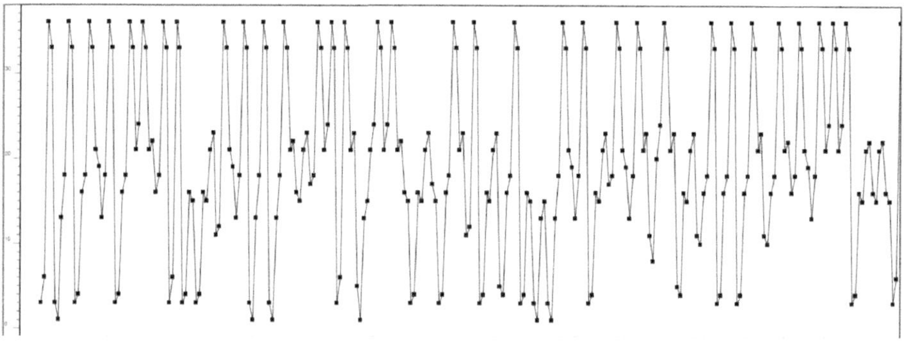

**Fig. 18.** Game dynamics resulting from the simulation with the maximum gain strategy. (Horizontal axis: moves; Vertical axis: cells 1 to 36).

**Choice of each agent using the "minmax" strategy:** In this last simulation, each agent chooses a line or a column, not only aiming towards a value that is in his favor, but also trying minimize his opponent's gain. As a result, the game only runs through 18 cells of the matrix, thus half of the 36 cells (see Fig. 19). One line *(III)* and one column *(IV)* are never chosen. As each agent tries to minimize his adversary's gain, conductor $Y$ never chooses column

*IV* to prevent conductor *X* from choosing line *III* and getting the maximum gain (5 points). Concerning the trajectory, it is possible to detect a group of irregularly repeating patterns: $\{36 \rightarrow 35 \rightarrow 5 \rightarrow 6 \rightarrow 36\}$, $\{11 \rightarrow 12 \rightarrow 36 \rightarrow 35 \rightarrow 11\}$, $\{1 \rightarrow 25 \rightarrow 29 \rightarrow 11 \rightarrow 9 \rightarrow 21 \rightarrow 23 \rightarrow 11 \rightarrow 8 \rightarrow 2 \rightarrow 1\}$, $\{11 \rightarrow 8 \rightarrow 20 \rightarrow 23 \rightarrow 11 \rightarrow 12 \rightarrow 36 \rightarrow 35\}$, $\{11 \rightarrow 9 \rightarrow 21 \rightarrow 23\}$, $\{5 \rightarrow 1 \rightarrow 7 \rightarrow 8 \rightarrow 20 \rightarrow 23\}$... (see Fig. 20).

| | I | II | III | IV | V | VI |
|---|---|---|---|---|---|---|
| I | 1 | 2 | 3 | 4 | 5 | 6 |
| II | 7 | 8 | 9 | 10 | 11 | 12 |
| III | 13 | 14 | 15 | 16 | 17 | 18 |
| IV | 19 | 20 | 21 | 22 | 23 | 24 |
| V | 25 | 26 | 27 | 28 | 29 | 30 |
| VI | 31 | 32 | 33 | 34 | 35 | 36 |

**Fig. 19.** Cells run through (in gray) by a simulation with a minmax strategy

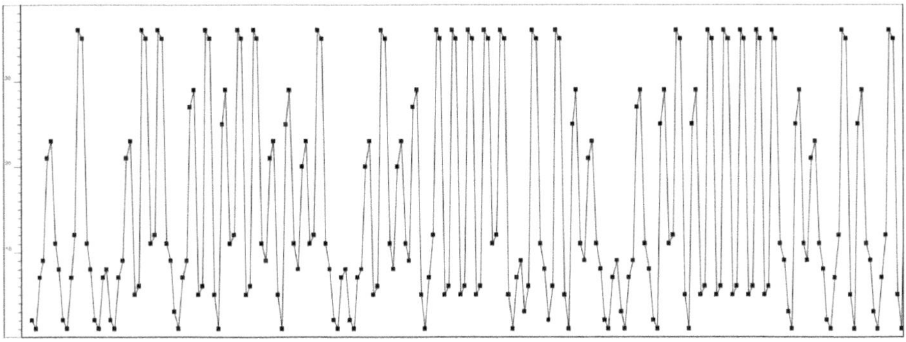

**Fig. 20.** Game dynamics resulting from the simulation with the "minmax" strategy. (Horizontal axis: moves; Vertical axis: cells 1 to 36.).

## 3.5 Practical Problems of Performance (II)

*Duel* is a challenge to perform according to the composer's wishes. Both conductors have to communicate with their orchestras to let them know which tactic is next and where they should start. They should avoid pauses resulting from hesitation or communication, as silence is a choice, and corresponds to tactic *VI*.

A preliminary observation could conclude that, according to the conductors' wishes, the strategy chosen leads to structural or formal issue. A completely random strategy and a weighted random strategy lead to a random structure. A "minmax", with or without gain, strategy leads to a game generating cycles and more or less repetitive patterns. A strategy combining weighted and "minmax" may generates a less rigid structure.

This means that the conductors need a tool to switch between strategies, in order to avoid the "degenerate" way of performing (see Sect. 3.3). For this reason *Duel* requires interfaces to display the game's results and to communicate.

Xenakis had conceived an electrical interface based on the use of switches, relays and brackets with electrical bulbs of various colors [9]. The swift resulting changes, according to the different choices and tactics, also emphasize to the need for Computer Aided Performance (CAP) interfaces to display the chosen score to the conductor and performers [8].

## 4 Conclusion

While trying to compare the two works and their respective requirements, we can find some similarities and differences. They are both musical discourse with two agents. Two *open* works, that is, two works where the final form is not pre-established, or set by the composer. However, the performers combine pre-established musical elements according to specific rules. Both pieces contain detailed instructions for smooth going, and require a specific preparation for rehearsal and performance. Both pieces are based on the manipulation of six ensembles, either instrumental ensembles, as is the case in *Domaines*, or ensembles of timbres as is the case in Duel.

On the other hand, in *Domaines* the ordering of the parts is determined by a combinatorial model based on permutations. In *Duel*, the evolution of the piece is based on a model originating in game theory. In *Domaines*, the order of the sequences is predetermined, while in *Duel*, the composer insists that sequences must be decided upon during the performance of the piece. In *Domaines* Boulez allows (or does not oppose to) a choice of permutations based on musical considerations, accepting the fact that "some permutations work better than others". In *Duel*, Xenakis demands that conductors do not bother with these musical considerations. He wants them to play according to the rules of the game matrix, and takes full responsibility for the resulting sound. *Domaines* is a sequential discourse with minimal superposition of events. The superposition is just a short overlapping between sections, which provides fluidity to the musical discourse. In *Duel* the superposition is the fundamental element. Finally, *Domaines* has a fixed length, with a final form that can be seen as a variation on the rondo. As for *Duel*, it has an indeterminate length with a form based on repeating sequences and on cycles resulting from the dynamics of the game matrix.

The $M_6(\mathbb{Z})$ algebra is, in various ways, at the foundation of both works. However, each composer uses this structure differently. In *Domaines*, we could see the underlying matrix structure as an array "temporal positions $\times$ musical groups", while in Xenakis' *Duel*, the matrix is an array "musical groups $\times$ musical groups", where the final sonorities are issued from a Cartesian product. In *Domaines*, the permutation matrix gives directly the structure of a section, while in *Duel*, the final structure is derived from the game dynamics.

The epistemological and historical matrix status shows us that this structure is an important element in the unification process of mathematical knowledge [14] (p. 41) that characterizes the beginning of the 20th century. It is due mainly to the fact that a matrix can represent different practices and concepts, with the same representation and operatory mode. The openness of the matrix concept is probably what, in these two cases, made it appropriate to represent two different musical esthetics based on different mathematical models, in an "open work" context.

## References


1. Max/MSP by Cycling'74, `http://cycling74.com/products/maxmspjitter/`
2. OpenMusic by Ircam, `http://repmus.ircam.fr/openmusic/`
3. Sluchin, B., Malt, M.: Play and game in Duel and Strategy. In: Xenakis International Symposium, London (2011)
4. Boulez, P.: Domaines. Universal Editions, London (1961-1968)
5. Boulez, P.: Domaines, Michel Portal. Ensemble Musique Vivante. Direction Diego Masson, Harmonia Mundi (1971, 2001)
6. Bosseur J. Y., Michel, P.: Musiques Contemporaines, Perspectives analytiques 1950-1985, Minerve (2007)
7. Jameux, D.: Pierre Boulez. Fayard/Fondation SACEM, Paris (1984)
8. Sluchin, B., Malt, M.: Interpretation And Computer Assistance In John Cage's Concert For Piano And Orchestra (1957-1958). In: Sound and Music Computing Conference, Barcelona (2010)
9. Xenakis, I.: Duel. Editions Salabert (1959)
10. Xenakis, I.: Musiques Formelles. Stock (1981)
11. Xenakis, I.: Formalized Music. Indiana University Press, Bloomington (1971)
12. Neumann, J.V., Morgenstern, O.: Theory of Games and Economic Behavior. Princeton University Press, Princeton (1944)
13. Webb, N.J.: Game Theory, Decisions, Interaction and Evolution. Springer, Heidelberg (2007)
14. Brechenmacher, F.: Les matrices, formes de représentation et pratiques opératoires (1850-1930), CultureMATH - ENS Ulm, Paris (2006)

# Exploding the Monochord: An Intuitive Spatial Representation of Microtonal Relational Structures

Nicholas Stylianou

Independent Researcher
nickstylianou@yahoo.co.uk

**Abstract.** Microtonality appears in a wide range of historical and ethnomusicological contexts, particularly in theoretical aspects of tuning systems and as intonation in performance. Theoretical concepts of microtonality can be inaccessible due to difficulties arising in the reconciliation of mathematical and musical approaches. The development of sophisticated geometrical representations of pitch cognition has largely been focused on the Western tonal tradition with limited incorporation of microtonality. This paper presents a spatial model of microtonal intervals and their relational structures. The model enhances accessibility of microtonal-theoretic concepts through a visually intuitive representation. It also acts as a unifying framework with respect to the comparative assessment of microtonal schemes and the integration of the different dimensions of pitch cognition. The integrative characteristics of the model demonstrate the psychological emergence of cognitive structures and their potential isomorphism with algorithmic approaches. The comparative features of the model may provide the basis for computational applications of broader scope than a culturally specific model can provide, while the intuitive spatial aspects may inspire improvements in the human-computer interaction of such applications.

**Keywords:** Music Theory, Microtonality, Visualization, Geometric Models, Pitch Cognition, Music Education.

## 1 Background

### 1.1 Microtonality

The microtone is defined as "any musical interval or difference of pitch distinctly smaller than a semitone" and appears in a wide range of historical and ethnomusicological contexts [1], particularly in theoretical aspects of tuning systems [2] and as intonation in performance [3].

Theoretical concepts of microtonality can be inaccessible, with difficulties arising in the reconciliation of mathematical and musical approaches [4].

C. Agon et al. (Eds.): MCM 2011, LNAI 6726, pp. 270–284, 2011.

## 1.2 Diagrams

The extensive use of diagrams in connection with the theory of music has reflected "the full range of historical associations of music theory and mathematics from number and proportion to logical and spatial representations of relations" [5]. This distinction between numerical and logical representations is of particular relevance to microtonality, for which the use of the inherently numerical monochord and its diagrammatic analog persisted over many centuries [6] despite well-established recognition of its limitations[1].

**Historical Developments.** In advancing the diagrammatic representation of microtonality beyond the monochord, two notable historical developments are of particular significance. Firstly, in his *Compendium Musicae* of 1618 (published in 1650) Descartes applies octave equivalence to transform a one-dimensional octave-section of the monochord into a circular arrangement [8]. While Descartes' division of the circle preserves the arithmetical division of the monochord, subsequent circular depictions such as those of Brouncker (1653), Newton (c. 1664-1665) and Salmon (1672) divide the circle logarithmically whilst retaining microtonal distinctions [9]. The second historical development was the 'Table of Relations' (or Riemannian "Tonnetz"), a two-dimensional grid originating from Euler's *speculum musicum* but here infinitely extending its horizontal axis of perfect-fifth relations and vertical axis of major-third relations [10].

A key feature of these developments is *dimensionality* whereby different components of pitch cognition are distinguished (octave-independent pitch chroma in the circular depictions, fifth- and third-relations in the Tonnetz). A second key feature is *uniformity* such that any given musical interval is "transpositionally invariant" (i.e. the same size regardless of where it appears in the circle or grid).

**More Recent Developments.** The above features have been described and integrated into a multi-dimensional geometrical model of pitch cognition [11] whose structure corresponds with the results of the analysis of experimental data [12]. Similar geometrical models, incorporating higher-level musical structures such as chords and keys, have been successfully applied to a range of applications such as key-finding [13], segmentation [14] and pitch-spelling [15]. In another proposed geometrical model the Tonnetz, in which triangles represent major and minor triads, is expanded into three dimensions such that tetrahedrons represent four-note chords (in particular the dominant- and half-diminished seventh chords)[16]. However, these models have been largely focused on the Western tonal tradition, and assume equal temperament with relatively little scope of application to microtonality.

A microtonal extension of the Tonnetz is Fokker's three-dimensional lattice of harmonics. This uses its three axes to represent the musical interval relations based on the prime numbers 3, 5, and 7, thus incorporating septimal relations and forming the basis of the Euler-Fokker genera of justly-intoned scales and

[1] Notably in Aristoxenus's assessment of the Harmonicists' diagrams as inadequate for the representation of the logical relations of tones and scale structures [7].

their 31-tone equal-tempered approximations. This three-dimensional lattice has been used in microtonal compositions such as those of James Tenney and Ben Johnston [17].

Other specifically microtonal representations include hexachordal diagrams with numerical annotations indicating the temperament of fifth- and third-relations [18], and graphs [19] and bar charts [20,21] for the representation of temperaments. Both types of visualization are useful in depicting and comparing particular aspects of musical temperament within their Western historical contexts, but neither are well-suited to representing melodic structures in a microtonal context.

A further three specific representational devices share important characteristics with the model proposed in this paper. Firstly, two-dimensional charts of performance data [22] utilise the vertical axis to clearly display intonational deviation from 12-tone equal temperament in a melodic context. Secondly, the "skeletal diagrams" presented in Japanese music theory [23], whilst not specifically microtonal, make highly effective use of two-dimensional spatial representations of musical intervals and their concatenation to depict melodic patterns. Thirdly, the notion of projection, as applied to Pythagorean tuning and equal temperament in two dimensions [24], acts as a powerful mechanism for relating the microtonal and categorical domains of musical pitch perception and cognition.

**Summary and Aims.** In summary, whilst microtonality is of broad musicological relevance, models of pitch cognition have tended to be confined to Western tonality. Furthermore the predominantly mathematical treatment of microtonality can make theoretical concepts inaccessible. Whilst diagrammatic representations have featured extensively in music theory, their specific application to microtonality has been limited. Where microtonality has been incorporated, the abstract nature of the representation is often detached from the melodic domain.

This paper presents a model of pitch cognition which explicitly incorporates microtonality. The model is spatial with an emphasis on providing an intuitive representation of microtonal intervals and their relational structures, thereby enhancing the accessibility of microtonal-theoretic concepts. Furthermore, the model shall act as a unifying framework with respect to the comparative assessment of microtonal systems and the integration of the psychological aspects of pitch cognition.

## 2 The Proposed Model

### 2.1 Numerical Conventions

The monochord is represented in the accessible context of the modern guitar (Fig. 1), where musical intervals are described by:

- (a) String length ratio (integer divisions measured from the guitar's bridge).
- (b) Frequency ratio (the reciprocal of the string length ratio).
- (c) The logarithm of the frequency ratio (in the range 1 to 2) yielding an octave value (0..1) which may be scaled to semitones (0..12) or cents (0..1200).

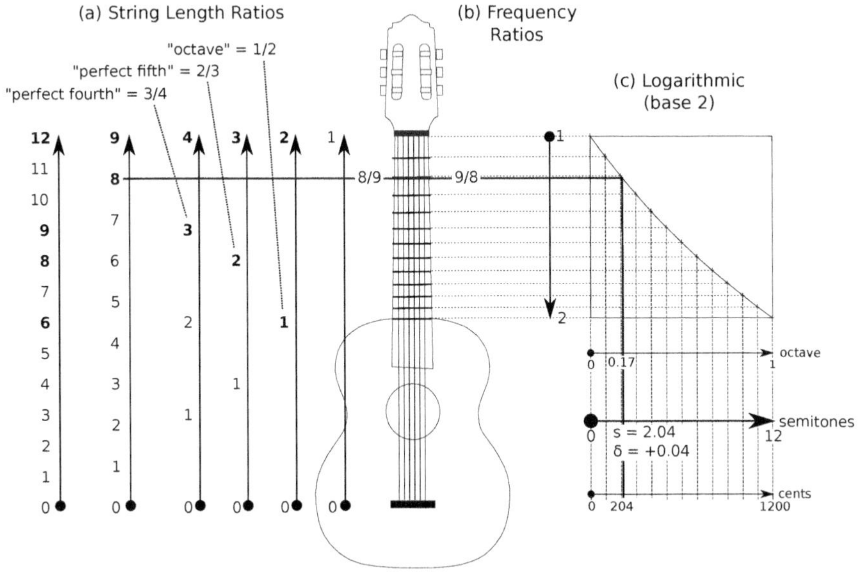


**Fig. 1.** The monochord in the context of the modern guitar, illustrating the relationship between (a) string length ratios, (b) frequency ratios, and (c) logarithmic measures

Assumptions are (i) 12-tone equal temperament as the default frame of reference[2], and (ii) octave equivalence in the ratio 2/1, such that frequency ratios are transposed into the range 1..2 and the corresponding logarithms are taken to base two[3].

The logarithmic semitone range forms the basis of the proposed spatial representation, as this provides the characteristic of uniformity described above and also allows straightforward numerical identification of microtonal variations (as fractional differences from the whole-numbers representing the equal-tempered divisions).

**Definition 1.** *Reference Deviation*
*Given a frequency ratio $r$ such that $1 \leq r \leq 2$ then:*

- *the semitone value* $s = N \log_2(r)$ such that $0 \leq s \leq N$ where $N = 12$
- *the deviation value* $\delta = s - round(s)$ such that $-0.5 \leq \delta \leq +0.5$

In the example shown in Fig. 1, the interval of a whole-tone (with string length ratio $d = 8/9$) has a frequency ratio $r = 9/8$, whose base-2 logarithm (value 0.17) scales to a semitone value of $s = 2.04$; this constitutes a microtonal deviation of $\delta = +0.04$ semitones (or 4 cents) from its closest equal-tempered semitone (the integer value $s = 2$).

[2] Equal divisions other than 12 are briefly considered later in this paper.

[3] Octave bases other than 2 are possible [25] but their consideration is beyond the scope of this paper.

## 2.2 Displacement in Two Dimensions

A simple, one-dimensional representation may be constructed by deriving a single (horizontal) coordinate $x$ in direct proportion to the semitone value $s$. Under this uniform scheme, a relatively small microtonal deviation ($\delta$) is represented by a correspondingly small horizontal displacement. This does not provide an effective visual representation, particularly for the distinction of the magnitudes of microtonal intervals and their structural relations.

A two-dimensional representation may be constructed by adding a second (vertical) coordinate $y$ in direct proportion to the microtonal deviation $\delta$.

**Definition 2.** *Vertical Displacement*

*Given the semitone value $s$ and deviation value $\delta$, a two-dimensional representation is defined by the coordinates ($x \propto s, y \propto \delta$).*

This provides increased visual emphasis of the direction and magnitude of the microtonal interval's deviation from its closest equal-tempered counterpart. Examples for the whole-tone frequency ratios 9/8 and 10/9 are shown in Fig. 2.

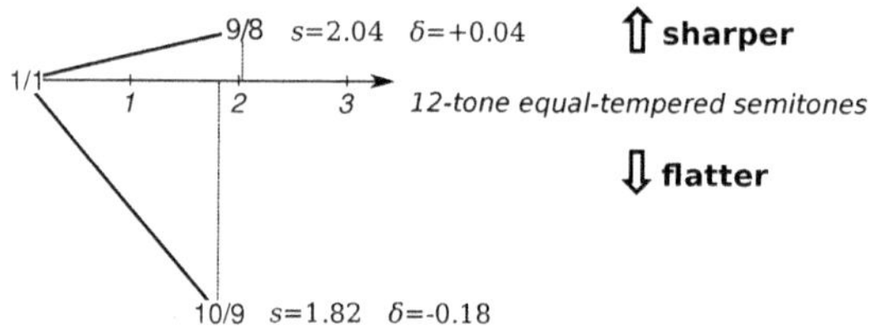


**Fig. 2.** The wholetone frequency ratios 9/8 and 10/9 are vertically displaced according to their deviation from the equal-tempered wholetone, providing increased visual emphasis of the direction and magnitude of deviation

In the following sections, this two-dimensional scheme is applied to examples of historical, speculative and ethnomusicological tuning systems, namely those of Pythagorean, Just, Septimal, Arabic 17-tone, and Turkish 24-tone intonation. Each system is presented with an overview of the theoretical basis of the system, a table of data values, a figure depicting the spatial representation (with musical annotations), and a summary of the structural features of the system.

Where musical stave representations are given, the starting note or tonic (represented by the ratios 1/1 and 2/1) is typically shown as D. This allows the structural symmetry to be reflected in the accidentals for each scale degree. It is hoped that this symmetry assists memory, provides structural clarity and allows straightforward comparison across the systems described[4].

**Pythagorean Intonation.** The intervals of Pythagorean intonation may be derived by taking ±6 perfect fifths (frequency ratio 3/2) from a reference tonic and

[4] The only exception to this is the Arabic 17-tone system whose asymmetric structure is more clearly presented starting on C; explanatory footnotes for this and the related Turkish 24-tone system are given.

normalising the resulting frequency ratios to fall within a single octave. Table 1 shows these normalised frequency ratios ($r$) with their semitone values ($s$) and deviation values ($\delta$). The corresponding two-dimensional spatial representation is shown in Fig. 3.

**Table 1.** Semitone ($s$) and deviation ($\delta$) values for Pythagorean ratios ($r$)

| $i$ | -6 | -5 | -4 | -3 | -2 | -1 | 1 | 2 | 3 | 4 | 5 | 6 |
|---|---|---|---|---|---|---|---|---|---|---|---|---|
| $r$ | $\frac{1024}{729}$ | $\frac{256}{243}$ | $\frac{128}{81}$ | $\frac{32}{27}$ | $\frac{16}{9}$ | $\frac{4}{3}$ | $\frac{3}{2}$ | $\frac{9}{8}$ | $\frac{27}{16}$ | $\frac{81}{64}$ | $\frac{243}{128}$ | $\frac{729}{512}$ |
| $s$ | 5.88 | 0.90 | 7.92 | 2.94 | 9.96 | 4.98 | 7.02 | 2.04 | 9.06 | 4.08 | 11.10 | 6.12 |
| $\delta$ | -0.12 | -0.10 | -0.08 | -0.06 | -0.04 | -0.02 | +0.02 | +0.04 | +0.06 | +0.08 | +0.10 | +0.12 |

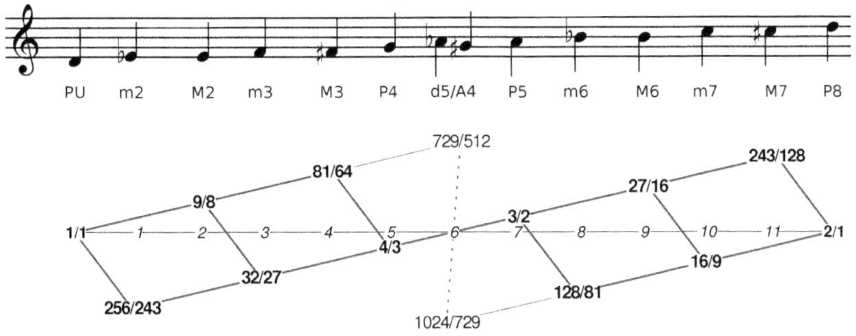


**Fig. 3.** A two-dimensional representation of Pythagorean intonation, illustrating the framework of tetrachords, wholetones and semitones, and the Pythagorean comma

Important features which are concealed in a one-dimensional representation are now clearly and intuitively apparent in the two-dimensional representation:

- The direction and relative magnitude of each Pythagorean interval's variation from its equal-tempered counterpart.
- The structural relations between the intervals, particularly the symmetrical division of the octave into a pair of disjunct tetrachords, each composed of repeated wholetone (9/8) and semitone(256/243) intervals.
- The Pythagorean Comma (between the intervals of 1024/729 and 729/512).

**Just Intonation.** A set of core intervals for Just intonation may be derived from the frequency ratios of the major and minor thirds (5/4 and 6/5), their respective inversions the minor and major sixths (8/5 and 5/3), and their respective perfect fourths and fifths, yielding minor and major seconds (16/15 and 10/9) and minor and major sevenths (9/5 and 15/8). Values for these ratios are shown in Table 2, while Fig. 4 shows the resulting two-dimensional plot superimposed on the Pythagorean system.

Again the direction and relative magnitude of each Just interval's variation from its equal-tempered counterpart is clearly presented, as are the structural relations between the intervals: in this case each Just tetrachord is composed of 16/15 semitones as well as 10/9 and 9/8 wholetones.

**Table 2.** Semitone ($s$) and deviation ($\delta$) values for Just ratios ($r$)

| $r$ | $\frac{16}{15}$ | $\frac{10}{9}$ | $\frac{6}{5}$ | $\frac{5}{4}$ | $\frac{8}{5}$ | $\frac{5}{3}$ | $\frac{9}{5}$ | $\frac{15}{8}$ |
|---|---|---|---|---|---|---|---|---|
| $s$ | 1.12 | 1.82 | 3.16 | 3.86 | 8.14 | 8.84 | 10.18 | 10.88 |
| $\delta$ | +0.12 | -0.18 | +0.16 | -0.14 | +0.14 | -0.16 | +0.18 | -0.12 |

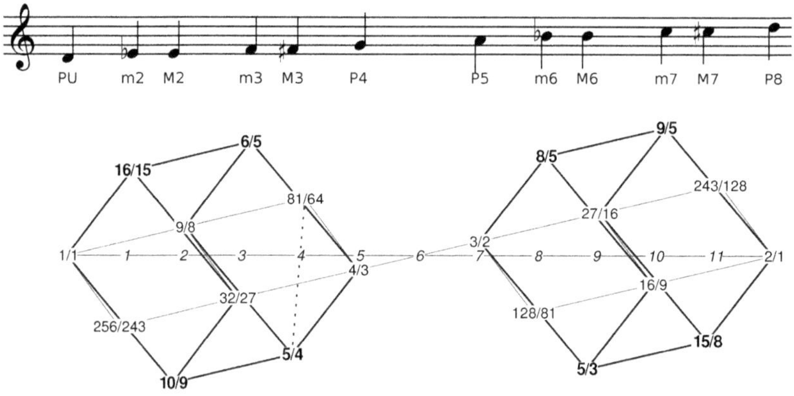


**Fig. 4.** A two-dimensional representation of Just intonation, illustrating the framework of tetrachords, wholetones and semitones, and the Syntonic comma

Furthermore, visual comparison between the Just and Pythagorean systems illustrates the spatial relation between the Just and Pythagorean major thirds (5/4 and 81/64) representing the Syntonic Comma, with multiple occurrences (eight in total) clearly discernible (256:243→16:15, 10:9→9:8, etc.)

**Septimal Intonation.** Dividing the octave ratio 2/1 as 8:7:6:5:4, the number 7 yields the septimal ratios 7/4, 7/5, 7/6 with respective inversions 8/7, 10/7 and 12/7. Furthermore the perfect fifth 3:2 may be divided as 9:8:7:6 yielding the additional ratio 9/7 with inversion 14/9[5]. These septimal ratios are shown in Table 3 with the resulting two-dimensional plot shown in Fig. 5.

**Table 3.** Semitone ($s$) and deviation ($\delta$) values for septimal ratios ($r$)

| $r$ | $\frac{8}{7}$ | $\frac{7}{6}$ | $\frac{9}{7}$ | $\frac{7}{5}$ | $\frac{10}{7}$ | $\frac{14}{9}$ | $\frac{12}{7}$ | $\frac{7}{4}$ |
|---|---|---|---|---|---|---|---|---|
| $s$ | 2.31 | 2.67 | 4.35 | 5.83 | 6.17 | 7.65 | 9.33 | 9.69 |
| $\delta$ | +0.31 | -0.33 | +0.35 | -0.17 | +0.17 | -0.35 | +0.33 | -0.31 |

Notable features of the septimal intervals are:

- They deviate considerably more from 12-tone equal temperament than the Pythagorean and Just intervals (i.e. the Septimal system is not as well-represented by 12-tone equal temperament as the previous systems).

[5] The remaining septimal ratios 11/7 and 13/7 are not included as they involve prime numbers greater than 7.

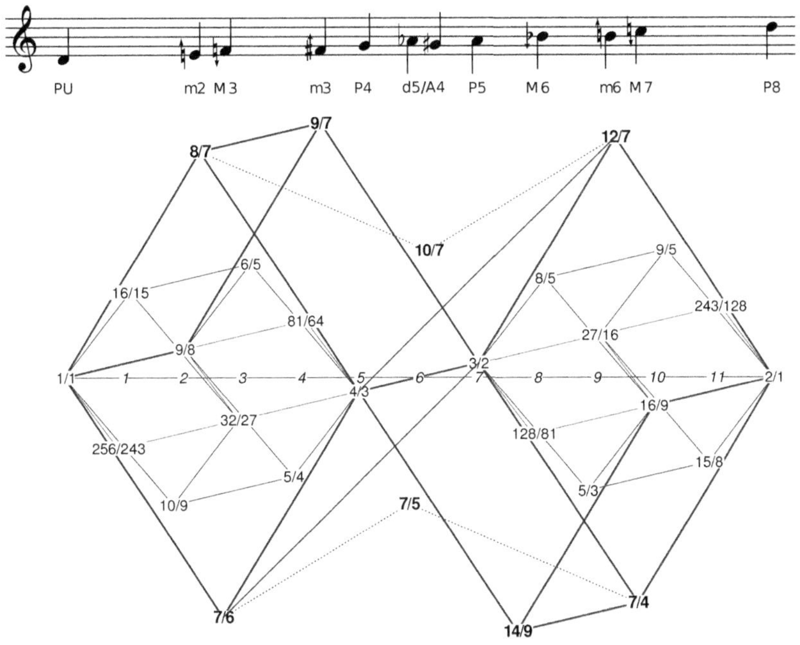


**Fig. 5.** A two-dimensional representation of septimal intervals and the framework of septimal tetrachords and pentachords in relation to the Pythagorean and Just systems

- They form two disjunct tetrachords (each spanning a perfect fourth) or two overlapping pentachords (each spanning a perfect fifth) which may form the basis of pentatonic, hexatonic or heptatonic octave species.
- Their relational structures incorporate Pythagorean relations (such as 9/8) and Just relations (such as the major triad at $7/6 \to 7/5 \to 7/4$) suggesting that conventional triadic harmonies may be used in a septimal context.
- The division of the perfect fourth 4/3 as 8:7:6 may be considered a dual of the more conventional division of the perfect fifth 3/2 as 6:5:4, suggesting the possibility of a system of specifically septimal harmonies.

**Arabic 17-tone Intonation.** The Arabic 17-tone intonation of the 13th century theorist Safi al-Din is based on an extension of the Pythagorean scheme such that it runs from $i = -12$ to $i = +4$ perfect fifths from a reference tonic [26]. The additional intervals are shown in Table 4 with the resulting two-dimensional plot of the complete scheme shown in Fig. 6[6].

This provides a clear visual representation of the extended Pythagorean system and its hierarchical decomposition of the octave into two tetrachords and a wholetone, each tetrachord into two wholetones and a semitone, and each wholetone into two semitones and a comma [26].

[6] For structural clarity, the stave representation is transposed to a tonic of C, such that each of the five instances of the Pythagorean comma (on the M2, M3, P5, M6 and M7 scale degrees) is indicated by the modified accidental denoting “comma flat”.

**Table 4.** Semitone ($s$) and deviation ($\delta$) values for the additional Arabic 17-tone ratios ($r$)

| $i$ | -12 | -11 | -10 | -9 | -8 | -7 |
|---|---|---|---|---|---|---|
| $r$ | $\frac{1048576}{531441}$ | $\frac{262144}{177147}$ | $\frac{65536}{59049}$ | $\frac{32768}{19683}$ | $\frac{8192}{6561}$ | $\frac{4096}{2187}$ |
| $s$ | 11.77 | 6.78 | 1.80 | 8.82 | 3.84 | 10.86 |
| $\delta$ | -0.23 | -0.22 | -0.20 | -0.18 | -0.16 | -0.14 |

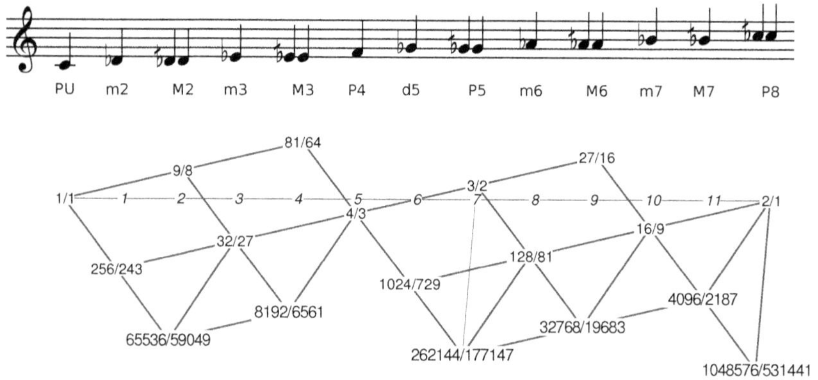


**Fig. 6.** A two-dimensional representation of Arabic 17-tone intonation, illustrating its Pythagorean-based framework of tetrachords, wholetones, semitones and commas

**Turkish 24-tone Intonation.** A further extension of the Pythagorean scheme forms the basis of the Turkish 24-tone system in which the system of perfect fifths is extended to run from $i = -12$ to $i = +11$ from the reference tonic [27]. The additional intervals are shown in Table 5, and the complete scheme is depicted in Fig. 7[7].

**Table 5.** Semitone ($s$) and deviation ($\delta$) values for the additional Turkish 24-Tone ratios ($r$)

| $i$ | +7 | +8 | +9 | +10 | +11 |
|---|---|---|---|---|---|
| $r$ | $\frac{2187}{2048}$ | $\frac{6561}{4096}$ | $\frac{19683}{16384}$ | $\frac{59049}{32768}$ | $\frac{177147}{131072}$ |
| $s$ | 1.14 | 8.16 | 3.18 | 10.20 | 5.22 |
| $\delta$ | 0.14 | 0.16 | 0.18 | 0.20 | 0.22 |

The adaptations of the stave notation reflect the theoretical division of the whole-tone into 9 commas, with the full set of 10 accidentals denoting 1, 4, 5, 8 and 9 commas sharp or flat respectively (as presented in [27]).

7 Yarman describes the Arel-Ezgi-Uzdilek system starting on C and the equivalent Yekta-24 system starting on D; the stave notation shown uses the latter in order to preserve structural and notational symmetry and facilitate straightforward comparison with the Pythagorean and Just representations.

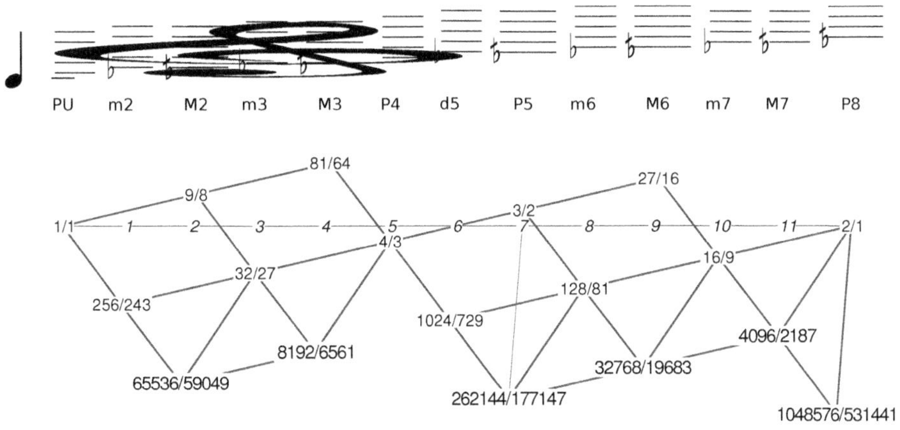


**Fig. 7.** A two-dimensional representation of the Turkish 24-tone system, illustrating its Pythagorean-based framework of tetrachords, wholetones, semitones and commas

Of particular significance is the correspondence of some of these extended Pythagorean intervals with those of the Just intonation system (as is also partially the case for the Arabic 17-tone system [26]). This is immediately apparent in the visual representation of Fig. 8 where the eight Just intervals (introduced in Table 2) each have a very close correspondence to intervals from the extended Pythagorean series (for example the Just major third 5/4 and the Pythagorean interval 8192/6561).

## 2.3 Projection in Two Dimensions

Particular conceptions of pitch are attainable by projecting the two-dimensional spatial representations in specific directions. These are depicted in Fig. 9 where the diatonic scale is obtained by projecting along the direction between the

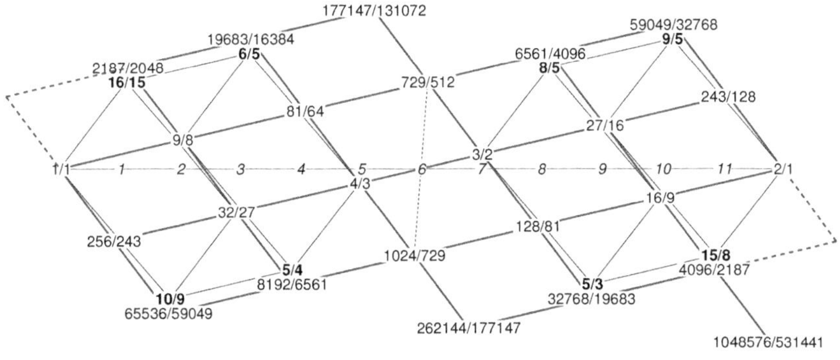


**Fig. 8.** The extended Pythagorean intervals of the Turkish 24-tone system, and their correspondence with the intervals of the Just intonation system

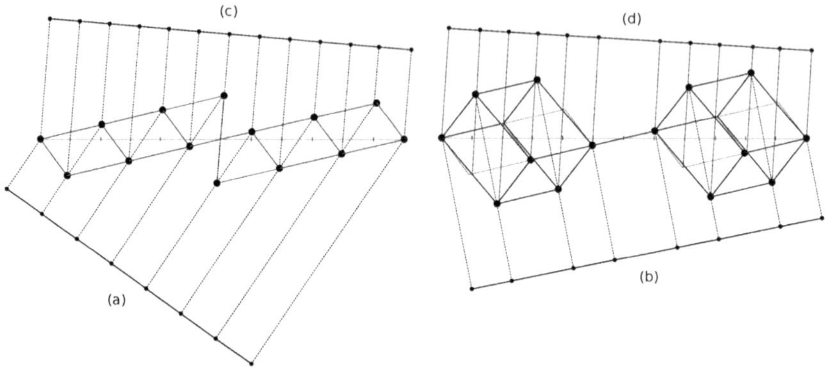


**Fig. 9.** Two-dimensional projection of the diatonic scale for (a) Pythagorean and (b) Just intonation, and the chromatic scale for (c) Pythagorean and (d) Just intonation

major- and minor-second intervals for (a) Pythagorean and (b) Just intonation; likewise the chromatic scale is obtained by projecting along the direction of the respective comma (Pythagorean or Syntonic) for (c) Pythagorean[8] and (d) Just intonation. This links the numerically-derived spatial model with the more familiar one-dimensional categorical models of the diatonic and chromatic scales.

### 2.4 Displacement and Projection in Three Dimensions

The cube-like appearance of the tetrachords of Just intonation suggest that Fig. 4 is a projection of a three-dimensional structure. To reflect this, the existing two-dimensional model may be extended with the addition of a third coordinate $z$ as a function of the deviation value $\delta$.

**Definition 3.** *Depth Displacement*

*Given the semitone value $s$ and deviation value $\delta$, a three-dimensional representation is defined by the coordinates $(x \propto s, y \propto \delta, z \propto f(\delta))$, where:*

– *let* $i = \delta_{3:2}$ , $j = \frac{\delta_{8:5}+\delta_{6:5}}{2}$ *and* $k = \frac{j}{2icos(atan(cos30))}$
– *then for Just Intonation let* $c = j(k + tan(\frac{1}{j}))$ *and:*
  - $f(\delta) = k\delta - c$ *for* $\delta > 0$,
  - $f(\delta) = k\delta + c$ *for* $\delta < 0$,

  *while for all other intervals* $f(\delta) = k\delta$.

Effectively the $z$-coordinate is scaled by $k$ while Just intervals are displaced by $c$ such that they are no longer embedded in the plane $y \propto z$.

The three-dimensional model may now be freely rotated, thus allowing projections from different viewpoints. In particular, the model may be viewed along the central equal-tempered reference axis (the $x$ axis) yielding a view isomorphic to the two-dimensional 'Tonnetz', as depicted in Fig. 10.

[8] This is a non-equal-tempered equivalent of Haluška's Pythagorean-chromatic projection [24].

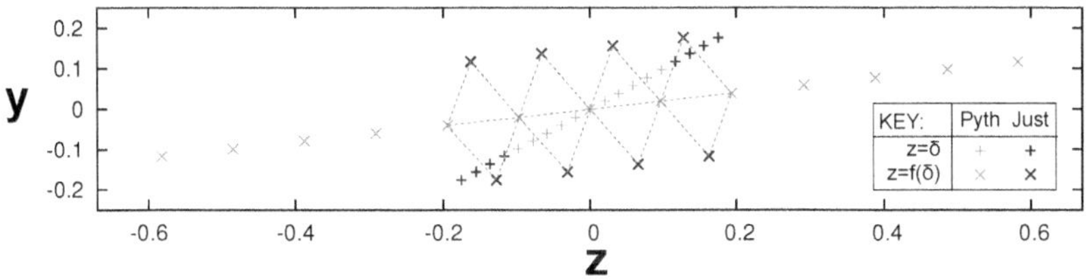


**Fig. 10.** The three-dimensional spatial representation projected along the $x$ axis, yielding the two-dimensional Tonnetz

# 3 Implications

The presented model exhibits *intuitive*, *comparative* and *integrative* characteristics, each of which has respective educational, theoretical, musicological, psychological and computational implications.

**Educational.** The visually intuitive nature of the representation enhances the accessibility of microtonal-theoretic concepts and clearly depicts the relational structures which form the basis of musical scales. This allows the specialized notational adaptations for microtonality to be placed in systematic context, and is therefore of practical as well as theoretical use.

**Theoretical.** By representing microtonal intervals as deviations from an equal-tempered reference, the model integrates the numerical and categorical domains of music theory. This yields a fresh and accessible perspective on individual historical tuning theories, and allows them to be placed in a contemporary context alongside speculative theories.

**Musicological.** The representational consistency of the model provides a comparative framework for assessing microtonal systems in ethnomusicological and historical contexts.

**Psychological.** The model integrates the different dimensions of pitch cognition through projection. This relates to the notion of "emergence" and the interaction of microscopic and macroscopic cognitive structures as encountered in the field of synergetics [28].

**Computational.** The integrative aspects of the model highlight the parallels between human and artificial intelligence, and in particular the potential isomorphism between cognitive and algorithmic structures. The comparative aspects of the model may provide the basis for computational applications of broader scope than a culturally specific model can provide, while the intuitive aspects may inspire improvements in human-computer interactional aspects of such applications.

**Further Possibilities.** Some further possibilities beyond the immediate scope of this paper include:

- The application of the model to the visualization and representation of non-equal temperaments (such as mean-tone or irregular temperaments) in addition to the ratio-based pure-tone systems presented above.

- The investigation of reference bases other than 12-tone equal temperament, such as the 19-tone (depicted in Fig. 11(a)) and 31-tone equal-temperaments adopted by Yasser and Fokker respectively [1].
- The extension of the idea of the two-dimensional Tonnetz projection (as in Fig. 10) to the three-dimensional septimal Tonnetz (Fig. 11(b)).

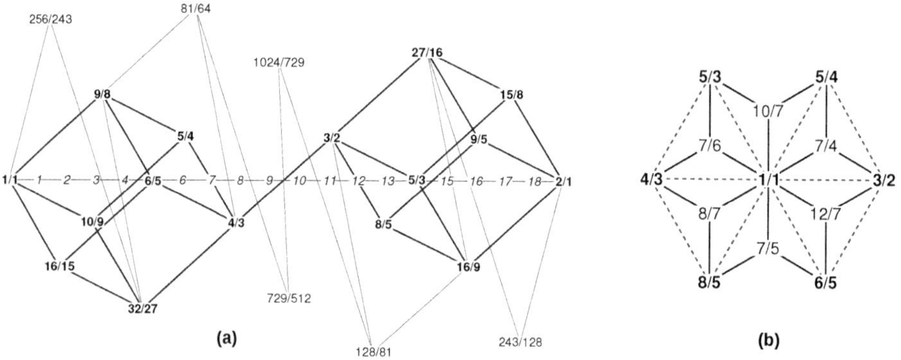


**Fig. 11.** Further possibilities: (a) 19-tone equal temperament preserving the relational structure of Just intonation while that of Pythagorean intonation breaks down, and (b) a three-dimensional Tonnetz projection incorporating septimal intervals

# 4 Conclusion

This paper has presented a spatial model of microtonality in two and three dimensions. The model provides enhanced accessibility of microtonal-theoretic concepts through its visually intuitive representation of microtonal intervals and their relational structures. Application of the model to various historical, speculative and ethnomusicological tuning systems has demonstrated a comparative framework in which these systems can be assessed. The model also integrated different dimensions of pitch cognition through projection, illustrating the interactions and potential emergence of hierarchical cognitive structures. The characteristics of the model have computational implications with regards to human-computer interaction, breadth of scope of application, and related algorithmic structures. It is hoped that the model is an example of the application of spatial reasoning which complements formal mathematical and computational treatments of music.

# References

1. Griffiths, P., Lindley, M., Zannos, I.: Microtone. In: Sadie, S., Tyrrell, J. (eds.) The New Grove Dictionary of Music and Musicians, vol. 16, pp. 624–625. Macmillan, London (2001)
2. Lindley, M.: Tuning. In: Sadie, S., Tyrrell, J. (eds.) The New Grove Dictionary of Music and Musicians, vol. 25, pp. 884–885. Macmillan, London (2001)

3. Leedy, D., Haynes, B.: Intonation (ii). In: Sadie, S., Tyrrell, J. (eds.) The New Grove Dictionary of Music and Musicians, vol. 12, pp. 503–504. Macmillan, London (2001)
4. Lindley, M.: Temperaments. In: Sadie, S., Tyrrell, J. (eds.) The New Grove Dictionary of Music and Musicians, vol. 25, pp. 248–268. Macmillan, London (2001)
5. Nolan, C.: Music Theory and Mathematics. In: Christensen (ed.) The Cambridge History of Western Music Theory, pp. 272–304. Cambridge University Press, Cambridge (2002)
6. Adkins, C.: Monochord. In: Sadie, S., Tyrrell, J. (eds.) The New Grove Dictionary of Music and Musicians, vol. 17, pp. 2–4. Macmillan, London (2001)
7. Mathiesen, T.J.: Greek Music Theory. In: Christensen (ed.) The Cambridge History of Western Music Theory, pp. 109–135. Cambridge University Press, Cambridge (2002)
8. Klumpenhouwer, H.: The Cartesian Choir. Music Theory Spectrum 14(1), 15–37 (1992)
9. Gouk, P.: Music, Science and Natural Magic in Seventeenth-Century England. Yale University Press, New Haven (1999)
10. Mooney, M.K.: The 'Table of Relations' and Music Psychology in Hugo Riemann's Harmonic Theory. Ph.D. diss., Columbia University (1996)
11. Shepard, R.N.: Geometrical Approximations to the Structure of Musical Pitch. Psychological Review 89(4), 305–333 (1981)
12. Krumhansl, C.L.: Cognitive Foundations of Musical Pitch. Oxford University Press, New York (1990)
13. Chew, E.: Modeling Tonality: Applications to Music Cognition. In: Proceedings of the 23rd Annual Meeting of the Cognitive Science Society (2001)
14. Chew, E.: The Spiral Array: An Algorithm for Determining Key Boundaries. In: Anagnostopoulou, C., Ferrand, M., Smaill, A. (eds.) ICMAI 2002. LNCS (LNAI), vol. 2445, pp. 18–31. Springer, Heidelberg (2002)
15. Chew, E., Chen, Y.C.: Real-Time Pitch Spelling Using the Spiral Array. Computer Music Journal 29(2), 61–76 (2005)
16. Gollin, E.: Some Aspects of Three-Dimensional "Tonnetze". Journal of Music Theory 42(2), 195–206 (1998)
17. Polansky, L.: Paratactical Tuning: An Agenda for the Use of Computers in Experimental Intonation. Computer Music Journal 11(1), 61–68 (1987)
18. Lindley, M.: Well-Tempered Clavier. In: Sadie, S., Tyrrell, J. (eds.) The New Grove Dictionary of Music and Musicians, vol. 27, pp. 274–277. Macmillan, London (2001)
19. Wolf, A.: Graphic Representation and Functional Systematics of Historic Musical Temperaments. ISO Information 21, 41–70 (1980)
20. Kellner, H.A.: A Visualization of Organ Tunings for Perception of their Structure. Das Musikinstrument 44(10), 76 (1995)
21. Sandoz, R.: Les problèmes posés par les tempéraments. Revue Musicale de Suisse Romande 35(3), 140–145 (1982)
22. Sundberg, J., Friberg, A., Frydn, L.: Rules for Automated Performance of Ensemble Music. Contemporary Music Review 8, 89–109 (1989)
23. Haruko, K., Mihoko, N.: Theory and Notation in Japan. In: Nettl, B., Stone, R. (eds.) The Garland Encyclopedia of World Music, vol. 7, pp. 565–584. Routledge, New York (2002)
24. Haluška, J.: Equal Temperament and Pythagorean Tuning: A Geometrical Interpretation in the Plane. Fuzzy Sets and Systems 1148, 261–269 (2000)

25. Sethares, W.A.: Tuning, Timbre, Spectrum, Scale, 2nd edn. Springer, London (2005)
26. Wright, O.: Safi al-Din. In: Sadie, S., Tyrrell, J. (eds.) The New Grove Dictionary of Music and Musicians, vol. 22, pp. 86–87. Macmillan, London (2001)
27. Yarman, O.: A Comparative Evaluation of Pitch Notations in Turkish Makam Music. Journal of Interdisciplinary Music Studies 1(2), 43–61 (2007)
28. Stadler, M., Kruse, P.: Gestalt Theory and Synergetics: From Psychophysical Isomorphism to Holistic Emergentism. Philosophical Psychology 7(2), 211–226 (1994)

# Music between Hearing and Counting (A Historical Case Chosen within Continuous Long-Lasting Conflicts)

Tito M. Tonietti

Dipartimento di matematica, Università di Pisa, Italy
tonietti@dm.unipi.it

**Abstract.** Here is shown Bernhard Riemann's reaction to Helmholtz's *Lehre von den Tonempfindungen*. Then I recall how Joseph Joachim and Johannes Brahms valued Helmholtz's "natural" tuning. In the end, Planck's experiments with a particular new harmonium, and an *a cappella* choir concerning "natural" or tempered tuning are described.

**Keywords:** Mathematics, Physics, Music, History, 19th Century, Tuning, Ear.

## 1 The Conflicts

As always, Leibniz was too optimistic with his "Music is a hidden arithmetical exercise of the soul who is unaware that he/she is counting"[1]. In history, often, the numbers offered by the theoreticians and natural philosophers did not agree with the notes chosen by musicians following their ears. At once, the conflict began between the Pythagorean sects and Aristoxenus who did not divide the *diapason* [octave] by numbers into intervals, but he listened to them [2]. Concerning the same problem, the lute player and composer Vincenzio Galilei (father of Galileo) opposed the *senario* calculated by Gioseffo Zarlino, who used ratios between the numbers 2, 3, 4, 5, 6. In fact, the former followed a good approximation of the equal temperament [3]. Among other things, the ear of the Florentine was very good, because he invented rules to tune the strings changing the tension and to tune the pipes of the organs changing the volume: rules which were mistaken at that time.

With the evolution of music and science during the centuries, one could think that this conflict was going to be smoothed out. But that did not happen.

The numerical side of the conflict was strengthened by natural philosophers by the theory of *ictus* [beats, strikes against the ear] first, and then by that of harmonic tones. While, to allow a better moving among notes by transposing and modulating, the hearing side spontaneously glided toward temperament, which

[1] "Musica est exercitium arithmeticae occultum nescientis se numerare animi" [1].

C. Agon et al. (Eds.): MCM 2011, LNAI 6726, pp. 285–296, 2011.

gained ground among musicians. Like Vincenzio Galilei, Simon Stevin was an exception among natural philosophers; while many others invented their own ways to support the old ratios, which musicians did not follow. Descartes, Newton, and Leibniz were very unwilling to accept the temperament; they suffered it for practical reasons only. As if he were more a theoretician than a composer, even Rameau converted to the temperament in a second moment only [4]. To protect his theory based on prime numbers, Euler went so far as to hypothesize that, whatever tempered or "natural" or Pythagorean intervals were tuned, the ear should hear *his* "right" numbers [5].

All that is known[2], and I do not need to give here details or new evidence again, and to stress them. Anyway, the conflict between hearing, and counting went on during the 19th Century. Between Georg Simon Ohm, and Ludwig Wilhelm Augustus Seebeck arose an opposition concerning the relative intensity of overtones in the sequence of the harmonics of a note. Here again Seebeck's ear was the source of criticism against Ohm:

> "Under the restrictive conjectures, which Ohm assumed to simplify considering the strikes of siren, one gets results which are not corroborated by experience, and at least on this side one must deal with theory in a more general way"[3].

Ohm's reaction was appealing to "facts" [Thatsachen]:

> "... since in this matter I cannot do anything by ear, because nature totally refused me a musical ear." "Those disagreements perceived by Seebeck rest on a hearing deceit, in which our ear is trapped, ... Because I assume exactly that our ear unwittingly perceives the fundamental note stronger than it actually is, and its harmonics weaker than they actually are ..."[4].

And now I will tell less known episodes of this long story. In this article, I neglect the different numerical theories of music, and the different ways of judging music by the ear [15][5].

---

[2] Cf. [6,7,8,9,10,11,12].

[3] "Unter den beschränkenden Voraussetzungen, welche Ohm in Betreff der Sirenestöße der Einfachheit wegen angenommen hat, gelangt man zu Resultaten, welche durch die Erfahrung nicht bestätigt werden, und man muß wenigstens von dieser Seite her die Theorie allgemeiner behandeln." [13], pp. 480-481. All English translations are by myself.

[4] "... da ich mit dem Ohre in dieser Sache nichts zu thun vermag, weil mir die Natur ein musikalisches Gehör ganz und gar versagt hat." "Jene von Seebeck wahrgenommenen Widersprüche beruhen auf einer Gehörstäuschung, in welcher unser Ohr befangen ist, ... Ich nehme nämlich an, dass unser Ohr unwillkührlich den Haupton für stärker ansieht, als er wirklich ist, und seine Beitöne für schwächer, als sie wirklich sind, ..."[14], pp. 7, 15. [10] Section 10.2.

[5] Cf. [10]. For an other not historical approach Cf. [15].

## 2 Hermann von Helmholtz, Bernhard Riemann, Johannes Brahms, and Joseph Joachim

It is well known that Hermann von Helmholtz wrote the book which dominated the landscape of acoustics during the 19th Century: *On the Sensations of Tone as a Physiological Basis for the Theory of Music*[6].

In short, he offered a theory of musical scales and consonances based on harmonic sounds (or overtones) and beats together with a tentative physiological analysis of how the ear worked. The general result was a justification on physical grounds of these intervals which are set by integers numbers: the multiples of the fundamental frequency. Helmholtz wrote that he had discovered "... the physiological basis of the mysterious law of the numerical ratios found by Pythagoras." Thus, "... consonant intervals are in the ratio of the smallest integers"[7].

Notwithstanding its success mainly among physicists, the reactions to Helmholtz's theory were not entirely positive. The mathematician Bernhard Riemann left a manuscript, unfortunately unfinished because he untimely died before, to be published only posthumous: "Mechanics of the Ear" [19].

> "In the book *On the Sensations of Tone as a Physiological Basis for the Theory of Music* by Helmholtz is collected the progress which, in more recent times, has been made in the establishment of the extremely difficult facts concerning the perception of musical notes, and made in particular by Helmholtz himself. Since I am often obliged to oppose the results which Helmholtz draws from his experiments and observations, I believe I must express here even more how much I acknowledge the great merits of his work on our subject. In my view, however, those are not to be looked for in his theories of the mouvements of the ear, but in improving the experimental basis for the theory of these movements"[8].

Reading Riemann's paper, as far as it breaks off, these criticisms are not always easy to assess explicitly. Let us explain a few. Helmholtz experimented with Seebeck's sirens which had 8, 12, 16 holes. Therefore he tested the numerical

[6] *Die Lehre von den Tonempfindungen als physiologische Grundlage für die Theorie der Musik* [16]. See also [17] and [18].

[7] [17], pp. 2, 6-7, 20-21. Cf. [10], Section 10.2. I admit that the way I depict here Helmholtz may be too dry and rough, but this article is not focused on him.

[8] "In dem Buche *Die Lehre von den Tonempfindungen als physiologische Grundlage für die Theorie der Musik* von *Helmholtz*, findet man die Fortschritte zusammengestellt, welche in der so äusserst schwierigen Ermittelung der Thatsachen, die die Wahrnehmung der Töne betreffen, in neuerster Zeit gemacht worden sind und zwar vorzüglich von *Helmholtz* selbst. Da ich den Folgerungen, welche *Helmholtz* aus den Versuchen und Beobachtungen zieht, entgegen zu treten vielfach genöthigt bin, so glaube ich um so mehr gleich hier aussprechen zu müssen, wie sehr ich die grossen Verdienste seiner Arbeiten über unsern Gegenstand anerkenne. Sie sind aber meiner Ansicht nach nicht in seinen Theorien von den Bewegungen des Ohres zu suchen, sondern in der Verbesserung der erfahrungsmässigen Grundlage für die Theorie dieser Bewegungen" [19], p. 373.

ratios of intervals confirming the scale of Ptolemaeus and Zarlino. He justified the Pythagoreans, and stressed Aristoxenus' mistakes. Euler's *Tentamen* [20], he wrote, had to be completed because it lacked overtones and beats: he, Helmholtz, himself did it in his *Lehre*. On the contrary, no numbers have to be found in Riemann's *Mechanik*. Here the argument was kept inside the ear.

In the present article, I have enough room for a very brief sketch only. Riemann tried to explain why the sense of hearing was so accurate. At the beginning, the German mathematician faced a problem of method. Is it better to start from the anatomy of the ear to study the causes which explain its functioning, and performances? Or perhaps, on the contrary, would it be preferred to begin from the results achieved by the ear to come back to the conditions of its functioning?

> "With Newton and Herbart one can call the approach of the first synthetic and of the latter analytic." "The author is Herbartian in Psychology and Theory of knowledge (Methodology and Eidolology). With Herbart's Philosophy of nature, and the relative metaphysical disciplines (Ontology and Synechology), he [Riemann] does not agree most of the time"[9].

Riemann took the second procedure. He therefore looked for physical properties of the parts constituting the organ which could justify its performance. The German mathematician described experiments that bring to light the extreme sensitivity displayed by the ear in detecting sound. He inquired into which means made it possible to transform a sound wave into the refined perception of the *Klang*, while they keep intact the harmonic characteristics, notwithstanding they had to amplify the sound strength scattered in the space. This *Mechanik* is organized in the following scheme: "On the method to apply in the physiology of the most refined sense organs. [...]Synthetic approach. [...] Analytic approach. [...] Cavity of the tympanum"[10].

Of the ear, with every attention, was analysed the continuous linking together of the several parts from the air to the acoustical nerve:

> "... they must be continuously linked to one another with an exactitude more than microscopic, ... So that the apparatus can transmit the smallest modifications of the pressure of the air to the water of the labyrinth, amplifying it in an always equal ratio, ..."[11]

[9] "Man kann mit Newton und Herbart den ersten Weg den synthetischen, den zweiten den analytischen nennen." "Der Verfasser ist Herbartianer in Psychologie und Erkenntniss-theorie (Methodologie und Eidolologie). Herbart's Naturphilosophie und den darauf bezüglichen metaphysischen Disciplinen (Ontologie und Synechologie) kann er meistens nicht sich anschliessen" [19], pp. 307, 508.

[10] "Ueber die in der Physiologie der feineren Sinnesorgane anzuwendende Methode. [...] Synthetischer Weg [...] Analytischer Weg [...] Paukenhöhle." [19] pp. 370-371, 374.

[11] "... fortwährend mit mehr als mikroskopischer Genauigkeit in einander greifen müssen, ... Damit der Apparat die kleinsten Druckänderungen der Luft, in stets gleichen Verhältniss vergrössert, dem Labyrinthwasser mittheilen könne, ..." [19], pp. 345, 349.

Riemann considered the sensitivity of the ear as much more important than the mechanics. "In fact is not completely improbable that by the same truly are transmitted sound-motions which are so thin that they could not be perceived by the microscope"[12]. Instead, even though he ascribed a similar sensibility to the ear, Helmholtz made acoustical phenomena visible. He let a tuning fork trace a $sin - curve$, and used the vibration-microscope invented by Jules Lissajous. Of course, he did not trust the judgement of the ear enough.

The German mathematician would have wished to yield a "Klang Curve" [harmonic curve]. We do not know either the way he would have followed to calculate it, or what it would have looked like. Creativity cannot be continued as an analytic function does, because it has too many singularities. His idea of space was embodied in the famous "Ueber die Hypothesen, welche der Geometrie zu Grunde liegen". To these anti-Newtonian Hypothesis, Helmholtz later replied with philo-kantian facts: "Ueber die Thatsachen, die der Geometrie zum Grunde liegen".

Riemann had begun to face a subjective, and conscious listening to music. "Then, our research on listening keeps its validity for an ear which is purpose-designed towards an exact hearing, ..."[13]. Even in the works of art, Helmholtz was looking for "... a similar order, which rules the universe where everywhere law and reason dominate"[14]. In a later addition to his theory, he wrote:

> "1. Theoretical intervals, which in my book I called natural, really are so for a not perverted ear; 2. The mistakes of the tempered scale actually are remarkable and disagreeable for a just ear; 3. Notwithstanding little difference between these intervals separately taken, is much easier to sing just intonation following the natural scale than following the tempered scale"[15].

The German physicist and physiologist looked for endorsements among musicians. Did he get them? *Das Rheingold* [Rhinegold] by Richard Wagner was created later, in 1869 only. At that time, he could not know that the first chord after a while glided into "natural" tuning. Anyway, Helmholtz criticized Beethoven:

> "...Mozart and Beethoven were yet at the commencement of the reign of equal temperament. ... Beethoven eagerly and seized the wealth offered

[12] "In der That ist es durchaus nicht unwahrscheinlich, dass durch denselben Schallbewegungen treu mitgetheilt werden, die so klein sind, dass sie mit dem Mikroskop nicht wahrgenommen werden könnten" [19], p. 375.

[13] "Es bleibt dann unsere Untersuchung für das lauschende, zum genauen Hören absichtlich vorgerichtete Ohr giltig [sic! gültig], ..." [19], p. 380.

[14] "... un ordre semblable, qui règne dans l'universe où dominent partout la Loi et la Raison" [17], p. 481.

[15] "1. Les intervalles théoriques, que j'ai appellés intervalles naturels dans mon livre, le sont bien effectivement pour une oreille non pervertie; 2. Les erreurs de la gamme tempérée sont en réalité appréciables et désagréables pour une oreille juste; 3. Malgré le peu de différence des intervalles pris isolément, il est beaucoup plus facile de chanter juste suivant la gamme naturelle que suivant la gamme tempérée" [17], pp. 537-538.

> by instrumental music; ... But he used the human voice as a mere handmade...And after all, I do not know that it was so necessary to sacrifice correctness of intonation to the convenience of musical instruments"[16].

By J & P. Schiedmayer of Stuttgard, Helmholtz let modify a great harmonium, with two keyboards tuned in his "natural" way. Playing it, he began doing experiments. By him, an episode was told. "Mr. Joachim, the famous artist, was so kind to play the scale on his violin, beside my harmonium. He made thirds and sixths just perfectly"[17].

But this is one side of the history only. What was the point of view of the musicians? We are lucky enough to be aware of another evidence. Joseph Joachim was a close friend of Robert Schumann and Johannes Brahms. With the latter, a day he visited Helmholtz's home.

> "Once Joachim and I [Brahms] were at Helmholtz', who demonstrated us his discoveries and 'pure' harmonies on the instruments invented by him. He claimed that the seventh should sound a little higher, the third lower than normally. Joachim, who certainly is a a very polite person, should have received a quite particular impression of those intervals, and acted as if he heard them as Helmholtz did. But I told him that the matter were too earnest to be decided by politeness; I [Brahms] always heard the opposite of what Helmholtz claimed. Then he [Joachim] admitted that quite the same thing happened to him too. Helmholtz had several keyboards, and I pointed out to him that the notes of the second sounded much sharper than the others; he had to admitt this. In musical matters, as it turns out, he is just an awful *dilettante*"[18].

---

[16] "...Mozart et Beethoven sont encore au début de la période où commence la domination du tempérament égale. ... Beethoven a pris hardiment possession des richesses que pouvait lui offrir développement de la musique instrumentale; ... Mais il a traité la voix humaine comme une humble servant, ... Et, dans tout cela, je ne vois pas qu'il soit nécessaire de sacrifier la justesse de la voix aux commodités de la musique instrumentale." [17], p. 432.

[17] "Mr. Joachim, le célèbre artiste, a eu la bonté de jouer la gamme sur son violon, à côté de mon harmonium. Il faisait les tierces et les sixtes parfaitement justes" [17], p. 429.

[18] "Einmal waren Joachim und ich bei Helmholtz, der uns seine Entdeckungen und die reinen Harmonien auf den von ihm erfundenen Instrumenten vorführte. Er behauptete, die Septime müsse etwas höher, die Terz tiefer klingen als gewöhnlich. Joachim, der ja ein sehr höflicher Mann ist, wollte erst einen ganz eigentümlichen Eindruck von der Intervallen empfangen haben und tat so, als ob er sie gerade so höre wie Helmholtz. Da sagte ich ihm, die Sache sei doch zu ernst, als daß auch hier die Höflichkeit entscheiden könne; ich hörte immer das Gegenteil von dem, was Helmholtz behauptete. Da gab er denn zu, daß es eigentlich auch bei ihm der Fall sei. Helmholtz hatte mehrere Klaviaturen. Ich machte ihn darauf aufmerksam, daß die Töne der zweiten viel schärfer klängen als die der andern; er mußte es zugeben. Er selbst ist eben in musikalischen Dingen ein entsetzlicher Dilettant" [21], p. 279.

Even on its high, the scientific theory of sound collided with the ear of musicians, and what a composer and pianist was that person!

## 3 Max Planck

Because of his other success in physical researches, while he was looking after harmoniums tuned in his way, Helmholtz reorganized German scientific institutions, and became the outstanding personality of German science. So, in Berlin to the Institute for Theoretical Physics, was assigned a new harmonium tuned in "natural" scale, following the system of Carl Eitz, and built by the same Schiedmayer's factory of pianos already mentioned. This experimental harmonium met Max Planck when he was called to the University of Berlin. The famous theoretical physicist played it, and studied the problem of tuning which still laid in the dilemma between "natural" or tempered one.

In 1893, during a session of the Physikalische Gesellschaft zu Berlin, Planck announced how the instrument worked. The keyboards covered four octaves and one half, each could tune 104 different notes. The keys were colored in green, blue, white, and red. Max Planck "... then discussed some experiments of musical relevance collected upon studying the instrument, with whose further development he is still occupied, and supported these with several examples"[19].

In the same year, 1893, the theoretical physicist accomplished a much longer work, "The natural tuning in modern vocal music", but it was published in a musical journal, now [23][20]. A student also of Helmholtz during his youth, Planck followed the master's idea about the "natural" tuning and checked whether he could find any case for which the "natural" one worked better than the usual tempered.

Our physicist was very skillful in music, more than an average German *amateur*. During his youth, he committed himself very much to music. He composed *Lieder*, and even an *operetta*. Many times he sang parts in lyric performances. At the beginning, he wavered whether he had to choose this road. He played the piano, and had absolute pitch. Often his home was open to perform music with friends, with whom Albert Einstein played violin, and even Joseph Joachim sometimes participated [25]. He conducted a little choir, which Otto Hahn remembered:

> "Planck loved domestic gatherings. During the years before the First World War, a number of young ladies and gentlemen who enjoyed singing —and to whom also I belonged—gathered in his house in Berlin Grunewald every two weeks. We formed a 4-voices choir with Planck

[19] "... besprach sodann einige beim Studium des Instruments gesammelte Erfahrungen von musikalischer Bedeutung, mit deren weiterer Ausarbeitung er gegenwärtig noch beschäftigt ist, und belegte dieselben durch verschiedene Beispiele" [22], B. I, p. 436.

[20] "Die natürliche Stimmung in der modernen Vokalmusik". The article does not appear in the *Physikalische Abhandlungen und Vorträge*, as the previous one, and in general it has been neglected by historians of science. See [10] Chapter 10.3. Cf. Barbour [24].

> conducting, and we sang Brahms and others. In one of Brahms' *Zigeunerlieder*, I could even sing a small solo"[21].

It was in this context that the physicist carried out musical experiments concerning his problem: why did musicians prefer the tempered tuning to the "natural", "pure", or "just" scale, as it was theorized by Helmholtz? Planck's article is crowded with observations, examples, statements, difficulties, and his results were heavy depending on all these circumstances. Among them I choose the following:

> "One can straight adjust his hearing to a higher or lesser degree of adaptability. The ear of a person who follows with great attention a concert is much more adaptive when this person is attending the concert for pleasure or to act as a critic"[22].

Musicians would get notes the more "pure", the more they had prevented their ear to adjust itself. "... The possibility of obtaining as much purity as possible ..."[23] should have effect on timbre, and on sensations felt by the audience. The famous physicist described his own sensations:

> "To me, this moment of transition [toward "natural" intervals] is an immediate source of artistic pleasure: it is as if certain last earthly remainders faded away, and at the same time a veil was raised, which opens to the phantasy an insight into a new world, which reaches into infinite distances."

He got a rule:

> "... the adaptation is achieved the easier and can be driven the furthest, the less the interval is consonant. ... This clearly shows how our ear gets used to the tempered tuning"[24].

[21] "Planck liebte die häusliche Geselligkeit. In den Jahren vor dem ersten Weltkrieg versammelten sich in seinem Hause in Berlin-Grunewald alle 14 Tage eine Anzahl gesangsfreudiger jüngerer Damen und Herren, zu denen auch ich gehörte. Wir bildeten einen vierstimmigen Chor unter Plancks Leitung, sangen Brahms und anderes. Bei einem der Brahmsschen *Zigeunerlieder* durfte ich sogar eine kleine Solostelle singen" [22], B. III p. 421.

[22] "Man kann sein Gehör geradezu einstellen auf einen höheren oder geringeren Grad von Akkommodationsvermögen. Das Ohr des Konzertbesuchers ist, bei gespanntester Aufmerksamkeit im Übrigen, akkommodationsfähiger, wenn er als Genießender, wie wenn er als Kritiker erscheint" [23], p. 423.

[23] "... die Erzielung größtmöglicher Reinheit ..."

[24] "Für mich ist dieser Übergangspunkt geradezu eine Quelle künstlerichen Genusses: es ist, als ob ein gewisser letzter Erdenrest verschwände und zugleich sich ein Schleier lüfte, welcher der Phantasie den Einblick in eine neue, bis in unendliche Fernen reichende Welt eröffnet." "... die Akkomodation um so leichter erfolg um sich so weiter treiben läßt, je weniger konsonant das Intervall ist. ... Hier zeigt sich deutlich die Gewöhnung unseres Ohrs an die temperirte Stimmung" [23], pp. 423-426.

Like Helmholtz, Planck preferred the "natural" tuning, and to justify the habit of temperament among musicians he gave the fault to the musical context.

> "Therefore the ear finds itself in a peculiar antagonism. On the one hand, the fresh third in the usual pitch as wished by the ear, together with the hammering beats, on the other hand the third in peace, but however too deep according to the needs of the moment"[25].

He thought inconsistent to mix "pure" octaves with "impure" thirds. The core of the test was the following. Planck himself wrote a sequence of chords in $C$ major. Then he let his choir of friends sing them *a cappella* with him conducting. Thus he discovered that the choir sang the final $c$ lower of five sintonic *comma* ($\frac{81}{80}$) than the initial $c$: they were flattened more than a semitone. Planck drew a conclusion from that:

> "If at the end the choir came back to the initial $c$, this would be a compelling proof of the fact that the "natural" tuning had not the least practical meaning. From the point of view of modern art, it would then only be idle speculation to continue to care about the theory of "natural" tuning"[26].

He kept the singers in the dark about the meaning of the performance, in order not to perturb the result. He repeated it, and did a counter experiment at the end. The choir now sang the sequence in reverse order. Would he get a higher $c$? "The final result did neither yield a decrease, nor the longed-for raising of the pitch"[27]. The rigorous physicist devised some explanation, and concluded that more experiments would have been necessary.

> "These cases prove that, under some circumstances, the differences of the tempered and the "natural" tuning are more significant than the deviations from intervals commonly accepted in practical music. They therefore give to every conductor the task to form an opinion on this issue"[28].

[25] "Dabei befindet sich das Ohr in einem eigenthümlichen Zwiespalt: auf der einen Seite die Terz in der vom Ohre gewöhnten und begehrten frischen Höhe, zugleich mit den hämmernde Schwebungen, auf der anderen Seite die Terz in Ruhe, aber doch für das augenblickliche Bedürfniß zu tief" [23], p. 429.

[26] "Wenn dann am Schluß der Chor wieder auf dem Ausgangs-$c$ anlangen sollte, so wäre das ein zwingender Beweis dafür, daß für ihn die natürliche Stimmung auch nicht die geringste praktische Bedeutung hat. Dann wäre es, vom Standpunkt der modernen Kunst betrachtet, lediglich müßige Spekulation sich noch einen Augenblick länger mit der Theorie der natürlichen Stimmung zu beschäftigen."

[27] "Das schließliche Resultat ergab kein Sinken, aber auch nicht das erwartete Steigen der Tonhöhe."

[28] "Diese Fälle beweisen, daß die Differenzen der natürlichen und der temperirten Stimmung die in der praktischen Musik zulässigen Schwankungen der Intervalle unter Umständen merklich überschreiten, und stellen daher jeden Dirigenten vor die Aufgabe, sich darüber ein Urtheil zu bilden."

To the many dilemmas of the conductor performing a musical piece, Planck added the tuning, but he did not offer any solution. He gave the last word to the composer, and when it was impossible to get, he referred to the artistic effect one wished.

> "Because art finds its justification in itself, and no theoretical system of music, even if it was logically founded, and developed in a consistent way, is in the position to fulfill all the requirements of art which is in continuous exchange with the human spirit. In this spirit, the "natural" system has indeed not advantage to the tempered one..."[29]

Planck wrote that the main results of the research were the followings. Everybody tunes tempered because the ear adjusts itself to it. However, in some case, it could be better to tune "natural" to obtain particular aesthetic effects. The prospects of the "natural" system in future times would depend upon the birth of a genius who, by using it, was able to compose better than by the other one. Only this man could resolve those dilemma [23] (p. 439-440).

I guess that Planck had his own personal dilemma. How to reconcile the pleasure of playing and listening to music with the profession of a theoretical physicist, and of being a student of Helmholtz who had made "natural" Zarlino's scale by beats and overtones? Later, in his "scientific autobiography", Planck remembered:

> "... I had the task to study the "natural" tuning on this instrument [Eitz's harmonium]. I did this with great interest, in particular with relation to the issue about the role played by "natural" tuning in our modern vocal music, without instruments. In this process I obtained the unexpected result that our ear prefers the tempered tuning to the "natural" one in any circumstance. Even in a chord with harmony in a major tonality, the "natural" third sounds slack and without expression in comparison to the tempered one. Without doubt, this fact goes back to habits which have been developed over many years and generations"[30].

[29] "Denn die Kunst findet ihre Begründung in sich selbst, und kein theoretisches System der Musik, wäre es noch so logisch begründet und konsequent durchgeführt, ist im Stande, alle Forderungen der zugleich mit dem menschlichen Geiste ewig wechselnden Kunst ein für alle Mal zu fixiren. In dieser Beziehung hat das natürliche System durchaus keinen Vorzug vor dem temperirten, ..." [23], pp. 434-439.

[30] "... ich hatte die Aufgabe, an diesem Instrument Studien über die natürliche Stimmung zu machen. Das tat ich denn auch mit großem Interesse, besonders in bezug auf die Frage nach der Rolle, welche die natürliche Stimmung in unserer modernen, von Instrumentalbegleitung freien Vokalmusik spielt. Dabei kam ich zu dem mir einigermaßen unvermuteten Ergebnis, daß unser Ohr die temperierte Stimmung unter allen Umständen der natürlichen Stimmung vorzieht. Sogar in einem harmonischen Durdreiklang klingt die natürliche Terz gegenüber der temperierten Terz matt und ausdruckslos. Ohne Zweifel ist diese Tatsache in letzter Linie auf jahre- und generationenlange Gewöhnung zurückzuführen" [22], B. III, p. 383-384.

By dint of playing the piano, and of conducting domestic choirs, the serious Planck in the end convinced himself that "in any circumstance" the tempered tuning was the best for music, in his times.

## 4 The Moral

To draw a moral from our story: the closer theoreticians and scientists were to practical music and musicians, the more they trusted their ears. And today? What happens when theoretical studies are more easily done with the help of the screens of the computers? Is the need still felt to stay close to the ears of the audience and of the musicians? I would like to hear other points of view. Anyway, we must remember the privileged role of musicians played in history of science, at least in acoustics. The musical context should never be forgotten because it has always the final word. Predating Helmholtz, another instance is given by Hector Berlioz's *Traité de l'instrumentation et d'orchestration moderne* of 1844: "The ear quite neglects imperceptible differences, in spite of mathematicians' opinion" [26]. I regret that modern history of science disregards or overlooks too often the important contributions of music in the past [10,11,12]. Of course, the conflicts considered here also depend on how musical reality, or even "reality" in itself, is conceived; but in this paper, I neither have enough place to sketch this very general and large theme nor to begin a discussion about it.

I thank the reviewers and the editors to help me improve the paper. I acknowledge the help of Emiliano Rago for my problems with the computer, and of Chiara Letta for finding some texts.

## References

1. Leibniz, W.G.: Epistolae ad diversos. In: Kortholt, C. (ed.), vol. I, p. 241. Breitkopf, Lipsiae (1734)
2. Da Rios, R. (ed.): Aristoxenus: Elementa Harmonica. Istituto Poligrafico dello Stato, Roma (1954)
3. Galilei, V.: Dialogo della musica antica et della moderna. Marescotti, Firenze (1581); Rep. Fano F. (ed.) Reale Accademia d'Italia, Roma (1934)
4. Rameau, J.-P.: Génération harmonique, ou Traité de musique théorique et pratique. Prault Fils, Paris (1737)
5. Euler, L.: Conjecture sur la raison de quelques dissonances généralement reçues dans la musique. In: Opera Omnia, Ser. III, vol. III, pp. 508–515. Teubner, Leipzig (1926)
6. Cohen, H.F.: Quantifying Music. Reidel, Dordrecht (1984)
7. Gozza, P.: La musica nella rivoluzione scientifica del Seicento. Il Mulino, Bologna (1989)
8. Bailhache, P.: Sciences et musique: quelques grandes étapes en théorie musical. Littérature, Médicine, Société (13) (1996)
9. Knobloch, E.: Rapports historiques entre musique, mathématique et cosmologie. In: Quadrivium, Musique et Sciences, pp. 123–167. ipmc, Paris (1992)
10. Tonietti, T.M.: Eppur si ode: Storia, musicale, multilingue e policulturale, delle scienze matematiche (unpublished manuscript)

11. Tonietti, T.M.: The Mathematical Contributions of Francesco Maurolico to the Theory of Music of the 16th Century. Centaurus 48, 149–200 (2006)
12. Barontini, M., Tonietti, T.M.: Umar Al-Khayyam's Contribution to the Arabic Mathematical Theory of Music. Arabic Sciences and Philosophy 20, 255–279 (2010)
13. Seebeck, L.W.A.: Ueber die Sirene. Annalen der Physik und Chemie 60, 449–481 (1843)
14. Ohm, G.S.: Ueber die Definition des Tones, nebst daran geknüpfter Theorie der Sirene und ähnlicher tonbildender Vorrichtungen. Annalen der Physik und Chemie 59, 513–565 (1843); Noch ein Paar Worte über die Definition des Tones. Annalen der Physik und Chemie 62, 1–18 (1844)
15. Sethares, W.A.: Tuning Timbre Spectrum Scale, `http://sethares.engr.wisc.edu/ttss.html`
16. Helmholtz, H.: Die Lehre von den Tonempfindungen als physiologische Grundlage für die Theorie der Musik, Brunswick (1863)
17. Helmholtz, H.: Théorie Physiologique de la Musique. Masson, Paris (1868)
18. Helmholtz, H.: On the Sensations of Tone as a Physiological Basis for the Theory of Music, London (1885); Repr. Dover, New York (1954)
19. Riemann, B.: Mechanik des Ohres. In: Weber, H., Dedekind, R., Narasimhan, R. (eds.) Gesammelte mathematische Werke und wissenschaftlicher Nachlass und Vorträge. Collected Papers, pp. 370–382. Springer, Berlin (1990)
20. Euler, L.: Tentamen novae theoriae musicae, St. Petersburg (1739); Repr. Opera Omnia, Ser. III, vol. III1, pp. 197–427. Teubner, Leipzig (1926)
21. Kalbeck, M.: Johannes Brahms, Band I. Hans Schneider, Tutzing (1976)
22. Planck, M.: Physikalische Abhandlungen und Vorträge. In: Laue, M. (ed.) Vieweg & Soh., Braunschweig (1958)
23. Planck, M.: Die natürliche Stimmung in der modernen Vokalmusik. Vierteljahrsschrift für Musikwissenschaft 9, 418–440 (1893)
24. Barbour, J.M.: Tuning and Temperament. Dover, New York (1951)
25. Grassmann, G.: Max Planck. Mitteilungen aus der Max-Planck-Gesellschaft zur Förderung der Wissenschaften, pp. 91–104 (1973)
26. Berlioz, H., Strauss, R.: Instrumentationslehre. Peters, Leipzig (1904)

# Mazzola's Model of Fuxian Counterpoint

Dmitri Tymoczko

Princeton University
dmitri@princeton.edu

**Abstract.** This paper critiques Guerino Mazzola's derivation of traditional counterpoint rules, arguing that those principles are not well-modeled by pitch-class intervals; that Mazzola's differential treatment of fifths and octaves is not supported musically or by traditional counterpoint texts; that Mazzola's specific calculations are not reproducible; that there are a number of intuitive considerations weighing against Mazzola's explanation; that the fit between theory and evidence is not good; and that Mazzola's statistical arguments are flawed. This leads to some general methodological reflections on different approaches to mathematical music theory, as well as to an alternative model of first-species counterpoint featuring the orbifold $\mathbb{T}^2/\mathcal{S}_2$.

**Keywords:** Counterpoint, Guerino Mazzola, Fux, Palestrina, Parallel Fifths, Antiparallel Fifths, Orbifolds, Methodology of Mathematical Music Theory.

## 1 Introduction

For an American music theorist, Guerino Mazzola's *The Topos of Music* (henceforth *Topos*) can be a forbidding book: dense, intricately systematic, and more complex in its mathematics than the writings of Allen Forte, John Clough, or David Lewin. And where American theorists are often apologetic in their invocations of advanced mathematics, offering simplified tutorials for untrained readers, Mazzola can seem almost aristocratic in his disdain for nonmathematicians. If you can't learn algebraic geometry, he sometimes seems to be saying, then you have no business trying to understand Mozart.

Confronted with this attitude, some theorists might feel tempted to dismiss *Topos* as a mathematical fantasy with no real-world relevance. But to do so is to ignore the fact that Mazzola makes claims of the utmost centrality to music theory. Perhaps foremost among these is the "Counterpoint Theorem," located 650 dense and technical pages into the tome: here Mazzola proposes that Fux's first-species counterpoint rules—rules which are basic to Renaissance composition and tonal pedagogy more generally—are in fact reflections of deep mathematical symmetries inherent in the twelve-tone universe. This proposal is important enough that the conscientious music theorist must take it seriously, even if that means working through some technical details.

The purpose of this paper is to use Mazzola's proposal to open a dialogue between American and European approaches to mathematical music theory. In doing so, I want to make it clear that my intent is not hostile: every writer deserves critics, and every book—even a great book—deserves careful scrutiny.[1] Like many American readers, I

[1] It may be relevant that Guerino encouraged me to write this paper, precisely in order to facilitate dialogue between different strands of mathematical music theory.

C. Agon et al. (Eds.): MCM 2011, LNAI 6726, pp. 297–310, 2011.

am fascinated by *The Topos of Music;* indeed, my shock upon first encountering it is comparable to my shock on first encountering the philosophy of Hegel or the music of Cecil Taylor. Here, I felt, was something new, powerful, and yet utterly beyond my comprehension: was it a great intellectual achievement or a majestic shaggy-dog story?

My initial goal in writing this paper was to try to answer this question for myself. I am motivated to try to publish it by the thought that, although *Topos* has had many authors (twenty names are listed on the front cover), it has had relatively few critics—that is, independent, established, and probing music theorists who are able to follow the argument in all its details.[2] This is on one level unsurprising: if it takes an extraordinary degree of mathematical sophistication simply to read Mazzola's book, it also takes a good deal of *musical* sophistication to evaluate its boldest theoretical claims. Thus an algebraic geometer, though perhaps comfortable with Mazzola's mathematics, may find it difficult to evaluate the statements about Fux. (Of course, a specialist in Renaissance music has no hope of understanding Mazzola's at all.) For this reason, I suspect that there are relatively few readers who are truly able to evaluate this extraordinary and challenging work. All the more reason, then, for someone such as myself to give it a try.

## 2 Mazzola's Description of Fux's First-Species Rules

Mazzola proposes to use symmetry to explain Fux's first-species counterpoint rules. But before we can evaluate this idea, we need to know whether he has characterized Fux's rules correctly. (Roughly speaking, the rules are the data that his theory is meant to describe.) There are essentially three separate questions here. First, can Fux's rules be usefully modeled by an approach that abstracts away from pitches in favor of pitch-class intervals? Second, if we accept this abstraction, is Mazzola justified in focusing on parallel fifths to the exclusion of octaves? And third, are Mazzola's claims technically correct on their own terms?

Contemporary theorists may be somewhat surprised to learn that Fux's discussion of first-species counterpoint contains no specific prohibition against parallel perfect intervals; instead, it forbids *all similar motion into perfect consonances* ([2], p. 22). Thus for Fux the successions (C4, G4)→(D4, A4) and (C4, E4)→(D4, A4) are equally bad, even though only the former involves forbidden parallels.[3] Conversely, successions like (C4, G4)→(G3, D5) would seem to be acceptable, even though they involve "antiparallel fifths"—a fifth moving to a twelfth by contrary motion (or vice versa). Fux also seems to permit "registral parallels" such as (C4, G4)→(C5, F4), in which crossed voices articulate a pair of perfect fifths in register.[4] Thus he is concerned with a *very specific kind of melodic motion,* rather than a more general harmonic situation in which one seven-semitone pitch-class interval follows another. Figure 1 shows that Fux is here being reasonably faithful to his Renaissance sources, which often contain registral parallels or "antiparallels" unusual in later music.

[2] John Roeder has written a valuable critique of Mazzola's early work [1].

[3] Throughout this paper, I follow [3] in using $(x, y)\to(w, z)$ to indicate that note $x$ moves to $w$ and $y$ moves to $z$.

[4] Fux ([2], p. 36) permits voice crossings.

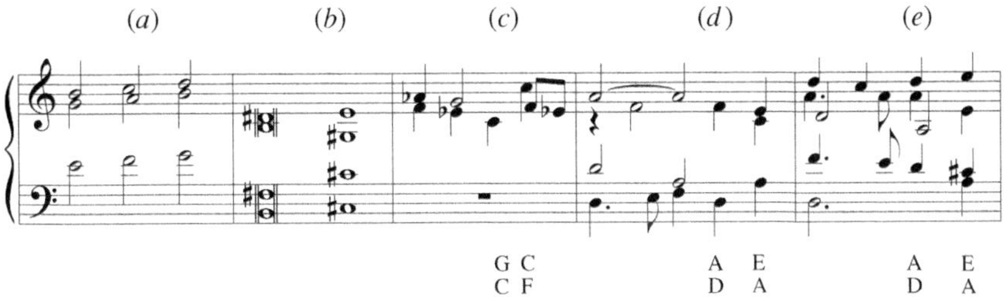


**Fig. 1.** Renaissance composers often articulated consecutive fifths registrally (*a–c*) or by contrary motion (*d–e*). (*a*) Palestrina, Missa "In te Domine speravi," Credo; (*b*) Lassus, *Prophetiae Sybillarum,* Prologue, mm. 3–4; (*c*) Josquin "Gaude Virgo, Mater Christi," m. 51, (*d*) Palestrina, Missa "Sanctorum meritis," Agnus, m. 26, (*e*) Palestrina, Missa "Vestiva i Colli," Kyrie, m. 78. Mazzola's first-species counterpoint rules forbid antiparallel fifths such as those in the last two examples.

All of which suggests that the conception of "forbidden parallels" has changed over time. I believe that passages such as Fig. 1 are more characteristic of Renaissance than Baroque music. If so, then this may reflect an increasing awareness of the *harmonic* dimension of music, according to which chords are entities in themselves, ordered registrally, and not simply the byproducts of melodic lines. From this point of view, Fig. 1(b) is unsatisfactory since it produces a strong *sense* of parallelism even though no pair of voices articulates parallel perfect intervals. Similarly, an increased sensitivity to antiparallels (Fig. 1(c–e)) may reflect a growing awareness of *pitch class* as an important musical parameter. If this is right, it provides a new window into the development of harmonic thinking in the seventeenth and eighteenth centuries: the increasing sensitivity to harmony is reflected not just in the increasing use of certain chord progressions and cadential formulae, but also in the gradual broadening of the prohibitions on forbidden parallels. Progressions unobjectionable from a strictly contrapuntal point of view start to seem problematic on other grounds.

For this reason it is probably somewhat anachronistic to model Fux using intervals modulo the octave. But since abstraction is the theorist's prerogative, let's permit Mazzola this relatively innocent simplification. The next question is why Mazzola permits progressions from unisons to octaves (or vice versa), while prohibiting the analogous progressions between fifths and twelfths. Most pedagogues instead treat perfect intervals as a unified class subject to the same fundamental constraints: in general, theorists either warn against *both* antiparallel octaves and fifths or permit both sorts of notion.[5] It is unclear why Mazzola decides to prohibit only antiparallel-fifths: in a move that is emblematic of *Topos*'s challenges, detailed explanations are found not in the book itself, but in an unpublished paper.[6]

[5] See, e.g. textbooks by Kostka and Payne ([4], p. 84) or Gauldin ([5], p. 136).

[6] Item 342 in Mazzola's bibliography is "Mazzola G and Muzzulini D: Deduktion des Quintparallelenverbots aus der Konsonanz-Dissonanz-Dichotomie. Accepted for publication in: *Musiktheorie*, Laaber 1990."

The important point here is that Fux's "counterpoint rules," rather than being clearly and explicitly stated, are only discoverable through interpretation. For example, though neither Fux nor Jeppesen directly prohibits antiparallels, both prohibit unisons within the phrase while limiting the inter-voice distance to the interval of a tenth.[7] As a result, antiparallels can occur only in unusual circumstances—octaves at the very beginning of a phrase, and fifths when there is unusually wide separation between the voices. Did Fux and Jeppesen mean to permit antiparallels in these specific circumstances, or were they more concerned with providing general rules that would help students write well? Did they reject antiparallel fifths *only* because of the wide registral separation, or were they simply uninterested in providing a comprehensive list of the multiple ways in which this particular progression was bad? The answers to these questions are by no means clear: one could certainly argue that Fux did reject first-species antiparallels, if only because there are no Fuxian examples in which they appear. (This is further supported by the fact that other theorists allow antiparallels *only when there are a large number of voices.*[8]) However, this is not the only possible conclusion one could draw. The bottom line is that there is a serious gap between the pedagogical remarks in traditional counterpoint texts and rigorous models of the sort found in contemporary theory. And this is why we might hope for more careful argumentation in favor of Mazzola's unusual decision to treat fifths and octaves so differently.

Finally, some technical details. Mazzola writes: "within an ecclesiastical mode, there are 287 a priori possible progressions. According to the consonance-dissonance counterpoint theorem, 37 of them are forbidden. Among them, 21 coincide with the 54 Fux-inadmissible progressions." Despite having corresponded with Mazzola, I am able to reproduce only two of these four numbers. The number 287 refers to the number of distinct progressions of the form $(0, x) \rightarrow (y, z)$ where (*a*) $x$, $y$, and $z$ are pitch classes in the range 0–12; (*b*) each dyad is a consonance (that is: unison, third, perfect fifth, or sixth); and (*c*) the entire progression fits within some diatonic scale. Mazzola states that 54 of these 287 progressions are "Fux inadmissible." I take this to mean that 54 progressions either involve parallel perfect fifths or contain a tritone in one of the voices.[9] However, by my count, the number of Fux-inadmissible progressions should be 55 or 65: 45 progressions contain tritones, 10 contain parallel fifths, and additional 11 contain parallel octaves, one of which also has tritone leaps (Fig. 2).[10]

[7] See Jeppesen ([6], p. 112) and Fux ([2], p. 38). Fux does not explicitly require that voices stay within a tenth, though all his examples obey this restriction.

[8] Vincentino ([7], book 2, ch. 3, f. 41v.) allows anti-parallel fifths in five parts and anti-parallel octaves in eight parts. Note that this suggests that antiparallel octaves are *more* problematic than antiparallel fifths, contrary to Mazzola's claim. (Thanks here to Peter Schubert, who also supplied some of the examples in Fig. 1.) On the other hand, Jeppesen ([6], p. 113) provides a first-species example with antiparallel octaves.

[9] This is because "Fact 16" on page 657 of *Topos* states "in the reduced strict style, only the rule of forbidden fifth parallels and the tritone rules have an unrestricted validity."

[10] Note that I do not consider repetitions, such as (C4, G4)→(C4, G4), to be parallel fifths. Mazzola forbids all such repetitions, no matter what interval they involve (Fig. 3).

(*a*)

| The 21 diatonic progression-classes involving parallel fifths or octaves | |
|---|---|
| **octaves** | **fifths** |
| (0, 0)→(1, 1) | (0, 7)→(1, 8) |
| (0, 0)→(2, 2) | (0, 7)→(2, 9) |
| (0, 0)→(3, 3) | (0, 7)→(3, 10) |
| (0, 0)→(4, 4) | (0, 7)→(4, 11) |
| (0, 0)→(5, 5) | (0, 7)→(5, 0) |
| (0, 0)→(6, 6)* | (0, 7)→(7, 2) |
| (0, 0)→(7, 7) | (0, 7)→(8, 3) |
| (0, 0)→(8, 8) | (0, 7)→(9, 4) |
| (0, 0)→(9, 9) | (0, 7)→(10, 5) |
| (0, 0)→(10, 10) | (0, 7)→(11, 6) |
| (0, 0)→(11, 11) | |

(*b*)

| The 45 consonant diatonic progression-classes involving a tritone in at least one voice | | |
|---|---|---|
| (0, 0)→(10, 6) | (0, 3)→(0, 9) | (0, 4)→(10, 10) |
| (0, 0)→(11, 6) | (0, 3)→(2, 9) | (0, 4)→(2, 10) |
| (0, 0)→(2, 6) | (0, 3)→(5, 9) | (0, 4)→(6, 2) |
| (0, 0)→(3, 6) | (0, 3)→(6, 1) | (0, 4)→(6, 6) |
| (0, 0)→(6, 1) | (0, 3)→(6, 10) | (0, 4)→(6, 9) |
| (0, 0)→(6, 10) | (0, 3)→(6, 3) | (0, 4)→(7, 10) |
| (0, 0)→(6, 2) | (0, 3)→(6, 6) | (0, 8)→(10, 2) |
| (0, 0)→(6, 3) | (0, 3)→(9, 9) | (0, 8)→(2, 2) |
| (0, 0)→(6, 6)* | (0, 9)→(0, 3) | (0, 8)→(5, 2) |
| (0, 0)→(6, 9) | (0, 9)→(3, 3) | (0, 8)→(6, 1) |
| (0, 0)→(9, 6) | (0, 9)→(6, 2) | (0, 8)→(6, 10) |
| (0, 7)→(1, 1) | (0, 9)→(6, 6) | (0, 8)→(6, 3) |
| (0, 7)→(10, 1) | (0, 9)→(6, 9) | (0, 8)→(6, 6) |
| (0, 7)→(5, 1) | (0, 9)→(7, 3) | (0, 8)→(7, 2) |
| (0, 7)→(6, 2) | | |
| (0, 7)→(6, 6) | | |
| (0, 7)→(6, 9) | | |

**Fig. 2.** (*a*) Consonant, diatonic progression-classes containing parallel fifths or octaves. (*b*) Consonant, diatonic progression-classes containing a tritone in at least one voice. A label like (0, 0)→(1, 1) refers to a class of progressions related by transposition; thus it could stand for (E, E)→(F, F) as well as (B, B)→(C, C). Each element of each class can be embedded within any diatonic scale by a suitable choice of pitch-class 0. Note that the starred progression contains both melodic tritones and parallel octaves.

Mazzola states that 21 of the 37 progressions forbidden by the Counterpoint Theorem are also Fux-inadmissable. The number 37 refers the number of diatonic successions in the bottom cell of the table on page 654.[11] But in my view, the number of Fux-inadmissable progressions should be 19, with the remaining 17 not specifically prohibited by Fux (Fig. 3).[12] It is noteworthy that the disagreements involve Fux's prohibitions, underscoring the point that it is difficult to agree about them. This may bode ill for the project of modeling the prohibitions mathematically, since it might seem unlikely that a precise mathematical formula would exactly capture this inherently vague and subjective set of principles.

[11] Mazzola's table contains a few other progressions that articulate augmented triads, diminished seventh chords, and other nondiatonic sets.

[12] Mazzola says that Fux prohibits the "battuta," and (C, E)→(D, D) appears on his list of forbidden progressions. But this is a subtle issue: Fux's counterpoint-master notes that the progression is traditionally prohibited when it moves from tenth to octave, but not when it moves from sixth to octave; furthermore, he says that he can find no reason for this asymmetry, leaving it to the student to decide whether to use the progression. Rather than articulating a strong prohibition, he concludes the traditional rule is "of little importance."

(*a*)

| Forbidden by Mazzola and Fux | |
|---|---|
| **Tritone Motion** | **Parallel Fifths** |
| (0, 0)→(6, 6) | (0, 7)→(1, 8) |
| (0, 3)→(0, 9) | (0, 7)→(2, 9) |
| (0, 3)→(6, 3) | (0, 7)→(3, 10) |
| (0, 4)→(2, 10) | (0, 7)→(4, 11) |
| (0, 4)→(6, 6) | (0, 7)→(5, 0) |
| (0, 4)→(6, 2) | (0, 7)→(7, 2) |
| (0, 4)→(10, 10) | (0, 7)→(8, 3) |
| (0, 9)→(0, 3) | (0, 7)→(9, 4) |
| (0, 9)→(6, 9) | (0, 7)→(10, 5) |
| | (0, 7)→(11, 6) |

(*b*)

| Forbidden by Mazzola but not by Fux | | |
|---|---|---|
| **Repetitions & Exchanges** | **Diminished Triad** | **???** |
| (0, 0)→(0, 0) | (0, 3)→(3, 6) | (0, 4)→(0, 0) |
| (0, 3)→(0, 3) | (0, 3)→(9, 0) | (0, 4)→(2, 2) |
| (0, 4)→(0, 4) | (0, 9)→(3, 0) | (0, 4)→(4, 4) |
| (0, 7)→(0, 7) | (0, 9)→(9, 6) | (0, 4)→(2, 6) |
| (0, 8)→(0, 8) | | (0, 4)→(10, 2) |
| (0, 9)→(0, 9) | | |
| (0, 3)→(3, 0) | | |
| (0, 4)→(4, 0) | | |
| (0, 9)→(9, 0) | | |

**Fig. 3.** (*a*) Consonant, diatonic progression-classes forbidden by both Mazzola and Fux. These include 9 of the 45 classes containing a tritone motion in one voice, and all eleven examples of parallel fifths. (*b*) Consonant, diatonic progression-classes forbidden by Mazzola but not Fux. These include repetitions, exchanges in which two voices swap notes, four progressions outlining a diminished triad, and five unproblematic progressions that begin with a major third. In all of these progressions, the cantus is listed first.

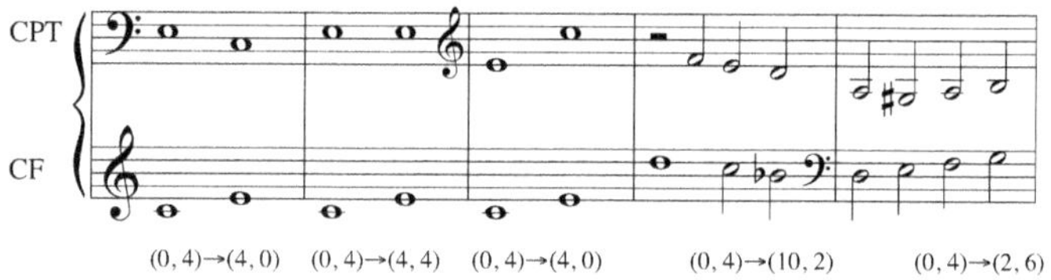


**Fig. 4.** Mazzola's forbidden progressions in Fux, Lassus, and Palestrina. (*a–c*) are from Fux ([2], 39–40); (*d*) is from Lassus *Cantiones duarum vocum* 2-5, m. 28; (*e*) is from Palestrina, Missa *Lauda Sion,* Gloria. In some of these examples, the cantus appears above the counterpoint, but Fux gives no indication that this is at all relevant to the progressions' acceptability.

## 3 Mazzola's Derivation of the Rules

So far, we have confined our discussion to Fux's rules: accepting Mazzola's decision to work with pitch-class intervals, we have found the 65 progressions in Fig. 2, rather than the 54 progressions mentioned in Mazzola's text. I now want to turn to Mazzola's *explanation* of these rules, focusing less on the mathematical details of his model than on its actual output. Mazzola begins with the fact that traditional counterpoint enjoins a careful balance between perfect and imperfect consonances, arguing that consonances cannot progress completely freely. Instead, there are "dissonances within consonances":

> The first elementary rule of counterpoint "note-against-note" says that we are not allowed to take other intervals than the consonances [...] This seems evident, but

> it imposes a strong obstruction against another, more hidden directive: the idea of creating a tension between each interval and its successor. More precisely, the meaning of "contra" is not only that of a vertical opposition between cantus firmus and discantus. As Sachs has remarked [...], the preposition "contra" equally means a horizontal opposition between successive intervals in the given sequence. This requirement is not very explicit, but it is reflected in the distinction between perfect consonances (prime, fifth, octave) and the others, the imperfect sixths and thirds, and the idea of changing between perfect and imperfect consonances in order to create tension. This conceptual distinction seems to evoke a dissonant ingredient in the consonant character, although it does not really abolish the consonance, it is a kind of coloring effect ([9], p. 646).

Mazzola proposes to use symmetry to capture this "dissonant ingredient," arguing (roughly speaking) that a progression between consonances $A \rightarrow B$ is allowable only if there is some affine transformation that sends $A$ into a consonance and $B$ into a dissonance (or vice versa).[13]

Having taught and studied counterpoint for some years, I admit that I was initially suspicious of this idea, largely because I could not imagine a plausible mechanism that would serve to connect the mathematics to real-world musical practice. In particular, Mazzola's explanation—like many other contemporary accounts of diatonic music—anachronistically assumed a diatonic collection embedded within a twelve-tone chromatic scale; whereas in the Renaissance just intonation and even nineteen-tone equal temperament were active subjects of exploration [7]. Furthermore, the form of Mazzola's argument seemed to involve a subtle logical leap, beginning with the indisputable fact that counterpoint requires blending perfect and imperfect intervals, but shifting to a mathematical model that provides *no guidance* about how to move between these two kinds of consonance. Instead the model delivers a whole collection of rules (avoiding parallel fifths, tritone melodic intervals, and note repetitions) that have nothing to do with the motivating idea of "creating tension" by mixing consonances: the model encodes the idea of "moving from consonance to dissonance and back" only at an abstract mathematical level, with these abstractions being connected *by analogy* to the manifest practice of mixing perfect and imperfect consonances. Thus, Mazzola's argument seems to depend on a prior conviction that there should be analogical similarities between underlying mathematical relationships and seemingly unrelated surface procedures.

Most importantly, Mazzola proposes a single explanation for three very different musical practices. The prohibition on tritone melodic motion has typically been explained, at least in part, by reference to the fact that tritones are *very difficult to sing*.[14] (On this view, the prohibition on tritones is analogous to the prohibition on major sevenths.) Parallel perfect intervals, by contrast, are quite singable; this prohibition is instead explained by the supposition that parallel perfect intervals lead to auditory fusion, thus decreasing voice independence. (This fusion, in turn, has been explained

[13] For the details see [8] or [9]. Here I am mainly concerned with the model's claims, rather the mathematical means by which they are derived.

[14] See Fux [2], p. 35 ("hard to sing and sounds bad").

both by the acoustic properties of perfect intervals, and also by the structure of the diatonic scale—which ensures that parallel diatonic octaves are always parallel chromatic octaves, and that parallel diatonic fifths are mostly parallel chromatic fifths [3].) Mazzola's counterpoint rules also include a host of non-Fuxian prohibitions on note repetition—prohibitions that are presumably justified by reference the rhythmic peculiarities of first-species writing. Intuitively, it seems highly unlikely that these three independent and unrelated considerations would come together such that they could all be modeled by interesting mathematical machinery. Were this to happen, it would be as if a bunch of pottery shards had spontaneously leapt up to form a beautiful unbroken vase.

To be sure, these are defeasible worries: if it were true that Mazzola's model could exactly reproduce Fux's rules—even as understood by later theorists—then this would provide evidence that my initial suspicions had been wrong. If, on the other hand, there was only an approximate match between Mazzola and Fux, then this would tend to confirm my intuitions about the model's irrelevance. And here it must be said that the results are not very good: of my sixty-five prohibited Fuxian motions, Mazzola prohibits only nineteen, or less than a third; at the same time, roughly half of his model's prohibitions are acceptable according to Fux. (Figure 4 provides a few examples of these progressions, drawn from Fux, Lassus, and Palestrina.[15]) This is on the face of it a strikingly bad fit between model and reality. A scientist with no prior bias in favor of Mazzola's hypothesis might well discard it for this reason.

There is also the important fact that Mazzola's treatment of thirds and sixths is *not symmetrical under registral inversion.* Thus we cannot always take a passage in which the counterpoint appears in thirds and sixths *above* the cantus, and move it *below* the cantus to produce another acceptable progression: according to Fig. 3, (E4, C5)→(C4, E5) is a permissible (0, 8)→(8, 0) progression, while (E4, C3)→(C4, E3) is an impermissible (0, 4)→(4, 0). I am aware of nothing in the contrapuntal literature that supports such a sharp distinction between thirds and sixths: indeed, virtually every contrapuntal theorist, including Fux [10], explicitly discusses "double counterpoint at the octave," whose central principle is that a melody can appear either above or below the cantus, so long as there are no fifths or fourths between the voices. Ironically, then, Mazzola ends up *breaking* one of the style's manifest symmetries, the invertibility of imperfect consonances, in his attempt to construct a model that uses symmetry at a deeper mathematical level.

Considerations of symmetry lead toward some fascinating methodological issues, which we unfortunately cannot pursue in full detail. In *Topos,* Mazzola explicitly rejects an approach in which intervals (or more generally, harmonies) are modeled as unordered sets:

> "Our requirements would not be satisfied if we thought of an interval as being a set of pitch events. This would in particular not do justice to the concept of voices. The cantus firmus could not be distinguished from the discantus, and the crossing of voices could not be conceived" ([9], p. 619).

---

[15] My assumption is that Fux was trying to model the style of composers such as Lassus and Palestrina, and thus that their music provides a guide to what Fux would've tolerated—particularly in cases where Fux makes no explicit mention of the progression in question.

Recent developments suggest that these claims are at best overstated: voice leadings (which represent *how voices move the notes of one chord to those of another*) can be represented as continuous paths in the orbifolds representing unordered sets, just as voice crossings can be conceived as certain kinds of *paths that move through singularities.*[16] The Appendix shows that it is possible to interpret Fux's counterpoint rules as demarcating a set of "allowable moves" on the orbifold in which intervals are represented as unordered pairs of pitch classes. What results is a plausible model of counterpoint in which voices are equal partners, rather than Mazzola's hierarchical conception in which the discantus is a mere "melodic variation" on the cantus line ([9], p. 617).

This new model, however, is a matter for another paper. Let's instead turn to the statistics Mazzola offers in support of his view. Noting that his model's 37 prohibitions include 21 of Fux's, he observes "if somebody tries to hit at least 21 of the 54 inadmissible cases of the reduced strict style without knowing anything about counterpoint by 37 trials, the chance is less than $2.10^{-8}$." But this is not convincing, since there are countless mathematical formulae which, for purely coincidental reasons, correspond in surprising ways to real-world facts. (For example, there is an urban legend according to which U.S. presidents die in office, usually by assassination, when elected in years evenly divisible by 20.[17]) Mazzola's style of argument, if valid, could be applied equally to any of these cases. (It is, for instance, unlikely that a random sequence of numbers would reproduce the election years of presidents who die in office.) The problem here is that the predictions of Mazzola's theory are not independent trials, analogous to throwing a series of darts at a dartboard: Mazzola gives us a *single* theory which makes a *single* set of musical claims. (The same point also applies to the purported pattern of presidential deaths.) If there is an experiment being performed, it is the endlessly repeated *human being investigates some apparent connection between mathematics and the world*. Thus, rather than asking "what are the odds of getting 21 hits in 37 trials?" we should be asking "if we engage in something like $10^8$ experiments with 25–50 trials in each, what are the chances that at least one of them will contain 21 spurious successes?"[18] *These* odds are of course quite good, which is precisely why we require that mathematical explanations of real-world phenomena either be extraordinarily accurate, or else be accompanied by some plausible explanatory mechanism. Mazzola's theory, as we have seen, is problematic on both counts.

---

[16] See [11], [12], [3], and [13]. Hall and Tymoczko [13] explicitly note that voice crossing in the singular space $\mathbb{R}^n/S_n$ is represented by a line segment that "bounces off" the singularity.

[17] Harrison, elected in 1840, died of pneumonia; Lincoln, elected in 1860, was assassinated in 1865; Garfield, elected in 1880, was assassinated in 1881; McKinley, re-elected in 1900, was assassinated in 1901; Harding, elected in 1920, died of a stroke in 1923, and was said to have been poisoned; Roosevelt, re-elected in 1940, died of a stroke in 1945; Kennedy, elected in 1960, was assassinated in 1963; and Ronald Reagan, elected in 1980, was shot and nearly killed. Proponents of the theory sometimes argue that Reagan's survival broke the "curse."

[18] The number $10^8$ is made-up. Nobody knows what the actual number is, though we can be certain that it is large.

Finally, it is important to say that standards of accuracy depend crucially on context. In some cases, it could be useful to have a test that was 30% accurate while delivering a 50% rate of false positives.[19] The situation we are considering is not of this sort, largely because we *already have* a fairly satisfactory derivation of the fundamental Fuxian prohibitions: tritones are dissonant and difficult to sing and parallel perfect intervals are more likely to fuse, for reasons of both acoustics and scale structure. From this standpoint, Mazzola's explanation has to pass a relatively high crossbar: the crucial question is not whether it is better than *no explanation at all,* but whether it is better than the very good explanations we already have.

## 4 Conclusion

It might seem unfair to draw any conclusions about a 1300-page book on the basis of a single small section of a single chapter. But for some theorists, the value of models lies precisely in their real-world applications: do they, or do they not, teach us something new about music as it has actually been practiced? These readers will be prepared to accede to *Topos*'s extraordinary demands only insofar as they receive an extraordinary payoff in return. And in the single case we have examined, the payoff is anything but remarkable: we have found significant problems at almost every stage of Mazzola's argument, from the description of Fux's rules, to the accuracy of the model, to the baffling use of statistical arguments. Given the severity of these issues, some readers may be tempted to draw negative conclusions about *Topos* as a whole. Sometimes, problems with a part really do cast doubt upon the rest.[20]

From this perspective, the problem with *Topos* lies less in the complexity of its mathematics than in the quality of its applications. But some readers will not be so focused on applications. Confronted with a mismatch between theory and the world, we can always respond by saying "well, so much the worse for the world." And one could legitimately argue that Mazzola's system has a beauty all its own, a compelling logic that overrides the occasionally tenuous assertion about reality. For example, one could treat *Topos* as a radical and challenging artwork that adopts the guise of scholarship, but whose ultimate goal is beauty rather than truth. Another option is to allow *Topos* the sort of autonomy we grant to abstract mathematics: for just as we do not criticize higher-dimensional hyperbolic geometry for being inapplicable to ordinary 3D experience, so too can we declare that Mazzola is engaging in a kind of "theoretical music theory"—an activity more concerned with how we *might* think about music than with solving particular historical or analytical problems. Perhaps *Topos* is a kind

[19] Suppose there were some fatal disease that only struck middle-aged people, but which could be prevented by giving infants an expensive drug with no side effects: if a test could identify 30% of those who would be afflicted, then it might save numerous lives at reasonable cost.

[20] Roeder [1] critiques two other applications that appear throughout Mazzola's writing: his model of "cadences" and his reading of keys in the "crisis" passage in Beethoven's *Hammerklavier*. This leads him to declare that Mazzola's theory "is impressive but not viable."

of mathematics whose questions originate in familiar musical problems, but which has managed to transcend them.[21]

For me, these are the fundamental issues: Is music theory an applied science, whose value is largely determined by its practical results? Or is it something self-standing, whose worth lies in the intrinsic beauty of its constructions? How do we evaluate a work like *Topos*, which is trying not just to describe previous music, but also to open up new conceptual territory? How should we prioritize mathematical sophistication relative to musical insight? When doing music theory, should we aspire to the disinterested caution of the scientist, or the committed intensity of the visionary artist? These questions, ultimately, are not matters for intellectual debate, but rather opportunities to display our fundamental intellectual values. Different answers are hard to reconcile, precisely because there is little neutral ground on which impartial discussion can take place.

From this point of view, the very inadequacies of *Topos*, when viewed under the lens of practical, results-oriented science, strongly suggest that Mazzola is playing a different game altogether.[22] There are, indeed, many activities that go under the rubric of "music theory," and it may be that Mazzola and I are quite far apart in our conceptions of the discipline. If so, my criticisms should be taken as an invitation for two music-theoretical traditions—largely but not exclusively associated with Europe and America—to engage in a deeper and more sustained philosophical dialogue about fundamental goals and methods.[23]

## Appendix

Figure A1 shows the orbifold $\mathbb{T}^2/\mathcal{S}_2$, representing unordered pairs of pitch classes. The space is a Möbius strip whose singular boundary acts like a mirror. Though it is continuous, only diatonic consonances are labeled. Voice leadings are represented by

---

[21] It might also be reasonable to treat *Topos* as something like an autobiography. Guerino is, after all, an extraordinary individual: a jovial and genuinely talented mathematician, a gifted free-jazz pianist, and a European pioneer of mathematically informed music theory. *Topos* may perhaps be read as the record of a unique and extraordinary man's attempt to grapple with the mystery of music. From this point of view, there is something charming about the generous manner in which the book incorporates the work of Guerino's students and collaborators. A charitable lack of skepticism can sometimes be a virtue in a teacher.

[22] As Roeder writes: "Americans value music theories for their practical (analytical or compositional) application, and usually present them with little philosophical motivation. In contrast, Mazzola values his theory primarily according to how well it realizes a particular philosophy about the nature of music" ([1], p. 307).

[23] These issues are reminiscent of those separating "Anglo-American" and continental approaches to philosophy. Anglo-American philosophers typically prize conceptual clarity and rational argument, conceiving of their enterprise as being continuous with science, and ignoring philosophers such as Hegel and Schopenhauer. By contrast, continental philosophers often ask broader questions, conceive of their work in (partially) aesthetic terms, and take philosophers like Hegel quite seriously. Here, as in music theory, it is difficult to bridge the gap precisely because the dispute is (in part) about what sorts of intellectual activities are worth pursuing.

the images of directed line segments in the covering space $\mathbb{R}^2$, "generalized line segments" that may bounce off the singular boundary in the quotient. We can think of the space as a kind of gameboard on which composers of two-voice music necessarily ply their trade. Fux's first-species rules determine a specific game that can be played on the board; an alternate game, modeling $11^{th}$-century contrapuntal practice, is discussed in Chapter 6 of [3]. For simplicity, I will model Fux's rules in the case where one is composing both voices at once. Working with a precomposed cantus amounts to fixing one 45° diagonal component of each generalized line segment.

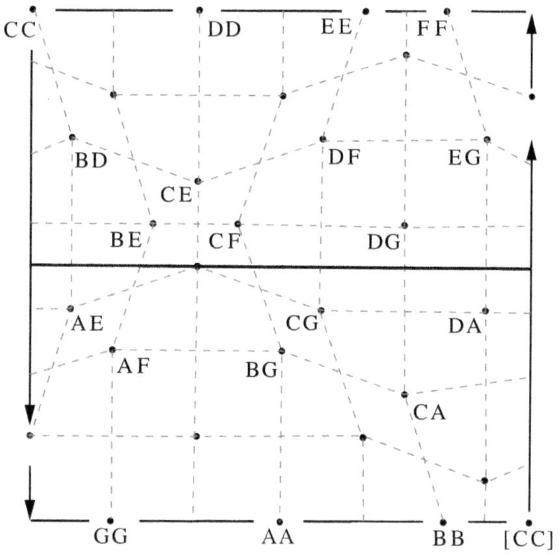


**Fig. A1.** The orbifold $\mathbb{T}^2/S_2$, with only diatonic consonances labeled. Unlabeled points correspond to diatonic dissonances, while dashed lines represent parallel and contrary stepwise motion within the diatonic scale. The left edge is glued to the right with a half twist. The top and bottom boundaries act like mirrors, while the dark line at the center of the space contains tritones.

Note that the following rules do not aspire to have the explanatory force of Mazzola's model. Instead, they consider traditional contrapuntal principles to be exogenous factors explained outside the theory (e.g. by considerations such as those in §3). The goal is to show that, contra Mazzola, one *can* represent Fux's counterpoint rules using the geometry of unordered sets, and furthermore that geometry provides an intuitive characterization of the possibilities available to composers—one in which the musically salient alternatives (involving efficient voice leading) are also *geometrically* salient (being represented by short paths on the orbifold). For more on the underlying analytical philosophy, see [3].

Rule 1 (*voice leadings*). Each move occurs along "generalized line segments" as defined above.

Rule 2 (*consonant intervals*). A move can begin and end only at the labeled points (representing octaves, thirds, sixths, and fifths). In the case of unisons and octaves, one must "desingularize" by beginning and ending infinitesimally close to the boundary, rather than exactly on it.

Rule 3 (*no tritones*). One cannot move along any path whose projection onto either 45° diagonal is six units long.[24]

Rule 4, v.1 (*no strictly parallel fifths or octaves*). One cannot move *horizontally* from one octave/unison to another, or from one fifth to another.

Rule 4, v. 2 (*no parallel or antiparallel fifths or octaves*). One cannot move *in any way* from one octave to another, or from one fifth to another.

Rule 4, v. 3 (*no similar motion into perfect consonances*). One cannot move into an octave or fifth along a generalized line segment whose slope is in the range (–1, 1).

Rule 5 (*no fourths, simple version*). Once you have landed on a fourth/fifth, you can only move to another fourth/fifth after an even number of total crossings of the tritone and unison lines.

Rule 5 (*no fourths, full version*). At the very beginning of the game, when one first chooses a point on the Möbius strip, one chooses an integer, the *tritone parameter,* representing the integer part of the distance between voices, measured in tritone units from lower note to higher note.

This number is incremented or decremented whenever the music (*a*) bounces off the mirror boundary; or (*b*) crosses the horizontal line of tritones at the center of the strip. When the number is even, decrement when crossing the line of unisons and increment when crossing the line of tritones; when the number is odd, increment when crossing the line of unisons and decrement when crossing the line of tritones. When decrementing from zero, move to "–0" before moving to –1. Similarly, when incrementing from –1, move to –0 before 0.

One can land on a point representing a fourth/fifth only if the tritone parameter is odd (counting both 0 and –0 as "even"). A voice crossing occurs when the tritone parameter changes sign.

Together the rules demarcate the complete set of Fux-admissible paths in $\mathbb{T}^2/\mathcal{S}_2$. That is, any geometrical motion in accordance with the five rules can always be realized in pitch space ("lifted") so that the resulting progression conforms to Fux's rules; conversely every Fux-admissable passage of first-species counterpoint (in the sense of Fig. 2) projects from pitch-space into $\mathbb{T}^2/\mathcal{S}_2$ as a series of motions in accordance with the rules. Thus, Fux-admissible paths have a recognizable shape even when we represent intervals as unordered sets of pitch classes.

---

[24] More precisely: one cannot move along any generalized line segment that is the image, in the covering space, of a path either of whose diagonal projections have length 6 (mod 12).

A few last points. First, the Fux-admissible paths can be sorted into three classes: *invertible paths* which contain no fifths, so that every pitch-space realization ("lifting") conforms to Fux's rules; *crossing-free paths*, in which there is a lower bound to the distance between voices (e.g. for one voice starting on C4, the other can start on any G above); and *finitely constrained counterpoint*, in which there are only a finite number of possible distances between voices. The simple version of Rule 5 models the first two, while the full version is needed for the third. Second, we can use the model to represent a range of prohibitions on motion into perfect consonances, including the rule actually enunciated by Fux. Finally, if we identify one note in a pair as corresponding to the cantus firmus, this distinction can be carried along any continuous path, such that we can always identify the cantus throughout the rest of the sequence.

## References


1. Roeder, J.: A MaMuTh Achievement. Perspectives of New Music 31(2), 294–312 (1993)
2. Fux, J.: The Study of Counterpoint. Ed. and Trans. Alfred Mann. Norton, New York (1971) [1725]
3. Tymoczko, D.: A Geometry of Music. Oxford, New York (2011)
4. Kostka, S., Payne, D.: Tonal Harmony, 4th edn. Knopf, New York (2000)
5. Gauldin, R.: A Practical Approach to Sixteenth-Century Counterpoint. Waveland, Long Grove (1985)
6. Jeppesen, K.: Counterpoint: The Polyphonic Vocal Style of the Sixteenth Century. Trans. Glen Haydon. Prentice Hall, Englewood Cliffs (1939)
7. Vincentino, N.: L'antica Musica Ridotta alla moderna prattica. Antonio Barre, Rome (1555)
8. Agustín-Aquino, O.: Counterpoint in 2*k*-tone equal temperament. Journal of Mathematics and Music 3(3), 153–164 (2009)
9. Mazzola, G., et al.: The Topos of Music. Birkhauser, Basel (2002)
10. Mann, A.: The Study of Fugue. Dover, Mineola (1987); contains passages from Fux that discuss double counterpoint
11. Tymoczko, D.: The Geometry of Musical Chords. Science 313, 72–74 (2006)
12. Callender, C., Quinn, I., Tymoczko, D.: Generalized Voice Leading Spaces. Science 320, 346–348 (2008)
13. Hall, R., Tymoczko, D.: Submajorization and the Geometry of Unordered Collections. Under Review (2010)

# Introduction to Scale Theory over Words in Two Dimensions

Marek Žabka

Katedra hudobnej vedy, Univerzita Komenského, Bratislava, Slovakia
zabka@fphil.uniba.sk

**Abstract.** Recently, an interaction between the mathematical discipline of combinatorics on words and musical scale theory has led to various interesting results. So far, the focus was mainly on scales generated by a single interval. The paper proposes an extension of word scale theory to tone systems of higher dimensions, i.e. generated by more than one interval. It is shown that the number of specific varieties for any non-zero generic interval in $n$-dimensional comma-demarcated generated tone systems is between 2 and $2^n$. Therefore, generating patterns in two-dimensional systems are words over a four-letter alphabet. A concept of *quasi pairwise well-formed* words is introduced as a weakening of Clampitt's pairwise well-formedness. The main result of the paper is that a four-letter word is a generating pattern in a comma-demarcated two-dimensional system if and only if it is quasi pairwise well-formed.

**Keywords:** Word Scale Theory, Interval Variety, Generated Tone System, Quasi Pairwise Well-Formed, Product Word.

## 1 Preliminaries

### 1.1 Words

In a series of papers, Clampitt, Noll and Domínguez laid down a solid basis for scale theory over words [1,2,3,4]. They have applied various concepts and results of combinatorial word theory to problems of musical scales. The formal framework summarized below follows their approach, especially, the introductory work [1].

A finite set $A$ is called *alphabet* and its elements *letters*. A finite string (series) $w = a_1 \ldots a_n$ of $n$ (possibly repeating) letters is called a *word* of *length* $n$ over alphabet $A$. We write that $|w| = n$ and the letter on the $i$-th position is denoted by $w(i)$, $i = 1, \ldots, |w|$. An empty string is also considered a word (an *empty word*). The set of all words over $A$ is denoted by $A^\star$.

*Concatenation* of two words $v = a_1 \ldots a_n$ and $u = b_1 \ldots b_m$ is the word $w = a_1 \ldots a_n b_1 \ldots b_m$ and we write $w = v \cdot u$ or $w = vu$. The concatenation is an associative binary operation on $A^\star$ and $(A^\star, \cdot)$ is a monoid (the empty word is the unit). Two words $w_1$ and $w_2$ over alphabets $A_1$ and $A_2$, respectively, are called *isomorphic* if there exists a monoid isomorphisms $f : A_1^\star \to A_2^\star$ such that $f(w_1) = w_2$.

C. Agon et al. (Eds.): MCM 2011, LNAI 6726, pp. 311–325, 2011.

We say that two words $w_1$ and $w_2$ are *conjugate* if there exist words $v$ and $u$ such that $w_1 = vu$ and $w_2 = uv$. This binary relation is an equivalence relation and therefore it factorizes the set $A^\star$ into equivalence classes, which are called *conjugacy classes*. The conjugacy class of a word $w$ is denoted by $[w]$.

Assume a two-letter alphabet $A = \{a, b\}$ and two positive integers $m$ and $n$, $m \leq n$. We define *rational mechanical word* $\mathsf{RMW}(m:n)$ over alphabet $A$:

$$\mathsf{RMW}(m:n)(i) = \begin{cases} a, \text{ if } \lfloor \frac{im}{n} \rfloor = \lfloor \frac{(i-1)m}{n} \rfloor, \\ b, \text{ if } \lfloor \frac{im}{n} \rfloor > \lfloor \frac{(i-1)m}{n} \rfloor, \end{cases}$$

for $i = 1, \ldots, n$. Any conjugate of a rational mechanical word is called *maximally even* word. Furthermore, if $\gcd(m, n) = 1$ then $\mathsf{RMW}(m:n)$ is called *Christoffel word* and any conjugate of a Christoffel word is called *well-formed*[1] word.

Consider a concatenation $w$ of $k$ copies of a word $v$. Then we say that $w$ is a *power* of $v$, $v$ is a *root* of $w$, and we write $w = v^k$. A non-empty word $w$ is *primitive* if it is not a power of any other word, i.e. $w = v^k$ implies $k = 1$.[2] Finally, a word $v$ is called the *primitive root* of $w$ if $w$ is a power of $v$ and $v$ is primitive. The primitive root exists and is unique for any non-empty word.

If $w_1$ and $w_2$ are conjugate then $w_1$ is primitive if and only if $w_2$ is primitive. It is also easy to see that the mechanical word $\mathsf{RMW}(m:n)$ is primitive if and only if $\gcd(m, n) = 1$. Therefore, the well-formed words are exactly the primitive maximally even words. Denote $d = \gcd(m, n)$. Then, the primitive root of a maximally even word $w$ belonging to $[\mathsf{RMW}(m:n)]$ is a well-formed word $v$ which is a conjugate of the Christoffel word $\mathsf{RMW}(\frac{m}{d}:\frac{n}{d})$ and $w = v^d$.

*Example 1.* The usual Ionian scale can be represented as a series of tones and semitones: *ttsttts*. This word is well-formed and is conjugate of the Christoffel word $\mathsf{RMW}(2:7)$ over the alphabet $\{t, s\}$, which is *tttstts*. □

*Example 2.* The hexatonic scale is usually interpreted as a series of alternating semitones and minor thirds or augmented seconds: 131313. This is a maximally even word and also a rational mechanical word $\mathsf{RMW}(3:6)$. Its primitive root is 13, which is a well-formed (Christoffel) word. □

### 1.2 (Unpitched) Generated Tone Systems

The concept of a generated tone system is taken from [6]. The only difference is that the approach here is purely structural: pitch is not reflected at all and, therefore, it is left out from the theoretical framework. For related theoretical concepts compare also [7,8,9,10].

Let $T$ and $X$ be non-empty finite sets and $K$ be a subset of the group $\mathbb{Z}[X]$ such that the elements of $X$ and $K$ have same cardinality $n$. Let $I = \mathbb{Z}[X]/\langle K \rangle$

[1] This definition encompasses the non-degenerate well-formed words over two-letter alphabets and the words of length 1 (when $m = n$). The latter case has not been excluded by purpose: such words may participate in product words representing the generating patterns in comma-demarcated GTS's.

[2] Compare [5].

denote the factor group of $\mathbb{Z}[X]$ modulo subgroup $\langle K \rangle$ generated by $K$ and let $I$ be finite. Further, assume two mappings:

1. $\mathsf{int} : T \times T \to I$ satisfies the following conditions:
   (a) $\mathsf{int}(t, u)\mathsf{int}(u, v) = \mathsf{int}(t, v)$ for any $t, u, v \in T$,
   (b) for any $t \in T$ and $i \in I$ there exists unique $u \in T$ such that $\mathsf{int}(t, u) = i$; we will write $u = t + i$;
2. $\mathsf{spec} : T \to \mathbb{Z}[X]$ satisfies the following condition:
   (a) $[\mathsf{spec}(t)] + \mathsf{int}(t, u) = [\mathsf{spec}(u)]$ for any $t, u \in T$;

The quintuple $(T, X, K, \mathsf{int}, \mathsf{spec})$ is called $n$-dimensional *(unpitched) generated tone system* (*GTS*). The elements of $T$, $X$, $K$, and $I$ are called *tones*, *generators*, *commas*, and *(generic) intervals*. The mappings spec and int are called *specifying function* and *interval function*.

Conditions (1a) and (1b) imply that $(t, i) \mapsto t + i$ is a simply transitive action of the group of generic intervals $I$ on the set of tones $T$. Therefore, the triple $(T, I, \mathsf{int})$ is a Lewinian Generalized Interval System and, in particular, cardinalities of $T$ and $I$ are equal.

Let $K = \{\kappa_i \mid i = 1, \ldots, n\}$. Assume an element $\varepsilon \in \mathbb{R}[X]$ such that for any tone $t \in T$:

$$\mathsf{spec}(t) = \varepsilon + \sum_{i=1}^{n} r_i \kappa_i, \qquad (1)$$

for some $r_i \in \mathbb{R}$, $0 \leq r_i < 1$, $i = 1, \ldots, n$. Then we say that $S = (T, X, K, \mathsf{int}, \mathsf{spec})$ is a *comma-demarcated* GTS. The element $\varepsilon \in \mathbb{R}[X]$ is called *extremity* and elements $\varepsilon + \sum_{i=1}^{n} z_i \kappa_i \in \mathbb{R}[X]$ for any $z_i \in \{0, 1\}$ are called *limits* of the GTS $S$. Moreover, if $\varepsilon \in \mathbb{Z}[X]$ then we say that $S$ is *strictly comma-demarcated.*

*Example 3.* Equation (1) means that the specifying function maps the tones into elements of $\mathbb{Z}[X]$ contained in a parallelogram given by the comma vectors. The left diagram in Figure 1 depicts a comma-demarcated generated tone system which models the diatonic scale in just intonation. Generators are the perfect fifth ($x_1 = (1, 0)$) and the major third ($x_2 = (0, 1)$). Thus, the free commutative group $\mathbb{Z}[X] = \mathbb{Z}^2$ generated by set of generators $X = \{x_1, x_2\}$ corresponds to usual Euler lattice.[3] Commas are $\kappa_1 = (-1, 2)$ and $\kappa_2 = (-3, -1)$. Extremity $\varepsilon$ coincides with the specific value of tone $D$. Values of specifying function are given by a parallelogram demarcated by commas, e.g. $\mathsf{spec}(C) = [0, 0]$, $\mathsf{spec}(G) = (0, 1)$, $\mathsf{spec}({}_1B) = [1, 2]$ etc. The interval function can be defined as $\mathsf{int}(t, u) = [\mathsf{spec}(u) - \mathsf{spec}(t)]$. This GTS is *strictly* comma-demarcated as $\varepsilon = (2, 0)$ belongs to the lattice.

The diagram on the right side of Figure 1 depicts a strictly comma-demarcated GTS modeling the hexatonic scale. □

[3] In this and similar diagrams, subscripts and superscripts preceding tone letters may be thought of as denoting (negative and positive) corrections of the Pythagorean values by syntonic commas. However, this is not included in formal framework and we use the notation in diagrams only as a mnemonic aid.

*Example 4.* Figure 2 shows two variations of the first GTS from the previous example. The GTS's shown here share sets of generators and commas. They differ in selection of extremity, i.e. in definition of specifying function. Resulting GTS's model the harmonic minor scale and the Hungarian scale in just intonation. Both are comma-demarcated and neither is strictly comma-demarcated. □

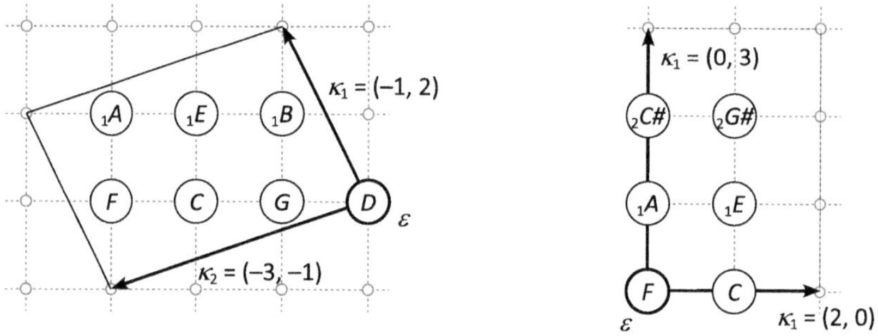


**Fig. 1.** The diatonic scale and the hexatonic scale as two-dimensional comma-demarcated GTS's

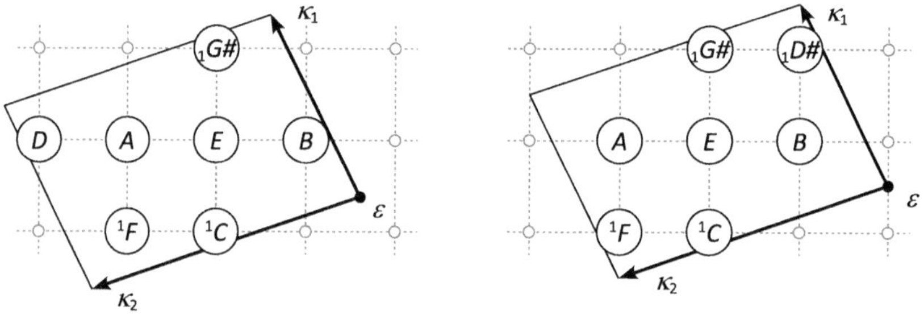


**Fig. 2.** The harmonic minor and the Hungarian scale as two-dimensional comma-demarcated GTS's

## 2 Conceptual Framework

### 2.1 Quasi Pairwise Well-Formed Words

The concept of pairwise well-formedness was introduced by Clampitt [11,12]. We propose a weakening of the concept: quasi pairwise well-formed words seems suitable for investigating the generating patterns appearing in two-dimensional comma-demarcated GTS's.

Assume a four-letter alphabet $A = \{a, b, c, d\}$ and a two-letter alphabet $X = \{x, y\}$. A monoid homomorphism $\pi : A^\star \to X^\star$ is called *pairwise projection* if $\pi(a') = \pi(b') = x$ and $\pi(c') = \pi(d') = y$ for some $\{a', b', c', d'\} = A$ and $\{x', y'\} =$

$X$.[4] We say that a word $w \in A^\star$ is *pairwise well-formed* if $\pi(w)$ is well-formed for any pairwise projection $\pi$ and $w$ contains[5] exactly three distinct letters.

Further, we say that pairwise projections $\pi_1$ and $\pi_2$ are distinct if there are $s, t \in A$ such that $\pi_1(s) = \pi_1(t)$ and $\pi_2(s) \neq \pi_2(t)$. Finally, a word $w \in A^\star$ is called ***quasi pairwise well-formed*** if it is primitive and there are at least two distinct pairwise projections $\pi_1$ and $\pi_2$ such that $\pi_1(w)$ and $\pi_2(w)$ are maximally even words.

The concept of quasi pairwise well-formed words is a generalization of the concept of pairwise well-formed words. The next lemma formalizes this fact. (It follows directly from the definitions of pairwise well-formed and quasi pairwise well-formed words.)

**Lemma 1.** *If a word is pairwise well-formed than it is also quasi pairwise well-formed.*

*Example 5.* Clampitt uses the classical Indian scale *ma-grāma* as an illustration of the pairwise well-formedness. However, the related scale *sa-grāma* is not pairwise well-formed. It turns out that quasi pairwise well-formedness encompasses both basic scales.

*Ma-grāma* and *sa-grāma* are selections of seven tones out of the universe of 22 *śrutis.* In both scales, there are step intervals of three different sizes: 2-, 3- and 4-*śruti* large. Structural details of the scales are the following:

| | | | | | | | | | | | | | | | |
|---|---|---|---|---|---|---|---|---|---|---|---|---|---|---|---|
| *SA-grāma*: | | SA | | RI | | GA | | MA | | PA | | DHA | | NI | |
| step sizes: | 4 | | 3 | | 2 | | 4 | | 4 | | 3 | | 2 | | |
| *MA-grāma*: | | MA | | PA | | DHA | | NI | | SA | | RI | | GA | |
| step sizes: | 4 | | 3 | | 4 | | 2 | | 4 | | 3 | | 2 | | |

If we replace step intervals with the letters $a$, $b$ and $c$ then *ma-grāma* is represented by the word *abacabc*. It can be directly verified that all three possible pairwise projections result in well-formed words.

On the other hand, *sa-grāma* is represented by the word *abcaabc*, which is not pairwise well-formed. However, two out of three possible pairwise projections still give a well-formed word:

$\pi_1 : a \mapsto x,\ b \mapsto x,\ c \mapsto y;\quad abcaabc \mapsto xxyxxxy$ – well-formed
$\pi_2 : a \mapsto x,\ b \mapsto y,\ c \mapsto x;\quad abcaabc \mapsto xyxxxyx$ – well-formed
$\pi_3 : a \mapsto y,\ b \mapsto x,\ c \mapsto x;\quad abcaabc \mapsto yxxyyxx$ – not well-formed

Therefore, *sa-grāma* is (only) quasi pairwise well-formed. □

[4] In general, the number of distinct letters in a word from $A^\star$ can be anywhere between 1 and 4. Therefore, pairwise projections for words with lesser than 4 distinct letters are also included in this definition. For instance, $\pi : a, d \mapsto x;\ b, c \mapsto y$ is a pairwise projection, which maps the three-letter word *abacaba* on the well-formed word *xyxyxyx*.

[5] It is necessary to add this condition explicitly. Otherwise, the definition would lead to an enlarged family of objects. Besides all Clampitt's three-letter pairwise well-formed words it would encompass also one-letter word $a$ (which in our framework is well-formed) and an infinite number of four-letter words.

*Example 6.* Let us explore the words representing the step patterns in the comma-demarcated GTS's from Examples 3 and 4:

| | | | | | | | | | | | | | | | |
|---|---|---|---|---|---|---|---|---|---|---|---|---|---|---|---|
| Diatonic scale: | $C$ | | $D$ | | $_1E$ | | $F$ | | $G$ | | $_1A$ | | $_1B$ | | (C) |
| (PWF) | | $a$ | | $c$ | | $b$ | | $a$ | | $c$ | | $a$ | | $b$ | |
| Hexatonic scale: | $C$ | | $_2C\sharp$ | | $_1E$ | | $F$ | | $_2G\sharp$ | | $_1A$ | | $(C)$ | | |
| (QPWF) | | $c$ | | $b$ | | $a$ | | $d$ | | $a$ | | $b$ | | | |
| Hungarian scale: | $A$ | | $B$ | | $^1C$ | | $_1D\sharp$ | | $E$ | | $^1F$ | | $_1G\sharp$ | | (A) |
| (PWF) | | $c$ | | $a$ | | $b$ | | $a$ | | $a$ | | $b$ | | $a$ | |
| Harmonic minor: | $A$ | | $B$ | | $_1C$ | | $D$ | | $E$ | | $^1F$ | | $_1G\sharp$ | | (A) |
| (QPWF) | | $b$ | | $a$ | | $c$ | | $b$ | | $a$ | | $d$ | | $a$ | |

We see that all the step patterns are quasi pairwise well-formed (QPWF) words and only some of them are also pairwise well-formed (PWF). For instance, the harmonic minor is QPWF as pairwise projections $\pi_1 : a, b \mapsto x;\ c, d \mapsto y$ and $\pi_2 : a, c \mapsto x;\ b, d \mapsto y$ map the 4-letter word *bacbada* to well-formed words and pairwise projection $\pi_3 : a, d \mapsto x;\ b, c \mapsto y$ maps it to a non-well-formed word. □

### 2.2 Product Words

Now a construction of words over two-dimensional alphabets will be suggested. Let $v$ be a word over alphabet $A$ and $u$ be a word over alphabet $B$. We define a word $w$ over the alphabet $A \times B$ of length[6] $k = \mathrm{lcm}(|v|, |u|)$ by the following formula:

$$w(i) = (v((i-1) \bmod |v| + 1), u((i-1) \bmod |u| + 1)),$$

for $i = 1, \ldots, k$. We say that the word $w$ is ***product word*** of $v$ and $u$ and write $w = v \times u$.

*Example 7.* Assume two words of same length:

$$w_1 = x_1x_1y_1x_1x_1y_1x_1, \qquad w_2 = x_2y_2y_2x_2y_2x_2y_2.$$

The product word $w = w_1 \times w_2$ will have the same length, i.e. 7, and will be constructed as a series of ordered pairs of corresponding letters of the words $w_1$ and $w_2$:

$$w = (x_1, x_2)(x_1, y_2)(y_1, y_2)(x_1, x_2)(x_1, y_2)(y_1, x_2)(x_1, y_2).$$

The word $w$ is isomorphic to the three-letter word *acbacab*, which modeled the diatonic scale in Example 6. □

*Example 8.* Now consider two words of different lengths:

$$w_1 = x_1y_1, \qquad w_2 = x_2y_2y_2.$$

[6] By $\mathrm{lcm}(a, b)$ we denote the least common multiple of $a$ and $b$.

Then the length of the product word $w = w_1 \times w_2$ will be the least common multiple of $|w_1|$ and $|w_2|$, which is $6 = \mathrm{lcm}(2, 3)$. To construct the product word both words are repeated in the respective dimension appropriately many times:

$$w_1^{6/2} = x_1 y_1 x_1 y_1 x_1 y_1, \qquad w_2^{6/3} = x_2 y_2 y_2 x_2 y_2 y_2,$$
$$w_1 \times w_2 = (x_1, x_2)(y_1, y_2)(x_1, y_2)(y_1, x_2)(x_1, y_2)(y_1, y_2).$$

The product word is isomorphic to the four-letter word *cbadab*, which modeled the hexatonic scale in Example 6. □

**Lemma 2.** *Let $v$ and $u$ be well-formed words. Then the product word $w = v \times u$ is quasi pairwise well-formed.*

*Proof.* Assume that $v \in A^\star$, $A = \{a_1, a_2\}$, $u \in B^\star$, $B = \{b_1, b_2\}$, and denote $k = |w| = \mathrm{lcm}(|u|, |v|)$. Consider the following mappings:

$$\pi_A : A \times B \to A, (a_i, b_j) \mapsto a_i, \qquad \pi_B : A \times B \to B, (a_i, b_j) \mapsto b_j.$$

Then $\pi_A$ and $\pi_B$ are distinct pairwise projections and $\pi_A(w)$ and $\pi_B(w)$ are maximally even words, whose primitive roots are $v$ and $u$, respectively. Therefore $|v|$ and $|u|$ divide the length of any root of $w$. Let $p$ be the length of the primitive root of $w$. Then $k = \mathrm{lcm}(|v|, |u|)$ divides $p$, which in turn divides $k$. Therefore, $p = k$ and the word $v \times u$ is primitive. □

**Lemma 3.** *Let $w$ be a quasi pairwise well-formed word. Then there exist well-formed words $v$ and $u$ such that their product $v \times u$ is isomorphic to $w$.*

*Proof.* Let $w$ be a quasi pairwise well-formed word over four-letter alphabet $A$, $X$ be a two-letter alphabet and let $\pi_1$ and $\pi_2$ be distinct pairwise projections of $A^\star$ into $X^\star$ such that $v' = \pi_1(w)$ and $u' = \pi_2(w)$ are maximally even. Let $v$ and $u$ be the primitive roots of $v'$ and $u'$, respectively. Then both $v$ and $u$ are well-formed and $w$ is isomorphic to $u \times v$ because $\pi_1$ and $\pi_2$ are distinct. □

The previous two lemmas assure that there is a one-to-one correspondence between the quasi pairwise well-formed words and the product words of well-formed words.

### 2.3 Interval Variety

In a classic work, Clough and Myerson [13] introduced the concepts of generic and specific intervals and investigated Myhill's Property as a key structural feature of diatonic scales. One of the basic results of musical scale theory is that non-degenerate one-dimensional scales are well-formed if and only if they have Myhill's Property (MP), i.e. every non-zero generic interval appears with exactly two specific values [14]. Inspired by Myhill's Property, Clampitt [11] introduced the concept of trivalence: every non-zero generic interval appears with exactly three specific values. He showed that pairwise well-formedness implies trivalence of a scale.

Usually, the concepts of specific and generic intervals rely on pitch realization of scale: the generic interval is given by its span, i.e. by the count of tones between the end tones of the interval represented in pitch domain, and the specific value of an interval is associated with its size, which is also a pitch related property. In their recent paper, Clampitt and Noll provided a lucid formalization of the generic/specific dichotomy. [4] Although they do explicitly involve the concept of pitch, in a certain way they approached the possibility to introduce the dichotomy even without it.[7] In the present approach, pitch is not formally introduced at all and the generic/specific dichotomy is explored on a more abstract level: the generic intervals are elements of the group $\mathbb{Z}[X]/\langle K\rangle$ and their specific values come from $\mathbb{Z}[X]$.

Let $S = (T, X, K, \mathsf{spec}, \mathsf{int})$ be a generated tone system. For any generic interval $g \in \mathbb{Z}[X]/\langle K\rangle$, consider the following set of elements of $\mathbb{Z}[X]$:

$$\mathsf{SPEC}(g) = \{\mathsf{spec}(t) - \mathsf{spec}(u) \mid t, u \in T, \mathsf{int}(t, u) = g\}.$$

The set $\mathsf{SPEC}(g)$ contains all *specific* values of the generic interval $g$. Assume that $\mathsf{SPEC}(g)$ has exactly $v$ distinct elements. We say that $v$ is the *variety* of the generic interval $g$ (the interval $g$ is $v$-varietal) and we write $v = \mathsf{var}(g)$.

*Example 9.* Assume the hexatonic scale as modeled in Examples 3 and 6. Although the hexatonic scale is usually thought of as having only two different "specific steps", in our model, its step pattern is represented by a word *cbadab* over four-letter alphabet. Therefore, the variety of the generic interval $[(-1, 2)] \in \mathbb{Z}^2/\langle\kappa_1, \kappa_2\rangle$ (the "generic step") is 4.[8] This can be interpreted that the diatonic semitone (e.g. ${}_1E - F$) and the chromatic semitone (e.g. $C - {}_2C\sharp$) are differentiated within our theoretic framework. And so are the minor third (e.g. ${}_2C\sharp - {}_1E$) and the augmented second (e.g. $F - {}_2G\sharp$). However, this differentiation is not based on acoustic considerations as the concept of pitch has not been introduced. Even if 12-tone equal temperament was considered, the step-pattern would involve four different species of step and, therefore, the generic step would be 4-varietal. □

## 2.4 Generating Patterns

Let $t \in T$ be a tone and $g \in I$ be a non-zero generic interval of order $k$, i.e. $k$ is the smallest positive integer for which $kg = 0$ in $I$. Consider a series of tones $\tau = (t_0, \ldots, t_k)$ such that $t_0 = t$ and $\mathsf{int}(t_{i-1}, t_i) = g$ for all $i = 1, \ldots, k$. We say that $\tau$ is a *generating series*. The conditions (1a) and (1b) from the definition of GTS imply that the generating series exists and is unique for any given tone and any given generic interval. Moreover, the generating series contains $k$ different tones and $t_k = t$.

[7] See their examination of the degenerate seven-tone scale.

[8] Moreover, Theorem 2 (see below) implies that the hexatonic scale cannot be modeled as a comma-demarcated two-dimensional GTS with two-letter step pattern *ababab* at all as this word is not quasi pairwise well-formed.

Let $v = \mathsf{var}(g)$ and let $\mathsf{SPEC}(g) = \{\gamma_1, \ldots, \gamma_v\}$. We define a word $\mathsf{GP}(g,t)$ of length $k$ over a $v$-letter alphabet $A = \{a_1, \ldots, a_v\}$ by the following rule:

$$\mathsf{GP}(g,t)(i) = a_j \quad \Leftrightarrow \quad \mathsf{spec}(t_i) - \mathsf{spec}(t_{i-1}) = \gamma_j.$$

The word $\mathsf{GP}(g,t)$ is called *generating pattern* of the generic interval $g$ starting from the tone $t$.

*Example 10.* Assume the harmonic minor scale as modeled through a comma-demarcated two-dimensional GTS in Example 4. Figure 3 illustrates how the generating pattern is constructed in this GTS for two generic intervals: "the generic step", $\text{STEP} = [(2,0)]$, and "the generic fifth", $\text{FIFTH} = [(1,0)]$. The generating pattern $\mathsf{GP}(A, \text{STEP}) = bacbada$ has been already computed in Example 6. It will be analyzed even more deeply below in Example 12. The generating pattern $\mathsf{GP}(A, \text{FIFTH}) = aabacba$ is an example of three-letter quasi pairwise well-formed word which is not pairwise well-formed. □

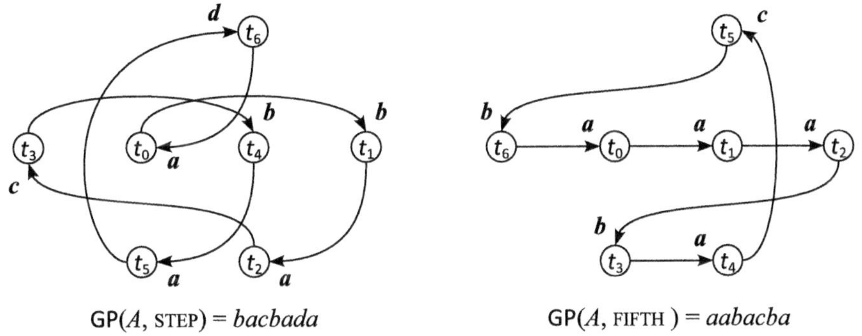


**Fig. 3.** Generating patterns $\mathsf{GP}(A, \text{STEP})$ and $\mathsf{GP}(A, \text{FIFTH})$ , both starting from the tone $A$ in $A$ harmonic minor

*Remark 1.* "Scale step pattern" and "folding pattern" are key concepts of the scale theory over one-dimensional words as developed by Clampitt, Noll and Domínguez. One of their most fundamental results concerns the duality between the two patterns. In the present approach, both patterns are viewed simply as "generating patterns". The investigation of the dual generating patterns remains beyond the scope of this paper.

# 3 Main Results

## 3.1 Generalization of Myhill's Property

As mentioned above, Myhill's Property is very strong in the one-dimensional case: a one-dimensional generated scale is non-degenerate well-formed if and only if every non-zero generic interval is 2-varietal. For the category of pairwise well-formed scales, a similar condition holds in one direction: every non-zero generic interval is 3-varietal. In the general case of two-dimensional comma-demarcated generated tone systems similar strong conditions are not valid. Instead, there are upper and lower bounds for the variety of generic intervals.

**Theorem 1.** *Let $S$ be an $n$-dimensional comma-demarcated GTS. Then the following inequality hold for the variety of any non-zero generic interval $g$:*

$$2 \leq \mathsf{var}(g) \leq 2^n.$$

*Furthermore, the upper limit and the lower limit are achieved for certain cases.*

*Proof.* Let $S = (T, X, K, \mathsf{spec}, \mathsf{int})$, $K = \{\kappa_1, \ldots, \kappa_n\}$, $\varepsilon$ be the extremity of $S$, and $g$ be a non-zero generic interval. Take tones $t$ and $u$ such that $\mathsf{int}(t, u) = g$. Comma-demarcatedness of $S$ yields existence of $r_i, s_i \in [0, 1)$, $i = 1, \ldots, n$ such that

$$\mathsf{spec}(t) = \varepsilon + \sum_{i=1}^{n} r_i \kappa_i, \qquad \mathsf{spec}(u) = \varepsilon + \sum_{i=1}^{n} s_i \kappa_i.$$

Denote:

$$\gamma = \sum_{i=1}^{n} (s_i - r_i)\kappa_i.$$

Then $\gamma \in \mathsf{SPEC}(g)$. Consider the following series of elements of $\mathbb{Z}[X]$:

$$\mathsf{spec}(t), \mathsf{spec}(t) + \gamma, \ldots, \mathsf{spec}(t) + j\gamma, \ldots$$

As $\gamma$ is non-zero this series contains infinite number of distinct elements and so it contains elements which do not belong to $\mathsf{spec}[T]$. Take the lowest $j$ for which $\mathsf{spec}(t) + j\gamma$ is not in $\mathsf{spec}[T]$. Thus, $\mathsf{spec}(t) + (j-1)\gamma \in \mathsf{spec}[T]$ and there are $v, w \in T$ such that $\mathsf{spec}(v) = \mathsf{spec}(t) + (j-1)\gamma$, $\mathsf{int}(v, w) = \gamma$, and $\mathsf{spec}(w) \neq \mathsf{spec}(t) + j\gamma$. Then $\mathsf{spec}(w) - \mathsf{spec}(v) \in \mathsf{SPEC}(g)$ while $\mathsf{spec}(w) - \mathsf{spec}(v) \neq \gamma$. Therefore $\mathsf{var}(g) \geq 2$.

To prove the second inequality, consider $t$, $u$, $g$, and $\gamma$ as defined in the previous paragraph and assume any tones $v, w \in T$ such that $\mathsf{int}(w, v) = g$. The comma-demarcatedness implies that there are coefficients $c_i \in \{0, \mathrm{sign}(s_i - r_i)\}$, $i = 1, \ldots, n$ such that:

$$\mathsf{spec}(w) - \mathsf{spec}(v) = \gamma - \sum_{i=1}^{n} c_i \kappa_i.$$

This yields up to $2^n$ potential combinations for values of $c_i$, thus, up to $2^n$ possible values for $\mathsf{spec}(w) - \mathsf{spec}(v)$. This implies the upper limit for the variety of the generic interval $g$: $\mathsf{var}(g) \leq 2^n$.

Now we will construct examples where the upper and lower limits are reached. For the case $\mathsf{var}(g) = 2$ consider the following vectors in $\mathbb{Z}^n$:

$$\begin{aligned}
\kappa_1 &= (2, 0, \ldots, 0), \\
\kappa_2 &= (0, 1, \ldots, 0), \\
&\ldots \\
\kappa_n &= (0, 0, \ldots, 1).
\end{aligned}$$

Denote $\gamma = (1, 0, \ldots, 0)$ and assume the generic interval $g = [(1, 0, \ldots, 0)] \in \mathbb{Z}^n/\langle K \rangle$ in the strictly comma-demarcated GTS defined by the set of commas $K = \{\kappa_1, \ldots, \kappa_2\}$ and the extremity $\varepsilon = [0, \ldots, 0]$. It is easy to see that $\mathsf{SPEC}(g) = \{\gamma, -\gamma\}$ and, therefore, $\mathsf{var}(g) = 2$.

For the upper limit $\mathsf{var}(g) = 2^n$ consider the series of first $n$ prime numbers $p_1 = 2, p_2 = 3, \ldots, p_n$ and define the following vectors in $\mathbb{Z}^n$:

$$\begin{aligned} \kappa_1 &= (p_1, \ldots, 0), \\ &\ldots \\ \kappa_n &= (0, \ldots, p_n). \end{aligned}$$

Assume the GTS defined by the set of commas $K = \{\kappa_1, \ldots, \kappa_n\}$ and the extremity $\varepsilon = [0, \ldots, 0]$. Denote $\gamma = (1, \ldots, 1)$ and consider the generic interval $g = [\gamma]$ and the tone $e = \mathsf{spec}^{-1}(\varepsilon)$. Finally, define:

$$\gamma_z = (1 - z_1 p_1, \ldots, 1 - z_n p_n),$$

for $z = (z_1, \ldots, z_n) \in \{0, 1\}^n$. We will show that any of the $2^n$ vectors $\gamma_z$ belongs to $\mathsf{SPEC}(g)$. Therefore, $\mathsf{var}(g) \geq 2^n$ and considering the inequality $\mathsf{var}(g) \leq 2^n$, which has already been proved in general, we obtain $\mathsf{var}(g) = 2^n$.

Assume any $z = (z_1, \ldots, z_n) \in \{0, 1\}^n$ and consider the tones:

$$t_{k-1} = e + (k-1)\gamma, \qquad t_k = e + k\gamma,$$

for $k = p_1^{z_1} \ldots p_n^{z_n}$. Then $\mathsf{spec}(t_k) - \mathsf{spec}(t_{k-1}) = \gamma_z$, which implies $\gamma_z \in \mathsf{SPEC}(g)$. This completes the proof. □

**Corollary 1.** *The variety of any non-zero generic interval in a one-dimensional comma-demarcated GTS is 2.*

**Corollary 2.** *The variety of any non-zero generic interval in a two-dimensional comma-demarcated GTS is 2,3, or 4.*

*Example 11.* The variety is invariant in both well-formed and pairwise well-formed scales: every non-zero generic interval has the same variety (2 or 3, respectively). Quasi pairwise well-formed scales do not exhibit such a property. Consider the generating patterns from Example 10. The generic step is four-varietal while the generic fifth is three-varietal. □

### 3.2 Necessary Condition for Generating Patterns

**Theorem 2.** *Let $S$ be a two-dimensional comma-demarcated GTS. Then any generating pattern in $S$ is a quasi pairwise well-formed word.*

*Proof.* Let $S = (T, X, K, \mathsf{spec}, \mathsf{int})$, $K = \{\kappa_1, \kappa_2\}$, $k$ be the cardinality of $T$, and $\varepsilon$ be an extremity. Consider any generic interval $g$ and any tone $t$ and denote $u = t + g$. As $\varepsilon$ is an extremity we have:

$$\mathsf{spec}(t) = \varepsilon + r_1\kappa_1 + r_2\kappa_2, \qquad \mathsf{spec}(u) = \varepsilon + s_1\kappa_1 + s_2\kappa_2,$$

for some $r_1, s_1, r_2, s_2 \in [0, 1)$.

One can assume that $s_1 \geq r_1$ and $s_2 \geq r_2$. Otherwise, we could consider a GTS $S' = (T, X, K', \mathsf{spec}, \mathsf{int})$ defined by the set of commas $K' = \{z_1\kappa_1, z_2\kappa_2\}$, where:

$$z_i = \begin{cases} 1, \text{ if } s_i \geq r_i, \\ -1, \text{ if } s_i < r_i, \end{cases}$$

for $i = 1, 2$. It is easy to see that $S'$ is also comma-demarcated, the generating patterns in $S$ and $S'$ are same and the required inequations for the comma coefficients hold.

Let $k$ be the order of $g$ in $I$. Then we have $kg = 0$ in the group of generic intervals. Thus, $m_1 = k(s_1 - r_1)$ and $m_2 = k(s_2 - r_2)$ are non-negative integers. Considered two-letter alphabets $X_1 = \{x_1, y_1\}$ and $X_2 = \{x_2, y_2\}$ and define the following words:

$$w_1(i) = \begin{cases} x_1, \text{ if } \lfloor r_1 + \frac{im_1}{k} \rfloor = \lfloor r_1 + \frac{(i-1)m_1}{k} \rfloor, \\ y_1, \text{ if } \lfloor r_1 + \frac{im_1}{k} \rfloor > \lfloor r_1 + \frac{(i-1)m_1}{k} \rfloor, \end{cases}$$

$$w_2(i) = \begin{cases} x_2, \text{ if } \lfloor r_2 + \frac{im_2}{k} \rfloor = \lfloor r_2 + \frac{(i-1)m_2}{k} \rfloor, \\ y_2, \text{ if } \lfloor r_2 + \frac{im_2}{k} \rfloor > \lfloor r_2 + \frac{(i-1)m_2}{k} \rfloor, \end{cases}$$

for $i = 1, \ldots, k$. The words $w_1$ and $w_2$ are maximally even:

$$w_1 \in [\mathsf{RMW}(m_1 : k)], \qquad w_2' in[\mathsf{RMW}(m_2 : k)].$$

Moreover, the product word $w_1 \times w_2$ is the generating pattern of $g$ starting from $t$, i.e. $\mathsf{GP}(g, t) = w_1 \times w_2$ over the alphabet $X_1 \times X_2$.

To finish the proof we only have to show that $w_1 \times w_2$ is primitive. Assume the opposite and let $w_1 \times w_2 = v^p$ for a word $v$ over the alphabet $A_1 \times A_2$ and a divisor $p$ of $k$, $p > 1$. Then $t + \frac{k}{p}g = t$, which is in contradiction with the minimality of $k$. This completes the proof. □

*Example 12.* Let us illustrate the main idea of the previous proof. The generating pattern of a two-dimensional GTS is replaced by a product of its projections in the one-dimensional GTS's given by the commas. These projections define pairwise projections and the resulting two-letter words are maximally even. Figure 4 shows the comma projections for the harmonic minor scale (compare Examples 4 and 10). The projections yield two maximally even (in this case even well-formed) words $w_1 = x_1x_1y_1x_1x_1y_1x_1$ and $w_2 = x_2y_2y_2x_2y_2x_2y_2$. The product word $w = w_1 \times w_2 = (x_1x_2)(x_1y_2)(y_1y_2)(x_1x_2)(x_1y_2)(y_1x_2)(x_1y_2)$ is isomorphic to the generating pattern *bacbada* of the generic step as computed in Example 10. □

*Example 13.* The acoustic scale is quite common in the folk music of certain ethnics (e.g. Slovak and Polish). It consists of the overtones 8 through 14 and is usually approximated by the Lydian-Mixolydian scale: $C - D - E - F\sharp - G - A - B\flat - (C)$. It is obvious that it can not be modeled as a one-dimensional fifth-generated (Pythagorean) scale as it selects a discontinuous subseries from the series of fifth. However, it can be represented by a contiguous selection on

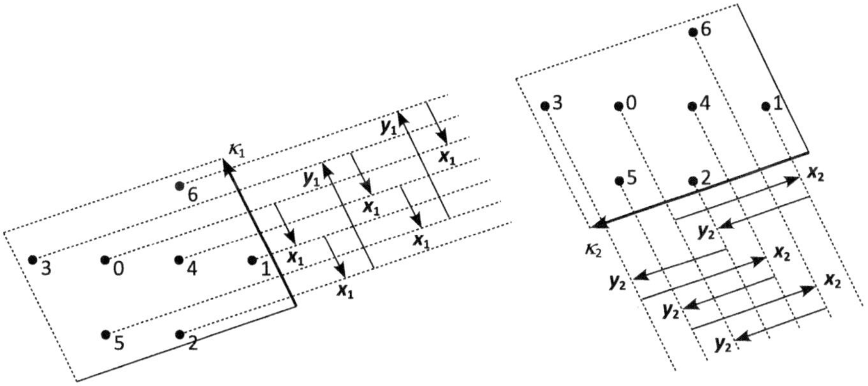


**Fig. 4.** The comma projections of the generating patterns of the generic step and the generic fifth in the harmonic minor

Euler's lattice. (See Figure 5.) Therefore, one may ask whether the natural scale can be modeled as a two-dimensional comma-demarcated GTS. The answer is no: The generating pattern of the generic step starting from $C$ is *aabcacb*, which is not pairwise well-formed. Theorem 2 implies that this pattern can not be a generating pattern of a two-dimensional comma-demarcated GTS.[9] □

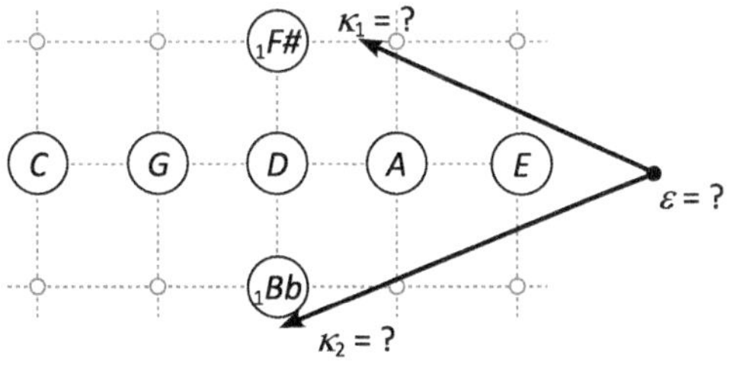


**Fig. 5.** The natural scale on Euler's lattice

### 3.3 Sufficient Condition for Generating Patterns

**Theorem 3.** *Let $w$ be a quasi pairwise well-formed word. Then there exists a two-dimensional comma-demarcated GTS, a generic interval $g$ and a tone $t$ such that the generating pattern* $\mathsf{GP}(g,t)$ *is isomorphic to $w$. Moreover, the example can be constructed in such a manner that the group of generic intervals is cyclic and $g$ is its generator.*

[9] On the other hand, the natural scale may be modeled as a comma-demarcated GTS of a dimension higher than 2. One obvious approach is to follow the acoustic origin of the scale and consider 5-dimensional GTS with the generators 3, 5, 7, 11, and 13, and the following set of commas: $K = \{(3,0,0,0,-1),(0,1,2,0,0),(2,0,1,0,0),(0,1,-1,1,0),(0,1,0,0,1)\}$.

*Proof.* Applying Lemma 3 we may assume that $w = w_1 \times w_2$ and the words:

$$w_1 \in [\mathsf{RMW}(m_1 : k_1)], \qquad w_2 \in [\mathsf{RMW}(m_2 : k_2)]$$

are well-formed. Therefore, we have $\gcd(m_1, k_1) = \gcd(m_2, k_2) = 1$. Denote $d = \gcd(k_1, k_2)$. Then also $\gcd(m_2, d) = 1$ and there exist $z, z' \in \mathbb{Z}$ such that $m_1 + m_2 z = z'd$. Define:

$$\kappa_1 = (\frac{k_1}{d}, 0), \qquad \kappa_2 = (\frac{k_2 z}{d}, k_2), \qquad \gamma = (\frac{m_1 + m_2 z}{d}, m_2).$$

Then $\kappa_1, \kappa_2, \gamma \in \mathbb{Z}^2$ and:

$$\gamma = \frac{m_1}{k_1}\kappa_1 + \frac{m_2}{k_2}\kappa_2.$$

Now let us consider the comma-demarcated GTS $S_0$ defined by set of generators $X = \{(1,0),(0,1)\}$, set of commas $K = \{\kappa_1, \kappa_2\}$, and extremity $\varepsilon_0 = [0,0]$. Then $g = [\gamma]$ is a generator of the group of generic intervals $I = Z^2/\langle K \rangle$ and the generating pattern of $g$ starting from $\varepsilon_0$ is:

$$\mathsf{GP}(g, \varepsilon_0) = \mathsf{RMW}(m_1 : k_1) \times \mathsf{RMW}(m_2 : k_2).$$

Finally, one can choose the extremity $\varepsilon \in [0, \kappa_1) \times [0, \kappa_2)$ in such a way that the generating pattern of $g$ is $w_1 \times w_2$. □

*Remark 2.* Previous theorems 2 and 3 characterize generating patterns in comma-demarcated two-dimensional GTS's: a word is a generating pattern in some comma-demarcated two-dimensional GTS if and only if it is quasi pairwise well-formed. This result can even be extended to higher dimensions. The key idea is that the quasi pairwise well-formed words are exactly the two-dimensional products of well-formed words (See Lemmas 2 and 3). It can be shown that generating patterns in $n$-dimensional comma-demarcated GTS's are $n$-dimensional products of well-formed words.

**Aknowledgements**

I am grateful to Thomas Noll and David Clampitt for valuable discussions which inspired the present research.

## References

1. Domínguez, M., Clampitt, D., Noll, T.: WF Scales, ME Sets, and Christoffel Words. In: Klouche, T., Noll, T. (eds.) MCM 2007. Communications in Computer and Information Science, vol. 37, pp. 477–488. Springer, Heidelberg (2009)
2. Clampitt, D., Domínguez, M., Noll, T.: Plain and Twisted Adjoints of Well-Formed Words. In: Chew, E., Childs, A., Chuan, C.-H. (eds.) MCM 2009. Communications in Computer and Information Science, vol. 38, pp. 65–80. Springer, Heidelberg (2009)
3. Noll, T.: Ionian Theorem. Journal of Mathematics and Music 3(3), 137–151 (2009)

4. Clampitt, D., Noll, T.: Modes, the Height-Width Duality, and Handschin's Tone Character. Music Theory Online (to appear)
5. Lothaire, M.: Algebraic Combinatorics on Words. Cambridge University Press, Cambridge (2002)
6. Žabka, M.: Well-Formedness in Two Dimensions: A Generalization of Carey and Clampitt's Theorem. Journal of Mathematics and Music 4(1), 1–30 (2010)
7. Fokker, A.D.: Selections from the Harmonic Lattice of Perfect Fifths and Major Thirds Containing 12, 19, 22, 31, 41 or 53 Notes. In: Proceedings of Koninklijke Nederlandse Akademie van Wetenschappen, Series B, vol. 71, pp. 251–266. Koninklijke Nederlandse Akademie van Wetenschappen, Amsterdam (1968)
8. Fokker, A.D.: Unison Vectors and Periodicity Blocks in the Three-Dimensional (3-5-7-) Harmonic Lattice of Notes. In: Proceedings of Koninklijke Nederlandse Akademie van Wetenschappen, Series B, vol. 72(3), pp. 153–168. Koninklijke Nederlandse Akademie van Wetenschappen, Amsterdam (1969)
9. Hellegouarch, Y.: Gammes naturelles I. Gazette des mathématiciens 81, 25–39 (1999)
10. Hellegouarch, Y.: Gammes naturelles (suite). Gazette des mathématiciens 82, 13–25 (1999)
11. Clampitt, D.: Pairwise Well-Formed Scales: Structural and Transformational Properties. Ph.D. diss., State University of New York at Buffalo (1997)
12. Clampitt, D.: Mathematical and Musical Properties of Pairwise Well-Formed Scales. In: Klouche, T., Noll, T. (eds.) MCM 2007. Communications in Computer and Information Science, vol. 37, pp. 464–468. Springer, Heidelberg (2009)
13. Clough, J., Myerson, G.: Variety and Multiplicity in Diatonic Systems. Journal of Music Theory 29, 249–270 (1985)
14. Carey, N., Clampitt, D.: Self-similar Pitch Structures, Their Duals, and Rhythmic Analogues. Perspectives of New Music 34, 62–87 (1996)

# The Planet-4D Model: An Original Hypersymmetric Music Space Based on Graph Theory

Gilles Baroin

Laboratoire LLA Creatis, Université de Toulouse
Gilles@Baroin.org

**Abstract.** Beside a geometrical part that has been calculated with the help of the graph theory, the Planet-4D model includes twelve ideograms that can either symbolize notes, chords or scales depending on the context. Based on symmetry principles, it presents the following innovations:

1. the hyper spherical environment grants each symbol an equivalent physical position, and involves more symmetries than any 3D model,
2. the concept of bi-dimensional ideograms provides an intuitive understanding of pitch relationships,
3. it contains implicitly the chromatic and fourth circles as well as the original Tonnetz.

NB: the pertinence of this model is effective when demonstrated in motion with colored CGI animations of the 4D Space including sound examples. Videos shown during this conference are available on the web at www.planetes.info.

**Keywords:** Symmetry, Hypersphere, Pitch Space, Tonnetz, Spectral Projections, Graph Theory, Animated GCI, Quaternions.

## 1 Symmetry

Nöthers theorem specifies that "to every local symmetry there corresponds a conservation law" [1,2] Interpreting symmetry as invariance to the transposition leads to a model where each pitch class plays the same role. That is actually the case for the Tonnetze describing the equal tempered space.

We will also use symmetry as invariance between musical interval and geometrical distance. The physical distance within the geometrical model for a defined interval will be constant i.e., each fifth will be drawn in the model with the same length. That condition is not present in a 3D model but can be respected on a 2D circle.

We apply finally the invariance principle to the ideographic system: since our ideograms are two-dimensional, moving into one direction will conserve one of the two parameters: color or form! That condition on the symbols is an enhancement to the merely mathematical decomposition.

C. Agon et al. (Eds.): MCM 2011, LNAI 6726, pp. 326–329, 2011.

## 2 Mathematical Background

Using different schemes, mathematicians and music theorists have demonstrated that the tempered twelve tone pitch space [3,4] can be considered as a combination of minor and major thirds [5,6,7]. We use the Cartesian product of two circular graphs C3 □ C4 to build the 12 edge graph called the Planet graph [8] (see Fig. 1).

The demonstration involves a step by step composition of 12 vertex graphs as a combination of simple sub elements. The Planet graph is an Abelian Cayley graph on $Z_{12}$ with neighborhood of 0 generated by (±3,±4).

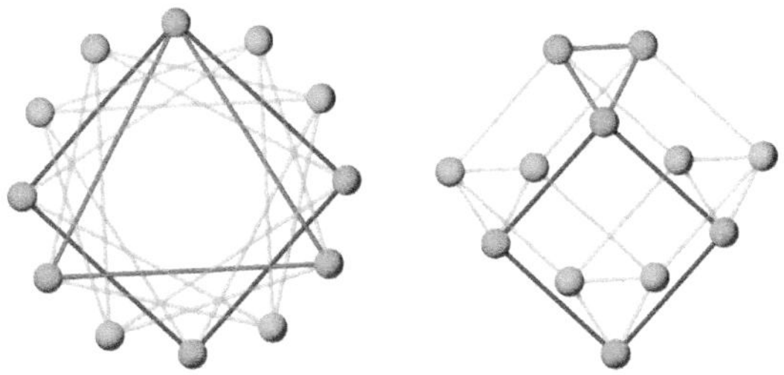

**Fig. 1.** Two isomorphic representations of the Planet graph: each Triangle graph (C3) and Square graph (C4) always have one unique and common node

A graph being a combinatoric object, in order to build a geometrical model, we perform a spectral analysis to determine its Eigen spaces and obtain geometrical coordinates [8]. The spectral projection gives 12 points equally dispatched on a Hypersphere. Since the 4D space is built up with two orthogonal spaces, each point is represented by a quaternion defined by a couple

$$(\frac{1}{\sqrt{3}}e^{i\frac{2\pi k}{3}}, \frac{1}{\sqrt{2}}e^{i\frac{2k'\pi}{4}}),$$

with $k = 0$ to 2 and $k' = 0$ to 3, where the couple $(k, k')$ determines the belonging to triangles and squares. If we consider $n = 0$ to 11 as the pitch index, the quaternions ensemble is defined by

$$(\frac{1}{\sqrt{3}}e^{i\frac{2n\pi}{3}}, \frac{1}{\sqrt{2}}e^{i\frac{2n\pi}{4}}).$$

We call "physical distance" the Euclidian distance in the geometric space; the "logical distance" is the number of rotations to perform to reach the next point, (distance on the graph). The acoustic distance is calculated in semi tones.

The table of distances (Table 1) shows three different physical distances whose values are noteworthy numbers. The distance between neighbor pitch classes is 1. For notes that are neighbor on the chromatic circle or on the circle of fourth or have a triton interval, we obtain $\sqrt{2}$. Finally, the biggest possible distance between two nodes: $\sqrt{3}$ (diagonal of the unity cube) applies for the whole tone.

**Table 1.** Table of distances

| Physical | Logical | Acoustic | Interval |
|---|---|---|---|
| 1 | 1 | 3;9;4;8 | M3;m6;m3;M6 |
| $\sqrt{2}$ | 2 | 1;11;5;7;6 | m2;M7;P4;P5;TT |
| $\sqrt{3}$ | 3 | 2;10 | M2;m7 |

## 3 Vizualisation

In conventional Tonnetz, pitch classes are represented with letters or numbers. Since the decomposition involves two sets and each pitch class being a unique combination of these two sub-groups, we use bi-dimensional ideogram. Each dimension of the pictogram is linked to a subgroup. The two dimensions (parameters) for the ideograms have been arbitrary chosen: form and color [9].

In order to see the fourth dimension and to feel its symmetry, we project the model into our 3D space [10] and let it rotate about its two main symmetry axes (see Fig. 2).

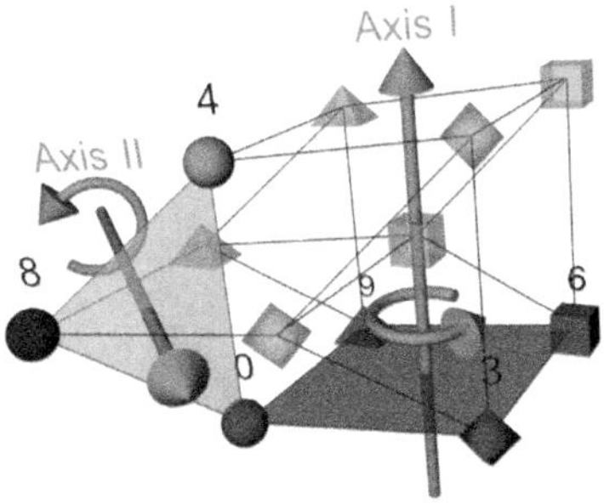


**Fig. 2.** Major axes of the Model

We must imagine that the first axis passes through the center of every square and the second axis through each triangle center. To feel that the model resides in true 4D space, it will be set in motion in our 3D space (see Fig. 3). In the computer animated model, the nearest node to the spectator represents the current position within the model. We control the camera focal distance in order to blur the distant nodes and focus the attention to the nearest nodes (one major or minor third).

As the rotation of a 3D cube on a screen looks like a 2D deformation within the plane, the rotation of the 4D model will be perceived as a deformation within our 3D space without affecting the hull. The feeling is analog to manipulating a Rubik™ cube. As for the torus, chromatic scale and circle of fourth are included within the Planet-4D model. They are obtained by rotating the Hypersphere in both directions simultaneously.

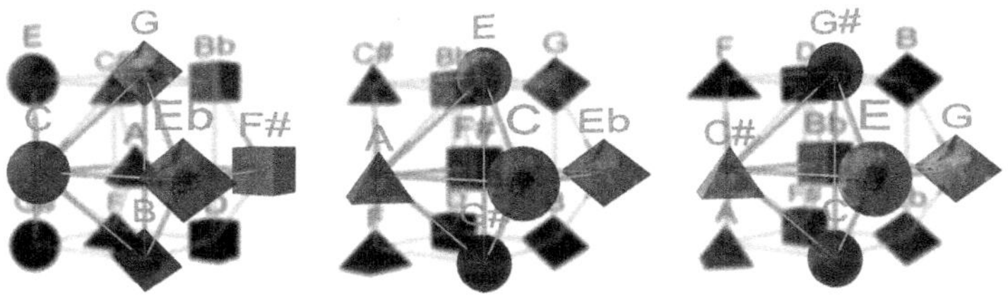


**Fig. 3.** 4D Rotations from start position C (center image)

## 4 Applications and Perspectives

Since the experiment raised curiosity of both mathematicians and musicians, we have applied the same mathematical principles (spectral graph projections) to other musical representation in order to let appear the inner structure of pitch space decomposition. The same principles of visualization in hyperspaces are experimented for a representation of the dual space of the Tonnetz on the Hypersphere. In this case, we illustrate visually and musically, some well known harmonic path, made of P,L,R relations [11,12]. Beside existing methods, these new techniques enable the layman and the music student to better understand proposals that are evident to experimented musicians.

## References

1. Byers, N.: E.Noether's Discovery of the Deep Connection Between Symmetries and Conservation Laws. IMC 12 (1999)
2. Lederman, L., Hill, C.: Symmetry and the Beautiful Universe. Prometheus Books (2004)
3. Andreatta, M., Agon, C.: Formalisation algébrique des structures musicales aide de la Set-Theory: aspects théoriques et analytiques. Actes des Journées d'Informatique Musicale (2003)
4. Forte, A.: The Structure of Atonal Music. Yale University Press, New Haven (1973)
5. Balzano, G.: The group-theoretic description of 12-fold and microtonal pitch systems. Computer Music Journal, 66–84 (1980)
6. Mazzola, G.: The Topos of Music. Birkhuser Verlag, Basel (2003)
7. Noll, T.: Geometry of Chords. In: Puebla, E.L. (ed.) Electronic Bulletin of the Sociedad Matematica Mexicana, vol. 1 (2001)
8. Baroin, G., Ferré, L.: Representation of musical pitch space using graphs, spectrum and hyperspaces. In: Proceedings of 8FCC Combinatorical Conference, Paris (2010)
9. Baroin, G.: De Newton à Riemann, Graphes et Graphisme: Interactions Mathématico-Musico-Plastiques. Litter@incognita (3), 41–57 (2010)
10. Leys, J., Ghys, E., Alvarez, A.: Dimensions: A walk through mathematics! (CGI movie) (2008), `http://www.dimensions-math.org`
11. Cohn, R.: Dramatization of hypermetric conflicts in the Scherzo of Beethoven's Ninth Symphony. 19th-Century Music 15, 22–40 (1992)
12. Crans, A., Fiore, T., Satyendra, R.: Musical Actions of Dihedral Groups. Amer. Math. Monthly 116(6), 479–495 (2009)

# Motivic Topologies: Mathematical and Computational Modelling in Music Analysis

Chantal Buteau[1] and Christina Anagnostopoulou[2]

[1] Department of Mathematics, Brock University, Canada
chrisa@music.uoa.gr
[2] Department of Music Studies, University of Athens, Greece
cbuteau@brocku.ca

**Abstract.** This paper discusses a mathematical model together with its computational realization, for the motivic analysis of a piece of music. Relations between the mathematical model (motivic topologies), computational counter-part (OM-Melos), and music analysis are presented in the light of general concepts of computational music analysis, stressing the importance of neutrality and scientific rigour in the modelling part, while preserving the freedom of the analyst.

**Keywords:** Motivic topologies, motivic analysis, mathematical modelling, computational music analysis.

## 1 Introduction

Mathematical modelling in music addresses topics ranging from traditional music theory to more contemporary theories of gesture and performance. A mathematical model however is an abstract construct, which needs a computational counterpart in order to be applicable to music analysis. Computational music analysis (CMA) is an area of research which uses computational means for music analysis purposes. Its aims are to produce musicologically interesting results and formalize the human analytical process, while assisting the analyst with data, calculations, and making explicit analytical choices. The whole modelling process, be it mathematical and/or computational, in order to be meaningful needs to remain close to the main questions of music analysis. In this paper, we exemplify some of the challenges in this enterprise.

## 2 The Mathematical and Computational Models

### 2.1 Motivic Spaces: The Mathematical Model

We briefly enumerate the main concepts of the model; see [1,2] for details and examples. A topological space on the infinite set $MOT$ of all theoretical motives is constructed: A **motif** $M$ of cardinality $n$ is a non-empty finite set of $n$ notes with all different onsets. A set mapping $t$ on $MOT$, called **shape type**,

C. Agon et al. (Eds.): MCM 2011, LNAI 6726, pp. 330–333, 2011.

a group $P$ action on motive shapes leading to **gestalts** (imitation classes), and pseudo-metrics (similarity functions) $d_n$ on motive shapes with fixed cardinalities $n$, retracted to motives as $gd_t^P(M,N) := \inf_{p,q\in P} d_n(p\cdot t(M), q\cdot t(N))$, are introduced. This leads to a topological space on each set $MOT_n$ of motives of cardinality $n$. But a crucial step for regrouping motives of different cardinalities is the introduction given a **similarity threshold** $\epsilon > 0$, of the sets: $V_\epsilon^{t,P,d}(M) := \{N \in MOT | N^* \subset N \text{ s.t. } gd_t^P(N^*, M) < \epsilon\}$. If the inheritance property is fulfilled [1], the collection of all $V_\epsilon^{t,d,P}(M)$ forms a basis for a topology $\mathcal{T}_{t,P,d}$ on $MOT$. In this space, the **$\epsilon$-variations** sets, $Var_\epsilon^{t,d,P}(M) := \{N \in MOT | N \in V_\epsilon^{t,P,d}(M) \text{ or } M \in V_\epsilon^{t,P,d}(N)\}$, conceptualize variations of motives.

The **motivic space of a piece** $S$ is the relativization of $\mathcal{T}_{t,P,d}$ to $MOT(S)$, an arbitrary finite collection of motives in $S$, and represents the motivic structure of $S$ [1]. Motivic analysis schemes are further modeled: (1) The **identification of germinal motives**, as proposed by Rudolph Réti [3], is modeled by quantifying $Var_\epsilon^{t,d,P}(M)$ sets through a weight function [1]; (2) **Paradigmatic categorizing**, the first stage of semiotic analysis as proposed by Nattiez [4], is realized using $Var_\epsilon^{t,d,P}(M)$ sets [5]; and (3) an **indirect temporal distribution of motivic paradigms**, is proposed by extending the weight function to notes [2].

### 2.2 OM-Melos: The Computational Model

Applying the mathematical model to motivic analysis means to explicitly construct a motivic space of a piece together with the analysis scheme models. The following describes the computational model implementation ('OM-Melos' [5]).[1]

1. **Data representation.** The manipulation of data is symbolic (using MIDI as the input format). The piece $S$ is reduced to its set of notes: (onset, pitch).
2. **Segmentation.** Given the segmentation preference (using large sections or a time window), the set $MOT(S)$ of motives in $S$ for the analysis is computed.
3. **Choice of knowledge representation of motives.** Given the shape type $t$ selected by the analyst, that is the musical parameter(s) the analysis will focus on, OM-Melos computes the shape of each motif in $MOT(S)$.
4. **Motivic grouping into gestalts.** Given the (paradigmatic) group $P$ selected by the analyst, motives are regrouped with their imitations (gestalts).
5. **Motivic similarity.** Given a similarity function $d$, the distance between any two gestalts of motives in $MOT(S)$ with same cardinality is calculated.
6. Futher motivic analysis procedures
   (a) **Calculation of weights.** Given a weight function, the weight of each gestalt in $MOT(S)$ and of each note in $S$ are calculated.
   (b) **Paradigmatic categorization procedure.** Given a collection $X \subset MOT(S)$ of motives in $S$, the paradigmatic categorization is calculated.
7. **Result production.** Intermediate and final results (of diverse types, e.g. numerical, graphic, and music) are returned to the analyst for interpretation.

[1] It is a complete-model, stand-alone version of MeloRUBETTE in RUBATO [6].

## 3 Related Issues in Computational Music Analysis

The above procedure describes the steps in the present computational approach. Below, some related key concepts of computational music analysis are discussed.

- **Segmentation:** Any set of motives in a piece could theoretically be set as $MOT(S)$; e.g. all motives in a piece. But, this would include potentially absurd note combinations, musically meaningless (e.g. the 2-note motif comprising the very first and last note of a piece). The choice of the segmentation is carried out manually and relies on the judgement of the analyst. Overlapping segmentation is allowed, as well as segments that span across voices.
- **The concept of motif:** It is a particularly general one. Theoretically, a motif can be any combination of notes (with different onsets) taken from the piece. However, in the realization of the piece, only selected motifs through the segmentation are taken into account (Step 2). A motif may also comprise non-consecutive notes in the piece.
- **Knowledge representation:** The representation of information (or knowledge) for use in intelligent problem solving lies at the core of any computational modelling approach, almost forming an independent area of study within Artificial Intelligence. In music processing, this defines not only what musical information is represented, but also how this is done. There can be several parameters extracted, at different levels of abstraction. In this case, they are: onset, pitch, duration, and loudness, which can be considered either alone, or in combinations. Further derived representations, including interval and contour functions are calculated on any of the parameters, defining thus the shape of the motif (Step 3). Knowledge representation allows motives with the same representation to be grouped together. In this case, the similarity of the musical surface (different motives) is based on the identity on the musical parameter level.
- **The concept of similarity and categorization:** This approach allows for several levels of motivic groupings and categorizations, based on various types of similarity relations and criteria: *Level 1* is realized through the different abstract knowledge representations, as explained above (Step 3). Motives with the same representation are considered identical for further analysis; *Level 2* takes place when constructing the gestalts (Step 4). Motives with the same gestalts are considered identical for further analysis; *Level 3* is related to distances between gestalts, according to a chosen similarity function and threshold (Step 5). In this level, which allows variations, only motives with the same cardinality are considered; *Level 4* is directly related to the formalization of variations of motives (as involved in Step 6a). Motives with different cardinalities are now considered and regrouped in $\epsilon$-variations $Var_\epsilon$, according to a similarity threshold $\epsilon$.

## 4 Theoretical Implications and Conclusions

In this paper we discussed connections between the mathematical and the computational model of motivic topologies, as applied to music analysis. While similar concepts exist in the two disciplines, mathematics and informatics, their relation is not always straightforward. To conclude, we bring up two issues that we consider important for this type of collaborative interdisciplinary research.

a. Scientific rigour and objectivity: A model is by definition rigorous and scientific. The analysis procedure should thus be reproducible as long as the parameters used by the analyst are taken into account. One related question is whether the process of music analysis, following a model, can also be thought of as "neutral" (in Nattiez' terms) and objective.[2] When looking at the described mathematical model, it first proposes an abstract construct on an infinite set of theoretical motives. It then carries the topological structure to "an arbitrary finite set of motives". Whether the actual choice of the "arbitrary" set of motives for the analysis of a piece makes sense, or the choice of analysis parameters is appropriate, one possible answer is that as long as the points of choice for the analyst are made explicit, the analytical process is formalized, and therefore perhaps closer to what Nattiez had originally envisaged.

b. Analytical freedom: It is clear that the freedom of choice given to the analyst is a significant aspect of the model in all its steps. Since the model is of generic nature, one would wish to vary the parameters for a more complete and context-free analysis. As in many computational music analysis approaches, the proposed model involves a similarity threshold in its procedures. Should this analytical parameter also be made accessible to the analyst (although it is not clear on which criteria the analyst would base his choice)? This turns out to be a key aspect of the approach: due to the diverse visualizations of results, the analyst does not have to select a similarity threshold a priori, but instead can use the overall motivic spectrum, e.g. the paradigmatic categorization dynamic plots, and select, a posteriori, which similarity thresholds should be considered based on the meaningfulness of their corresponding results.

## References


1. Buteau, C., Mazzola, G.: Motivic Analysis Regarding Rudolph Réti: Formalization Within A Mathematical Model. Journal of Mathematics and Music 2(3), 117–134 (2008)
2. Mazzola, G., et al.: The Topos of Music. Birkhäuser, Basel (2002)
3. Réti, R.: The Thematic Process in Music. Greenwood Press, Connecticut (1951)
4. Nattiez J.-J.: Fondements d'une sémiologie de la musique. Édition 10/18, Paris (1975)
5. Buteau, C., Vipperman, J.: Representations of Motivic Spaces of a Score in *OpenMusic*. Journal of Mathematics and Music 2(2), 61–79 (2008)
6. `http://www.rubato.org`
7. Anagnostopoulou, C., Buteau, C.: Can computational music analysis be both musical and computational? Journal of Mathematics and Music 4(2) (2010)


[2] In [7], the authors ask "*Can this type of analysis* [CMA] *be closer to what Nattiez originally thought about the neutrality, objectivity and scientific nature of music analysis? Researchers working in CMA are called to address the issue.*" (pp. 75-76).

# Exploring Rameau and Beyond: A Corpus Study of Root Progression Theories

Thomas Hedges[1] and Martin Rohrmeier[2]

[1] Trinity Laban Conservatoire of Music and Dance, London, UK
Thomas.H10@edu.trinitylaban.ac.uk
[2] Centre for Music and Science, Faculty of Music, University of Cambridge, UK
mrohrmeier@cantab.net

**Abstract.** This study empirically explores root progression theories as a differentiator between tonal and pre-tonal music with a statistical corpus analysis of Palestrina Madrigals and Bach Chorales. Results found some quantitative evidence in the corpora for Rameau's rule-based root progression theory and Meeus' symmetry between "dominant" and "subdominant" root progressions. Further investigation revealed statistically significant differences between the underlying structures of the corpora, suggesting the cycle of fifths as fundamental to tonal music.

**Keywords:** Harmony, Root Progression Theory, Rameau, *Basse Fondamentale*, Tonal and Pre-tonal Harmony, Corpus Analysis.

## 1 Introduction and Motivation

Root progression theories prescribe or categorise intervals between (implied) root transitions to explain harmonic progressions and can potentially be employed to investigate tonal and pre-tonal harmony. Dahlhaus [1] draws heavily on Rameau's *basse fondamentale* [2] to characterise the emergence of tonal harmony from pre-tonal music. Rameau explains tonal harmonic progressions through an implied bass line governing all chord transitions irrespective of chord inversions. Consonant and dissonant intervals prescribe a set of permissible root progressions; fourths/fifths progressions are preferred over thirds/sixths and dissonant intervals are prohibited except the rising second underpinning a deceptive cadence [2] (p. 60). Inversional equivalence of intervals and chords is argued based on the harmonic overtone series, after Helmholtz [3], implying a basic level of chord categorisation. Meeus [4] draws on Schoenberg [5] and chordal substitution (allowing chords a third apart to substitute one another) to create an explicit hierarchy for root progressions. According to Meeus, "dominant" root progressions of an ascending fourth may be substituted by falling thirds and ascending seconds. "Subdominant" root progressions (descending fourths) may be substituted by rising thirds and descending seconds.

This paper aims to extend previous research [4,6,7] and explore specifically the extent to which root progression theories differ in tonal and pre-tonal harmony. Meeus' [4] small-scale analysis suggests that around 90% of harmonic progressions

C. Agon et al. (Eds.): MCM 2011, LNAI 6726, pp. 334–337, 2011.

in tonal music are "dominant", whilst in pre-tonal harmony "dominant" and "subdominant" progressions are more equal. Tymoczko [6] further explores symmetry between dominant and subdominant root progressions. A small and simple corpus analysis of 186 Bach chorales and 17 Palestrina compositions collecting raw data from root progressions of all sequential triads and seventh chords (ignoring everything else) shows "dominant" and "subdominant" progressions to be less different in pre-tonal (65%:35%) than tonal (74%:26%) harmony. In the current study, an automatic, statistical corpus analysis assesses the following hypotheses derived from the literature: After Rameau [2], fourths and fifths should be the most frequent interval in root progressions, followed by thirds and sixths, followed by the ascending second. After [4,6] the dominant/subdominant symmetry is expected to be more pronounced in pre-tonal music than tonal.

## 2 Method

A tonal corpus of 343 Bach chorales and a pre-tonal corpus of 61 Palestrina madrigals in four parts were selected. A two-stage segmentation and labelling process adapted some of the methodology as chosen by [7]. Harmonic segments were sampled at the quarter note and with a dissonance based scoring system, a representative chord was chosen for each segment to filter dissonances and passing notes. After labelling, a second segmentation grouped adjacent and identically labelled segments. For labelling, pc sets were normalised and assigned a pitch class according to a template method (cf. [8]). Open fifths {0,7}, major {0,4,7} and minor {0,3,7} triads were assigned a root of 0, the dominant seventh, {0,3,6,8}, a root of 8, the major seventh, {0,1,5,8} and {0,3,7,8}, assigned roots of 1 or 8 and the minor seventh, {0,3,5,8}, a root of 5. Open fifths were not normalised and six-four chords were labelled in root position after [2] (p. 99). The method (Fig. 1) provides a fair representation of harmonic progression, although certain pc sets ({0,3,6}, {0,3,6,9} and {0,4,8}) were not labelled since they are ambiguous with respect to their roots. A shorter segmentation was avoided after Rameau's preference for slower root progressions [2].

**Fig. 1.** Segmentation and labelling of BWV 33.6: *Allein zu dir, Herr Jesu Christ* showing the fundamental roots and their progression in semitones

## 3 Results

To identify patterns in root progressions the frequency of each scalar interval (in semitones) was tabulated to create interval profiles for each corpus (Fig. 2). To quantify the extent to which root progressions followed Rameau's interval rankings, the scalar intervals were ranked by frequency. A Spearman's rank correlation test compared the observed rankings for the Palestrina/Bach corpora with Rameau's, returning very similar correlation coefficients of 0.530/0.574 ($p$ = .094/.063). The rising tone ranks a higher than expected third overall, accounting for 45.0%/56.6% of dissonant intervals. A paired t-test shows the rising tone to be significantly more frequent than the closest ranked dissonant interval, the falling tone, in both corpora ($t(60/342) = 2.675/14.003$ and $p = .005/ < .0005$) confirming it as the most frequent dissonant interval.

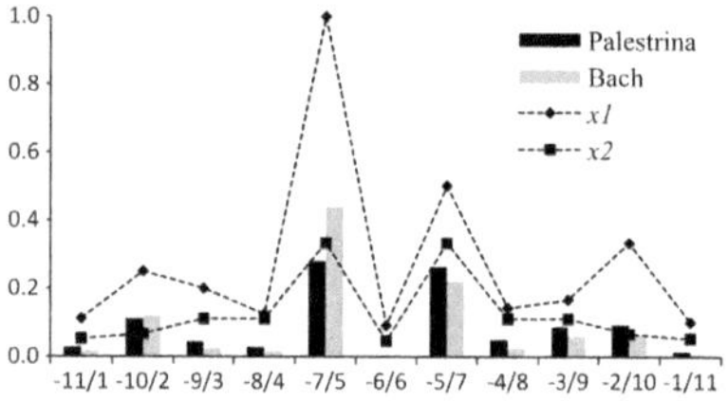


**Fig. 2.** Interval profiles for the two corpora including independent variables characterising the cycle of fifths ($x1$) and consonance ($x2$) used for multiple linear regression

Meeus [4] and Tymoczko [6] suggest the strength of dominant/subdominant symmetry to be a useful distinction between tonal and pre-tonal music. The "dominant" to "subdominant" ratios of 52.9%:42.9% / 64.5%:32.6% in the Palestrina/Bach corpora were less pronounced than Meeus' prediction of 65%:35% / 90%:10% prompting further investigation. A measure of symmetry was defined by (1) to assess overall symmetry around the tritone.

$$\frac{\Sigma_n^{i=1}|log(x_n) + log(y_n)|}{2n} \quad (1)$$

Intervals equidistant from the tritone were paired and the absolute difference of the logarithms of their frequencies was averaged. $x$ and $y$ represent the frequency of n paired intervals. As $S$ approaches 0, the observed pattern becomes increasingly symmetrical. $S$-values of 0.093/0.147 revealed both distributions to be relatively symmetrical, with the Palestrina being more so.

Since neither ranking nor symmetry could fully differentiate between the corpora, a multiple linear regression explored their underlying structures. Two independent variables (Fig. 2), the cycle of fifths ($x1$) and consonance ($x2$), were used to judge predictability of the scalar interval profiles. $x1$ characterises each scalar interval with the inverse rank of its distance along the cycle of fifths (preferring "descending" over "ascending"). $x2$ defines consonance in each scalar interval with the inverse sum of its two inversionally equivalent intervals' rankings for consonance (after [3]).

The analysis revealed significant adjusted $R^2$ values of 0.888/0.957 ($f(10) = 40.49/113.2$ and $p < .0005/.0005$). The Palestrina corpus returned $\beta_{1,2} = 0.156/0.502$ for $x_1/x_2$ with significance values of $t_{1,2}(10) = 2.189/7.040, p_{1,2} = .060/.025$ showing the consonance model as the only significant predictor. In contrast, the Bach corpus returned $\beta_{1,2} = 0.426/0.163$ for $x_1/x_2$ with significance values of $t_{1,2}(10) = 7.040/1.054, p_1 < .0005, p_2 = .323$ revealing the cycle of fifths model as the only significant predictor. These findings suggest that the impact of the cycle of fifths underlying the structural organisation of chord progressions differentiates tonal from pre-tonal music in the two corpora.

## 4 Discussion and Conclusions

Both corpora were found to have several mutual properties with respect to Rameau [2]; roots progressed predominantly by consonant intervals, although the rising tone was more prominent than expected and found to be more prominent than many consonant intervals. The findings of Tymoczko [6] with regards to Meeus' [4] theory on symmetry were reviewed with a larger, more inclusive corpus analysis, revealing the pre-tonal corpus as only marginally more symmetrical. Symmetrical patterns were found to extend beyond the fifth, third and second to complete scalar interval profiles.

It was found that the cycle of fifths as a predictor of root progressions reflects an interesting difference between the corpora, providing some empirical backing towards the importance of syntactical root progressions in the emergence of tonal music as asserted by Dahlhaus [1]. In particular, its strong underpinning in the tonal corpus reflects the directional, goal-oriented nature of tonal harmony as well as its more complex syntactic organisation around fifth progressions.

## References


1. Dahlhaus, C.: Studies on the Origin of Harmonic Tonality. Princeton University Press, Princeton (1990)
2. Rameau, J.-P.: Treatise on Harmony. Dover Publications, New York (1971)
3. Helmholtz, H.: On the sensations of tone as a physiological basis for the theory of music, 2nd edn. Dover Publications, Inc., New York (1954)
4. Meeus, N.: Toward a Post-Schoenbergian Grammar of Tonal and Pre-Tonal Harmonic Progressions. Music Theory Online 6(1) (2000)
5. Schoenberg, A.: Structural Functions of Harmony. In: Stein, L. (ed.), W. W. Norton and Company, New York (1969)
6. Tymoczko, T.: Function Theories: A Statistical Approach. Musurgia 10(3-4), 35–64 (2003)
7. Rohrmeier, M., Cross, I.: Statistical Properties of Tonal Harmony in Bach's Chorales. In: Proceedings of the 10th International Conference on Music Perception and Cognition, pp. 619–627 (2008)
8. Temperley, D., Sleator, D.: Modeling Meter and Harmony: A Preference-Rule Approach. Computer Music Journal 23(1), 10–27 (1999)

# Melodic Morphing Algorithm in Formalism

Keiji Hirata[1], Satoshi Tojo[2], and Masatoshi Hamanaka[3]

[1] NTT/Future University Hakodate
hirata@fun.ac.jp
[2] Japan Advanced Institute of Science and Technology
tojo@jaist.ac.jp
[3] PREST, JST/University of Tsukuba
hamanaka@iit.tsukuba.ac.jp

**Abstract.** We introduce a feature structure, corresponding to a time-span tree of Lerdahl and Jackendof's *A Generative Theory of Tonal Music*, and represent the reduction of the tree by the subsumption among these feature structures. As the collection of them forms a lattice, we can define the *join* and *meet* operations. We show a melodic morphing algorithm based on these simple operations.

**Keywords:** Time-span Reduction, Feature Structure, Join, Meet.

## 1 Introduction

To facilitate composing music, we often render a pitch event, a chord, and so on in a formal representation, together with supporting tools. However, there exists a trade-off between descriptive power and simplicity in the formal representation, two of which are basically incompatible. Descriptive power is the capability as to how faithfully the composer's original thoughts and emotions are expressed. On the other hand, simplicity means how concise the description of the music is. In general, the more abstract a description is, the shorter it is, and the less the writing and reading costs are. For instance, a Standard MIDI File has high descriptive power yet low simplicity, while the chord symbol is the opposite. The trade-off can also be considered an issue of controllability in rendering music.

We argue that the key for being compatible is to separate the basic well-understood operations from creator's intention for combining them. Intuitively, these basic operations include the ones like set arithmetics; addition, subtraction, intersection, and union. Thus we are led to an algebraic framework, in which a creator assembles basic operations into a calculation process for a target task as the creator intends.

The aim of the paper is to provide the theoretical foundations for proving the theorem for the property of a musical operation, melodic morphing. We start with defining the subsumption relation among melodies and building a lattice of feature structures, each element of which corresponds to a melody. Using the *join* and *meet* operations in the lattice, we construct a melodic morphing algorithm in the formal way.

C. Agon et al. (Eds.): MCM 2011, LNAI 6726, pp. 338–341, 2011.

## 2 Time-Span Trees in Feature Structures

First, we design the feature structure, *f-structure* hereafter, for a "time-span tree" of a music piece [1]. An f-structure is a directed acyclic graph as used in [2]. Since the reduction is an intrinsic method to reflect musical structures, we can properly map the reduction of the time-span trees to the subsumption relation '$\sqsubseteq$' between f-structures.

The type of an f-structure is shown headed by '~' (tilde). An f-structure for a ~*tree* is shown on the left-hand side of Fig. 1, and that for an ~*event* on the right-hand side. A binary tree has left and right branches where there are trees recursively; we call these branches daughters (*dtrs*). Let $\sigma$ be a ~*tree* f-structure, then the left daughter and the right daughter, denoted by $\sigma.dtrs.left$ and $\sigma.dtrs.right$ respectively, have recursively ~*tree* type. The set notation $\{x, y\}$ means the choice either of $x$ or $y$. When $\sigma.dtrs$ is $\bot$ (empty), that is, there is no daughter in the tree, $\sigma.head$ must be a single pitch event.

The whole f-structure is referred to by a tag, which is shown by such a boxed number as $\boxed{i}$ or $\boxed{j}$. The head of a ~*tree* f-structure must be either the head of its left daughter tagged by $\boxed{i}$, that of its right daughter tagged by $\boxed{j}$, or another single pitch event. If $\sigma.head = \sigma.dtrs.left.head$, the node has the right-hand elaboration of shape $\wedge$ , and if $\sigma.head = \sigma.dtrs.right.head$, the left-hand elaboration $\wedge$ . As such, the *head* value at the parent level is recursively taken from either the left-hand or right-hand branch. As for ~*event* type, feature *pitch* include $C4$, $B\flat 6$, $F\sharp 3$, and so on. Feature *pos* stands for the start timing, and its value is an f-structure consists of feature *bar* (the $n$-th bar) and *meter* ($m$-th meter measured by a quarter note). Feature *duration* also has a *Length* value.

A type of an f-structure specifies a set of indispensable features. When no indispensable feature is missing, the typed f-structure is said to be *full-fledged.* For example, ~*tree* type requires the feature set of *head* and *dtrs*, and ~*event* type does *pitch*, *pos.bar*, *pos.meter*, and *duration*. The property "full-fledged" is concerned with whether or not rendering a real melody from a f-structure representation. Then, we can provide the formal definition of subsumption relation between f-structures, which allows the mechanical calculation.

$$\begin{bmatrix} \tilde{}tree & \\ head & \{\{\boxed{i},\boxed{j}\},[\tilde{}event]\} \\ dtrs & \left\{ \begin{bmatrix} left & \begin{bmatrix} \tilde{}tree & \\ head & \boxed{i}[\tilde{}event] \\ dtrs & \{\cdots\} \end{bmatrix} \\ right & \begin{bmatrix} \tilde{}tree & \\ head & \boxed{j}[\tilde{}event] \\ dtrs & \{\cdots\} \end{bmatrix} \end{bmatrix}, \bot \right\} \end{bmatrix} \qquad \begin{bmatrix} \tilde{}event & \\ pitch & Pitch \\ pos & \begin{bmatrix} bar & Integer \\ meter & Length \end{bmatrix} \\ duration & Length \end{bmatrix}$$

**Fig. 1.** Feature structures for a time-span tree (left) and a pitch event (right)

## 3 Calculus in Melody Lattice

Theorem 3.13 in Carpenter [3] defines that the unification of f-structures $A$ and $B$ is the least upper bound of $A$ and $B$. Therefore, we adopt the unification as the definition of *join* in this paper, and the intersection of the unifiable f-structures as that of *meet.*

**Definition 1 (*Meet and Join*).** *Let $A$ and $B$ be full-fledged f-structures representing the time-span trees of melodies $A$ and $B$, respectively.*

*If we can fix the greatest lower bound of $A$ and $B$, that is, the greatest $x$ such that $x \sqsubseteq A$ and $x \sqsubseteq B$ is unique, we call such $x$ the meet of $A$ and $B$, denoted as $A \sqcap B$.*

*If we can fix the least upper bound of $A$ and $B$, that is, the least $y$ such that $A \sqsubseteq y$ and $B \sqsubseteq y$ is unique, we call such $y$ the join of $A$ and $B$, denoted as $A \sqcup B$.*

**Definition 2 (Reduction Path).** *Suppose that each pitch event is given a beat strength as a result of metrical analysis [1]. For such two f-structures (melodies) $A$ and $B$ that $T_B \sqsubseteq T_A$, a reduction path from $A$ to $B$ is defined as a sequence of f-structures (melodies) obtained by removing a pitch event (a note) in $A \setminus B$ from $A$ one-by-one, according to the algorithm, as follows:*

*Step 1: $N := \#(T_A \setminus T_B)$, $i := 0$, and $T_0 := T_A$.*

*Step 2: Select a pitch event $p$ of the minimum beat strength in $T_i \setminus T_B$.*

*Step 3: Reduce $T_i$ to $T_{i+1}$ by removing $p$.*

*Step 4: Iterate Steps 2 and 3 $N$ times ($i = 0 \sim N-1$).*

*The resulting sequence $T_0$, $T_1$, $T_2$, $\cdots$, $T_N$ is the reduction path from $A$ to $B$.*

Note that at Step 2 the pitch events (notes) with the minimum beat strength are the least important notes in the time-span tree. Since such least important notes are multiple and one is chosen from them nondeterministically, there are more than one reduction path in general. The selection is justified by the time-span reduction preference rules: TSRPR1 (metrical position) and TSRPR5 (metrical stability) [1]. Hence the algorithm may automatically compute more than one reduction paths, for every melody $C$ on the reduction path from $A$ to $B$, $B \sqsubsetneq C \sqsubsetneq A$ holds.

## 4 Melodic Morphing Algorithm

Here, we present the morphing algorithm in the formal way [4,5].

**Definition 3 (Melodic Morphing Algorithm).** *The algorithm consists of the following steps (see Fig. 2):*

*Step 1: Calculate $T_A \sqcap T_B$ (meet).*

*Step 2: Select melody $T_C$ on the reduction path from $T_A$ to $T_A \sqcap T_B$, and select $T_D$ on the reduction path from $T_B$ to $T_A \sqcap T_B$.*

*Step 3: Calculate $T_C \sqcup T_D$ (join), and the result is morphing melody $\mu$.*

If $C$ close to $T_A$ is chosen, the characters of melody $A$ are reflected in output $\mu$. On the other hand, If $C$ close to $T_A \sqcap T_B$ is chosen, those of $A$ are less emphasized in $\mu$. The character of $D$, as opposed to $B$, behaves in the similar way.

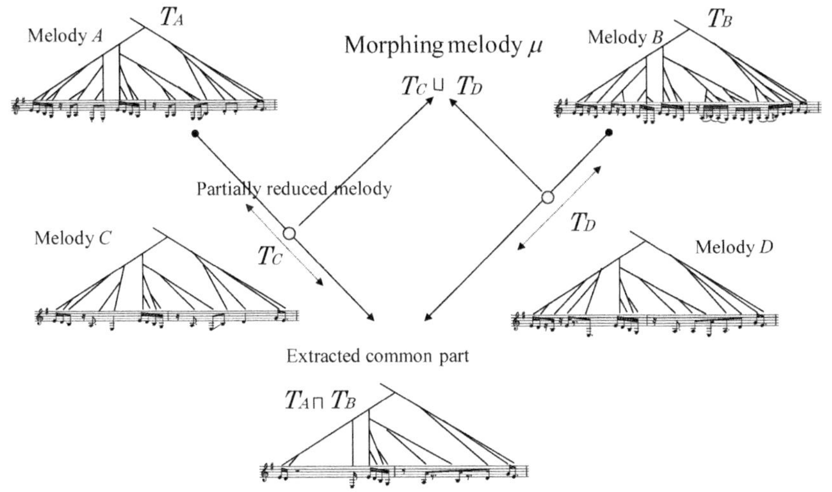


**Fig. 2.** Melodic morphing algorithm

# 5 Conclusion

We have provided an algebraic framework in which a music piece is represented by an f-structure, corresponding to its time-span tree. As these f-structures were ordered in terms of the subsumption relation, we could construct a lattice of melodies and could define *join* and *meet* operations. We have presented a morphing algorithm, which is a complicated calculation in general, from combinations of these simple algebraic operations. In the near future, we will prove the algorithm is exactly the interpolation of two given melodies.

# References

1. Lerdahl, F., Jackendoff, R.: A Generative Theory of Tonal Music. The MIT Press, Cambridge (1983)
2. Marsden, A.: Generative Structural Representation of Tonal Music. J. New Music Research 34(4), 409–428 (2005)
3. Carpenter, B.: The Logic of Typed Feature Structures. Cambridge University Press, Cambridge (1992)
4. Hamanaka, M., Hirata, K., Tojo, S.: Melody Morphing Method Based on GTTM. In: Proc. of ICMC 2008, pp.155–158 (2008)
5. Hamanaka, M., Hirata, K., Tojo, S.: Melody Extrapolation in GTTM Approach. In: ICMC 2009, pp. 89–92 (2009)

# From 2D to 3D: Using Geometry and Group Theory to Model Motivic Structure in Musical Composition

Jocelyn Ho

Department of Music, Stony Brook University
Stony Brook, New York, USA
jocelyn@jocelynho.com

**Abstract.** In this paper, I propose to model motivic development using the concept of manifolds. Compositional space is represented by a manifold that consist of musical "charts", and the music itself is represented by a path. This concept forms the basis of my two compositions, *Torus* and *12 Variations on a Dodecahedron*. In *Torus*, the idea of a path on a two-dimensional manifold in three dimensions is used to manifest different levels of circularity. In *12 Variations on a Dodecahedron*, the elements in the group of rotational symmetries of a dodecahedron are represented musically by exploiting its isomorphism with the alternating group on five elements.

**Keywords:** Dodecahedron, Geometry, Group Theory, Manifolds, Motivic Development, Rotations, Symmetry, Torus.

## 1 Motivation

In my research, I have been interested in representing thematic or motivic development,[1] and have proposed the concept of modelling the compositional space of a piece of music into a manifold $M$, generally embedded in more than two dimensions. The experience of the music in time is modelled as a path $\gamma(t)$ on $M$. Manifolds capture the concept of the local and the global; this is useful since, in music, there is always a sense of a global scheme while being always situated locally in time. Also, manifolds can capture recurrence more generically: while the circle depicts one thing repeating, a higher dimensional manifold can represent circularity on different levels.

This musical manifold has attributes that yield interesting observations. The k-dimensional manifold is composed of charts that represent motifs and their development. These charts, having coordinates on them, can capture the development of motifs precisely. They also intersect to represent motifs that are heard simultaneously. The musical composition as a path has observable mathematical properties. For instance, its derivative represents the perceived flow of

[1] I use this term in the sense of Rudolph Reti's transformation of themes and motifs [1].

C. Agon et al. (Eds.): MCM 2011, LNAI 6726, pp. 342–345, 2011.

time. Also, the path's distance to a particular motif gives information about the memory of that motif.

Manifolds could be used in the reverse process in a flexible way. The manifold created by a composer becomes infinitely interesting: it is a virtual compositional space with charts that he or she chooses. There are an infinite number of choices that a path could potentially take; however, the music is exactly the specific path that exists, not any other one. The compositional space can thus be a model of the choices that are faced by a composer when he or she is composing at any given time. In the following two sections, I will explain the structures of my two compositions *Torus* and *12 Variations on a Dodecahedron* that stem from this theoretical basis.

## 2 *Torus*

In *Torus*, the compositional space, that is, all possible combinations and developments of three themes A, B and C, is represented on the surface of the torus as sets and set intersections, shown in Fig. 1. The piece itself is represented by the black dotted path, such that each time the path enters a new region, an instance of musical material corresponding to that region is heard, shown as the labelled dots on the path.

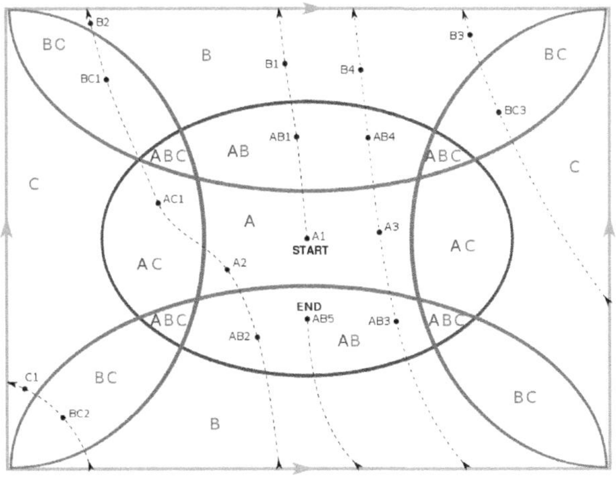


**Fig. 1.** Compositional space of *Torus*

*Remark.* The specific nature of the path, sets and set intersections chosen around the torus captures the concept of circularity in time in two ways. The torus is circular in both its vertical and horizontal cross-section. The path on this specific torus loops around it "vertically" almost four times and "horizontally" once (Fig. 2).

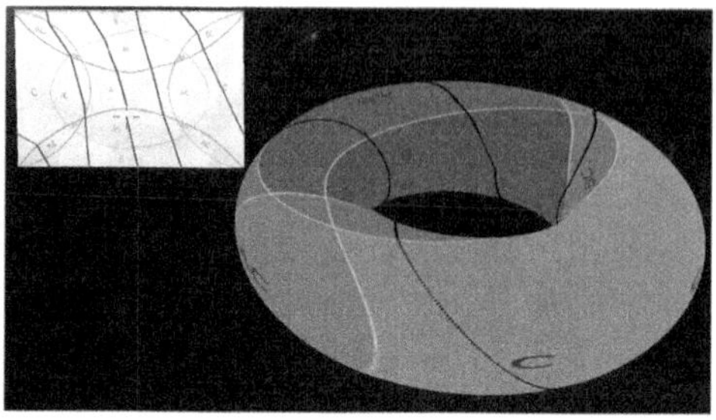

**Fig. 2.** Frame captured in multimedia video of *Torus*

# 3 *12 Variations on a Dodecahedron*

## 3.1 The Group of Rotational Symmetries of the Dodecahedron

In contrast, *12 Variations on a Dodecahedron* for percussion duo fully explores the dodecahedron, by representing all the rotations in the group of symmetries of the dodecahedron, or $S(D)$. The group isomorphism between $S(D)$ and the alternating group on five elements, $A_5$, is exploited to do so.

There are 5 different representative phrase lengths in the music – a, b, c, d and e, in eighth-notes units. If a 1-1 correspondence between $S(D)$ and $A_5$ is established, one can rotate the dodecahedron, look for its corresponding permutation and use this permutation to shape the phrase lengths in the music. This can be done by establishing an edge labelling of the dodecahedron that reflects the isomorphism between $S(D)$ and $A_5$.

## 3.2 Establishing an Edge Labelling of the Dodecahedron

We need to set up the bijection so that if we apply a rotation to an edge and the corresponding permutation to its label, we get the label of the resulting edge. We approach this problem by considering elements of orders 5 and 2 in separate cases.

**Considering elements of order 5:** It is clear that the 5 edges of one pentagonal face must be distinct. This is because a rotation around an axis through a face centre is order 5, which corresponds to a permutation of order 5. A permutation of order 5 does not leave any labels fixed, thus, its corresponding rotation must not leave any edge labels fixed.

**Considering elements of order 2:** Order 2 elements of $S(D)$ are the rotations of 180 degrees about the axis through the midpoints of 2 opposite edges. Order 2 elements of $A_5$ are the permutations where 2 disjoint pairs of labels are transposed. The only edges fixed by order 2 rotations are the 2 edges through which the axis passes. Hence as only one label is fixed, these two edges must be given the same label. Therefore, we can conclude that opposite edges have the same labels $(*)$.

Thus, the labelling of the dodecahedron is as follows: one face is labelled where edges are all different, and all the other edges are determined using $(*)$.

### 3.3 Representing the Dodecahedron in *12 Variations*

We now have a labelling of the dodecahedron that shows the isomorphism between $S(D)$ and $A_5$. With this labelling, we can rotate the dodecahedron, obtain the corresponding permutation,[2] and translate it literally as the phrase lengths in the music: for instance, the permutation $[b\ d\ c\ a\ e]$ will correspond to a metaphrase consisting of five phrases of lengths $b$, $d$, $c$, $a$, and $e$ eighth notes heard in that order.

Now we can methodically enumerate all the 60 rotations and permutations of phrase lengths simultaneously, such that in the music, one variation corresponds to one pentagonal face, and each variation will have five metaphrases corresponding to the five rotations around that face. These five metaphrases will be heard in the order of enumeration. In this way, the music carries with it geometrical meaning literally in every phrase, variation and in the whole piece, as shown in the correspondence table below:

| | | |
|---|---|---|
| Twelve variations | $\sim$ | The whole dodecahedron |
| One variation | $\sim$ | One pentagonal face |
| One metaphrase | $\sim$ | One rotation |
| One Small phrase length | $\sim$ | $a$, $b$, $c$, $d$ or $e$ |

The piece *12 Variations* consists of two players, each separately exploring the dodecahedron's symmetries in the procedures explained above. The two percussionists actually walk around this dodecahedron in different paths, only coming together on the same face thrice (the 1st, 7th and 9th variations).

## 4 Conclusion

In *Torus* and *12 Variations on a Dodecahedron*, the geometrical constructions enrich as well as concretise the process of transforming beautiful visual objects into sound. While the concept of manifolds provides a fresh way to look at motivic structures, it may also prove to be useful to composers who are always searching for methods to structure their musical framework. The use of mathematics, specifically three-dimensional geometry that is tangible to the audience, not only makes the connection from visual to aural concrete, but also provides the audience with a new accessible perception strategy.

## Acknowledgements

Thanks to Avinash Krishnan for implementing the video of *Torus* in Fig. 2 and Nicolo Davis for computer-graphics assistance. I would also like to thank Professor Guerino Mazzola for his ongoing guidance.

## Reference

1. Reti, R.: The Thematic Process in Music. Faber, London (1961)

[2] The details of obtaining the permutation is omitted here.

# Clustering and Classification of Music by Interval Categories

Aline Honingh and Rens Bod

Institute for Logic, Language and Computation, University of Amsterdam
A.K.Honingh@uva.nl, Rens.Bod@uva.nl

**Abstract.** We present a novel approach to clustering and classification of music, based on the concept of interval categories. Six interval categories exist, each with its own musical character. A piece of music can be represented by six numbers, reflecting the percentages of occurrences of each interval category. A piece of music can, in this way, be visualized as a point in a six dimensional space. The three most significant dimensions are chosen from these six. Using this approach, a successful visual clustering of music is possible for 1) composers through various musical time periods, and 2) the three periods of Beethoven, which illustrates the use of our approach on both a general and a specific level. Furthermore, we will see that automatic classification between tonal and atonal music can be achieved.

**Keywords:** Interval Category, Pitch-class Set, Classification, Clustering.

## 1 Introduction

Comparing music can be accomplished on several levels. Finding similarities and dissimilarities is important for the development of many music informationretrieval systems. Since the amount of digitized music has increased enormously over the past few years, the need for tools to organize, order and cluster music has increased as well. Automatic music classification has been investigated on the basis of different ideas, such as compression distance [1], Hidden Markov Models [2], and n-gram models [3].

In this paper we present a new approach to music clustering and classification on the basis of interval categories (IC's). With this new approach, we can not only cluster and classify music according to certain genres or styles, but we may also be able to explain the differences and commonalities in terms of interval and pitch content.

## 2 Interval Categories

It has been shown that pitch class sets (pc-sets) can be grouped into six interval categories [4,5]. Every interval category is associated with an interval class. IC1 corresponds to intervals of 1 (or 11), IC2 corresponds to intervals of 2 (or 10),

C. Agon et al. (Eds.): MCM 2011, LNAI 6726, pp. 346–349, 2011.

IC3 corresponds to intervals of 3 (or 9), and so on, meaning that a pitch class set belonging to IC$n$ contains a dominating number of intervals $n$. Every pitch class set can be classified into one of these categories, and it will belong to the category it is most similar to. For example, pitch class sets $\{0,1,2\}$, $\{0,2,3,4,5\}$, $\{0,1,2,3,5\}$ will all be grouped into IC1 since the number of semitones in each set dominates the set.

An entire piece or corpus of music can be represented by a distribution of interval categories. A piece of music is segmented and the notes in every segment form a pc-set. Of every pc-set it can be calculated to which IC it belongs. In this way, a piece of music can be represented by its IC distribution, listing for each category the percentage of occurrence in the piece or corpus. For a more elaborate explanation of these interval categories, as well as the detailed method to represent a piece of music in terms of interval categories, see [5,6].

## 3 Composer Clustering

Western art music, also referred to as classical music, is generally grouped into several musical periods (listed in Table 1) and is known to express different characteristics for every period. Using the IC distributions, we have visualized several composers and hope to see that the various musical periods can be distinguished in clusters. The composers we have used are listed in Table 1. For each composer we have selected around five pieces of music. The Medieval period was treated as one item, since for most medieval music the composers are unknown. Since the musical mode has been shown to influence the IC distribution [6], we have chosen to only use music in major mode for the present visualization purpose. Each composer is represented by an IC distribution that was calculated using the selected music. The six interval categories give rise to six dimensions, but we only chose three of these (IC3, IC4 and IC5) to visualize the composers—see Fig. 1(a). The Baroque music as well as the atonal music (represented by composers Schoenberg and Webern) are found as separate clusters, while the Romantic and Classical music are difficult to distinguish in the figure.

**Table 1.** Periods of European art music and the composers we used for Fig. 1(a)

| **Period** | **date** | **composers (abbreviation)** |
|---|---|---|
| Medieval | 500-1400 | |
| Renaissance | 1400-1600 | Palestrina (Pal) |
| Baroque | 1600-1750 | Bach (Ba), Händel (Ha), Vivaldi (Vi) |
| Classical | 1750-1830 | Haydn (Hay), Mozart (Mo), Beethoven (Be), Schubert (Schu) |
| Romantic | 1830-1900 | Brahms (Bra), Mahler (Ma), Tchaikovsky (Tchai), Debussy (De), Mendelssohn (Men), Sibelius (Sib) |
| Modern | 1900-... | Ravel (Ra), Stravinsky (Stra), Schoenberg (Schoe), Webern (Web) |

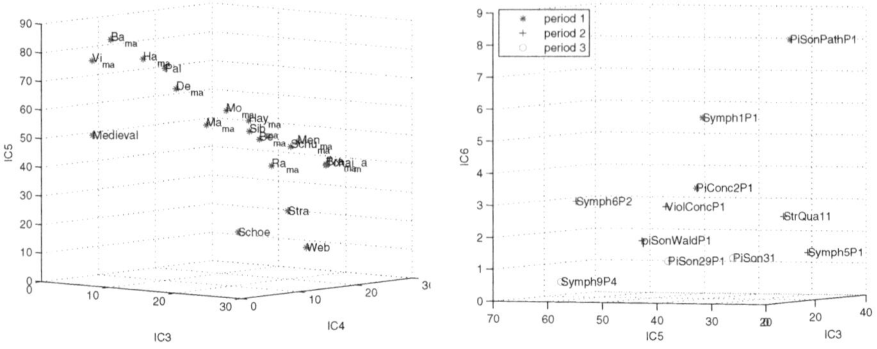


(a) (b)

**Fig. 1.** (a) 3-D plot of music from composers listed in Table 1. The label "ma" indicates the major mode. (b) 3-D plot of music from the three periods of Beethoven.

## 4 Tonal – Atonal Classification

On the basis of the results above, we have tried to automatically separate tonal from atonal music. For every piece of music, if the ratio $category5/category1$ was higher than a certain threshold, the piece was classified as tonal, and otherwise as atonal. To evaluate our results, we compared them to the results of a baseline algorithm, which reads: for every bar, the diatonic scale is selected that matches most of the notes in the bar, after which the number of notes that are not elements of that scale are counted. An average of those "atonal notes" is calculated over the whole piece. If this number is higher than a certain threshold, the piece is classified as atonal and otherwise as tonal.

The results can be found in Table 2. We conclude that the overall classification on the basis of IC's is quite successful (94.7 % correct).

**Table 2.** Results of classification of tonal and atonal music

| | Correctly classified **atonal** pieces | Correctly classified **tonal** pieces | Total number of correctly classified pieces |
|---|---|---|---|
| IC algorithm | 19 | 53 | 72 (94.7 %) |
| baseline algorithm | 14 | 53 | 67 (88.2 %) |
| total number of pieces | 20 | 56 | 76 |

## 5 Beethoven's Three Periods

Beethovens compositional career is usually divided into three periods [7]. From each period, we have chosen a number of works, and calculated the IC distribution of each piece. From period 1 we chose the piano sonata *Pathetique* part 1 (PiSonPathP1), piano concerto 2 part 1 (PiConc2P1), and Symphony 1 part 1 (Symph1P1). From period 2, we chose Symphony 5 part 1 (Symph5P1), piano sonata *Waldstein* part 1 (PiSonWaldP1), string quartet no 11 (StrQuar11), violin concerto part 1 (ViolConcP1), and symphony 6 part 2 (Symph6P2). From period 3, we chose symphony 9 part 4 (Symph9P4), piano sonata 29 part 1 (PiSon29P1), and piano sonata 31 (PiSon31).

IC3, IC5 and IC6 have been chosen as dimensions to visualize the pieces in—see Fig. 1(b). We see that the pieces from the three different periods can be distinguished on the basis of dimension IC6. Fig. 1(b) surprisingly shows that Beethoven has been using relatively many tritones (IC6) in his early works, and relatively few tritones in his late works. Of course, these differences in tritones do not fully characterize the three periods. However, the fact that a visual classification on the basis of this category is possible, suggests that the use of tritones contributes to the well-known classification of Beethoven's music.

## 6 Conclusions

We have presented a new approach to algorithmically cluster music based on interval categories (IC's). We have seen that, with this approach, 1) composers can be clustered according to the musical period they belong to, 2) tonal music can be automatically distinguished from atonal music, and 3) music from Beethoven can be clustered according to Beethoven's three different periods.

We should keep in mind that these IC clusterings are only based on pitch intervals and therefore can never capture all of the differences and commonalities of the investigated music. However, it is striking that a classification *can* be made with the use of this simple method and thus the approach provides a simple and powerful representation of music that can be used to investigate musical similarity on the basis of tonalness, composer, musical period and possibly more.

## References


1. Cilibrasi, R., Vitanyi, P., De Wolf, R.: Algorithmic Clustering of Music Based on String Compression. Computer Music Journal 28(4), 49–67 (2004)
2. Chai, W., Vercoe, B.: Folk music classification using hidden Markov models. In: Proceedings of ICAI (2001)
3. Hillewaere, R., Manderick, R., Conklin, D.: String Quartet Classification with Monophonic Models. In: Proceedings of ISMIR, Utrecht, the Netherlands (2010)
4. Quinn, I.: Listening to similarity relations. Perspect. New Music 39, 108–158 (2001)
5. Honingh, A.K., Weyde, T., Conklin, D.: Sequential Association Rules in Atonal Music. In: Chew, E., Childs, A., Chuan, C.-H. (eds.) MCM 2009. Communications in Computer and Information Science, vol. 38, pp. 130–138. Springer, Heidelberg (2009)
6. Honingh, A.K., Bod, R.: Pitch Class Set Categories as Analysis Tools for Degrees of Tonality. In: Proceedings of ISMIR, Utrecht, the Netherlands (2010)
7. Kerman, J. et al.: Beethoven, Ludwig van. In Grove Music Online. Oxford Music Online, `http://www.oxfordmusiconline.com/subscriber/article/grove/music/40026pg11`

# Plactic Classification of Modes

Franck Jedrzejewski

CEA Saclay, France
Franck.Jedrzejewski@cea.fr

**Abstract.** Classification of scales began to take shape in the nineteenth century through the works of Camille Durutte, Hoëne Wronski, Anatole Loquin and some others, but it really took a new start in the twentieth century. The aim of this paper is to study a new classification of modes based on the plactic congruences. These congruences mimic a small perturbation from one mode to the other by the move of only one note. Two modes are in the same plactic class if they are related by a path of modes which are pairwise linked by plactic congruences. In this paper, a *mode* is an ordered series of musical intervals (or steps). A *scale* is an ascending or descending series of notes, representing a class of modes under circular permutations. In traditional Western music, the C major scale represents the circular permutations of the seven usual modern modes (Ionian, Dorian, Phrygian, etc.)

**Keywords:** Plactic Monoid, Modes Classification.

## 1 The Plactic Congruences

Musical scales considered here are patterns of five notes or more that moves by steps. Since each kind of steps can be encoded by a letter of a totally ordered alphabet $A \subset \mathbb{Z}_n = \{0, 1, ..., n-1\}$, the structure of a scale is represented by a word $\omega$ over this alphabet such that the sum of each step is equal to the total number $n$ of notes in the $n$-tone equal temperament.

In the usual tempered system, the Dorian mode (D, E, F, G, A, B, C) is encoded by the word 2122212, or *babbbab*, with $a = 1,\ b = 2,\ a < b$ and the major scale or Ionian mode is represented by 2212221 or *bbabbba*. If for some musical reasons, a note is altered, the encoding of the mode changes. In this paper, we consider that two modes are in the same class if they are related by a chain of words in plactic relations. These relations have been introduced by Knuth in [1] and the plactic monoid of Schützenberger and Lascoux appears in [2]. The plactic monoid over some totally order alphabet $A = \{a, b, c, ...\}$ with $a < b < c < ...$ is the monoid whose generators are the letters of the alphabet verifying for any three arbitrary letters $x, y, z$ in $A$ the Knuth congruence relations:

$$\begin{cases} yzx \equiv yxz \text{ whenever } x < y \leq z \\ xzy \equiv zxy \text{ whenever } x \leq y < z \end{cases} \tag{1}$$

The plactic monoid on the alphabet $A$ is the quotient of the monoid of words $A^*$ by the Knuth relations $Pl(A) = A^*/\equiv$. The plactic monoid of modes is

C. Agon et al. (Eds.): MCM 2011, LNAI 6726, pp. 350–353, 2011.

the restriction of this set to the modes. If the alphabet has only two letters $A = \{a, b\}$ with $a < b$, the plactic relations reduce to $bab \equiv bba$, $aba \equiv baa$. When $a$ encodes a semitone and $b$ a whole tone ($a = 1$ and $b = 2$), two modes are in the same class if there exists a chain of modes that differ by a pattern of four consecutive notes such that their steps are $bab \equiv bba$ or $aba \equiv baa$. For example, the Dorian mode ($babbbab$) and the Mixolydian mode ($bbabbab$) are in the same class, because the move of the third note in the pattern (D, E, F, G) to (D, E, F♯,G) is allowed by the plactic congruence. But there is no direct plactic relations between the Carnatic Vachaspati mode ($bbbabab$) and the Rishabhapriya mode ($bbbaabb$). Vaschapati is linked to the Mixolydian mode ($bbabbab$), which is linked to the Carnatic Charukesi mode ($bbababb$), and this last mode is linked to Rishabhapriya.

If the alphabet has three letters $A = \{a, b, c\}$ with $a < b < c$, the plactic relations $bca \equiv bac$, $acb \equiv cab$ describe the switch of two steps ($ac \sim ca$) if and only if this pattern is preceded or followed by the step $b$ ($a < b < c$). In other words, in a four consecutive notes pattern, the second or the third note can be raised or lowered if the first step is $b$. For the simplest values ($a = 1$, $b = 2$, $c = 3$), the plactic relations mean that in a tritone 213, (F, G, G♯, B) the third note can be raised (F, G, A♯, B) but the second note cannot move (F, F♯, G♯, B). And if the last step is $b = 2$, in a tritone 312, (F, G♯, A, B), the second note can be lowered (F, F♯, A, B), but the third note cannot move (F, G♯, A♯, B).

In 1961, C. Schensted [4] gave an algorithm to find the length of the longest nondecreasing subwords of a given word $w \in A^*$. His method associates with each word $w$ a semistandard Young tableau $t = P(w)$. The words producing a given Young tableau $t$ form a *plactic class*. A. Lascoux and M.P. Schützenberger [2] have stated that the plactic classes $[w]$ and $[w']$ of two words $w$ and $w'$ uniquely determine the plactic class of their concatenation $[ww']$. And this gives the set of all plactic classes.

Schensted stated that the length of the first row of $P(w)$ is equal to the maximal length of nondecreasing subwords and the length of the first column of $P(w)$ is equal to the maximal length of the decreasing subwords. Some years later, Greene introduced the equivalence relations on two words of $A^*$ if they have the same Schensted tableau. Moreover, in 1970, Knuth [1] proved that this equivalence between two words coincides with the plactic congruence.

Another tool for studying the plactic monoid is the RSK (Robinson-Schensted-Knuth) correspondence. The RSK map associates to each word $w \in A^*$ over the linerarly ordered alphabet $A = \{1, 2, ...\}$, the pair $(P(w), Q(w))$ where $P(w)$ is the Schensted semistandard Young tableau and $Q(w)$ is a standard Young tableau. This map $w \mapsto (P(w), Q(w))$ is one-to-one. A standard tableau is a semistandard tableau where each $k \in A$ appears exactly once. By definition, $Q(w)$ is the standard tableau encoding the chain of shapes in the construction of $P(w)$. Let us denote $m_1, m_2, ..m_r$ the multiplicities of the letters $a, b, c, ...$ in the word $w$. By labeling from left to right each occurrence of $a$ with label 1 to $m_1$, and each occurrence of $b$ with labels $m_1 + 1$ to $m_1 + m_2$, and so on, we define a new word called the standardization $std(w)$ of the word $w$. For example

$std(142531) = 153642$. If two words are in Knuth relations $w \equiv w'$, then they have the same standardization $std(w) = std(w')$. Therefore, the standardization commutes with the Schensted tableau.

## 2 The Pentatonic Modes

As an application of the results of the previous section, we will classify all the pentatonic modes with the plactic congruences on the alphabet $A = \{a, b\}$. Recall that the *evaluation* of a word $w$ over the alphabet $A = \{a_1, a_2, ..., a_n\}$ is the vector whose components are the number of occurrences of each letter of $A$, namely $ev(w) = (|w|_{a_1}, |w|_{a_2}, ..., |w|_{a_n})$ where $|w|_{a_1}$ indicates the number of occurrences of $a_1$ in $w$. For any word $w = w_1....w_n$, a descent is a position $i$ where $w_i > w_{i+1}$. The major index maj$(w)$ is the sum of all descent positions. For example, the word 23232 has descents in position 2 and 4. Its major index is maj$(w) = 2 + 4 = 6$. The generating polynomial of a subset $R \subset A^*$ is the sum over $w \in R$ of $x^{\mathrm{maj}(w)}$. For a plactic classes $R$ related to a fixed pair $(a, b)$ this polynomial $T$ decomposes into cyclotomic polynomials and $T(1)$ is the number of modes or the cardinality of the set $R$. To study the pentatonic modes, we distinguish between classes according to their evaluation. 1) If $ev(w) = (4, 1)$, we get two plactic classes in the 12-tone equal temperament. The first class has only one element *aaaab* and the second class has four elements *aaaba* $\sim$ *aabaa* $\sim$ *abaaa* $\sim$ *baaaa*. 2) If $ev(w) = (3, 2)$, we get three plactic classes. The first class has only one element *aaabb*. In the second class of four elements, the pentatonic minor mode (22323) is in relation with the Blues major mode (23223). In the third class of five elements, the Egyptian mode (23232) is linked to the pentatonic major mode (32232) which is related by plactic congruence to the Blues minor mode (32322).

## 3 The Plactic Modal Classes

The graph of a plactic class is defined as follows. The vertices of this graph are the words representing the modes of the class. Two vertices are connected by an edge if and only if the two words representing the vertices are related by Knuth relations. A *plactic modal class* is defined to be a plactic class of five or more modes which has no trivial linear graph, otherwise the class is said *linear plactic class*. Most of the graphs have less than five vertices. Among the hexatonic modes, only two classes have nine modes, and among the heptatonic modes, two classes have more than five elements. The first heptatonic class is the class of the Phrygian mode (*abbbabb*). In the 12-tet, this class is obtained for $a = 1, b = 6$ or $a = 1, b = 2$. Its graph is linear and it is not a plactic modal class. It includes Locrian (*abbabbb*), Phrygian (*abbbabb*), Natakapriya (*abbbbab*) and Kikilapriya (*abbbbba*) modes. The second class is the plactic class of the major mode (*bbabbba* with $a = 1, b = 2$). These modes appears in John Foulds's classification under the Vth class, and three of them are used in the *Essays in the modes* for piano (1920-1927). The plactic heptatonic class has 14 elements and its graph is as follows:

| | | | | | | | | |
|---|---|---|---|---|---|---|---|---|
| *babbbba* <br> *Gourimanohari* | – | *babbbab* <br> *Dorian* | – | *babbabb* <br> *Eolian* | – | *bababbb* | – | *bbaabbb* |
| │ | | │ | | │ | | | | |
| *bbabbba* <br> *Major* | – | *bbabbab* <br> *Mixolydian* | – | *bbababb* <br> *Charukesi* | – | *bbbaabb* <br> *Rishabhapriya* | | |
| │ | | │ | | | | | | |
| *bbbabba* <br> *Mechakalyani* | – | *bbbabab* <br> *Vaschaspati* | – | *bbbbaab* | | | | |
| │ | | | | | | | | |
| *bbbbaba* <br> *Aug.Lydian* | – | *bbbbbaa* | | | | | | |

If we consider all modes with less than 12 notes, a computation shows that there are *only ten plactic modal classes*, that is to say ten plactic classes with non trivial graph. The plactic classes are unchanged if we replace the 12 tone system by the $N$-tone equal temperament (except for the values of $a$ and $b$). In this tuning, new classes appear. For example, in the 12-tet, there are two linear plactic classes of modes of 11 notes: a class with one element $a^{10}b$ and another class with 10 elements linked to $ba^{10}$. In the 13-tet, the modes of 11 notes have a new set composed of 3 classes: a trivial class with one element $a^9b^2$, a linear class of 10 elements with representative $ba^9b$, and a modal class of 44 elements with representative $b^2a^9$. Louvier's heptatonic mode 4334343 is used in its *Clavecin non tempéré* [3]. It belongs to a new (nonlinear) class of 14 elements. In the 24-tet, Wyschnegradsky's diatonicized chromatic scale associated to the mode 1222221222222 belongs to a linear class of 12 elements. A new modal class of 65 elements appears in this tuning with representative $b^{11}a^2$.

Since plactic congruences defined modulations from one mode to another, the plactic modal classes define the admissible domain of modal modulations. These modulations, also called sometimes *metaboles*, could be generalized to microtones. As mentioned above, microtonal combinatorics, as well as plactic modes classification, give new resources yet unexplored.

## References


1. Knuth, D.E.: Permutations, matrices, and generalized Young tableaux. Pacific Journal of Mathematics 34, 709–727 (1970)
2. Lascoux, A., Schützenberger, M.P.: Le monoïde plaxique. In: Non-commutative structures in Algebra and Combinatorics, Quaderni della Ricerca Scientifica del CNR 109, Roma (1981)
3. Louvier, A.: Recherche et classification des modes dans les temp éraments égaux. Musurgia IV(3), 119–131 (1997)
4. Schensted, C.: Longest increasing and decreasing sequences. Canadian Journal of Mathematics 13, 179–191 (1961)

# Feature Extraction Using Pitch Class Profile Information Entropy

Maximos A. Kaliakatsos-Papakostas,
Michael G. Epitropakis, and Michael N. Vrahatis

Computational Intelligence Laboratory (CI Lab),
Department of Mathematics, University of Patras, Greece
{maxk,mikeagn,vrahatis}@math.upatras.gr

**Abstract.** Computer aided musical analysis has led a research stream to explore the description of an entire musical piece by a single value. Combinations of such values, often called global features, have been used for several identification tasks on pieces with symbolic music representation. In this work we extend some ideas that estimate information entropy of sections of musical pieces, to utilize the Pitch Class Profile information entropy for global feature extraction. Two approaches are proposed and tested, the first approach considers musical sections as overlapping sliding onset windows, while the second one as non-overlapping fixed-length time windows.

**Keywords:** Pitch Class Profile, Information Entropy, Global Features, Composer Identification.

## 1 Introduction

Pitch Class Profile has previously been used for exploring information entropy of sections of musical pieces [1]. In the paper at hand, we utilize these ideas to produce global features from musical pieces. A *global feature* is a numerical descriptors that "encapsulates information about a whole piece into a single value" [2]. In this work we investigate whether the *tonal information entropy* contained in subdivisions of musical pieces is suitable as a feature for composer identification.

## 2 The Proposed Approaches

***Overlapping sliding onset windows:*** A musical piece can be considered as a set of temporally ordered note events, $\{e_1, e_2, \ldots, e_n\}$. Each note event, $e_i$, is characterized by an onset value, i.e. its relative time position in the piece. A note event can be either *monophonic* or *polyphonic* depending respectively on whether a single note or multiple notes are sounded simultaneously.

An onset window of length $l$ is a musical section $\{e_i, e_{i+1}, \ldots, e_{i+l}\}$. We can compute the PCP[1] of this section of length $l$ which is referred as $\text{PCP}^l_{\text{onset}}(i)$. If

[1] The PCP of a section of a musical piece, is the distribution vector of the 12 pitch classes within this section. The interested reader can find more information in [3].

C. Agon et al. (Eds.): MCM 2011, LNAI 6726, pp. 354–357, 2011.

we slide this window by a single onset, we obtain a new window $\{e_{i+1}, \ldots, e_{i+l+1}\}$ and its PCP is denoted by $\text{PCP}^l_{\text{onset}}(i+1)$. If a musical piece is constituted by $n$ onsets, we can obtain $n - l$ overlapping sliding onset windows as well as their respective PCPs. Since every PCP is a distribution over the discrete variable of the 12 pitch classes, we can compute its Shannon information entropy. Thus, for the PCP of the $i$-th onset window, $\text{PCP}^l_{\text{onset}}(i)$, we can compute the value of the Shannon information entropy, $S^l_{\text{onset}}(i)$. Two global features can be extracted for every piece by the aforementioned information entropies, the mean value $\mu_{\text{onset}}$ and the standard deviation $\sigma_{\text{onset}}$. Intuitively, the mean value indicates the overall *tonal uncertainty* of this piece, while the standard deviation indicates whether there exist sections of tonal stability as well as how steeply they are interchanged with unstable ones.

***Non-overlapping, fixed-length time windows:*** The above approach does not consider the temporal structure of a musical piece, since a window is forwarded to the next onset regardless of its time distance. This constitutes a drawback, as distant time events are less influential to the memory of tonal structure. To overcome this problem, we consider pitch information in a number of bars of the musical piece, thus in fixed-length time windows. We denote as $\text{PCP}^m_{\text{bar}}(i)$ the PCP of the $i$-th musical segment of length $m$ bars. Let $C^m_{\text{bar}(i)}$ denote the Shannon information entropy of the aforementioned *current* segment's PCP. Additionally, we consider the *movement information entropy* between the $(i-1)$-th and the $i$-th segments. This is the Shannon entropy of the absolute difference vector of the PCPs of the respective segments, i.e. $|\text{PCP}^m_{\text{bar}}(i-1) - \text{PCP}^m_{\text{bar}}(i)|$. We denote this entropy as $M^m_{\text{bar}}(i)$. For every musical piece, four global features can be extracted by *current* and *movement* information entropies. Those are, the mean value and standard deviation of current $(\mu^{\text{curr}}_{\text{bar}}, \sigma^{\text{curr}}_{\text{bar}})$ and movement $(\mu^{\text{move}}_{\text{bar}}, \sigma^{\text{move}}_{\text{bar}})$ entropies.

## 3 Results

We employ the $k$th Nearest Neighbor ($k$NN) supervised classifier on a data set that consists of 50 pieces by J.S. Bach (Ba), and 50 movements of string quartets by each of the composers Haydn (H), Mozart (M) and Beethoven (Be), thus a total of 200 pieces. Composer identification simulations are performed in pairs and with a leave one out strategy, as suggested in previous works in data sets of similar size [4,5].

***Overlapping sliding onset windows:*** In this approach, each musical piece is represented by two global features described in Sect. 2, the mean value ($\mu_{\text{onset}}$) and the standard deviation ($\sigma_{\text{onset}}$). In Fig. 1(a–c), we observe that smaller window sizes exhibit better results for separating pieces of J.S. Bach from the string quartets. Similar results are obtained by a range of $k$ values below 40. Identification success decreases with the increase of window size. This reveals that transition information from onset to onset is crucially different between those pieces. On the other hand, as we observe in Fig. 1(d–f) identification success increases with the increase of window size for the separation of the string quartets.

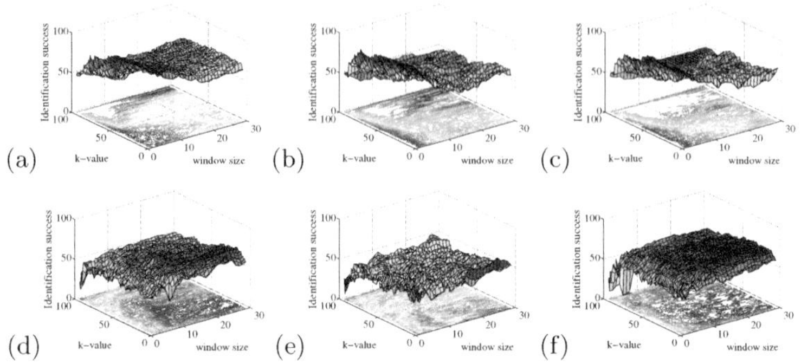


**Fig. 1.** Identification success between all composers for different overlapping sliding onset windows and different $k$ values. Identification simulations: (a) Ba – Be, (b) Ba – H, (c) Ba – M, (d) Be – H, (e) H – M and (f) M – Be

***Non-overlapping fixed-length time windows:*** In this approach, each piece is represented by four global features, the mean values and standard deviations of *current* ($\mu_{\text{bar}}^{\text{curr}}$, $\sigma_{\text{bar}}^{\text{curr}}$) and *movement* ($\mu_{\text{bar}}^{\text{move}}$, $\sigma_{\text{bar}}^{\text{move}}$) information entropies within different consecutive time windows. In Fig. 2(a–c) we observe that identification success between pieces of J.S. Bach and the string quartets presents relatively stable behavior for windows of multiple bars. This indicates that information entropy within a time frame of less bars does not reveal differences between composers.

The latter comment does not come into contradiction with the respective results presented in Fig. 1(a–c) but as complementary to them. Fig. 1(a–c) presents that 2 to 4 consecutive onsets are ideal for separating those composers. Identification success drops dramatically for windows of 10 onsets, which is near the expected number of onsets per bar. Thus Fig. 1(a–c) and Fig. 2(d–f) indicate that differences in information entropy between the pieces of J.S. Bach and the

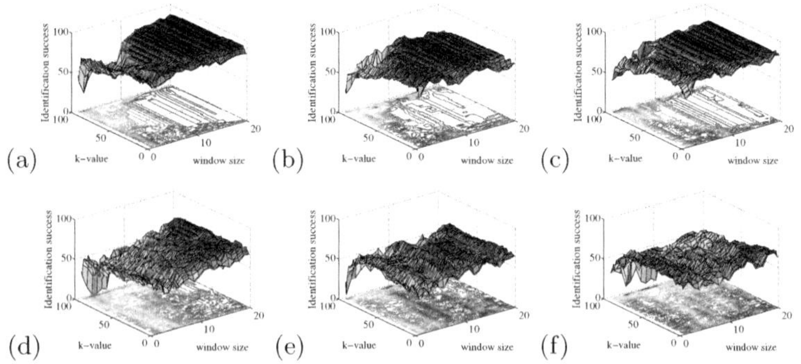


**Fig. 2.** Identification success between all composers for different non-overlapping fixed-length time windows and different $k$ values. Identification simulations: (a) Ba – Be, (b) Ba – H, (c) Ba – M, (d) Be – H, (e) H – M and (f) M – B

string quartets, are obvious either in a very small or a very large scale. It should also be noted that the maximum achieved identification success for the string quartets of Haydn and Mozart, 72%, is comparable or better than results of previous works, where more global features have been used [4,5]. String quartets of Haydn and Mozart have also been modelled as Markov Chains and their variations [4,6,7], with identification successes ranging between 65% to 80%.

## 4 Concluding Remarks

For the data set used in this work, both models exhibited better results for the separation of J.S. Bach's pieces by the string quartets. This exhibits that epoch or genre classification could be performed more safely by these features. Even though the separation of the string quartets did not yield impressive results, they were comparable to results obtained by previous works. The features proposed in this study could be combined with other features proposed in literature and hopefully improve music classification.

## References


1. Madsen, S.T., Widmer, G.: Towards a computational model of melody identification in polyphonic music. In: Proceedings of the 20th International Joint Conference on Artificial Intelligence (IJCAI 2007), Hyderabad, India. Morgan Kaufmann Publishers Inc, San Francisco (2007)
2. Hillewaere, R., Manderick, B., Conklin, D.: Global feature versus event models for folk song classification. In: Proceedings of the 10th International Society for Music Information Retrieval Conference (ISMIR 2009), Kobe, Japan, pp. 729–733 (2009)
3. Kaliakatsos-Papakostas, M.A., Epitropakis, M.G., Vrahatis, M.N.: Musical composer identification through probabilistic and feedforward neural networks. In: Di Chio, C., Brabazon, A., Di Caro, G.A., Ebner, M., Farooq, M., Fink, A., Grahl, J., Greenfield, G., Machado, P., O'Neill, M., Tarantino, E., Urquhart, N. (eds.) EvoApplications 2010. LNCS, vol. 6025, pp. 411–420. Springer, Heidelberg (2010)
4. Hillewaere, R., Manderick, B., Coklin, D.: Melodic models for polyphonic music classification. In: Proceedings of the 2nd International Workshop on Machine Learning and Music (MML 2009), held in conjunction with (ECML-PKDD 2009), Bled, Slovenia, pp. 19–24 (2009)
5. Kranenburg, P.V., Backer, E.: Musical style recognition - a quantitative approach. In: Parncutt, R., Kessler, A., Zimmer, F. (eds.) Proceedings of the Conference on Interdisciplinary Musicology (CIM 2004), Graz, Austria, pp. 1–10 (2004)
6. Kaliakatsos-Papakostas, M.A., Epitropakis, M.G., Vrahatis, M.N.: Weighted Markov Chain model for musical composer identification. In: Di Chio, C., Brabazon, A., Di Caro, G.A., Drechsler, R., Farooq, M., Grahl, J., Greenfield, G., Prins, C., Romero, J., Squillero, G., Tarantino, E., Tettamanzi, A.G.B., Urquhart, N., Uyar, A.Ş. (eds.) EvoApplications 2011, Part II. LNCS, vol. 6625, pp. 334–343. Springer, Heidelberg (2011)
7. Liu, Y.W.: Modeling music as markov chains: Composer identification, `https://www-ccrma.stanford.edu/~jacobliu/254report.pdf`

# Catastrophe Theory: An Enhanced Structural and Ontological Space in Music Composition

Fani Kosona[1] and Leontios Hadjileontiadis[2]

[1] Dept. of Music, Ionian University, Corfu, Greece
fkosona1@yahoo.com
[2] Dept. of Electrical & Computer Engineering,
Aristotle University of Thessaloniki, Greece,
State Conservatory of Thessaloniki, Greece
leontios@auth.gr

**Abstract.** The application of catastrophe theory in music composition offers a solid conceptual frame for handling discontinuity, resulting to an enhancement of the structural space, by converting the music work into a dynamical system. In this frame, the structural stability of the form is put under strain by forces as multiple attractors, consequently enlarging the ontological space of the work to contain indeterminist, de-autocorrelative and deconstructive aspects. A case study is briefly discussed.

**Keywords:** Catastrophe Theory, Music Composition Modeling.

## 1 Introduction

After 1950, composers' research in the domain of form has resulted to various approaches, converging to the quest for a logical organisation, but also to a—varying—tendency of disputing the deterministic symmetries of the past forms. In this direction, catastrophe theory [1,2] is adopted here as a means for enhancing the structural space in music composition, by integrating the concept of discontinuity and converting the music work into a dynamical system, where forces like multiple attractors put under strain its structural stability. Consequently, the ontological space of the music work could be enlarged to contain innovating situations of indeterministic, de-autocorrelative, deconstructive aspects. The liturgical potential of this theoretical approach is briefly presented in a case study (a piece for flute solo).

## 2 Methodological Background

The structural space of the music work can be considered as a parametric space with an elementary basis of reference—pitch, rhythm, timbre, dynamics organization.[1] The parametric handling allows mathematical modeling in a multi-dimensional space. In an enhanced structural space, deconstructed aspects of

[1] A list which can be extended with complex parameters at will.

C. Agon et al. (Eds.): MCM 2011, LNAI 6726, pp. 358–361, 2011.

musical parameters, as "refracted" pitch, complexified rhythm, topologically organized timbre, diversified dynamics and, generally, distorted elements of structure could be introduced (see Fig. 1). These structural decisions have serious effects on the informational "autocorrelation" [3] (p.135) of the form, which is associated with the determinist aim for internal coherence. Introducing indeterminism in music composition leads to deconstructive attitudes, which act against autocorrelation.

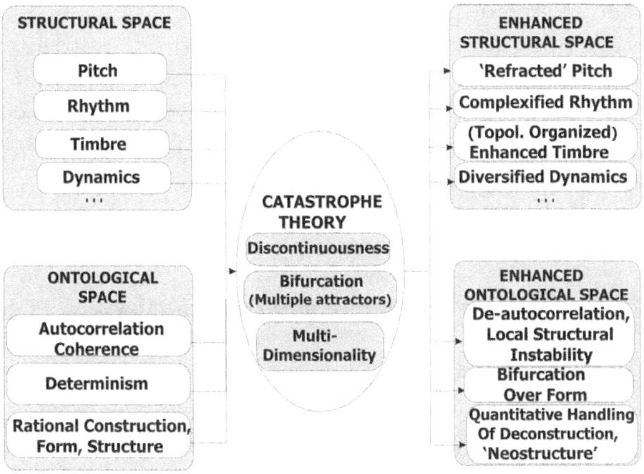


**Fig. 1.** Consequences of the application of catastrophe theory on the structural and ontological space of music composition

Catastrophe theory models the mechanism of sudden and discontinuous change, placing it in the frame of a dynamical system, where the abrupt changes of state are the result of smooth alterations of the control parameters. The models of catastrophe theory deal with objects and terms as bifurcations, attractors, critical points, singularities, structural stability/instability, multidimensional spaces and manifolds. The mapping of "catastrophic" (discontinuous) changes into a music work has a deconstructive effect. Contradiction and conflict is represented in catastrophe theory by the concept of multiple attractors, when operating in the range of the bifurcation set. Introducing this concept in the musical form leads to zones of divergence from the established musical material, where the uniformity, unanimity and stability of the musical structure are undermined. By the use of the mathematical tools of catastrophe theory, deconstruction can be handled not only qualitatively, but also quantitatively, by degrees, thus gaining precision and diversity in the compositional process. Moreover, deconstruction as a mathematical object can be introduced in different levels of the work of music, from the micro to the macro-level, thus generalizing its action in every aspect, from the sensual (sonic) to the structural and semantic fields. Complexity is also contributing to the aspect of deconstruction through diversion and informational overload. From the catastrophe theory aspect, complexity is introduced in the music work: (a) through multiple attractors

(b) caused by the mapping of multidimensional spaces and manifolds into the space of the music work. Non-smoothness, discontinuity and structural instability are related to indeterminism. In catastrophe theory, a locally indeterminist situation of abrupt change is incorporated to a determinist mathematical model, by transposition to a more complex, multidimensional space. This contradiction, concerning the indeterminist and determinist aspects of catastrophe theory, is interpreted in terms of musical form as ambiguity concerning the issue of its "wholeness" [4] (p. 16), which can be described as a "bifurcation over form", where two attractors are in action: one of disintegration and one of an "upgraded" completion to a "neostructure" [5] (p. 9 and pp. 140-142) incorporating a higher degree of complexity. This is a charming self-referential situation: the catastrophe model, applied to a parametric subspace of the musical work, creates a macroscopic copy of itself on the ontological level of the work. The block diagram depicted in Fig. 1 features the consequences of this methodology on the structural and ontological space.

## 3 Case Study: *Diathlassis* for Flute Solo

The work *Diathlassis* (= the Greek word for "refraction") for flute solo was composed by Fani Kosona in 2009 [6], based on the cusp catastrophe model. The mathematical formula that decribes the catastrophic manifold (see Fig. 2 of the cusp model is:

$$t^3 + Yt + X = 0 \tag{1}$$

In the equation (1), $X$ and $Y$ are the control variables and $t$ is the state variable.

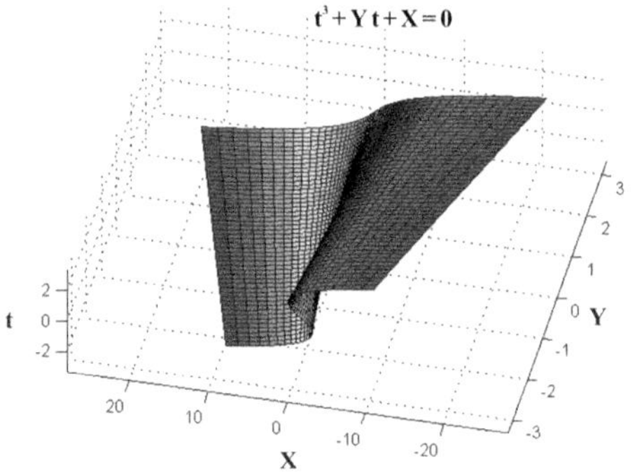


**Fig. 2.** The cusp catastrophe manifold

The state variable $t$ of the cusp catastrophe model is mapped to the musical parameter of pitch, thus interpreting the bifurcation as pitch "refraction". The control variable $X$ is associated with the rhythmic structure of the piece, being a factor of rhythmic fragmentation ("complexified rhythm"). The second control variable $Y$ is mapped to the timbre, which is enhanced to contain several intermediate stages of the evolution from pure air to normal flutistic sound (see Fig. 3) and from normal sound to an amalgam of normal sound with singing notes.

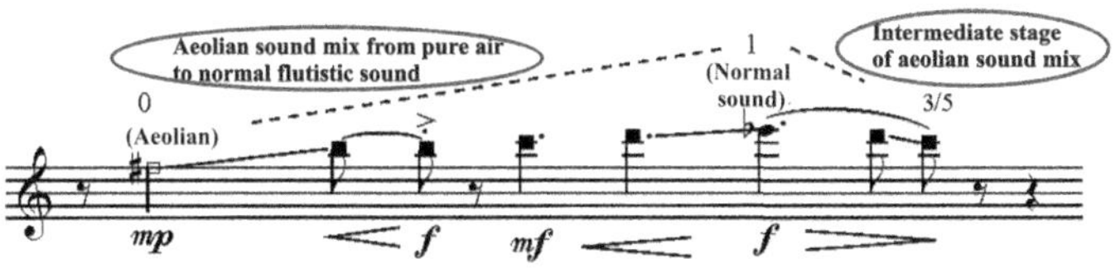


**Fig. 3.** Enhanced timbre usage

The "normative" case, where the autocorrelation function of the work would be based, is represented by smooth and continuous melodic contours. Bifurcations are represented by "refracted" melodic lines (see Fig. 4), creating "zones of deconstruction", where the material used is in conflict with the "zones of normativity".

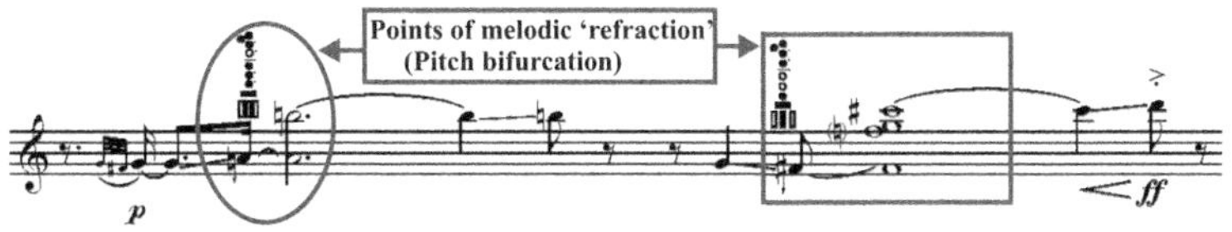


**Fig. 4.** "Refracted" melodic line, in a "zone of deconstruction"

## 4 Conclusion

From the analysis presented here it is deduced that catastrophe theory application in music composition is a promising tool, creating indeterministic, de-autocorrelating, deconstructive effects on the structure and the ontology of the music work.

## References

1. Thom, R.: Structural Stability and Morphogenesis: An outline of a general Theory of Models. Westview Press, USA (1989)
2. Poston, T., Stewart, I.: Catastrophe Theory and its Applications. Dover, USA (1996)
3. Moles, A.: Théorie de l'information et perception esthétique. Traduction grècque, Institut d'Education de la Banque National de Grèce, Athènes (2005)
4. Georgescu, C., Georgescu, M.: A System Approach to Music. Journal of New Music Research 19, 15–52 (1990)
5. Stroe, A. Georgescu, M., Georgescu, C.: Morphogenetic Musics: a catastrophe thermodynamic approach to the musical composition (1985) (unpublished) (copy kindly provided by Sever Tipei)
6. Zenz, K.: Flute Recital (Greek Flute Music of the 20th and 21st Centuries). [Cd], Naxos, Greek Classics, Cat. No: 8.572369 (2010)

# Enriched Score Access for Computer Assisted Composition in PWGL

Mika Kuuskankare

Sibelius Academy, Department for Doctoral Studies in Music and Research, Helsinki, Finland
STMS : IRCAM/CNRS/UPMC, Paris, France
mika.kuuskankare@siba.fi

**Abstract.** PWGL is a visual composition environment that can be used to, among other things, solve musical constraints problems. The constraints system within PWGL, PWGLConstraints, allows us to write rules using a special pattern-matching language. Typically, the assignments use as a starting point a score prepared with the help of Expressive Notation Package (ENP). In this paper we present an extension to the PWGLConstraints pattern-matching language which allows us to access information from ENP to assist with the compositional process. ENP provides a rich library of standard and user-definable expressions called ENP-expressions. They range from standard articulation markings (such as staccatos and slurs) to fully interactive multi-purpose graphical expressions. A special syntax is developed which allows us to retrieve information about and contained by the expressions. In this paper, the syntax and the present state of the system are illustrated using a working example.

**Keywords:** Constraint-based Computer-assisted Composition, Visualizing Musical Constraints, Computer-assisted Music Notation.

## 1 Introduction

There exists a wide range of general-purpose rule-based systems that have been used to solve musical constraint satisfaction problems, such as Situation [1], Arno [2], OMClouds [3], and Strasheela [4]. PWGLConstraints [8] is a rule-based pattern-matching language within PWGL [5] which can be used to solve musical problems. The main advantage of our system when compared to the aforementioned systems is that PWGLConstraints is closely integrated with a flexible music notation front-end, the Expressive Notation Package, or ENP [6]. Typically, the PWGLConstraints assignments use as a starting point a score skeleton prepared with the help of ENP. ENP allows us to enrich scores with additional attributes called ENP-expressions [7], which are Lisp-based multipurpose graphical objects that can be used to represent different kinds of information as part of a musical texture. The traditional expression markings, such as articulations and dynamic markings, form a subset of ENP-Expressions. The extension of

C. Agon et al. (Eds.): MCM 2011, LNAI 6726, pp. 362–365, 2011.

PWGLConstraints presented in this paper allows us to access the information about and contained by the ENP-expressions, using a simple but powerful syntax. The new syntax can be seen as a supplement to the existing accessor scheme reported in [8]. However, it allows us to write rules that access the score information in a completely new manner. Throughout this paper we will refer to the PWGLConstraints syntax which will not be presented here in detail. Instead, the reader is advised to study [8] for more information.

## 2 The New Expression Access Scheme

Access to the information contained by the expressions is provided through a new primitive called `e`. It takes as an argument a score object and a collection of keywords (for example, :pos, :first?, :last?, :sample, :at, :id) and finally returns either an object or a property or a list of objects or properties.

Example 1 shows the basic syntactic components of the system (the symbol `?1` represents a single note object in the score). The form `(e ?1 :accent)` is of interest here. It is a condition stating that the rule is executed only if there is an accent present in the note referenced by `?1`.

*Example 1:* A rule executed only for accented notes. The ellipsis denotes the rule part where the user defines the actual constraints.

```
(* ?1 (e ?1 :accent) (?if ...))
```

The e-syntax also allows us to "sample" an expression at any given point in time (obviously the ENP-expression in question has to contain time-varying information, such as a breakpoint function). In Example 2 an expression is sampled and the melodic movement is matched against the values retrieved from the expression using the combination of two keywords `:sample` and `:at`. The keyword `:at` defines the point in time that we want to sample. The argument can either be absolute time in seconds or a score object. Thus, `:at ?1` means that we want to access the value contained by the expression at exactly the onset-time of the note ?1.

*Example 2:* A rule forcing the melody to follow `:object3` exactly.

```
(* ?1 (?if (let ((ref-pitch (e ?1 :object3 :sample :at ?1)))
             (= (m ?1) ref-pitch))))
```

## 3 An Example of Basic Expression Access

In this section we provide a concise example using our new access scheme. Here, contrasting texture profiles are defined with the help of accents and slurs. Slurred texture uses small intervals and the accented texture, in turn, uses leaps larger than an octave. To make the example more interesting, we also use a collection of supplementary rules where, for example, the pitches inside each beat

**Fig. 1.** Contrasting textures created with the help of ENP-expressions: the slurred notes use small intervals and the accented ones use large intervals

(i.e., non-interleaving four note groups) must result in a pitch-class set 4-7. However, due to space limitations these rules are not discussed here. The resulting score can be seen in Fig. 1.

The two expression access rules (Examples 3 and 4) are used to define the two contrasting textures. The variables `?1` and `?2` in the pattern-matching part (line 1) give access to every two consecutive note objects in the score. The form `(and (e ?1 :accent) (e ?2 :accent))`, in turn, is a condition indicating that the rule is executed only if there is an accent on both of the notes. Finally, the rule simply states that the absolute value of the interval between the notes must be greater than an octave (12 semitones).

*Example 3:* Use big leaps on accented notes.

```
(* ?1 ?2
   (and (e ?1 :accent) (e ?2 :accent))
   (?if (> (abs (- (m ?1) (m ?2))) 12)))
```

Our second rule (Example 4) is almost identical to the first one but, instead of the `:accent` keyword, uses the `:slur` keyword to indicate that the rule is to be executed only on slurred passages. In the rule part, we specify that all of the intervals inside the slurred passages must be between the minor second and the major third intervals (the intervals must also be descending).

*Example 4:* Small descending intervals on slurred notes.

```
(* ?1 ?2
   (and (e ?1 :slur) (e ?2 :slur))
   (?if (<= 1 (- (m ?1) (m ?2)) 4)))
```

Finally, Example 5 defines the transition from one texture to another. Here, we indicate (lines 2-3) that this rule is active only when we are going from an accented note to a slurred one (line 2), or vice versa (line 3). In both cases, the interval between the two sections is constrained in such a way that it is either a minor or a major second (see line 4).

Note, that without this rule, the interval between the two contrasting textures would be random, since, by default, there is no rule that would be applied in this case.

*Example 5:* When going from one texture to another use a stepwise movement.

```
(* ?1 ?2
   (or (and (e ?1 :accent) (e ?2 :slur))
       (and (e ?1 :slur) (e ?2 :accent)))
   (?if (<= 1 (abs (- (m ?1) (m ?2))) 2)))
```

# 4 Conclusions

This paper presents a new scheme which allows us to access score information for the purposes of computer-assisted composition in the visual programming language PWGL. ENP, the music notation system of PWGL, is used to prepare the starting point of the search. The scores can be enriched with standard or user-definable markings called ENP-expressions. The information about and contained by the ENP-expressions can then be used by our rule-based compositional system to guide the search process. The underlying expression system itself is quite powerful and also user-extendable. Basically, any expression can be given an arbitrary meaning by the user. The scheme described in this paper is capable of expressing and solving highly sophisticated musical problems.

## Acknowledgments

The work of Mika Kuuskankare has been supported by the Academy of Finland (SA137619).

## References

1. Rueda, C., Lindberg, M., Laurson, M., Bloch, G., Assayag, G.: Integrating constraint programming in visual musical composition languages. In: ECAI 1998 Workshop on Constraints for Artistic Applications, Brighton (1998)
2. Anders, T.: Arno: Constraints programming in common music. In: Proceedings of the International Computer Music Conference (2000)
3. Truchet, C., Assayag, G., Codognet, P.: OMClouds, a heuristic solver for musical constraints. In: MIC 2003, Metaheuristics International Conference, Kyoto, Japan (2003)
4. Anders, T.: Composing Music by Composing Rules: Design and Usage of a Generic Music Constraint System. Ph.D. thesis, Queen's University, Belfast (2007)
5. Laurson, M., Kuuskankare, M., Norilo, V.: An Overview of PWGL, a Visual Programming Environment for Music. Computer Music Journal 33(1), 19–31 (2009)
6. Kuuskankare, M., Laurson, M.: Expressive Notation Package. Computer Music Journal 30(4), 67–79 (2006)
7. Kuuskankare, M., Laurson, M.: ENP-Expressions, Score-BPF as a Case Study. In: Proceedings of International Computer Music Conference, Singapore, pp. 103–106 (2003)
8. Laurson, M., Kuuskankare, M.: Extensible Constraint Syntax Through Score Accessors. Journées d'Informatique Musicale, 27–32 (2005)

# Historical Development of Tonal Syntax: Counting Pitch-Class Sets in 13th-16th Century Polyphonic Vocal Music

Richard Parncutt[1], Fabio Kaiser[1], and Craig Sapp[2]

[1] Centre for Systematic Musicology, University of Graz, Austria
{richard.parncutt,fabio.kaiser}@uni-graz.at
[2] CCARH, Stanford University, USA
craigsapp@gmail.com

**Abstract.** The evolution of tonal-harmonic syntax in European notated music, from the beginnings of polyphony to the emergence of major-minor tonality, has been the subject of intense historical study. Several authors have also attempted statistical analyses of the frequency of occurrence of specific pitch-time patterns in specific periods or composers. But no-one has compared prevalence profiles across different periods. Here, we estimate the frequency of occurrence of pitch-class sets of cardinality three in small samples of vocal polyphony from the 13th, 14th, 15th and 16th centuries. Throughout this period, sonorities that were later identified as major and minor became more prevalent (major more than minor). The rank order of sonorities was more variable in earlier music, where chords such as CDF or CE♭F were quite prominent; in later music, the third and fourth most common chords were suspended and diminished.

**Keywords:** Pitch-class set, Tn-type, Triad, Tonality, Major, Minor.

## 1 Introduction

Major-minor tonality "emerged" in the 16th-17th centuries as major and minor triads became commonplace, not only within chord progressions but also at phrase endings. That may be seen as the result of a long historical process in which compositional trial and error interacted with traditions and rules of contrapuntal writing [1,2]. Here, we investigate this historical and perceptual process by counting the frequency of occurrence of all pitch-class sets of cardinality three in representative music of different historical periods. It is difficult to compare scores from different historical periods, because several important parameters can vary simultaneously (reliability of written records, notations system, *musica ficta*). We have attempted to minimize this variation by confining our investigation to 4-part vocal music that is frequently performed and recorded today. Most of the music in the sample is sacred.

C. Agon et al. (Eds.): MCM 2011, LNAI 6726, pp. 366–369, 2011.

# 2 Method

## 2.1 Corpus and Transcription

We had planned to analyse diverse examples of representative notated music from the 13$^{th}$ to the 16$^{th}$ century. This criterion could not yet be fulfilled for the 13$^{th}$ and 14$^{th}$ centuries for which relatively little notated music exists. Most files were downloaded in MIDI format from internet sources[1] and then transfered to Kern format to enable processing (automatic counting) in Humdrum (www.music-cog.ohio-state.edu/Humdrum).

The analyzed works were: **13$^{th}$ century:** Perotin: *Viderunt omnes*, *Sederunt*; **14$^{th}$ century:** Guillaume de Machaut: *Messe de nostre Dame* (Kyrie, Gloria); **15$^{th}$ century:** Dufay: *Omnes amici*, *Missa Ave regina coelorum.* John Dunstable: *Hymn "Veni creator" for Whitsunday*, *Veni sancte spiritus.* Josquin Desprez: *Ave Maria*, *Salve regina*, *El grillo* (madrigal). Johannes Ockeghem: *Missa l'homme armé* (Kyrie). Jacob Obrecht: *Parce Domine* (Motet); **16$^{th}$ century:** Orlando de Lassus: *Matona mia cara* (madrigal). Palestrina: *Dies sanctificatus*, *Adoramus Te*, *Hodie Christus natus est*, *Missa brevis (Credo). Missa aeterna Christi munera.*

In the computer transcriptions, all pitches have been assigned to the 12-note chromatic scale, which allows us to count pitch-class sets.[2] The problem of musica ficta is not addressed in this preliminary analysis.

## 2.2 Analysis

Our analysis is confined to pitch-class sets of cardinality three (sets of three pitch classes). According to Forte [3], there are twelve such sets that do not map onto each other by transposition. Five of these sets are symmetrical in the sense that their intervallic inversions are the same (e.g. the set 3-1, which comprises intervals 0, 1 and 2 semitones above a reference pitch). The other seven are not symmetrical, so they comprise two different chord types (e.g. 3-11, which comprises the minor and major triads). The different inversions are commonly referred to Tn-types. There are 19 Tn-types of cardinality three.

In our analyses, a given Tn-type was considered to occur when either (i) the onsets of the tones sounded simultaneously, called "onset" in the figures below, or (ii) one or two of the tones sounds and the other one or two sounds are held over from the previous sonority, called "sonor".

# 3 Results

Results from our 13$^{th}$-century sample are shown in Fig. 1(a). The numbers on the vertical axis make it clear that the sample is too small to claim statistical

[1] Choral Public Domain Library (www2.cpdl.org); Kern library (kern.ccarh.org); Petrucci Music Library (imslp.org).

[2] See for a list of pc-sets http://solomonsmusic.net/pcsets.htm

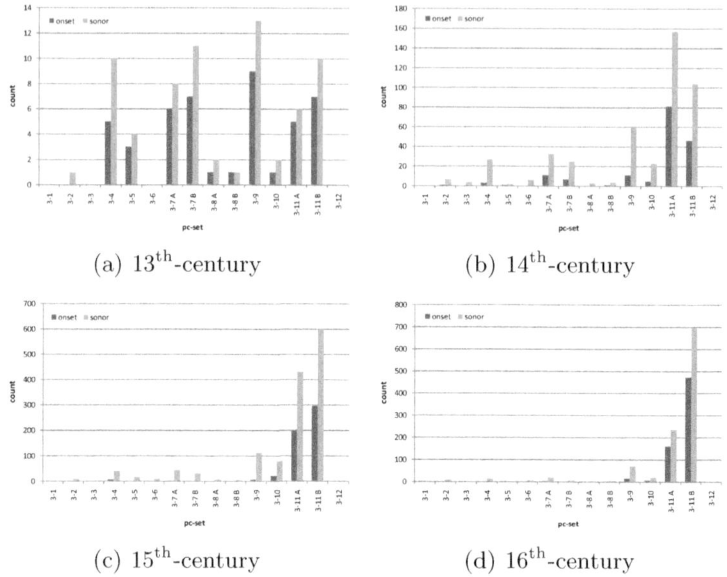


(a) 13th-century (b) 14th-century

(c) 15th-century (d) 16th-century

**Fig. 1.** Pitch-class set counts for the different time periods. The dark bars ("onset") show the number of times the pc-set occurs with three simultaneous onsets. The grey bars ("sonor") additionally include vertical occurrences of the pc-set in which one or two of the three tones are held from the previous sonority.

significance. Moreover, the sample can hardly be regarded as representative since it includes only two pieces by a single composer, Perotin (see above list). Tentatively we can say that the most common pc-sets of cardinality three in these pieces are 3-7 (in semitones: 025 or 035, e.g. CDF or CE♭F) and 3-11 (today's major and minor). There are also several examples of 3-9 (suspended triad 027) and 3-4 (015 or 045).

Our 14th-century sample is confined to extracts from a Mass by Machaut Fig. 1(b). The difference between the Figs. 1(a) and 1(b) reflects that Machaut wrote 3-11 sonorities more often than Perotin, and fewer of 3-4 (015, 045), 3-7 (025, 035) and 3-9 (027). In our Perotin sample, several examples of 3-5 had occurred in passing on weak beats, but 3-5 is generally avoided by Machaut.

Fig. 1(c) is the first for which counts are so high that we can make clear stylistic claims. The dominance of major and minor triads over all other possibilities is clearer in the 15th than the 14th century. There are also more major than minor triads—similar to Eberleins [1] analysis of 18th- and 19th-century music. Next in rank order are 3-9 (027, suspended) and 3-10 (036, diminished).

In the 16th century (Fig. 1(d)), the stylistic differences to the 15th-centuries lie not so much in the individual sonorities, but how they are joined together (voice-leading patterns). Consistent with the hypothesis of [2] that the emergence of major-minor tonality in the 16th-17th centuries was only possible after the sound of major-minor triads had become highly familiar to Western ears in the

$15^{th}$- $16^{th}$ centuries. The rank order of the four most common Tn-types in Fig. 1(d)—major, minor, suspended, diminished—is consistent with the theory [4] that the main psychological components of consonance/dissonance are fusion and roughness, of which fusion is more important.

## 4 Conclusions

This paper is exploratory and has primarily confirmed predictions that could have been made on the basis of intuitive theoretical and practical knowledge of the repertoire. Sonorities that were later identified as major and minor became more prevalent throughout the four centuries under consideration, with major more prevalent than minor. The dominance of major and minor triads in root position in the $15^{th}$-$16^{th}$ centuries cannot be explained purely on the basis of older conventions of voice leading, even though the composers in question consciously applied those conventions. Presumably, the new principles of chord construction that were discovered and explained by the music theorists of the $17^{th}$ century were already being applied intuitively by composers of the $15^{th}$ and $16^{th}$ centuries. This hunch is consistent with an approach to music history that emphasizes the role of hearing and psychology by comparison to contrapuntal theories, rule systems and Pythagorean ratios.

Since major-minor tonality now dominates world music, it is reasonable to ask how the system emerged, which will help us to gain a new, psychologically based understanding of its inner workings and apparently universal appeal. The striking dominance of major and minor in the 16th century is consistent with the idea that they are the foundation upon which major-minor tonality was built. These sounds had to be familiar to western ears before they could become tonic triads in the emerging tonal system. The increasing stability of the rank order of sonorities over these four centuries is consistent with the idea that a more stable tonal system was emerging by a process of perceptual trial and error. Composers and improvisers were constantly trying out new pitch combinations and testing them by ear. Sounds that listeners, composers and performers liked were more likely to "survive".

## References

1. Eberlein, R.: Die Entstehung der tonalen Klangsyntax. Frankfuft/Main, Peter Lang (1994)
2. Parncutt, R.: The tonic as triad: Key profiles as pitch salience profiles of tonic triads. Music Perception 28(4), 333–366 (2011)
3. Forte, A.: The Structure of Atonal Music. Yale University Press, New Haven (1977)
4. Parncutt, R., Hair, G.: Consonance and dissonance in theory and psychology: disentangling dissonant dichotomies. Journal of Interdisciplinary Music Studies (accepted)

# Surveying Musical Form through Melodic-Motivic Similarities

Atte Tenkanen

Department of Musicology, University of Turku, Finland
attenka@utu.fi

**Abstract.** The aim of this study is practical: we want to afford useful compositional schemas and insights, for instance, for students who apply counterpoint in order to construct larger musical forms. For that, we inspect by computer the melodic hierarchies in classical contrapuntal textures. The current model is based on mapping the frequencies of melodic-motivic repetitions throughout an entire piece. Our application creates schemas that illustrate how commonly the melodic segments occur in the piece. The results seem to correspond well to our intuitive impressions of thematic hierarchies.

**Keywords:** Computational Music Analysis, Musical Form, Compositional Process, Music Composition Pedagogy.

## 1 Introduction

The original question behind this study was, whether we can deduce something about the compositional process order in tonal music by computer. To do that, our current model searches for similar melodic structures in classical polyphonic pieces. Through the similarities, we attempt to map the large-scale musical structures and to propose hypotheses about the compositional process order between simultaneous voices. There are several computer-assisted studies in which melodic similarities in compositions are investigated, for example in order to find segmentation boundaries or patterns for music information retrieval (e.g. [1,2]). The novelty of the approach is its application and aim, rather than the development of the model. Consideration of what the composer might have written first, and what was subsidiary in the compositional process, benefits composition students especially, since such observations reveal something about the construction of the musical form. Counterpoint textbooks, for instance, seldom inform about the overall organisation of a piece [3].

The compositional process includes several phases. In classical fugal pieces, for example, the main theme, with its variants and fixed countersubjects, is most probably written before the "freer parts" that occur at the same time (see Fig. 1(a)). At least, our algorithm should distinguish between these themes and the freer filling voices. The method, which is described in Section 2, is tested in Section 3 by using the fugue from J.S. Bach's *Orchestral Suite in B minor* BWV 1067.

C. Agon et al. (Eds.): MCM 2011, LNAI 6726, pp. 370–373, 2011.

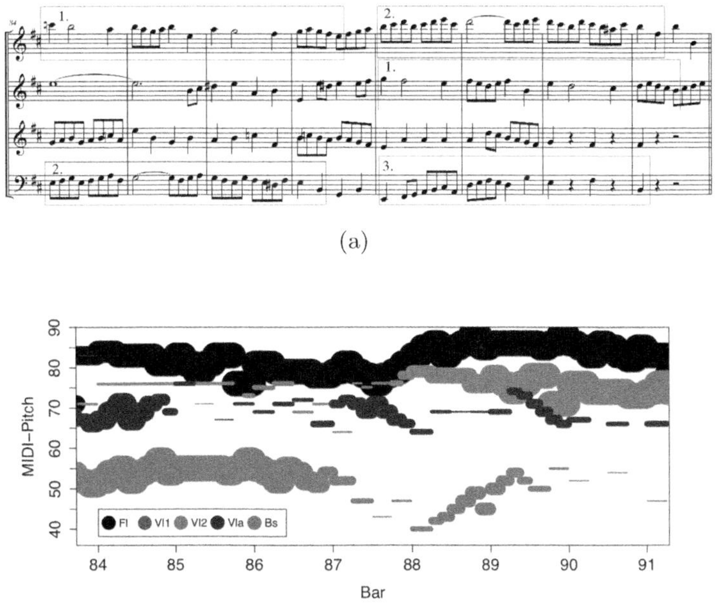


(a)

(b)

**Fig. 1.** (a) A manual analysis of the four-part fugue excerpt from J.S. Bach's *Orchestral Suite No. 2 in B minor* BWV 1067, first movement, bars 84–90. The main theme has been marked with the order number 1 and its fixed counterpoint with number 2, representing the supposed compositional process order. A bass voice has a crucial role in modal and tonal music; in bars 86-89 it is assumed to be primary in comparison with the viola part (order number 3). (b) Melodic priorities according to the weighted graph of the same bars.

## 2 The Method

We have defined only one simple hypothesis for the basis of our current application: *those melodic-motivic patterns that occur more than once are seen as more important and might have been written before those that occur only once in the piece.* In addition, we give more weight to those melodic patterns that occur more often.

The model estimates similarities between melodic patterns of equal length. A piece is considered to be a list of separate voices. The melodic "segments" are derived from the sequence of chromatic intervals between the consecutive pitches in a voice, that is to say, the representation is transposition-invariant. For that purpose, continuous musical data is automatically segmented into overlapping segments of the same cardinality (length), that is to say, n-grams. The only musical parameter hereby considered is intervals. Since the durations and rests are ignored, the melodic intervals interrupted by rests are also computed. As an example, the first violin pattern $s_1$ of cardinality 4 derived from the fugue theme (c.f. Fig. 1(a)) is defined with pitch-interval representation $(+1, -1, -2)$, the second one with $s_2 = (-1, -2, +2)$ and so on (see Fig. 2). After the segmentation of all parts, we compare all objects of the same length with each other. Only the strict identity of motive representations is accepted for the similarity of motives. However, we consider the chromatic intervals -1 and -2, and 1 and 2 to be similar. This emphasises the importance of the pattern contour and note directions, independently of the scale structure.

The resulting graph reveals which segments are more commonly found in the piece, at a particular moment in the piece. Each note in the "weighted" piano-roll score is marked in bold according to how many times it belongs to segments that occur at least twice (Figs. 1(b) and 3). Let us say that we segment the voices of a composition twice, using the segmentation cardinalities 4 and 5. One note can now belong to many segments: if all of the segments marked in Figure 2 occur more than once in a piece, the fourth note, for instance, belongs to eight different segments, and is weighted in the piano-scroll representation according to its number of hits' (4+4). Different parts in a piano-roll representation are coloured using different colours.

**Fig. 2.** First four sequential patterns of cardinality 4 and 5, segmented from the fugue theme of BWV 1067

## 3 Sample Analysis

We demonstrate our approach with the fugue from Bach's *Orchestral Suite No. 2 in B minor* BWV 1067 for solo flute, strings and basso continuo. The piece thus consists of five voices. In tutti sections, the flute doubles the first violin part. These doublings are omitted, because it is not desirable that the procedure takes into account melodic repetitions caused by doublings.

The piece was segmented seven times using the segmentation cardinalities 4 to 10. The values of cardinality 'thresholds' were chosen intuitively. The schematic representation of the score clearly reveals some basic structures, like repeated sections of the piece (see Fig. 3). The exposition seems to be based, for the most part, on the fugue theme and its fixed countersubject. Other material from the exposition is written more freely. We may also suggest some straightforward hypotheses based on the graph. For example, the bass part was probably first written in bars 18–25, 105–110 and 167–176 (marked in Fig. 3). This hypothesis is supported by the regular motives which make up the bass part. The weighted figure corresponds well to our intuitively proposed hierarchy in bars 84–90: compare also Fig. 1(a) with "zoomed" Fig. 1(b).

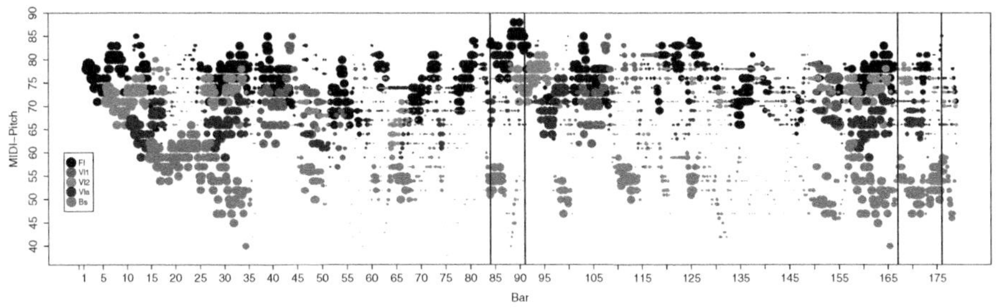


**Fig. 3.** Melodic similarities in the fugue of the first movement, BWV 1067. A weighted graph, based on the segmentation cardinalities 4–10.

## 4 Future Directions

The results seem promising. However, the approach presented here is only a starting point on the way to a more subtle application, that would evaluate priorities between simultaneous voices by taking the interplay between them into account. For that, we might, for example, measure the information transfer between different voices [4], and emphasise the less dissonant voices and longer notes.

To analyse a single composition of a particular composer is just the first stage in utilising this kind of application. The next step would be to analyse several compositions of the same type by the same composer in order to draw more general conclusions about the composer's formal conventions. This might be an even more fruitful way of improving what the method has to offer than refining the method itself. Finally, we can go further and make comparisons between the formal techniques of different composers.

## References


1. Hewlett, W.B., Selfridge-Field, E. (eds.): Melodic Similarity – Concepts, Procedures, and Applications. Computing in Musicology, vol. 11. MIT Press, Cambridge (1998)
2. Lartillot, O., Toiviainen, P.: Motivic matching strategies for automated pattern extraction. Musicae Scientiae, Discussion Forum 4A, 281–314 (2007)
3. Schubert, P.: Modal Counterpoint, Renaissance Style. Oxford University Press, New York (1999)
4. Schreiber, T.: Measuring Information Transfer. Physical Review Letters 85, 461–464 (2000)

# Author Index

GPSR Compliance

*The European Union's (EU) General Product Safety Regulation (GPSR) is a set of rules that requires consumer products to be safe and our obligations to ensure this.*

*If you have any concerns about our products, you can contact us on ProductSafety@springernature.com*

In case Publisher is established outside the EU, the EU authorized representative is:

Springer Nature Customer Service Center GmbH
Europaplatz 3
69115 Heidelberg, Germany

Batch number: 10844619

Printed by Printforce, the Netherlands